D0064971

PROCEEDINGS OF A CONFERENCE
Dartmouth College, June 24–29, 1973

CRITICAL EVALUATION of *Chemical and Physical Structural Information*

Sponsored by the
Committee on Chemical Crystallography
Division of Chemistry and Chemical Technology
National Research Council

with support from the
National Science Foundation

Edited by
DAVID R. LIDE, JR.
and
MARTIN A. PAUL

NATIONAL ACADEMY OF SCIENCES Washington, D.C. 1974

NOTICE: The project that is the subject of this report was approved by the Governing Board of the National Research Council, acting in behalf of the National Academy of Sciences. Such approval reflects the Board's judgment that the project is of national importance and appropriate with respect to both the purposes and resources of the National Research Council.

The members of the committee selected to undertake this project and prepare this report were chosen for recognized scholarly competence and with due consideration for the balance of disciplines appropriate to the project. Responsibility for the detailed aspects of this report rests with that committee.

Each report issuing from a study committee of the National Research Council is reviewed by an independent group of qualified individuals according to procedures established and monitored by the Report Review Committee of the National Academy of Sciences. Distribution of the report is approved, by the President of the Academy, upon satisfactory completion of the review process.

Supported by Contract No. NSF-C310, Task Order No. 254.

Library of Congress Cataloging in Publication Data
Main entry under title:

Critical evaluation of chemical and physical structural information.

Proceedings of a conference held at Dartmouth College, June 24–29, 1973, and sponsored by the Committee on Chemical Crystallography, Division of Chemistry, National Research Council.

Includes bibliographical references.

1. Chemical structure–Congresses. I. Lide, David R., 1928- ed. II. Paul, Martin Ambrose, 1910- ed. III. National Research Council. Committee on Chemical Crystallography.

QD471.C74 541'.225 74-4164
ISBN 0-309-02146-4

Available from
National Academy of Sciences, Printing and Publishing Office
2101 Constitution Avenue, N.W., Washington, D.C. 20418

Printed in the United States of America

Preface

The Conference on Critical Evaluation of Chemical and Physical Structural Information, the proceedings of which are herewith presented. was conceived by the Committee on Chemical Crystallography of the National Research Council. This group took note of the growing specialization in science and the difficulty experienced by a practitioner of one experimental technique in comparing his own results with those obtained by other methods. They felt that it would be useful to bring together experimentalists and theorists from various areas to address the question of the accuracy and reliability of structural information derived from different types of measurements. They outlined a tentative program and nominated members of an organizing committee to plan and administer the conference in detail.

In planning this conference, the word *structure* was taken to include the following types of information:

1. The geometric arrangement of atoms in a free molecule or crystal (symmetry, interatomic distances, bond angles);
2. The description of the forces between these atoms (bond-stretching and bending-force constants, barriers to internal rotation, and other large amplitude motions); and
3. Molecular parameters related to the electronic charge distribution

(electric and magnetic moments, chemical shifts, coupling constants, charge density maps).

The interest in these parameters derives from the manifold applications of quantitative structural information in physics, chemistry, and related disciplines. The development of modern concepts of chemical bonding has drawn very heavily on structural information, and it is likely that refinements in our understanding of bonding will require more extensive and more accurate structural data. Simple, but extremely useful, concepts such as conjugation, hyperconjugation, hybridization, and electronegativity have been correlated with various types of structural information. A large number of correlations between structure and macroscopic physical and chemical properties have been proposed. For instance, chemical reactivity has been correlated with interatomic distances, force constants, and other structural parameters; similarly, such properties as surface tension and vapor pressure have been correlated to dipole moments. In some cases, the direct calculation of macroscopic properties from structural data is a well-established procedure; for example, ideal-gas thermodynamic functions can be calculated with high accuracy from information on the geometric structure and the vibrational force constants of molecules. Over the past decade, as *ab initio* quantum mechanical calculations have become more and more sophisticated, the ability to reproduce a variety of experimentally measured structural parameters has provided a sensitive test of the computational procedure and of the approximations that are involved. Thus, the accuracy and the reliability of experimental structural information is of concern to a variety of scientists in a variety of disciplines.

The question of how accurately a given structural parameter need be known is difficult to answer in a general way. Most would agree that a measurement of the interatomic distance in the HCl molecule to the nearest 0.00001 Å is mainly a *tour de force* that has limited bearing on the correlation of this distance with other properties. On the other hand, it is easy to find examples in the literature where very shaky conclusions have been drawn from structural data that were not as reliable as implied by the original investigator. In particular, when one tries to interpret *differences* in a certain structural parameter—variations from one molecule to another—the question of accuracy becomes critical. The one point that cannot be disputed is that a realistic assessment of the accuracy of a structural parameter is just as important as the value of that parameter itself.

In all techniques for determining structural parameters, a considerable amount of data analysis is required in deriving parameters from the primary experimental measurement. The propagation of errors through this chain of analysis is often quite complex. Someone who is not intimately familiar with the method of analysis usually has a great deal of difficulty in developing a good feeling for the reliability of the final numbers. This problem is accentuated by the need to consider not only the uncertainty in the primary measurement—including both the random errors of measurement and possible systematic errors from calibration of the equipment, impurities in the sample, and other sources—but also the uncertainties associated with the model on which the data analysis is based. It is probably fair to generalize that these model errors are usually as significant or more significant, and certainly more difficult to assess, than the primary measurement errors.

When one tries to compare what is ostensibly the same quantity as determined by two different experimental techniques, ambiguities in the models used to reduce the data become particularly crucial. This is a problem that has become more serious since the advent of the high-speed electronic computer, because it is now feasible to use more elaborate models and more sophisticated techniques in the data analysis. The disadvantage is that the investigator is further removed from direct contact with the data reduction process and has fewer opportunities to insert his own judgment into the interpretation. A least squares fit to a large number of parameters can lead to a printout with standard deviations for each of the constants; however, unless the limitations of the model are thoroughly understood, these standard deviations bear little relation to the true accuracy of the structural parameters.

Another area of concern that arises when published structural information coming from a discipline with which we are not familiar is used is the possibility of a real blunder in the interpretation of the data. Whatever the technique employed in a structure determination, special situations usually exist where there is danger of a false conclusion on even the qualitative structural features, and some of these gross misinterpretations get into the literature. An expert in a particular experimental technique learns to recognize these pitfalls, but they may not be obvious to a person in a different discipline.

This publication includes the papers presented at the conference and summaries of the discussions that took place. It is hoped that these summaries, prepared by conference participants who agreed to accept the assignment, accurately reflect the consensus reached or, as the case may be, the disagreements outstanding in each discussion.

Although it is difficult to summarize the conclusions of a conference of this nature, some general observations can be made. In most of the fields discussed, the statistical treatment of experimental data cannot be regarded as completely satisfactory, in spite of elaborate fitting procedures made possible by modern computers. The following points emerged from these discussions:

1. Least squares fitting techniques now in use are not necessarily the best way of handling large data sets. The lack of resistance to small changes in the data (or in the weights and constraints) is a significant drawback.
2. New methods of data analysis are being developed that should provide more useful feedback to the person conducting the experiment.
3. No fitting technique can be expected to replace completely the judgment of the experimentalist in distinguishing random errors from systematic errors or model deficiencies.

Pitfalls still remain in determining the geometry of molecules and crystals, even for relatively simple systems. Pairs of interatomic distances that are not well separated cause problems in diffraction techniques, and atoms near a principal inertial axis are difficult to locate by spectroscopic measurements. On the other hand, quantitative structure determinations on large molecules have made major strides in recent years, and further progress can be expected. It is hoped that NMR, Raman, and infrared spectroscopy will prove valuable in complementing the information derived from x-ray diffraction.

A general problem that was noted throughout the conference was the effect of vibrations on derived structural parameters. Although vibrational corrections have been pursued to a more sophisticated level in some fields than in others, it is fair to say that the situation is not completely satisfactory for any of the techniques considered at the conference. The picture is particularly discouraging because of the difficulty in finding a level of approximation at which the influence of the large number of higher order terms can be safely neglected. Perhaps the best hope is to discover patterns in which certain anharmonic terms consistently dominate while others are negligibly small. There is a clear need for more extensive data to test model anharmonic potential functions.

One of the encouraging points emerging from the conference was the great strides made in *ab initio* quantum mechanical calculations of structural parameters. Sufficiently good wave functions can now be obtained, even on molecules of moderate complexity, to yield structural parameters precise enough for a meaningful comparison with experimental values. Chemical shifts, magnetic susceptibilities, electric field gradients, and dipole moments, as well as geometric parameters, are now being calculated with considerable success. The interactive coupling between theorists and experimentalists that has developed in the past few years is particularly encouraging. Perhaps it is not unrealistic to hope that theoretical calculations will someday provide a way out of the problems encountered (for example, with anharmonic potential constants) when the available experimental data are insufficient to determine all of the relevant structural parameters.

DAVID R. LIDE, Jr.
Co-Chairman

CONFERENCE ORGANIZING COMMITTEE

CARROLL K. JOHNSON, Oak Ridge National Laboratory, *Co-Chairman*
DAVID R. LIDE, Jr., National Bureau of Standards, *Co-Chairman*

SIDNEY C. ABRAHAMS, Bell Laboratories
BENJAMIN P. DAILEY, Columbia University
HAROLD N. HANSON, University of Florida
BRUCE R. McGARVEY, University of Windsor
WALTER H. STOCKMAYER, Dartmouth College

NATIONAL RESEARCH COUNCIL

DIVISION OF CHEMISTRY AND CHEMICAL TECHNOLOGY

Cheves Walling, University of Utah, *Chairman*

Martin A. Paul, Executive Secretary
Peggy J. Posey, Administrative Assistant

Contents

I

Analysis of Experimental Data

John W. Tukey

Introduction to Today's Data Analysis

Chemists and physicists are routinely miseducated about least squares, usually by being given some correct facts and incomplete–hence unsound–evaluations. As a result, they both make less of their data than they should and, too often, come to think better of their results than deserved. Statisticians must take most of the blame for this state of affairs; in many cases, indeed, it is their teaching, perhaps handed down from generation to generation, that is responsible. In all cases, however, they are responsible for not having developed better methods and not having explained just how things stand.

LEAST SQUARES IN PRACTICE

As a data analyst, what good things can I say about least squares? Only these few:

1. It can be computationally efficient. This was once far more important than today; computers have made a difference.
2. Much is now known about the numerical stability or instability of many specific ways of calculating its "solution." It is not clear to what extent these results are practically vital; it does not matter how close you come when the solution is statistically so imprecise as to be

worthless. There may be little loss from the fact that not much has been done in practice about these recent results.

3. It does a perfect job of insulating the value found for any parameter from any small change in the data that exactly corresponds to a small change in some other parameter(s) that appears in the model actually fitted. This is a property of any method that can be specified uniquely in terms of residuals—of differences between observed and fitted. At least an approximate form of this property is very important and is usually provided by any reasonable method specifiable in terms of residuals, even if not uniquely so.

With computers still speeding up and the costs of their use still going down, only the last of these three—shared, to an adequate approximation, by many other techniques—seems important. Some would like to add a fourth to this list, namely:

4. Once the "weights" are fixed, setting up the mills of least squares—as these are fixed by what is to be minimized—is purely a routine matter. Some would call this "objectivity"; others—the writer included—would prefer "thoughtlessness" and would not like to encourage its occurrence.

LEAST SQUARES IN THEORY

As a theoretical statistician, I would list a different set of good features that can be delineated, out of which only point 3 above would overlap:

5. If the parameters enter additively into the observations, say as

$$y = X\beta + \text{hash},$$

where X, the matrix of x's, is known, β is a vector of parameters, y the vector of the observations, and "hash" one of perturbations or errors, if the components of hash have zero covariances and variances in a known ratio, and if we restrict our attention to estimates b of β that are linear in the observations to those that can be written

$$b = Gy$$

for some G, then least squares with properly chosen weights (weights involving proportional variances) offers minimum variance estimates. This is the famous Gauss–Markov theorem, to which the best responses include "But what if we do not accept such a restriction? What about estimates that are not linear in the data?"

6. If we use only approximately proper weights, if

$$a \leqslant (\text{weight})\,(\text{variance}) \leqslant a(1 + \delta)$$

for some δ, then the fractional increase in variance is not more than $\delta^2/4(1 + \delta)$, which is 1/8 for $\delta = 1$. In practice, the increase for $\delta = 1$ is usually only a very few percent; we have to distort the weights very carefully to approach 1/8 while keeping $\delta \leqslant 1$. Weighting, once reasonably done, is almost never a crucial issue in least squares. This was proved for the trivial case in 1948[4]; the general proof now exists but is still unpublished.

7. If the perturbations (errors) follow a (joint) Gaussian distribution, the best estimate of the parameters will be linear in the data and can then be found by least squares. To this, the best reaction is: "And what if it isn't Gaussian?" What, indeed.

8. If the assumptions of point 5 hold, we can estimate the variances and the covariances of our estimates of the various parameters by a simple routine calculation. Note that among these assumptions is one that the model fitted is either the right model, or contains the right model as a submodel.

These results are theoretically nice–and with the exception of point 6, classical. Similar, though less polished, results can be had in the practically important situation where the parameters do not enter additively. Practically, though, these results tend to raise more questions than they answer, with the exception of point 6.

WHAT TO USE AND WHEN

Unfortunately, to date little information is available on general replacements for least squares. That which can be offered is in the nature of anecdotes. What specific methods, believed to be fairly reasonable, do when applied to certain specific sets of data can be shown. The next few years will see methods with reasonably understood properties that will be safer than least squares and will usually glean a little more out of the data than least squares would.

These methods will take more computer time. More than one of them is likely to be plausible as a means of analyzing a specific set of data. Choices will have to be made. But computer time and thoughtful choice are small investments if more useful information is then extracted from the same data.

THE TRIVIAL CASE

The simplest case of data depending additively on parameters must involve only one parameter. It takes the form

$$y_i = \mu + \epsilon_i \qquad i = 1, 2, \ldots, n,$$

where the $\{\epsilon_i\}$ are a sample from some distribution "located at" zero. We now know[1] a lot about the case where the distribution is symmetric around zero. How can we understand this case in a way that will suggest to us what is likely to be true in more complicated situations?

Mechanical Analogy

Least squares is usually described in terms of a "potential,"

$$\Sigma(y_i - m)^2,$$

that is minimized when its gradient vanishes, when (suppressing a factor of 2)

$$\Sigma(y_i - m) = 0,$$

when the "forces" balance. We can, indeed, regard $(y_i - m)$ as the pull exerted by y_i on m and say that the "energy" is minimized when the "forces" are in equilibrium.

If one y_i were sufficiently far away from the others, every one of us would find some reason for removing it from the calculation. The "pull" exerted by a very discrepant observation needs to be very small—conveniently, zero. (To recognize that this is so is to recognize the inadequacy of any Gaussian assumption for the distribution of the ϵ_i.)

We want, it turns out, to alter the "pull" of y_i on m so that:

- For reasonably small $y_i - m$ the "pull" is roughly proportional to $y_i - m$ (and may as well be taken as roughly equal to $y_i - m$);
- For large $|y_i - m|$ the "pull" is at least very small (and may as well be zero);
- For intermediate $|y_i - m|$ the transition from one regime to another is reasonably smooth.

All really high-performance techniques—this does not include least squares, which is not high performance in any broad sense—satisfy the first two of these criteria. It is probable that those that also satisfy the third perform slightly better statistically. It is certain that those satisfying the third show a kind of "continuity" closely related to another important characteristic to which we turn.

Resistance

We now say that a technique for dealing with data to produce one or more numerical answers is *resistant* if changing a small fraction—no matter how much this small fraction is changed—of the numbers making up the data has only a small effect on the answers. The arithmetic mean is *not* resistant. The median—the middle value once the values are ordered—*is* resistant, as are a variety of summary measures that perform even better than the median under what seem to be practically diverse circumstances.

We want resistant summaries—in general, resistant estimates of parameters—for a variety of reasons:

- It is aesthetically—and I believe scientifically—wrong for our answers to depend unduly upon any small part of the data *unless we know that this is happening.* Cases where straight lines are fitted to many points close together "over there" and a single point "over here" do arise, but it is important that we are clearly aware that they have arisen.
- No one knows more keenly than a good experimenter that accidents do happen to data. It is wise to decrease our vulnerability to accidents whenever we can.
- The changes in technique that give summaries good properties in a wide variety of situations also, almost inevitably, make such summaries much more resistant.
- Resistant summaries and estimates must, in their nature, have decreased difficulty with numerical instabilities. Roundoff in the middle of the calculation, for example, usually corresponds to a change in only a few input values. When this is so, resistance means that roundoff in the middle of the calculation will not greatly affect the answer. The sort of continuity discussed above is also helpful.

Only tradition and a desire to keep calculations a little simpler argue against using better, more resistant as well as more broadly effective, summaries and estimates.

Robustness of Efficiency

How are we to specify what we mean by "broadly effective"? If we specify an unrealistically narrow probability model, such as a Gaussian

(Maxwellian) distribution, we can compare the performance of any estimate we like (by Monte Carlo if not by algebra) with the best performance an estimate can have for that (unrealistically narrow) situation. Broadly effective ought reasonably to mean "close to the unrealistic optimum for each of a wide variety of narrowly specified probability models." This is just what it will mean to us. (Statisticians may choose to say "robust of efficiency" as a technical term; we need not bother here.)

A Crystallographic Note

Alert crystallographers will have drawn a simple lesson from this discussion. Separate measurement of equivalent reflections is a usual thing, with the intensities being combined before the main analysis. Such combination is still possible, but we now know that we cannot always expect to make these combinations by forming the arithmetic mean of the presumably equivalent observations and that we can sometimes gain by fitting before combination.

SOME EXAMPLES

It is natural to be attracted by the idea of a potential. Today it seems as though we can always find a summary (or more general estimate) based on a potential that performs rather like any reasonable summary (or more general estimate), except that it may have to be harder to compute. It is, however, easier to think of summaries and, with reasonable certainty, of estimates in terms of "forces" rather than in terms of "energy." I suspect this is also true in thinking about molecules. Even though we expect to leave potentials behind, it is both natural and reasonable to begin by looking at them.

If we take

$$\Sigma|y_i - m|$$

as the "energy," then the "force" is (up to a common multiplicative constant)

$$\operatorname{sgn}(y_i - m) = \begin{cases} -1, & y_i - m < 0 \\ 0, & y_i - m = 0 \\ +1, & y_i - m > 0 \end{cases}$$

For large $|y_i - m|$ this deviates from the "force" $y_i - m$ corresponding to $\Sigma(y_i - m)^2$ in a desirable direction, but does not go far enough. For small $|y_i - m|$ it is wasteful, being constant, rather than being proportional to $y_i - m$.

The estimate thus generated—in the trivial case, for which we have written our expression—is easily seen to be the median. We ought not to be surprised that, although inferior to the arithmetic mean when the errors follow a Gaussian distribution, the median is quite superior when the errors follow a rather long-tailed distribution. As a summary, the median is much safer than the arithmetic mean, though it will not perform as well under unrealistically ideal situations.

Recently, seemingly sound evidence has been offered[3] to indicate that the behavior of the estimates that minimize

$$\Sigma|\text{observed} - \text{fitted}|,$$

where the "fitted" depends additively on the parameters, is quantitatively like the behavior of the median of a simple sample when the perturbations (errors) are samples from a distribution of identical shape. (When the x-variables, whose coefficients are the parameters, are themselves spread out in a long-tailed way, and when the errors are themselves so long-tailed as to make the median considerably better than the arithmetic mean, this analogy somewhat overstates the performance of the median, but only by part of its apparent advantage over the arithmetic mean.) It is reasonable for us to believe that this will also turn out to be so—or nearly so—for estimators derived from other potentials—or even for estimators developed without regard to a potential.

A wide variety of summaries that (a) start at the median and (b) make an adjustment that would be the first step of a Newton–Raphson sequence tending to the minimum of some more reasonable potential are known to be broadly effective in the simple case.[1] For how complicated a situation will taking just one step do so well? As yet we have little evidence.

A related family of summaries takes the form

$$\frac{\Sigma w_i y_i}{\Sigma w_i},$$

where

$$w_i = \frac{1}{u}\,\psi\,(u_i)$$

with

$$u_i = \frac{y_i - T_0}{\text{scale}},$$

where T_0 is a first summary, apparently best taken as the median, and "scale" is an estimate of scale (spread) calculated from the y_i. As summaries, these appear to perform just about as well as the previous family.

This suggests the use of such a weighting process, perhaps with

$$w_i = \left(\left(1 - \frac{u_i^2}{U^2}\right)^+\right)^2,$$

where U is a measure of scale based on all the u_i and the + indicates that negative values should be replaced by zero, as a way of handling more general problems. We have only to insert an additional weighting factor into our least squares programs and use these programs iteratively—one big loop outside all loops now present—calculating w_i from the last iteration and using it for the next.

We cannot yet say exactly what can soon be recommended to improve on least squares, but we can suggest certain possibilities.

DISTORTION OF RESIDUALS

There is a major difficulty with least squares that we must mention. If there are a few bad values—perhaps only one—least squares will do all it can to "paper over the cracks," to choose a fit whose residuals at the bad points are not nearly as large as the original perturbations or errors. To do this, least squares will make the residuals at good points larger than they should be, thus further concealing the badness of the bad points.

If our analysis is going to direct the experimenter's attention adequately to those points that ought to be regarded with suspicion, we need to use fitting methods other than least squares as a means of defining the residuals. We need resistant/robust methods here, too.

SYSTEMATIC ERRORS

The words "systematic error," though harsh, are often a reality. If we understand something of how systematic errors are likely to behave—as we now do in x-ray crystal structure analysis, for example—we need to *put this knowledge* into our analysis, *into our least squares.*

The model for least squares needs to be reasonably close to what actually goes on, not just to what might be hoped to be going on. If there are likely to be systematic errors of (roughly) known form, they ought to go into the model. Least squares can only deal with what we tell it may be there. We owe it to ourselves to tell it something like the truth, else how can we believe what it tells us in return?

The remarks of Zachariasen[5] in closing the 1968 Conference on Intensities and Structure Factors can only be taken as an indictment of how the data had usually been analyzed in this field:

> It is evident that the positional parameters are reasonably good—even though the actual errors are about five times greater than the estimated ones. However, the thermal parameters are all nonsense. . . .

The results of the International Union of Crystallography project,[2] though much less extreme, still point in the same direction.

We need to note that the phenomenon that Zachariasen indicates could perfectly well come from not including appropriate terms for systematic error in the model fitted. Since it is important to understand why this is so, we will illustrate something of what can happen.

MODELS AND SUBMODELS

Again we will, for simplicity, write down our calculations for the case where parameters enter additively; essentially similar results hold, in terms of derivatives and series expansions, for the more general case, but we can hope that simplicity will lead to clarity. Similarly, we will stick to least squares, on grounds both of simplicity and of familiarity.

Suppose first that the model is

$$y \sim \gamma u + \beta_1 x_1 + \beta_2 x_2 + \ldots + \beta_k x_k$$

$$= \gamma u + \text{submodel } B.$$

The deviations of the y's from the expression in terms of $u, x_1, x_2, \ldots, x_k$, when the proper parameter values are inserted for $\gamma, \beta_1, \beta_2, \ldots, \beta_k$, are just a sample of perturbations (errors) without further structure.

When a model like this is written down for physical or chemical purposes, it will be rare that the experiment is so conducted as to make the covariances between u and the x_i—or the covariances among the x_i—zero. If we were to submit u to least squares analysis, as if it were a response, with submodel B as the model, we would find

$$u = B_1 x_1 + B_2 x_2 + \ldots + B_k x_k + v,$$

where v has zero covariance in this particular experiment with x_1, $x_2, \ldots, x_k$. This means that we can write

$$y \sim \gamma v + (B_1 x_1 + B_2 x_2 + \ldots + B_k x_k + \text{submodel } B).$$

Two things should be carefully noticed: γ is the same as before, and the whole parenthesis is submodel B again with a somewhat different interpretation of the parameters appearing in it.

Since the parenthesis contains only linear combinations of x_1, x_2, $\ldots, x_k$, it follows that the least squares estimate $\hat{\gamma}$ of γ can be obtained directly as

$$\hat{\gamma} = \frac{\Sigma y_i v_i}{\Sigma v_i^2}$$

so that, if each y_i has variance σ^2,

$$\text{var}\{\hat{\gamma}\} = \frac{\Sigma v_i^2 \sigma^2}{\left(\Sigma v_i^2\right)^2} = \frac{\sigma^2}{\Sigma v_i^2}.$$

Since the least squares solution depends only on the model and not on how we write it, the last formula applies as well to $\hat{\gamma}$ in

$$y \sim \gamma u + \text{submodel } B.$$

For comparison consider

$$y \sim \gamma u,$$

where

$$\hat{\gamma} = \frac{\Sigma y_i u_i}{\Sigma u_i^2}$$

$$\text{var}(\hat{\gamma}) = \frac{\sigma^2}{\Sigma u_i^2}.$$

The result of adding submodel B has been to decrease the denominator in var $\{\hat{\gamma}\}$ from Σu_i^2 to Σv_i^2. The more closely u can be fit with submodel B, the greater the increase of variance consequent on its inclusion. If we did not care what γ meant, we would determine $\hat{\gamma}$ most precisely by leaving submodel B out.

A similar result holds if we have to deal with

$$y \sim \gamma u + \text{submodel } B + \text{submodel } C$$

compared with

$$y \sim \gamma u + \text{submodel } B.$$

Adding submodel C to the model increases the variance of $\hat{\gamma}$ further. Instead of being σ^2 over

$$\Sigma\ (\text{residuals of } u \text{ from submodel } B)^2,$$

the variance is now σ^2 over

$$\Sigma\ (\text{residuals of } u \text{ from submodels } B \text{ and } C \text{ combined})^2.$$

The latter denominator is never larger than the first and, in fact, is often smaller.

Systematic Error Case

If there are systematic errors of a known form, including them in the model will increase the variance of $\hat{\gamma}$ (where γ is any one of the original parameters). This addition will also remove that part of the bias of $\hat{\gamma}$ due to the absence of these systematic error terms in the model. To rid $\hat{\gamma}$ of the systematic error (due to their omission), we must add the systematic error terms to the model, thus raising the variance of $\hat{\gamma}$. We cannot have it both ways: To exclude these systematic errors from $\hat{\gamma}$, we must accept a larger variance. A better $\hat{\gamma}$ has to be harder to estimate.

Notice that we have not said that the estimated variance of $\hat{\gamma}$ will go up (though it often will). If the systematic error terms, or other terms, are not in the model but are really present, their effects will contribute to our estimate of σ^2, biasing this upward. Thus adding systematic error terms will reduce both our estimate of σ^2 and the denominator that divides σ^2 to form the variance of $\hat{\gamma}$. Thus the estimated variance of $\hat{\gamma}$ can change in either direction.

If Zachariasen was right about magnitudes and if neglect of systematic errors were the sole cause, we ought to add enough systematic error terms to our models to increase the estimated variance of positional parameters by a factor of 25 and to increase the estimated variance of thermal parameters by a much larger factor. The International Union of Crystallography project[2] would suggest factors of 2 to 4 and 8 to 16, respectively.

Generality

We have concentrated upon x-ray crystal structure analysis in this discussion of what systematic errors can and ought to do. The same considerations apply, to a greater or lesser degree, to any application of least squares—or of its as yet incompletely specified successors—to experimental data.

To serve us well, the model has to adequately portray the behavior of the measurements as they really are. It is not enough to represent how we wish the measurements had been (but were not).

Prepared in part in connection with research at Princeton University sponsored by the Atomic Energy Commission.

REFERENCES

1. D. F. Andrews, P. J. Bickel, F. R. Hampel, P. J. Huber, W. H. Rogers, and J. W. Tukey, *Robust Estimates of Location: Survey and Advances,* Princeton University Press, Princeton, N.J. (1971).
2. W. C. Hamilton and S. C. Abrahams, *Acta Cryst.,* **A26**, 18 (1970).
3. B. Rosenberg and D. Carlson, *Econometrica,* **39**, 401 (1971).
4. J. W. Tukey, *Ann. Math. Stat.,* **19**, 91 (1948).
5. W. H. Zachariasen, *Acta Cryst.,* **A25**, 276 (1969).

Albert E. Beaton and John W. Tukey

Comments on the Fitting of Power Series

We are all used to writing down power series, but not all of us have ever bothered to think about why we do it. Once someone has written down any function, someone else is likely to say, "let's fit it to these data." Somehow the idea has gotten abroad that things are just that simple. Our purpose here is to review some of what is known—usually not widely enough—and to illustrate certain points, algebraically or numerically, in the context of the experimental methods discussed at this conference.

NATURE OF FITTING

Let us investigate the situation in molecular spectroscopy. As the basic example we take the paper on the emission spectrum of hydrogen fluoride by Mann, Thrush, Lide, Ball, and Acquista,[1] in which a given vibration/rotation band is analyzed with the relation

$$\nu_m = \text{wave number} = \sum_0^N c_k m^k.$$

Here $m = J + 1$ for the R branch and $m = -J$ for the P branch; the c_k's are related to the usual spectroscopic constants as follows:

$$c_0 = \nu_0 \qquad c_1 = B' + B''$$
$$c_2 = (B' - B'') - (D' - D'') \qquad c_3 = -2(D' + D'') + (H' + H'')$$
$$c_4 = -(D' - D'') + 3(H' - H'') \qquad c_5 = 3(H' + H'') + 8Y_{04}$$
$$c_6 = (H' - H'') \qquad c_7 = 8Y_{04}$$

How are these relations to be interpreted? Suppose first that we know the wave numbers with infinite accuracy. The major problem involves the choice of N, which will change the definitions of the c_k. (We deal with integer m; we cannot even think of defining the c_k in terms of derivatives. Our "power series" are not mathematicians' usual power series.) Clearly, the hope is that we are to choose N so large as not to make its choice affect the values of the c_k. Two points are now vital:

1. It is not clear that there is any range of N where this is so.
2. It is unlikely that imperfect knowledge of the wave numbers will leave us free to choose N in such a range if it exists.

The question of imperfect knowledge for a further moment aside, let us look at the even part of the expression for the wave number; it can be written as

$$\nu_0 + (B' - B'')m^2 - (D' - D'')(m^4 + m^2) + (H' - H'')(m^6 + 3m^4) + \ldots .$$

From the point of view of the usual spectroscopic constants, then, if there are natural even functions of m in which to expand the even part of the wave number, they are $1, m^2, m^4 + m^2, m^6 + 3m^4, m^8 + 8m^6 + m^4, m^{10} + 10m^8 + 5m^6, \ldots$; they are not the monomials $1, m^2, m^4, m^6, m^8, m^{10}, \ldots$.

Those who argue that expansions in these two sets of polynomials are equivalent miss two important points: (1) In practice we must cut off the expansions, and cutting off the two expansions at the same degree never gives the same result; (2) the coefficients in one expansion are not the same as those in the other. Thus,

$$3 + 2(m^2) + 5(m^4 + m^2) \equiv 3 + 7(m^2) + 5(m^4),$$

but $2 \neq 7$.

Both of these remarks indicate that fitting a polynomial through m^6, for example, must be different when the next term to be fitted is $m^8 +$

$8m^6 + m^4$ from when the next term is m^8. Presumably, at least so far as data analysts go, the usual spectroscopic constants have been so chosen as to make cutting off their sequence the best that we can do.

A fitting process cannot know what else we would have fitted had we fitted more than we did. The only way to protect a fit against the possible appearance of some functional form is to include that functional form in the fit. If we want to fit

$$\nu_0 + (B'-B'')m^2 - (D'-D'')(m^4 + m^2) + (H'-H'')(m^6 + 3m^4)$$

in a way that will not be perturbed by the existence of a term in $m^8 + 8m^6 + m^4$—or by terms in $m^8 + 8m^6 + m^4$ and $m^{10} + 10m^8 + 5m^6$—the only solution is to include these terms in the fit.

We now see that there are physical reasons why we are trying to fit a section of a polynomial series and not a section of a power series.

FITTING ONE HIGHER DEGREE

So much for the moment of what we are trying to fit. What are we actually fitting? To what does the internal workings of our fitting process correspond? A helpful way to understand more of these answers is to look at the effects of fitting one higher degree.

We have estimates $\hat{\nu}_m$ of ν_m for a certain set of integers (always omitting $m = 0$ and often skipping over other integers). Suppose we fit polynomials of degrees n and $n + 1$ to the given data. Let the fits be $q_n(m)$ and $q_{n+1}(m)$. What is the nature of

$$q_{n+1}(m) - q_n(m),$$

the difference between the two fits? This difference is a polynomial of degree $n + 1$ in m. It is all too natural to suppose that it resembles the added monomial

$$(m)^{n+1}$$

but easy to see that this never happens. We can write

$$q_{n+1}(m) = \text{a polynomial of degree } n + c r_{n+1}(m),$$

where $r_{n+1}(m)$ is any polynomial whatever of degree $n + 1$ in which the coefficient of m^{n+1} is not zero. Any polynomial of this degree now seems as natural as the monomial m^{n+1} as a pattern for the difference. Surely we can say more than this.

To do so we will have to think harder about what a fitting process is, of which there are three essentials:

1. A collection of alternate possibilities (essential, vital, containing almost all the physics and chemistry of the problem);
2. A coordinate system to describe these alternate possibilities (this may be important for the calculation, but is open to our choice, i.e., to facilitate or make numerically stable the calculation); and
3. A rule by which one of the alternate possibilities is to be selected (here the statistical properties of the data have to be faced realistically).

Our main concern here will be with the last of these. We leave discussion of it aside for the moment, however, since our concern is still with $q_{n+1}(m) - q_n(m)$.

What coordinate systems—one for polynomials of the nth degree and one for polynomials of the $n + 1$st degree—will make the study of this difference simplest? The answer will depend on the rule by which we fit. For least squares, the rule with which we are all fortunately or unfortunately most familiar, the answer is a suitable set of orthogonal polynomials

$$Q_0(m), Q_1(m), \ldots Q_n(m), Q_{n+1}(m),$$

where $Q_i(m)$ is of degree i in m and

$$\sum_{\text{data}} (\text{weight})\, Q_i(m) Q_j(m) = 0 \text{ for } i \neq j.$$

There will always be such orthogonal polynomials, and if we write our possible alternatives in terms of them, as

$$\alpha_0 Q_0(m) + \alpha_1 Q_1(m) + \ldots + \alpha_n Q_n(m)$$

and

$$\beta_0 Q_0(m) + \beta_1 Q_1(m) + \ldots + \beta_n Q_n(m) + \beta_{n+1} Q_{n+1}(m),$$

the least squares estimates will satisfy

$$\hat{\alpha}_i = \hat{\beta}_i \text{ for } i = 1, 2, \ldots, n.$$

Thus, for the least squares case, the difference between the fitted polynomials will be a multiple of $Q_{n+1}(m)$.

It is easy to see that $Q_{n+1}(m)$ can never look like $(m)^{n+1}$. In particular, it will have at least $n + 1$ changes of sign. In most cases it will be far smaller (at the actual data points) than the term $a \cdot m^{n+1}$ (by which $Q_{n+1}(m)$ differs from a linear combination of the Q's of lower degrees) because near (and especially at) the data points there will be much internal cancellation of the contributions. (In the special case of Chebyshëv polynomials, which arise for unrealistically continuous m and weights somewhat concentrated toward the ends of an interval, the largest value over the interval is 2^{-n} times the largest value in the interval of the term in $(m)^{n+1}$.) It is dangerously misleading to think of fitting one more degree as "adding a term in m^{n+1}." It is far safer to think of this as adding "a rippling polynomial of degree $n + 1$ in m that is likely when written out in terms of monomials to involve all degrees $\leqslant n + 1$ and to be much smaller in value than its term in $n + 1$."

One message is that a comfortably large size for

$$c_{n+1} m^{n+1}$$

across the data need not mean that we can make an estimate of c_{n+1} with reasonable precision. What matters—exactly for least squares and approximately for other rules of fitting—is the size of what is left after $c_{n+1} m^{n+1}$ has been fitted as well as we can (using the other components of what we are fitting—in our polynomial case after fitting with a polynomial of degree n). The notion of size, of course, has to be adapted to the fitting rule.

Once we turn least squares—or any other fitting rule—loose to fit what we may think are sections of power series, it will fit what is best thought of as a section of another polynomial series. The nature of this polynomial series is fixed by the data locations, the assumed weights, and the selected fitting rule. No physics enters. There is no reason why this polynomial series should be the same as the physically meaningful polynomial series, and it probably never has been.

We can convert numerically from one coordinate system to another without great pain. We need to recognize in our minds that at least three coordinate systems for polynomials are being used in any such example, namely, those with basis vectors consisting of the physically meaningful polynomials such as $m^8 + 8m^6 + m^4$; the monomials like m^8; and, in some sense, orthogonalized polynomials such as $Q_k(m)$. An easy way to err is to try to use in one of these coordinate systems insights that are appropriate in another.

ARITHMETIC OF LEAST SQUARES

Those who have not had their faces rubbed in the practical difficulties tend to think that fitting

$$\sum_{0}^{N} c_k m^k$$

by least squares–using the monomial coordinate system–ought to be numerically simple. In truth, it is numerically hard, not easy. The most standard family of matrices for demonstrating both ultrasmall determinants and difficulties with many conventional numerical processes are the matrices of the normal equations for fitting

$$y(t) = \sum_{0}^{N} c_k t^k$$

for $0 \leqslant t \leqslant 1$ and equal weights. The matrices of the normal equations for

$$y(m) = \sum_{0}^{m} c_k m^k$$

and $1 \leqslant m \leqslant N$ are essentially as bad; only their slightly greater algebraic complexity keeps them from being the examples of choice.

There are three ways in which we can try to make the computational pains of least squares bearable:

1. We can choose a better coordinate system.
2. We can carry our arithmetic to double, or higher, precision.
3. We can use a wisely sophisticated computational procedure.

A small experiment exploring all three of these has been conducted using some of the HF measurements referred to earlier. Two computational procedures were used–about the worst and best easily available: WRYALG[2] standing for the "Worst Routine You Are Likely to Get"; and ORTHO, whose double precision version will here be called DORTHO, a carefully planned routine originated at the National Bureau of Standards.[3] There seemed no point to use WRYALG with double precision arithmetic, so only 3 of the 4 = 2 × 2 combinations of routine and precision were used.

For a better coordinate system–better, not best–we compromised on the Chebyshëv polynomials for an interval extending one half-integer beyond the data in either direction. Since we are, in fact, using equal weights (and, sometimes, because of holes in the data), these polynomials will not be very close to being orthogonal. They, however, will be much closer than the monomials. They also have the practical convenience of being generated by a simple two-term recursion relation, so that a table of their values is almost as easy for the computer to prepare as a table of values of monomials.

Table 1 shows the number of significant figures defined as

$$\log_{10}\left|\frac{\text{correct value}}{\text{error}}\right|$$

TABLE 1 Significant Figures Delivered by Various Combinations of Routines (WRYALG and ORTHO) and Coordinates (Monomials and Chebyshëvs) Used in Single Precision[a] on Wave Numbers for the (3, 1) Band of HF[b]

Degree Fitted	WRYALG Monomials	WRYALG Chebyshëvs	ORTHO Monomials	ORTHO Chebyshëvs
0	5.4	5.4	5.6	5.6
1	4.5	4.5	5.1	5.3
2	4.4	4.2	4.9	5.2
3	3.7	3.7	4.8	5.0
4	2.0	2.4	4.2	3.9
5	0.6	2.0	3.2	3.9
6	−2.1	2.3	0.3	0.5
7	−1.1	1.2	0.8	1.2
8	−1.6	0.3	1.4	2.1
9	−1.7	0.4	1.3	1.4
10	−2.5	−1.0	1.3	1.6
11	−1.7	1.7	2.0	2.2
12	−1.6	0.3	1.9	2.4
13				
14				
15				

[a] DORTHO (the double precision version of ORTHO) applied to either coordinate system gave the same first six leading digits for every coefficient for every degree from the zeroth to the fifteenth.

Entries are the smallest for any coefficient.

The DORTHO coefficients have been taken as the correct values for the calculation of the significant figures above (see text for formula).

[b] Data from Mann *et al.*[1, p. 422]

obtained for the worst of the power series coefficients when fitting polynomials of increasing degree to the observed wave numbers for the (3,1) band of HF. WRYALG applied to monomials gives a negative number of significant figures (error larger than correct value) by the sixth degree, while WRYALG applied to Chebyshëvs behaves considerably better. The single-precision versions of ORTHO do not give good accuracy; a more detailed inquiry, however, indicates that they would usually give accuracy in this example that would be acceptable, in that numerical inaccuracies would be small compared with statistical inaccuracies. (There is no reason for this to be expected to continue for more precise data.)

The performance of DORTHO, the double precision version of ORTHO, was a very pleasant surprise to the writers. Up through the fitting of fifteenth degree polynomials the first six digits of each coefficient were the same whether monomials or Chebyshëvs were used for a coordinate system. This means that the internal orthogonalization scheme built into ORTHO and DORTHO functioned very effectively and that the double-precision arithmetic avoided meaningful roundoff errors. From this, we conclude that

1. Unless the basic routine is known to be of the quality of ORTHO and DORTHO, the use of better coordinates than monomials, say Chebyshëvs, is indicated;
2. Double precision should be used; and
3. DORTHO should meet almost any reasonable need.

WHEN TO STOP?

Once we think of fitting a section of a series we can hardly fail to ask, "How long a section?" It is more usual to ask this question in terms of "What fit is best?" or "What residuals reflect the disturbances or errors best?" In our present context, we need to answer this question in terms of "When will our coefficients be closest to the 'true' values?"—a seemingly more difficult question.

The classic answer of statisticians has been to ask about the significance of each new coefficient in turn, as judged from

$$t = \frac{\text{fitted new coefficient}}{\text{its estimated standard deviation}}$$

usually in terms of the distribution of "Student's t." We have less frequently thought about the value of

$$\tau = \frac{\text{true value of the new coefficient}}{\text{its actual standard deviation}}$$

If our concern is with the quality of estimate of one or more of the old coefficients, however, it turns out that τ is of central importance. (As a noncentrality parameter, τ is usually denoted by δ.)

Adding one more term does two things for each old coefficient: It changes its meaning, shifting its mean value (its value averaged over all disturbances); and it increases its variance (possibly by zero) because the change decreases what would be left of the monomial or polynomial whose coefficient concerns us after fitting the other terms to that monomial or polynomial (see Tukey, this volume, p. 13). Presumably, we think well of the redefinition, especially if it is large, and ill of the increase of variance. (The estimated variance may well go down, because the old calculation had no alternative but to include any consistent effects described by the new term in its error estimate, but the actual variance can only go up.)

By analogy with the mean square error, it is natural to want to compare

$$(\text{variance for degree } n) + \frac{1}{A}(\text{shift in mean})^2$$

with

$$(\text{variance for degree } n + 1).$$

Here $A = 1$ seems conventional and $\frac{1}{2} \leqslant A \leqslant 2$ seems likely. A certain amount of algebra shows us that to make this comparison is, except for some correction terms that need not concern us here, essentially to compare τ^2 with A.

This is what we would do if we could. Because τ is generally not known, what can we do? For $\tau = 0$, the standard deviation of t is somewhat larger than unity. This standard deviation does not change rapidly as the value of τ changes. We can judge from t rather well how τ^2 compares with 100 and somewhat how τ^2 compares with 10. Comparing τ^2 with 1, or any nearby value, cannot be done closely and directly.

We are left with the classic approach of looking at each new degree by itself, or trying to make something of the pattern of the t values for various degrees. In the latter context, we have

$$\text{ave}\left\{ t_f^2 \mid \tau^2 \right\} = \frac{f}{f-2}(1 + \tau^2),$$

TABLE 2 Values of t for, and s^2 after, the Fitting of Each Successive Degree to the Wave Numbers for the (1, 0) and (3, 1) Bands of HF[a]

	(1, 0) Band[b]					(3, 1) Band[c]				
Degree	t	$t^2 - 2\frac{f}{f-2}$	s^2	C^d	$10^4 \times s^2/f$	t	$t^2 - 2\frac{f}{f-2}$	s^2	C	$10^4 \times s^2/f$
0	–	–	–	–	–	87.9	large	large	large	large
1	–	–	–	–	–	12.3	large	large	large	large
2	–	–	–	–	–	−101.7	large	491.	large	large
3	35.1	large	2.35	large	large	−39.7	large	12.1	large	large
4	68.6	large	0.01389	large	5.15	57.8	large	0.169	large	36.6
5	6.48	39.8	0.00552	15.4	2.12	17.1	large	0.0229	13.4	5.1
6	0.32	−2.1	0.00571	17.3	2.28	−0.05	−2.1	0.0234	15.4	5.3
7	−1.35	−0.4	0.00553	16.8	2.30	−1.05	−1.0	0.0234	16.1	5.4
8	−0.88	−1.4	0.00559	17.7	2.42	−1.00	−1.1	0.0234	16.9	5.6
9	−0.28	−2.1	0.00581	19.6	2.64	−0.47	−1.9	0.0238	18.6	5.8
10	2.05	2.0	0.00508	16.4	2.42	−2.24	2.9	0.0217	15.1	5.4
11	2.76	5.4	0.00386	11.1	1.83	2.26	3.0	0.0197	12.0	5.0
12	0.31	−2.1	0.00405	12.0	2.13	0.36	−2.0	0.0201	13.8	5.3
13	–	–	–	–	–	−1.72	0.8	0.0191	13.0	5.2
14	–	–	–	–	–	0.74	−1.6	0.0194	14.4	5.9
15	–	–	–	–	–	−0.64	−1.0	0.0197	14.0	5.6
16	–	–	–	–	–	program fails				

[a] Data from Mann *et al.*[1]
[b] $N = 32$ data points.
[c] $N = 51$ data points.
[d] $C = \frac{s^2 f}{s_L^2} - [N - 2(d+1)]$, where s_L^2 is the final s^2 calculated and d is the degree ($d + 1$ constants fitted).

where ave $\{ t_f^2 | \tau^2 \}$ is the average (expected) value of the statistic t_f^2 given a value of τ^2 and f is the degree of freedom involved.

It is thus not unnatural to look at the values of

$$t_f^2 - 2\frac{f}{f-2}$$

for successive degrees and to try to decide when its average value has become small or negative. (We know that this will be difficult, perhaps impossible, but we may as well try. If we like some value of $A \neq 1$, we would use $1 + A$ instead of 2.)

Table 2 shows, for least square fits of successive degrees to the wave numbers of the (1, 0) and (3, 1) bands of HF, t values for, and s^2 values after, the fitting of each additional degree up to 12 and 15, respectively, as well as certain associated quantities. (If we do the fitting up to each degree carefully enough, it will not matter in which terms we do the fitting.)

There is no doubt about fitting at least through degree 5: All values of t are notably large. For each band, the next four values of t are small; the two values associated with degrees 10 and 11 are substantial; and those beyond—with the possible exception of the thirteenth degree for the (3, 1) band—are small. If we look at the columns of t values, we get the same answer for both bands, i.e., either fit through degree 5 or through degree 11. (There is no sense in skipping powers, especially since the physically relevant polynomials are not, we repeat *not*, monomials.)

If only one of the t values for degrees 10 and 11 had been large, we would have been tempted to regard this as accidental, especially in view of the values of $t^2 - 2[f/(f-2)]$.

One rule for selecting a stopping place[4] is to calculate the quantities called C in Table 2 and seek for a (local) minimum. This leads to fifth, seventh (not competitive), or eleventh degree for (1, 0) and fifth, eleventh, or thirteenth (not competitive) degree for (3, 1). (This choice attempts to select a good fit.) Another rule for selecting a stopping place[5] is to look at s^2/f, where f is the number of degrees of freedom involved in s^2. This leads to the same choices—fifth, seventh (not competitive), or eleventh degree—for (1, 0) and to fifth, eleventh, or thirteenth (not competitive) for (3, 1). (This choice attempts to get clean residuals.)

Almost surely we would have to look at the other bands and, quite possibly, try more than one choice before reaching a decision between the fifth and eleventh degrees. It is far from clear that we should choose the same degree for each and every band for which we have data.

One important lesson, we hope, is made clear by Table 2: It is unwise to stop looking when one t is small, or even when two consecutive t's are small. Here there were four small ones before the large ones appeared.

We have not tried to specify a single definite way to decide where to stop fitting. We would feel that we had done a disservice to good analysis of molecular structure data if we had. Mechanical computations can usually, as here, shorten the list of reasonable competitors. Sometimes only one will be left; in general, however, there will be a few from which choice must be made carefully.

HOW CAN WE BETTER LEAST SQUARES?

So far we have stuck to least squares as the rule for selecting our fitted coefficients, even though we were aware of this rule's deficiencies. Most of the questions addressed so far (What series is physically meaningful? What series is implied by the fitting process? When to stop?) are all general and could as well be discussed in a least squares context as not; indeed, the algebra is simplest there. It is time now to consider anecdotes about what other rules can do for us.

Examples soon follow of the performance of two sorts of rules that are intended to be robust/resistant—to be robust in the sense of performing well when the disturbances (errors) come from a variety of distributions (not just the Gaussian) and to be resistant in the sense of changing the fit only a little when one or a few data values are changed, no matter how much. So far as we can see, making a rule robust/resistant means making it at least somewhat iterative. Our computational effort will be greater than for least squares, but our reward should also be greater.

Biweight

One approach is to use weighted least squares in which the weights for each iteration are a function of the sizes of the residuals:

$$\text{residual} = \text{observed} - \text{fitted}$$

provided by the previous iteration. We want the weights of small residuals to be nearly constant and the weights of very large residuals to be zero. It seems desirable for the weights to change smoothly. One way to do this is to choose

$$\text{weight} = \begin{cases} (1-u^2)^2, & \text{for } |u|<1 \\ 0, & \text{elsewhere} \end{cases}$$

with

$$u = \frac{\text{residual (last iteration)}}{\text{scaling value}},$$

where the scaling value is a function of all the residuals, perhaps a numerical multiple of the median absolute deviation. The term "BIWEIGHT" is sometimes associated with this type of weighting.

In the examples that follow, we used this functional form for the weight and a somewhat different form for the scaling value.

Smofit

A second approach is based on an effective smoothing procedure; this procedure, given a sequence of numbers, produces a smoother sequence in such a way that unusual values are not only smoothed down (or up) but have very little effect on the resulting smoothed sequence–i.e., it is resistant. A classic smoothing process is the use of running (or moving) linear combinations:

$$y(t) = \tfrac{1}{4}y(t-1) + \tfrac{1}{2}y(t) + \tfrac{1}{4}y(t+1).$$

More recently, we have come to appreciate the advantages of running (or moving) medians:

$$y(t) = \text{median of } \{x(t-1), x(t), x(t+1)\}.$$

Using first one smoothing process and then another is routine practice and can be very effective.

To produce a resistant smoother, it is convenient to use suitable combinations of running medians (to provide resistance and robustness) and running linear combinations (to provide greater smoothness). It is probably important to use the principle of "twice"–namely, to smooth once, form residuals from the first smoothing, smooth these residuals, and then to combine the two smoothings by term-by-term addition. Given such a smoothing process, how can we approach fitting

$$y \sim \Sigma b_i x_i,$$

where in our present case each carrier x_i is a polynomial in m? Smoothing processes work on sequences, so that it seems natural to work on one carrier (one x_i) at a time (just as essentially all equation-solvers, including essentially all least squares programs seem to work one variable or carrier at a time). Since we are gaining resistance by deliberately choosing operations where the result of the operation on the sum is not the sum of the results of the operations on the parts, we have no guarantee that doing whatever we do for a single carrier once is enough. Accordingly, for our examples, we have gone through the carriers in the order $x_1; x_2; x_3; x_1, x_2, x_3; x_1, x_2, x_3, x_4; x_1, x_2, x_3, x_4, x_5; \ldots$. If we feel a need, we do all this over again.

What we need to do for a single carrier is a version of what is often called an elementary sweep; this process calculates one constant (no constant term in the fit) or two constants (when, as here, a constant term is being fitted) and then transfers corresponding amounts from the previous remainders to the new fits, thus (we hope) decreasing the size of the remainders appropriately. A simple sweep associated with a smoothing process operates as follows, when we attend to the carrier x_i:

- Sort the data sets $(y, x_1, x_2, \ldots, x_k)$ on the carrier x_i.
- Smooth the values (in this sorted order) of y—here, the remainders of the observed values after previous transfers.
- Smooth the values (in the same order of data sets) of this one x_i.
- Do a fit $y \sim a + bx_i$, which probably may as well be by least squares using both smoothed y and smoothed x_i.
- Transfer an amount $a + bx_i$ from each y to the corresponding fitted value, subtracting $a + bx_i$ from y and adding a to the fitted constant term and b to the fitted coefficient of x_i.

In practice, we found it better (to our surprise) to omit the sorts and do all the calculation with the data sets in order of m.

Smogonalization

We are or ought to be keenly aware that the effectiveness of any sweeping procedure depends on the nature of the carriers, especially on their mutual relationship. To try the procedure just described on

$$(y, 1, m, m^2, \ldots, m^k)$$

would be both unwise and ineffective. We need to replace the monomials by an appropriate sequence of polynomials. The natural sequence to try first, as for any reasonably simple fitting process, is

residual of m after fitting a constant;
residual of m^2 after fitting a linear function of m; and
residual of m^3 after fitting a quadratic in m.

For obvious mnemonic reasons, we will denote these polynomials as $S_0(m), S_1(m), S_2(m), \ldots, S_k(m)$ and call them *smogonalized* polynomials. (*Smidual* might seem more precise, but not as suggestive.)

INSTANCES OF RESISTANT FITTING

It will be clear from the above that, while we do intend to explain the sort of calculation we have used as outlined above, we do not plan to give precise details in this account. We omit detail because we are keenly aware that we do not know what details to recommend. It is hoped that some reasonable recommendations will be available before too much time has passed.

We turn then to some examples, based on eighth degree fits to the data for the (1, 0) band of HF. Table 3 sets out the residuals from four fits to the actual data. The fitting procedures are DORTHO, high quality least squares; BIWEIGHT, sixth iteration of iterative fitting with weights, using DORTHO for each iteration; SMOFIT (first application), iterative sweeping based on the resistant smoothing process (see p. 33); SMOFIT (second application), results of applying the whole of SMOFIT again to the residuals from its first application. Judged by this one set of residuals, all four fitting procedures seem to be of rather similar quality.

Rather than make detailed comparisons, we shall turn next to an artificial example. Five of the 32 lines (data sets) for the (1, 0) band were selected by random numbers (see below) and the observed wave numbers drastically modified (by amounts also selected by random numbers). Strangely enough the lines selected for modification were at $m = -8, -6, -5, 8$, and 27 and included the two adjacent to the central gap. The random numbers really did it!

Table 4 shows the residuals from the fits to these artificially perturbed data. Least squares, as represented by DORTHO, has smeared lack of fit across the 27 correct values in a quite unacceptable way. Instead of three residuals larger than 100, we now have (skipping the five parenthesized residuals, which ought to be large) four residuals larger than 20000.

BIWEIGHT has done a relatively tasteful job. Setting aside the parenthesized residuals, only 3 of 27 are beyond 100 (compared with 4 of 32

TABLE 3 Residuals (in $10^{-3}\,cm^{-1}$) from Four Eighth Degree Fits to the Actual Data for the (1, 0) Band of HF[a]

m	DORTHO	BIWEIGHT (6th iteration)	SMOFIT (1st application)	SMOFIT (2nd application)
−15	6	−6	−49	−37
−14	15	27	12	48
−13	−52	−38	−46	−6
−12	−16	−11	−22	11
−11	65	54	41	60
−10	34	7	−6	−1
−9	−58	−95	−104	−111
−8	43	2	0	−16
−7	−72	−110	−103	−122
−6	36	9	26	8
−5	−4	−12	11	−1
8	92	202	152	195
9	−7	77	36	67
10	−61	−3	−33	−14
11	−104	−70	−88	−81
12	−64	−50	−58	−60
13	138	137	139	129
14	31	20	30	15
15	−52	−66	−53	−70
16	51	40	53	37
17	58	55	63	51
18	−35	−28	−26	−34
19	−7	10	4	2
20	−29	−4	−17	−14
21	−15	12	−3	2
22	−31	−10	−22	−17
23	−23	−19	−19	−19
24	19	−2	17	8
25	88	37	82	60
26	70	−10	62	25
27	−176	−268	−180	−229
28	65	−1	73	22
# beyond*[b] 50	15	10	13	12
# beyond* 100	3	4	5	5

[a] The data are given only to 10's in the units used here.

[b] # beyond* means "number (of 32) greater in absolute value than."

TABLE 4 Residuals (in 10^{-3} cm^{-1}) from Four Fits to the Artificially Perturbed Data[a]

m	DORTHO (unweighted)	BIWEIGHT (6th iteration)	SMOFIT (1st application)	SMOFIT (2nd application)
−15	−2454	2	− 5	−11
−14	2134	9	6	17
−13	1560	−36	−78	−61
−12	12	7	−58	−43
−11	−546	61	16	26
−10	409	−35	−7	−2
−9	2772	−222	−75	−74
−8	(16188)	(9763)	(10664)	(10664)
−7	9483	−475	−4	0
−6	(−121636)	(−134484)	(−133841)	(−133833)
−5	(97895)	(83375)	(84174)	(84191)
8	(−27121)	(−19904)	(−19783)	(−19754)
9	−2678	32	79	97
10	1931	−4	−7	−1
11	6046	−45	−78	−80
12	9137	−15	−58	−68
13	10785	172	132	117
14	10194	46	18	1
15	7621	−53	−66	−82
16	3456	39	41	28
17	−2016	44	55	46
18	−7892	−45	−31	−34
19	−12746	−6	3	6
20	−15339	−14	−15	−7
21	−14136	11	0	10
22	−7963	−3	−19	−11
23	3923	−8	−19	−17
24	21220	5	13	6
25	42011	33	73	52
26	61948	−29	49	14
27	(−159575)	(−233293)	(−233194)	(−233241)
28	65377	3	67	18
# beyond*[b] 50	26	5	11	7
# beyond* 100	26	3	1	1
# beyond* 1000	24	0	0	0
# beyond* 10000	9	0	0	0

[a] Actual data for the (1, 0) band of HF perturbed at 5 of the 32 points. Residuals for the five artificially perturbed values are in parentheses. The data were given only to 10's in the units used here.

[b] # beyond* means "number (of 32 − 5 = 27) not in parenthesis greater in absolute value than."

TABLE 5 Coefficients from Eight Eighth Degree Fits for the (1, 0) Band of HF

Term		DORTHO		BIWEIGHT (6th cycle)		SMOFIT (1st use)		SMOFIT (2nd use)		
		Act[a]	Pert	Act	Pert	Act	Pert	Act	Pert	Median[b]
0	$10^{0}\times$	3961.56	3961.26	3961.43	3962.24	3961.43	3961.21	3961.38	3961.15	3961.43
1	$10^{0}\times$	40.32	45.26	40.30	40.26	40.31	40.31	40.30	40.31	40.30
2	$10^{-1}\times$	−7.73	−7.35	−7.71	−7.85	−7.70	−7.67	−7.69	−7.66	−7.70
3	$10^{-2}\times$	−0.80	−10.38	−0.77	−0.69	−0.79	−0.80	−0.78	−0.79	−0.78
4	$10^{-4}\times$	0.55	−0.36	0.38	1.13	0.31	0.21	0.22	0.12	0.34
5	$10^{-6}\times$	−0.73	562.84	−1.63	−6.52	−0.31	0.21	−0.72	−0.02	−0.72
6	$10^{-8}\times$	2.57	−638.05	7.48	−0.87	7.79	8.01	10.55	10.23	7.64
7	$10^{-8}\times$	0.20	−107.34	0.25	1.26	−0.03	−0.06	−0.03	−0.07	0.08
8	$10^{-10}\times$	−0.54	259.51	−0.86	−2.50	−0.28	−0.26	−0.37	−0.33	−0.46

[a] Act, actual data; Pert, artificially perturbed data.
[b] Of the four fits to the actual values.

before perturbation), and only 5 of 27 are beyond 50 (compared with 8 of 32). The increases among the 27 caused by "ruining" five of the given values can only be considered strikingly small.

SMOFIT is also clearly much better than least squares. Whether applied once or twice, it does not make the small residuals quite as small—or the large residuals quite as large—as BIWEIGHT does.

Table 5 shows what happens to the coefficients of the monomials. Going from actual (Act) to artificially perturbed (Pert) values has some shockingly large effects for DORTHO and only uncomfortable but moderate effects for BIWEIGHT and SMOFIT. Clearly, the coefficients testify to SMOFIT's greatly reduced susceptibility to erratic values.

A DETAILED LOOK AT SOME RESIDUALS

When we look at the residuals carefully, we are likely to find unexpected appearances. This seems to be true in our example here. Table 6 gives us a rather hard look at two sets of residuals, both of which seem to show some tendency toward relatively regular clumping. Inquiry of Dr. Lide as to whether there was a "least count" of 0.04 cm^{-1} to 0.05 cm^{-1} (40 or 50 in the units of Table 5) in the wave numbers either in measurement or calculation brought a response that this was not to be ruled out. While the discovery of such an effect at the time when the data were being analyzed might have been quite helpful, neither the suggestion nor a possible proof of its existence today can be of importance.

The message of this example is not "See what we found!" but rather "See what can be found, even if it is entirely unexpected!" Residuals are meant to be calculated and looked at, usually in several different sorts of graphical display. There is never a valid excuse for omitting such examinations.

If we turn back to the middle column of Table 3, BIWEIGHT has pinpointed two residuals at −268 and +201 followed by three more at +137, −110, and −95; the remaining residuals are between −70 and +77. Clearly, there is ground for inquiry into the five largest residuals. Had this turned up during the actual data analysis, inquiry would certainly have been made.

A COMMENT

The more precise and careful the data gathering, the greater the certainty that something can be learned from the pattern of residuals. When all known sources of error are carefully controlled, the chance that one unexpected source will dominate all others is increased. The more careful the work, the harder one should look at the residuals.

TABLE 6 Display of Residuals[a] (in 10^{-3} cm^{-1}) after (Left) First Application of SMOFIT to Fit Eighth Degree Polynomial to the (1, 0) Band and (Right) Application of DORTHO to Fit Eighth Degree Polynomial to the (3, 1) Band

Last Digit(s) for (1, 0)	Leading Digits	Last Digit(s) for (3, 1)
		(3 higher)
	17	3
	16	8
	15	
4	14	1
4	13	
	12	1489
	11	
9	10	579
1	9	
2	8	2
	7	08
0	6	6
834	5	6
3	4	247
6	3	3
	2	
721	1	6
816	0	33
47	−0	
0	−1	99
57	−2	21
67	−3	88
804	−4	6
9	−5	
1	−6	4
	−7	86620
	−8	77
	−9	
06	−10	7
	−11	998
	−12	
	−13	
	−14	
	−15	
4	−16	8
	−17	
		(6 lower)

[a] The entries 4|14|1 refer to 144 for (1, 0) and 141 for (3, 1); the entry |16|8 refers to 168 for (3, 1); the entry 4|13| refers to 134 for (1, 0), and so on. The entry 834|5|6 indicates residuals of 58, 53, and 54 for (1, 0) and 56 for (3, 1). The 3 higher for (3, 1) are 427, 305, and 223. The 6 lower for (3, 1) are −273, −231, −225, −216, −211, and −200.

Prepared in part in connection with research at Princeton University sponsored by the Atomic Energy Commission.

REFERENCES

1. D. E. Mann, B. A. Thrush, D. R. Lide, Jr., J. J. Ball, and N. Acquista, *J. Chem. Phys.,* **34**, 420 (1961).
2. A. E. Beaton, D. B. Rubin, and J. L. Barone, *Res. Bull.,* RB-72-44. Educational Testing Service, Princeton, N.J. (1972).
3. P. J. Walsh, *Comm. Assoc. Comp. Mach.,* **5**, 511 (1962).
4. C. David and F. S. Wood, *Fitting Equations to Data,* Wiley-Interscience, New York (1971), 86*ff.*
5. J. W. Tukey, *J. R. Stat. Soc. B,* **29**, 47 (1967).

D. F. Andrews

Some Monte Carlo Results on Robust/Resistant Regression

Many techniques have been proposed for fitting the parameters of a model to data. Some of these seek to minimize some function of the residuals of differences between observed and fitted values. In the early nineteenth century, the methods of least absolute deviations and of least squares were used. These sought to minimize

$$\Sigma \mid \text{observed} - \text{calculated} \mid$$

or

$$\Sigma \mid \text{observed} - \text{calculated} \mid^2,$$

respectively. More recently, least squares has been more commonly used. Several reasons might account for this use. Among these are computational convenience and some optimality properties. For example, if the observations are assumed to be statistically independent with the same variance, the Gauss–Markov theorem states that the method of least squares yields estimates with the smallest variance among the class of unbiased estimates that are linear in the observations. If, in addition, the observations are assumed to be drawn randomly from a Gaussian distribution, the least squares estimates have the smallest possible variance.

But why should our procedures be restricted to linear and unbiased estimates if others tend to be more precise? Is there any evidence that experimental observations follow a Gaussian distribution rather than any other? The evidence points rather to longer tailed distributions (see, for example, ref. 6).

PRELIMINARY DEFINITIONS AND NOTATION

Consider n observations $y_1, \ldots, y_n$ and a polynomial $b_0 + b_1 x^1 + \ldots + b_k x^k$ with coefficients $b_0, \ldots, b_k$, chosen to fit the data. This collection may be written

$$y_i = b_0 + b_1 x_i^1 + \ldots + b_k x_i^k + r_i,$$

where r_i is the residual left over from fitting the ith observation. A probability model for this system might be expressed

$$y_i = \beta_0 + \beta_1 x_i^1 + \ldots + \beta_k x_i^k + \sigma e_i,$$

where the β_i are unknown regression parameters, σ is an unknown scale parameter, and the e_i are random variables usually assumed to be independent, identically distributed with a density function $f(e)$.

The problem of fitting polynomials is a special case of the more general linear model

$$y = \beta_0 x_0 + \beta_1 x_1 + \ldots + \beta_k x_k + e,$$

where the independent variables x_j are not restricted to powers of a single x. It is convenient to express such a system in matrix notation as

$$y = \mathbf{x}^t \beta + e.$$

A set of n observations may be expressed

$$\mathbf{y} = X\beta + \mathbf{e},$$

where X is a $n \times k$ matrix of independent variables. In this discussion the x's are considered as fixed known quantities. They also are assumed to have a reasonably concentrated distribution. All of the techniques discussed here are sensitive to observations corresponding to extreme values of the x vectors. These observations should be set aside for a resistant fit.

For any estimate $\hat{\beta}$, the residuals are given by

$$\mathbf{r} = \mathbf{r}(\hat{\beta}) = \mathbf{y} - X\hat{\beta}.$$

The least squares estimate $\hat{\beta}_{\mathrm{LS}}$ minimizes

$$\sum_{i=1}^{n} (y_i - \mathbf{x}_i{}^t\hat{\beta})^2$$

and is defined by the system of linear equations

$$\sum_i x_{ij}(y_i - \mathbf{x}_i{}^t\hat{\beta}_{\mathrm{LS}}) = \sum_i x_{ij}r_i(\hat{\beta}_{\mathrm{LS}}) = 0.$$

In the following we will need an estimate of scale. Since the properties of the regression estimate do not depend markedly on the efficiency of this estimate, we will always use the resistant estimate

$$s = \text{median}\left\{|r_i|\right\}.$$

M ESTIMATES

Maximum likelihood estimates of β are defined to maximize the likelihood

$$\prod_{i=1}^{n} f\left(\frac{y_i - \mathbf{x}_i^t\beta}{\sigma}\right)$$

or, equivalently, to maximize the log-likelihood

$$\sum_{i=1}^{n} \log f\left(\frac{y_i - \mathbf{x}_i^t\beta}{\sigma}\right).$$

This leads to the system of equations

$$\sum_{i=1}^{n} x_{ij}\Psi\left(\frac{y_i - \mathbf{x}_i^t\beta}{\sigma}\right) = 0$$

$$= \sum_{i=1}^{n} x_{ij}\Psi\left(\frac{r_i}{\sigma}\right),$$

where $\Psi(z) = \dfrac{d \log f(z)}{dz}$. These equations will be approximated by substituting $s = \text{median}\ \{|r_i|\}$ for σ.

If $f(z)$ is the Gaussian density function

$$f(z) = (2\pi)^{-1/2} \exp\left\{-\tfrac{1}{2} z^2\right\},$$

then $\Psi(z) = z$, and the equations reduce to those of least squares. While any density function f could be used to define such an estimator, we are, in fact, free to investigate the properties of the estimator that corresponds to any function Ψ without regard to its relation to any density function. Such estimates are called *M* estimates. (See Huber.[3]) Figure 1 illustrates several possible choices for the function Ψ. Figure 1a is the function $\Psi(z) = z$ that corresponds to the least squares estimate. The function is unbounded. One observation can greatly affect the equations and hence their solution. Figure 1c is the Ψ function corresponding to the maximum likelihood estimate for a Cauchy distribution. Figure 1b is the Ψ function investigated by Huber.[3] The form of this estimator

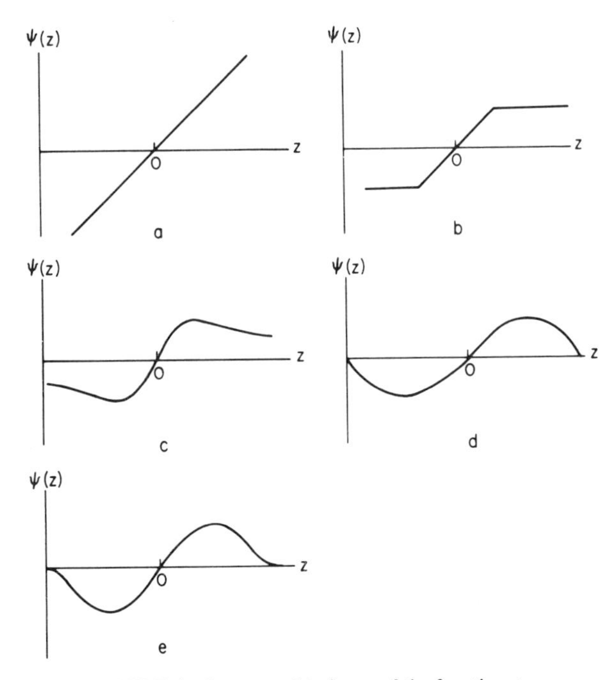

FIGURE 1 Some possible forms of the function ψ.

may be traced to Newcomb.[4] Figure 1d is the Ψ function of the SINE estimator studied by Andrews *et al.*[1] The function is defined by

$$\Psi(z) = \sin(z/c) \quad |z| \leqslant c\pi$$

$$= 0 \qquad \text{otherwise.}$$

The calculation of M estimates is similar in form to that of weighted least squares, where the weights are chosen iteratively on the basis of the current fit. The M estimates satisfy

$$\sum_i x_{ij} w_i^2 r_i = 0,$$

where $w_i^2 = \Psi\left(\frac{r_i}{s}\right)/r_i$. This last equation defines the weighted least squares estimate, where the ith observation is given weight w_i. Thus an estimate may be found by

1. Selecting an initial estimate $\mathbf{b}^0$;
2. Using this estimate, finding residuals r^0, a scale estimate s^0, and weights w^0;
3. Iteratively calculating $\mathbf{d}^{(m)}$, the least squares estimate of the incremental model

$$w^{(m)} r^{(m)} = w^{(m)} \mathbf{x} \mathbf{d}^{(m)} + r^{(m+1)}$$

and updating the estimate $\mathbf{b}^{(m+1)} = \mathbf{b}^{(m)} + \mathbf{d}^{(m)}$.

The results of this calculation will depend on the starting point $\mathbf{b}^0$ and M, the maximum number of iterations.

In the following we will discuss the SINE estimators denoted by $L\cdot\rightarrow(Sc)^M$ where $L\cdot$ is the starting point, c is the scaling constant, and M is the maximum number of iterations. The least squares estimate denoted by LS, and the least-absolute-deviations estimate denoted by L1 will be used as starting points. Similarly $L\cdot\rightarrow(Bc)^M$ will denote the corresponding estimates using the Ψ function suggested by Tukey and illustrated in Figure 1e,

$$\Psi(z) = z\left[1 - \left(\frac{z}{c}\right)^2\right]^2 \quad |z| \leqslant c$$

$$= 0 \qquad \text{otherwise.}$$

For the calculation of the Ll estimate see Claerbout.[2]

CALCULATIONS

The variance matrix of each of a collection of estimates was calculated for a variety of distributions using the Monte Carlo techniques of Andrews and co-workers[1] and Relles.[5] The distributions were all expressed in terms of mixtures of Gaussian distributions. G(0, 1) denotes a standard Gaussian distribution with mean 0 and variance 1. The distribution formed by contaminating this with 10 percent of a Gaussian distribution with mean 0 and variance 9 is denoted by 10% G(0, 9). A distribution with longer Lorentzian tails may be formed by dividing a standard Gaussian variate by an independent variate with a distribution uniform on the interval (0, 1). This distribution is denoted by G/U.

The estimation was of the parameters of a second-degree polynomial fitted to 20 equally spaced points. Thus three coefficients were calculated corresponding to the constant term and the linear and quadratic orthogonal polynomials. The coefficients, obviously uncorrelated, were rescaled to have variance 1 under least squares and the Gaussian distribution. The discussion will be in terms of the sum of these three variances.

DISCUSSION

The SINE estimate with

$$\Psi(z) = \sin(z/c) \quad |z| \leqslant c\pi$$
$$= 0 \quad \text{otherwise}$$

will be compared with the least-squares estimate LS and the least-absolute-deviations estimate Ll. Table 1 gives the sum of the variances for the Ll→(S2.1)[1] estimate.

The variance associated with this SINE estimate is slightly (16 per-

TABLE 1 Sum of Variances and, in Parentheses, the Percent This Sum Is in Excess of the Minimum Sum Attained for the Same Distribution

Estimate	G(0,1)	%	10% G(0,9)	%	G/U	%	Total %
L1	4.63	(54)	5.49	(28)	29.0	(21)	(103)
L1→(S2.1)[1]	3.47	(16)	4.28	(0)	23.9	(0)	(16)
LS	3.00	(0)	5.26	(23)	∞	(∞)	(∞)

TABLE 2 Sum of Variances for L1→(Bc)^1 for Various Values of c and, in Parentheses, the Percent This Sum Is in Excess of the Minimum Sum Attained for the Same Distribution

Estimate	G(0,1)	%	10% G(0,9)	%	G/U	%	Total %
L1	4.63	(54)	5.49	(31)	29.0	(24)	(109)
L1→(B5.25)^1	4.02	(34)	4.69	(12)	23.3	(0)	(46)
L1→(B6.3)^1	3.69	(23)	4.42	(5)	23.4	(0)	(28)
L1→(B7.4)^1	3.46	(15)	4.27	(2)	23.9	(3)	(20)
L1→(B8.5)^1	3.31	(10)	4.20	(0)	24.8	(6)	(16)
LS	3.00	(0)	5.26	(25)	∞	(∞)	(∞)

cent) larger than that of least squares for the Gaussian distribution. However, for a distribution quote close to the Gaussian, 10% G(0, 9), the variance is 19 percent smaller than the LS estimator. For the longer tailed distribution G/U, the LS variance is infinite and the SINE estimator has a smaller variance than that of L1.

The constant c in the expression for Ψ affects the efficiency of the estimates. Table 2 presents the sum of the variances for the L1→(Bc)^1 estimates for various values of c, again for the same three distributions.

For the Gaussian distribution, larger values of c correspond to smaller values of the variance. In the limit as $c \to \infty$ the estimate and its variance correspond to least squares. For distributions far from the Gaussian, such as G/U, the variance *increases* with c. Some compromise is required in selecting an estimate useful when the distribution is not known (the usual case).

TABLE 3 Sum of Variances for L·→(S2.1)^M and, in Parentheses, the Percent This Sum Is in Excess of the Minimum Sum Attained for the Same Distribution

Estimate	G(0,1)	%	10% G(0,9)	%	G/U	%	Total %
L1	4.63	(54)	5.49	(33)	29.0	(21)	(108)
L1→(S2.1)^1	3.47	(16)	4.28	(4)	23.9	(0)	(20)
L1→(S2.1)^2	3.34	(11)	4.18	(1)	24.7	(3)	(15)
L1→(S2.1)^4	3.31	(10)	4.17	(1)	26.1	(9)	(20)
L·→(S2.1)$^\infty$							
LS→(S2.1)^4	3.20	(7)	4.14	(0)	29.0	(21)	(28)
LS→(S2.1)^2	3.15	(5)	4.12	(0)	39.8	(67)	(72)
LS→(S2.1)^1	3.08	(3)	4.22	(2)	103.9	(334)	(339)
LS	3.00	(0)	5.26	(28)	∞	(∞)	(∞)

The number of iterations and the starting point also affect the variance. Table 3 presents the variance of the SINE estimator for 1, 2, and 4 iterations, again under the same distributions.

When L1 is the starting estimate, the variance decreases with the number of iterations for the Gaussian distribution and for 10% G(0, 9) but increases with the number of iterations for G/U.

When LS is the starting point, the variances increase with the number of iterations for the Gaussian distribution but decrease with the number of iterations for G/U. This is consistent with the view that LS is the optimal estimate for the Gaussian distribution from which further iteration will tend to move, while LS is a very poor estimator for G/U and further iteration can improve this.

CONCLUSION AND TENTATIVE RECOMMENDATIONS

There is little evidence to support the view that data from experiments have Gaussian distributions and much to suggest that the distributions have longer tails. The *M* estimators with Ψ functions satisfying

$$\lim_{|z|\to\infty} \Psi(z) = 0$$

are resistant to large deviations and have smaller variance for the longer tailed distributions and only slightly larger variance for the Gaussian distribution. These *M* estimates may be easily calculated iteratively. A small number of iterations 1 or 2 starting from the Ll estimate may yield estimates combining the resistant properties of the Ll estimate with the efficiency of the iterated *M* estimate.

The estimates $L1\to(B9)^1$ or $L1\to(S2.1)^1$ have standard errors less than 9 percent larger than the least squares estimate if the errors had a Gaussian distribution and have much smaller standard errors than least squares for the more typical longer tailed distributions. For *very* large data sets where the L1 estimate may be prohibitively expensive, the estimates $LS\to(B9)^6$ or $LS\to(S2.1)^6$ might be useful alternatives. While the properties of these fitting techniques have not yet been explored for situations of any substantial complexity, their performance is, so far, very encouraging. It would thus seem sensible to try them in more complex situations and see how they perform.

This Research was supported in part by the National Research Council of Canada and, while the author was on leave at the Department of Statistics, University of Chicago, by NSF Grant GP 32037x.

REFERENCES

1. D. F. Andrews, P. J. Bickel, F. R. Hampel, P. J. Huber, W. H. Rogers, and J. W. Tukey, *Robust Estimates of Location: Survey and Advances.* Princeton University Press, Princeton, N.J. (1972).
2. J. F. Claerbout, *Geophysics,* **38**(5), 826 (1973).
3. P. J. Huber, *Ann. Math. Stat.,* **35**, 73 (1964).
4. S. Newcomb, *Astron. Pap.,* **9**, 1 (1912).
5. D. A. Relles, *Technometrics,* **12**, 499 (1970).
6. "Student," *Biometrika,* **19**, 151 (1927).

W. L. Nicholson

Some Revised Approaches to X-Ray Crystal Structure Analysis Calculations

This note is a progress report on a reanalysis of the International Union of Crystallography (IUCr) Single-Crystal Intensity Measurement Project on *d*(+)-tartaric acid. X-ray structure factor data are currently being analyzed by classic least squares procedures with complex pseudo-theoretical model components to account for observed inadequacies of the classic model, a linear combination of Gaussian characteristic functions. The purpose of the present reanalysis is to illustrate that introduction of terms for systematic error into the classic model and use of fitting techniques not crucially dependent on the Gaussian error assumption may bring the results of supposedly discrepant collaborative studies into agreement. Specific results are not as yet available as the needed computer programs are still in the checkout stage.

X-ray structural calculations have been the subject of much study and careful development, essentially in the framework of classic least squares analysis. From the point of view of a modern data analyst, two sorts of increased realism appear vital: recognition of systematic error and introduction of methods that do not depend crucially on utopian assumptions.

The results of collaborative x-ray-diffraction studies have shown substantial systematic errors. There is no reason to believe such errors are absent from routine structure determinations. Therefore the expressions fit to observed intensities should include terms that allow for such systematic errors. Adding these terms will alter both fitted constants and the estimated variances of such constants. (If we can bring the internally estimated variances closer to the differences seen in collaborative studies, we will have gained invaluable realism.)

Behavior corresponding to belief in the exact Gaussian distribution of error has long been blamed on "mathematicians" by "scientists" and vice versa. Nowhere has careful examination of the consequences of such behavior shown it to be desirable. Techniques for robust/resistant fitting are just now being developed. X-ray diffraction is but one of many areas of molecular structure where such techniques are almost sure to be needed.

The structure factor data from the IUCr Single-Crystal Intensity Measurement Project on *d*(+)-tartaric acid were selected to illustrate the effect of incorporating the robust/resistive fitting techniques discussed in the Andrews and Tukey papers preceding. Part I of the IUCr project[1] identified systematic intensity, angle, and Miller indices effects between experiments. These could be partially, but not completely, accounted for by the instrument type and the incident radiation. As a result, it was concluded that sets of scaled structure factors measured by different experiments under the conditions of this project would most probably differ by about 6 percent, with much poorer agreement a distinct possibility. In Part II of the Project[2] each of the 17 sets of structure factors was subjected to a 10-atom least squares refinement using a single scale factor and three positional and six anisotropic thermal parameters for each of the oxygen and carbon atoms. A second refinement was attempted with three positional and a single isotropic thermal parameter for each of the six hydrogen atoms. In both these refinements the systematic effects noted in Part I of the project carried over into variability among parameter estimates from the 17 experiments; the variability was significantly greater than expected when measured by least squares estimated standard deviations. It was concluded that standard deviation estimates for the positional parameters were too small by a factor of $\sqrt{2}$ to 2 and that the thermal parameters were about twice as bad.

A least squares program, written by Larry W. Finger and modified by Edward Prince is being further modified to include:

1. The BIWEIGHT function $[1-(u/c)^2]_+^2$ for smooth, soft structure refinement;
2. The ability to look for systematic model discrepancies with appropriate computer plots; and

3. The ability to include and refine such systematic effects in further iterations.

At present, points (1) and (2) of these modifications are complete, while point (3) is still in the checkout stage. This work will continue with the goal of an efficient program for robust/resistive estimation of structural parameters using x-ray-diffraction data. If successful, inclusion of experiment-dependent systematic effects should allow agreement among different experiments on positional parameters when measured against the proper robust/resistive standard deviation estimates. If Zachariasen's[3] conjecture is correct, the thermal parameters will also agree but will have such large standard error estimates that the thermal portion of the model must clearly be "nonsense."

Possibly at some future data, a complete robust/resistive reanalysis of the IUCr project Part II data can be reported to interested crystallographers for their evaluation.

REFERENCES

1. S. C. Abrahams, W. C. Hamilton, and A. McL. Mathieson, *Acta Cryst.*, **A26**, 1 (1970).
2. S. C. Abrahams and W. C. Hamilton, *Acta Cryst.*, **A26**, 18 (1970).
3. W. H. Zachariasen, *Acta Cryst.*, **A25**, 276 (1969).

John W. Tukey

Data Analysis for Molecular Structure: Today and Tomorrow

My task here is to summarize where we seem to stand in regard to the diversity of data-fitting problems encountered in the experimental fields covered by this conference. As we have moved through this part of our discussion, the emphasis has shifted—sometimes insensibly—from doing a better job of fitting our temporarily final model to fitting as a part of diagnosis, as a part of an ongoing inquiry into the data. This is as it should be. If each of us moves a little further down this road, the data analysts and statisticians will feel that their efforts are more than repaid, and the chemists and physicists will practice their arts more effectively.

The marriage of data and model is always a trial marriage. None of our models dares to assert itself completely. All have omissions. Our task is to choose a model whose omissions are well below the level of failures and then to use this model to extract useful and communicable information—perhaps the length of a C–C bond—from our data. In doing this, our single most important task is to seek the incompatibilities of each trial marriage, to ask whether, and to what extent, the data call for modification of the model that we are trying. Diagnosis is of the greatest importance.

To diagnose a model means being prepared for behavior in the data not describable by that model. Any or all of systematic error, erratically measured points, or omitted higher order terms can contribute to such behavior, to behavior that the model we are diagnosing cannot represent. The principle is always the same; only the practical details vary.

Systematic errors and higher order terms are always better diagnosed by fitting a model whose parameters include, in addition to those of the model being diagnosed, parameters apt to absorb the appropriate kinds of systematic behavior. Erratically measured points and points that are almost alone in responding to some unmeasured parameter are better diagnosed by examining appropriate (not least-square) residuals. We need to expect to include extra constants, to choose a fitting procedure that is not dangerously effective in covering up otherwise illuminating residuals, and to examine our residuals both individually and collectively.

Two comments: It seems a little misleading to speak of the presence or absence of "model error" since all models are somewhat in error; hence, the use of "omissions" and "failures" above. As Kirchhoff[5] emphasizes, gaining accuracy only through having the "uncertainties for the derived parameters . . . diminish . . . in proportion to the square root of the number of observations" is not always very rewarding. Similarly, using a carefully tuned fitting process to reduce the variance a few percent—corresponding to taking a few percent more observations—is even less likely to be very rewarding. True, we can use all the data-analytical resolution we can get when we are concerned with diagnosis, but there are better ways to do this than to squeeze the variance down; our needs in postdiagnostic fitting are far less compelling.

THIRD TYPE OF FITTING

Three main types of fitting are involved in the papers discussed in this volume:

1. (Discrete) intensity fitting as in x-ray crystal structure analysis;
2. (Line) location fitting as in visible, infrared, and microwave band spectroscopy;
3. (Curve) shape fitting as in electron spin resonance and gas electron diffraction.

The practitioners of shape fitting may rightly claim some neglect: The data analyst/statistician papers here presented have emphasized the other two types of problems, not because it was felt that shape fitting offered no important data-analytical problems, but rather because the purely data-analytical problems of intensity fitting and location fitting seemed simpler, better defined, and easier to attack. (This says nothing about the relative difficulty of problems of numerical stability or of handling larger bodies of data.) To make really sensible statements about shape fitting would require experience with working through, and reworking, realistic examples involving real data, experience that those who write here from a data analytic/statistical viewpoint have not yet gained. So anything I say now about shape fitting ought to be quite general (else it might well prove misleading).

EVOLUTIONARY STAIRCASE

The history of shape fitting in any field seems to follow a standard pattern:

- Look at the output trace and count peaks (and valleys).
- Simulate shapes from assumed parameters and try to put the peaks and valleys in the places they were observed.
- Build the trial-and-error procedure just described into a least square fit (one that is often quite nonlinear).
- Learn to use a fitting procedure, eventually not a least square procedure, to diagnose the nature and, perhaps, to help in suggesting the causes of the model failures, of the inabilities of any of the possibilities from which the fit is selected to fit the actual situation.

Where on these stairs stand the various fields in which shape fitting is the main issue? Judging by the references to which some of those organizing this conference were kind enough to direct our attention, most such fields are on the second or third steps. We ought to expect them to move up to the fourth step soon, to face problems rather like those that intensity fitting and location fitting have already begun to face.

ESSENTIALS FOR ALL THREE TYPES OF FITTING

The same essentials will be there for shape fitting as for intensity fitting or location fitting:

1. Recognition of the fitting process as first and most importantly a *diagnostic* procedure. (Only after cycles of diagnosis, each involving fitting, examination of residual and extra-term behavior, reconsideration of models, and repair or discarding of data, can it become appropriate to make a data analysis that can be hoped to be terminal!)

2. Emphasis on *residuals* as the key tools in diagnosing the apparent difficulties of each trial marriage between data and model. (Nothing except well-selected additional fitted constants can be as revealing!)

3. Recognition that plain least squares does the most, among all methods likely to be considered, to make residuals relatively UNinformative, relatively UNsatisfactory guides to diagnosis. (Diagnosing with least-squares residuals is a little like diagnosing the illness of a supermodest patient, one the doctor is not allowed to see, touch, or collect specimens from!) If least-squares residuals are all we have, of course, we should use them for diagnosis as thoroughly as we can.

4. Recognition that other methods of fitting—methods that may (as a practical convenience) incorporate iteratively weighted least-squares fits as inner parts—can provide both more useful, more diagnostic residuals and estimates of parameters that are usually of higher quality than those found by plain least squares. (Fitting procedures that *appear* to deny "full weight" to some points are essential if only reasonable efficiency of fitting is to be attained, even when disturbances are random and model failures are absent!)

5. Recognition that every measurement procedure is subject to some kind of systematic error and that the only way to deal with such systematics is *to include them in the fits.* (A fit cannot take account of what you leave out of it. Even the best residuals are less sensitive indicators of systematic error than are fitted coefficients for systematic terms!)

6. Recognition that including additional flexibility in a fit ordinarily both alters the meaning of old parameters and increases the actual variability with which they are estimated—the good of the improved meaning very often outweighing the evil of the increased variability. (To measure better things is usually harder!)

7. Recognition that reducing the random errors of the results by only a few percent is rarely worth the trouble, especially since the randoms are likely to be dominated by the systematics. (Reducing a lesser cause of variation can only be worthwhile if the reduction is large!)

8. Understanding that if a final few percent improvement in quality of fitted coefficients is to be real, rather than fallacious, it must be based on a rather detailed understanding of the shape of distribution of the disturbances and the changes in variability from one part of the data to another. (It is not easy to do better than what, fortunately, is usually good enough!)

9. Understanding that near linear dependences in the modeled effect of changing different parameters leads to great uncertainty as to how the corresponding part of the observed effect is to be allocated among

TABLE 1 Ten Important Points Summarized

1. **Diagnosis** is the most important purpose for a **fit.**
2. **Residuals** are unexcelled for diagnosing the **unexpected.**
3. **Least-squares** residuals are relatively **uninformative.**
4. **Other fits** are better, i.e., more diagnostic, safer, and even more precise.
5. **Systematic errors** are present in your data.
6. **A fit** cannot take account of what you **omit.**
7. **Accordingly,** systematic-error collecting terms should be **fitted.**
8. Estimating **better chosen** quantities is usually **harder**; it is also often worthwhile.
9. Reducing the **internal** standard error the last few percent, which requires very detailed knowledge, is **rarely** worthwhile.
10. The presence of even near-linear **dependencies** should **always** be **looked for** and, if found, should be **flagged.**

these effects. (For molecular structure examples and approaches see, for instance, Curl[3] and Less.[6] This problem can be crucial, and the question of the correct approach to it deserves careful thought!)

These points are important enough to deserve recapitulation, as in Table 1.

DATA DISPLAY AND GATHERING

Back to Shape Fitting

There is one distinctive feature of shape fitting—one opportunity to choose for better or worse not shared by intensity or location fitting. This is the choice of how one "looks at the curve." In electron spin resonance, for instance, it is commonplace to look at a curve that comes close to representing the first derivative of the absorption. It would be possible to look at a wide variety of other presentations: suggestions of "sharpening"[1] only scratch the surface of the possibilities. All that should be said here is that this is not a simple problem—surely not one for which an *a priori* calculation can give a satisfactory answer—and that it is hard to see why something close to a first derivative is the best choice.

Data Gathering—Intensity Experiments

Insights suggested by data analysis and statistics are frequently even more helpful in rearranging the details of data gathering than they are in the improvement in the final analysis. Three possibilities deserve attention here; all seem applicable to intensity experiments (applications to location and shape experiments are more doubtful):

- Blocking, dividing the taking of data into distinct time intervals and allowing for changes from one time interval to another in the fitting;
- Degradation compensation, measuring each intensity at several epochs during the exposure of the specimen followed by extrapolation to zero exposure;
- Replication, measurement divided over two or more specimens so as to do some averaging out of the idiosyncrasies of individual specimens.

Blocking offers a method of detecting (if large) and at least partly compensating for (if small) effects that are actually—whether meaningfully or accidentally—associated with epoch. To obtain the most from blocking, the order of data taking should be such as to make the possible effect of each block as poorly fittable as possible by the effects of parameter change. This will minimize the loss of leverage in estimating parameters by the converse relation (fittability of parameter change effects by block effects).

If such choice of data order can indeed be made, there are smaller but definite advantages to randomizing the order of measurements within each block and to making the end of natural intervals in the data-taking process coincide with the ends of blocks.

The feasibility of degradation compensation depends on how well a consistent pattern of change of intensity with exposure can be developed.

Greater use of replication may seem to some counter to usual experimental practice, but can give better values for equal effort.

Kirchhoff's technique[4, p.355] of selecting lines for further measurement should not be overlooked.

Dissection Techniques

As techniques for measuring line profiles improve, the use of the techniques of shape fitting, both for locating centers of individual lines and for resolving overlaps, should come into play. Conversely, the use in shape experiments of careful fitting techniques for the unoverlapped portion of a line of modelable shape may allow us to see better the overlapping line after the fitted line shape of the first line is subtracted. This is not likely to be feasible as a terminal process, but may be useful at an early stage.

CARE AND FEEDING OF RESIDUALS

Looking at Residuals

Careful authors in fields like electron spin resonance[7] are now showing not only the observed and fitted curves but a curve of the residuals—of the differences observed minus fitted—at the same scale. This is good and deserving of praise, but it is not good enough. One virtue of residuals, once fitting is relatively close, is their ability to be shown on an enlarged scale. Most readers brought up in the United States know Linus, the comic-strip character who goes everywhere accompanied by his blanket—his symbol of security. A condensed picture of residuals is a "security blanket," an attempt only to show how close the fit appears, rather than an attempt to show forth the characteristics and detailed behavior of the lack of fit. It is of the latter that we need more. The failure of carefully attempted diagnoses to show anything meaningful is the best seal of experimental (and analytical) quality.

Intensity fitting and line fitting deal with residuals that can be much more reasonably rearranged into various orders. A careful examination of a set of their residuals is likely to need examination of their appearance when plotted against a variety of abscissas (chosen to illuminate more or less plausible classes of possibilities). Especially in intensity fitting, a reasonably large number of residuals will be available; as a result, merely plotting a cloud of points without partial summarization inevitably reduces the sensitivity of our examination. (I was brought up to think highly of just plotting the points. For more than a few points, I have learned better.) Techniques exist for drawing useful curves (smoothed broken lines[8]) not only near the center of the point cloud but also toward its upper and lower edges; the locally highest and lowest points will still need to accompany these curves and be plotted individually.

Making Residuals Comparable

Assuming constant variability for the disturbances when the analysis is specified and having this assumption correct are not enough to make the long-run size of all residuals the same. If, for instance, we fit a straight line by equally weighted least squares to five equally spaced points and the model applies precisely, we find the following:

	End Points	Intermediate	Center
Variance of fitted line	$0.6\sigma^2$	$0.3\sigma^2$	$0.2\sigma^2$
Average {residual^2}	$0.4\sigma^2$	$0.7\sigma^2$	$0.8\sigma^2$

Note that in each case, the sum is $1.0\sigma^2$ as it must be for least squares, where these two terms are uncorrelated. Factors even larger than the $0.8\sigma^2/0.4\sigma^2 = 2$ shown here can arise when, as is not as rare as we would wish, a single observation furnishes a large share of the information about one parameter. The consequence of least squares is a great squeezing together of the distribution of the residual corresponding to such a single observation.

If we want to work with residuals in a comparative way—if we want to go beyond naive and qualitative (but effective!) diagnosis—we need to rescale our residuals to make them really of comparable size.

The Least Squares Cases When the residuals come from linear least squares, not only the parameter estimates but even the fitted values are linear combinations of the observations with coefficients that can be calculated. If we know the value of

$$\frac{\partial \hat{y}_i}{\partial y_i},$$

where $\hat{y}_i$ is the fit at the place where y_i is observed, then the appropriately scaled-up residual might seem to be

$$\frac{y_i - \hat{y}_i}{1 - \partial \hat{y}_i / \partial y_i}$$

where the partial derivative only depends on the x's. Such an expression neglects the contributions to this residual from the disturbances in the other observations. An intermediate divisor is needed. Actually, for least squares, only the square root of this divisor should be applied; thus the correct scaled-up residual is

$$\frac{y_i - \hat{y}_i}{\sqrt{1 - (\partial \hat{y}_i / \partial y_i)}}.$$

When the weights used in a weighted least squares fit do not reflect the variances of the observations, the result is more complicated, but qualitatively similar.

Two lessons are to be drawn from this: Rescaling is likely to be essential; and residuals that have to be heavily rescaled are less informative about the actual disturbance at the point concerned because a larger fraction of them comes from the propagated effects of other observed values. In drawing inferences about disturbance distributions, for example, we need to be wary of heavily rescaled residuals—rescaling them was desirable, but their results may not be wonderful.

In nonlinear least squares we have the same relation. The only change is that $\partial\hat{y}/\partial y$ now depends on the general nature of the fit and only indirectly on the observed y's.

Diagnostically Satisfactory Fits We know what the response of any diagnostically satisfactory fit to the changing of one observation is qualitatively like: a linear response for y close to the fit (here y is being "measured"); a decreasing rate of response as y is more deviant (when this response is constant, the observation is being "counted" as high or low); and a decreasing response, eventually leaving the fit where it would have been had the observation been absent (once this stage has been reached the observation is being "forgotten"). This qualitative behavior applies as well to the fitted value $\hat{y}_i$ that corresponds to y_i as it does to each parameter or to any other linear combination of parameters. Accordingly, $\partial\hat{y}_i/\partial y_i$ is a maximum for $\hat{y}_i - y_i \sim 0$ and falls, first, to zero and, then, below zero, and then returns to zero. This means that the divisor required to rescale $y_i - \hat{y}_i$ now depends on the size of $y_i - \hat{y}_i$ as well as on the structure of the data. Qualitatively, this divisor is less than 1 for $y_i - \hat{y}_i \sim 0$, increases to a value greater than 1 as $y_i - \hat{y}_i$ increases in magnitude, and decreases back to something that is still greater than 1 as the residual increases still further.

A Statistician's Burden If we are to have satisfactory methods for inferring shapes of disturbance distributions from sets of residuals, we need to implement the sorts of rescaling and careful analysis just hinted at, making the implementation easy to use. While this can presumably be done without too much effort for the plain least squares case, it does not seem, in fact, to have been even considered for the diagnostically satisfactory procedures. Clearly, statisticians have an obligation to proceed apace with this implementation.

FITTING TOMORROW

Diagnostic Fits

From evidence here and elsewhere, I can now prescribe what I would recommend using in diagnostic fitting. To use the notation of Andrews (this volume, p. 41), if an L1 start—a fit minimizing the sum of absolute residuals—is conveniently available, I would recommend

$$\text{L1} \rightarrow \text{B9}^{\text{once}}$$

and for those who prefer a SINE-based polishing process, something like L1 → S2.6^{once}.

No guarantee is given that this procedure will give the best fit, none is needed, none is even appropriate. A guarantee can be given, and is, that the diagnostic properties of this fit will prove to be much better than any familiar fit and that the quality of estimation will be quite adequate. Of course, this recommendation will include a call, in the strongest terms, for inclusion in the fit of parameters corresponding to the likely sorts of systematic errors. Including systematics has three main effects:

1. Residuals will be made appropriately smaller, thus reducing the noise background against which diagnosis must take place;
2. Estimated variabilities for the parameters estimated will be much more realistic; and
3. Warnings will be issued about the apparent presence of certain kinds of systematic error.

As long as we are concerned with either the important aspect of fitting, namely diagnosis, or with the assessment of variability for our estimates, no valid argument speaks against the inclusion of systematics in our fitting.

The sort of procedures we have just presented will do—automatically and quite effectively—the things that Kirchhoff has described as manual processes for the case of microwave spectroscopy.[4] Whether the extreme disturbances are due to model failure, or merely to long tails on the distribution of actual fluctuations, matters not; fits of this kind will give the best diagnosis. The fact that a residual is large enough to be downweighted is a reason for suspicion, it will be worthwhile taking a harder look at that observation's provenance. However, even acquittal from suspicion would not cause us to want to change our downweighting. Especially for diagnosis—and to only a slightly lesser degree for high quality estimation—the downweighting of unusually deviant residuals is called for by all probability models that are realistic about the tails of the distribution of disturbances.

Terminal, Postdiagnostic Fits

Personally, I would use a diagnostically satisfactory fit for my terminal fit. I would probably leave my systematics in, also. Here, we need experience that I still lack before we can be clear about this choice. At a minimum, I would use error estimates for the parameters calculated as if these systematics were included.[2]

Some will insist on going back to least squares for their postdiagnostic fit. So much the worse for them. They still owe it to themselves to include their best judgment of systematics, at least in the calculation of error estimates for the parameters.

Before too long, if statisticians do our duty, we will have refined methods of postdiagnostic fitting, whose advantages over least squares will be clear and irrefutable to all those who are willing to read. Who will do what then in choosing a terminal fit? I suspect that I will continue to use the diagnostically satisfactory fits, as will some others. The use of plain least squares ought to die out in favor of a combination of diagnostically satisfactory and carefully tuned postdiagnostic fits. All this will be of lesser importance than the doing of more and better diagnosis.

Prepared in part in connection with research at Princeton University sponsored by the Atomic Energy Commission.

REFERENCES

1. L. C. Allen, H. M. Gladney, and S. M. Glarum, *J. Chem. Phys.,* **40**, 3135 (1964).
2. L. S. Bartell and H. Yow, *J. Mol. Struct.,* **15**, 173 (1973).
3. R. F. Curl, Jr., *J. Comput. Phys.,* **6**, 367 (1970).
4. W. H. Kirchhoff, *J. Mol. Spectrosc.,* **41**, 333 (1972).
5. W. H. Kirchhoff, *J. Mol. Spectrosc.,* **41**, 367 (1972).
6. R. M. Less, *J. Mol. Spectrosc.,* **33**, 124 (1970).
7. M. Sugié, T. Fukayama, and K. Kuchitsu, *J. Mol. Struc.,* **14**, 333 (1972).
8. J. W. Tukey, *Exploratory Data Analysis,* Addison-Wesley, Reading, Mass. (in press).

Discussion Leader: JOHN W. TUKEY

Reporters: WILSON H. De CAMP
ALBERT E. BEATON

Discussion

The concept that the best fit to a set of experimental observations is not necessarily provided by a least squares analysis was central to the opening session of the conference. The resulting discussion was so extensive and wide-ranging that its sense would have been difficult to communicate by direct quotations. In avoiding this approach, the reporters have elected to summarize the majority of the comments without, in general, identifying the speakers. While effort has been made to preserve the intent of the various speakers, their thoughts have been extended at places, where appropriate, in order to assure understanding on the part of the nonspecialist.

The problems associated with diagnosing errors by inspection of least squares residuals are particularly familiar to those who work with spectroscopic or diffraction data. The statement by Tukey that such values are uninformative was taken by some participants to be too strong. General agreement was reached that, because of the nature of the least squares fitting process, the residuals were not as informative as they could be. Conceivably, they could be sufficiently uninformative that significant points could be missed. With fitting methods more resistant than least squares, more information may be obtained since outlying values have less influence on the fit. More information may sometimes be obtained by examining the standardized residuals, which are defined as the residuals divided by their estimated standard deviations, whose values allow both for differences in variance from one observation to another, when known, and for differences in the degree of overfitting associated with the character of what is fitted (easily calculable). Least squares analyses are less informative than a more resistant method would be. A data point in poor agreement with the fitted value would be more likely to remain in disagreement according to a resistant method than it would according to least squares.

The universal presence of systematic error is exemplified by the Single Crystal Intensity Measurement Project of the International Union of Crystallography.* Seventeen laboratories throughout the world measured structure amplitudes for *d*(+)-tartaric acid with as high precision as possible. Seven of the sets of data appeared to be severely affected by systematic errors. The exact nature of these systematic errors was difficult to deduce under the experimental conditions. In a case such as this, a common interpretation is that an outlying measurement is necessarily faulty. However, it may be that the outlier is really the only good value, in which case statistical treatment alone is uninformative.

The consideration of systematic experimental errors involves semantic problems as well as questions of proper statistical treatment. Although random errors are well-defined in the mathematical sense, the same cannot be said for systematic errors. The effect may be to bias all values in the same direction. In other cases, effects are uniform over a group of data, but differ between groups, or the effect can be so variable in its influence on the data as to be nearly indistinguishable from a truly random error. Careful use of terms is essential to full understanding of this aspect of data analysis.

In practice, the distinction between random and systematic errors is the extent to which they are revealed by the data. Consistent, constant errors cannot be seen in the data, no matter how the experimenter looks for them. An essential consideration in data analysis is the finding and use of procedures that will best reveal errors one might call "systematic separable." Model-related errors of this type are frequently encountered.

The insight of an expert in the field in question, when suitably guided by good displays, is more effective than any purely statistical treatment when it comes to identifying a systematic separable error of unknown form. Plotting residuals as a function of various reasonable parameters can be informative. Other approaches that might be informative are Fourier-transform methods of varying the model on an empirical basis to see what happens. While it does not amount to a complete diagnosis of a defect in the model, unexpected or physically unrealistic behavior of parameters in the model being fitted (e.g., non-positive definite thermal ellipsoids in crystal structure determination, or imaginary interatomic distances in spectroscopic calculations) may result from defects elsewhere in the model. When replicate measurements of a given data point are available, it may be most prudent to fit to the individual values before averaging them. This approach may identify faulty measurements early in the fitting process, particularly if the model includes a parameter to account for variations between individual measurements or if resistant methods of fitting are used.

Standard deviations in the parameters that describe the model may be estimated from the square roots of the diagonal terms of the variance/covariance matrix. These standard errors are purely internal to the calculation in that they are correct only when consistent systematic errors are absent from the data. The values would

* S. C. Abrahams, W. C. Hamilton, and A. McL. Mathieson, *Acta Cryst.*, **A26**, 1 (1970).

be more meaningful if known sources of error could be included. TUKEY suggested that, for each source of consistent error, a variance/covariance matrix could be written to describe how that error propagated through the calculations. These matrices describing the consistent systematic errors could be combined with the variance/covariance matrix describing the random errors to produce a more physically reasonable estimate of the uncertainty in a quantity derived from the fitting process.

In some cases, constraints on the model exist, such as restrictions placed on atomic positional or thermal parameters for atoms in special positions in a crystal structure. The question was raised of the proper place for such information in the fitting process. The more certain the knowledge is, the more reasonable is its inclusion. Allowing the model more freedom to fit the data by omitting known constraints may be unwise, particularly when a variable in the model can be adjusted to account for systematic error.

Although it may well be true that the method of least squares is widely misused because of its apparent objectivity and general availability, it was clearly also true that much of the information obtainable from least squares is not used as completely as it could be. The problem of correlation between model parameters illustrates this clearly. High correlation between parameters amounts only to a statement about the data structure as opposed to the data values. The essential issue is the nature of the dependence of the parameter being determined on the data set. If two parameters have similar dependences, then their estimates are going to be correlated. Measuring more data points or a different set of data points would result in a different correlation matrix. The physical limitations of the experimental method, such as the inability to measure spectral characteristics of weak transitions or transitions that fall in inaccessible frequency regions, make it impractical to avoid correlations.

It is not generally necessary to avoid correlations, although techniques exist for reducing correlation coefficients by changing the structure of the data analysis. It is more important to reduce the standard errors of interesting quantities than to reduce the correlations.

PLÍVA noted that more understandable fits can be obtained for vibrational and rotational constants by fitting one constant and the difference between it and each of the others, rather than by fitting all the constants simultaneously. Although the constants for the lower state will be poorly determined, the differences will be obtained with much higher precision. Furthermore, even if the individual constants are correlated, the differences between various constants are generally roughly orthogonal to the comparison of the rotational constants themselves, thus reducing the correlation. RAMSAY responded with a comparison of the rotational constants for *cis*-glyoxal for a two-state fit* with those for which the ground-state constants were fixed. The differences between the two fits were significant, although the difference between the optical spectroscopic and microwave values for the ground-state constants was adequately accounted for by the standard errors. The use of orthogonal parameter sets in describing fits may be particularly important when fitting power series. Variability in parameters may be more clearly

* G. N. Currie and D. A. Ramsay, *Can. J. Phys.*, **49**, 317 (1971).

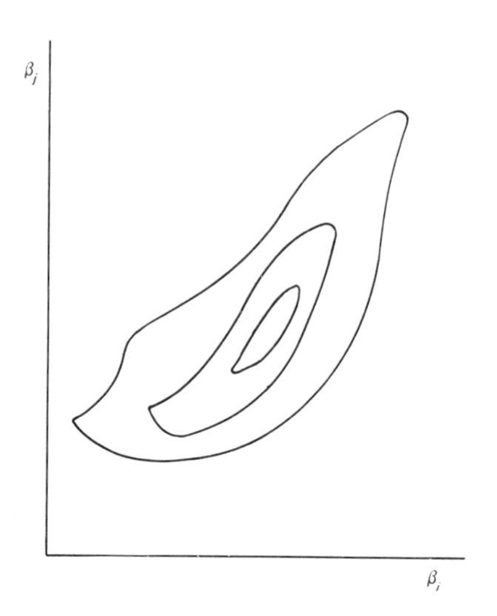

FIGURE 1 Contours (schematic) of function $(N_{\text{o}} - N_{\text{p}})\ln s^2(\bar{\beta})$ plotted against two parameters β_i and β_j.

identified, so that the user is more aware of which combinations of parameters are well- or ill-estimated.

A means by which the effects of correlations can be visualized was presented by ANDREWS, who suggested plotting the function $(N_{\text{o}} - N_{\text{p}})\ln s^2(\bar{\beta})$, where N_{o} and N_{p} are the numbers of observations and parameters, respectively, and $s^2(\bar{\beta})$ is a measure of the fit of the parameter set $\bar{\beta}$, such as the median residual or the mean square residual. Contours of this function, plotted against two individual parameters, β_i and β_j, which are allowed to range over several standard errors from the fitted values, graphically present the correlation between β_i and β_j (Figure 1). For the linear case, the mean square contours are ellipsoidal, but they have been known to vary widely from this shape in nonlinear problems.

Although correlation between parameters is a function of the data structure and has nothing to do with deficiencies in the model, it has implications for both the choice of the model and the design of the experiment. EVANS described his experiences with the determination of the crystal structure of tetragonal barium titanate ($BaTiO_3$).* The problem was simple in that it involved only three atomic positional parameters (one for Ti and two for O), plus nine thermal parameters. There was considerable interest in the details of the structure because of the ferroelectric properties of the material. The proposed model was essentially a simple cubic arrangement of atoms, but with Ti displaced slightly from the center of an octahedron. By ordinary x-ray standards, this distortion (which was expected to be on the order of 0.15 Å) could be measured with a standard error of 0.01–0.02 Å if

* H. T. Evans, Jr., *Acta Cryst.*, 14, 1019 (1961).

150 reflections were measured to very high precision, which was done on one of the earliest single-crystal diffractometers. Large correlations between atomic displacements and thermal amplitudes of vibration along the same direction prevented determination of either. Although this case is an example of a particularly ill-conditioned problem, it could have been solved by redesigning the experiment. For instance, the use of a single-domain crystal would have permitted distinguishing by anomalous dispersion between the *hkl* and $\overline{hkl}$ reflections, thus changing the nature of the correlation.*

The importance of weights to the fitting process was also discussed. The resistance of the least squares fit to errors in weights was shown in TUKEY's observation that, as long as the weights are not out of line with respect to each other by a factor greater than two, the loss in efficiency must be less than one ninth and probably will be considerably smaller. Larger deviations rapidly become more costly. A naïve approach, such as the use of unit weights, may be apparently successful. However, there is no assurance that a fit based on data naïvely weighted will yield estimates and residuals that are as informative as they could be.

Purely statistical weighting functions are not always effective. X-ray crystallographers have been aware for some time that fitting a model to a set of structure amplitudes using weights based only on counting statistics is not satisfactory. Terms dependent on factors such as the magnitude of the structure amplitude must be included in the weight. Some sort of parameter that is related to the fit may also be needed to adjust the weights, particularly toward the end of a refinement. In some cases, weights are adjusted on empirical grounds on the basis of the fit of a few individual observations. This amounts more to using a limited ignorance factor than to a real improvement in the weighting function, but the purpose is served of identifying seriously discrepant observations and reducing their effect on the fit. Further study is clearly needed on the inclusion of fitting factors in observational weights. Caution is essential in placing reliance on such a practice. Weighting schemes in which the weight of an observation is inversely proportional to a power of its deviation from the fit can lead to catastrophic results, since a point that fits very closely is given an extremely high weight.

The simplest resistant fitting procedures use weights determined from deviations from fits. The two kinds of weight—the observational weight and the fitting weight—should be used multiplicatively. Almost any weighting function that falls off smoothly yields a fitting weight more or less in line with good statistical technique. Functions such as the following:

$$w \propto (1 - u^2)^2 \tag{1}$$

$$w \propto \frac{\sin(\pi u)}{\pi u} \tag{2}$$

* See further an x-ray and neutron diffraction study of this compound by J. Harada, T. Pedersen, and Z. Barnea, *Acta Cryst.*, **A26**, 336 (1970).

in which w is the weight and u is a measure of the lack of fit in an observation, are quite satisfactory in that they reach a maximum at $u = 0$, decrease slowly, and are generally smooth. TUKEY reported that a wide variety of weighting functions have been tested and found to be effective in treating data both with Gaussian errors and with several other oppositely unrealistic error distributions. Monte Carlo experiments were run using a particular weighting function and data sets with different error distributions. For example, one set might be Gaussian, while the observations in another set would have 5 percent chance of drawing an error from a Gaussian distribution but with 10 times the standard deviation. The weighting function was judged effective if it yielded variances close to the best that one knows how to reach in both these and other cases.

Cases of linear dependence between variables present problems that are often difficult to handle. A frequent treatment of this situation is a compression along principal components, after which one or more of the smaller eigenvalues is dropped. It is not always recognized that this procedure has implications for the model parameters that may or may not be physically realistic and that it may be worthwhile in such a case to redesign the experiment.

This type of problem occurs for all fitting methods. The identification of linear combinations that are ill-determined is almost always important in a statement of what has been learned from an experiment. A more resistant fitting method is not necessarily more helpful in dealing with cases of exact or approximate linear dependence.

The personal commitment of the individual scientist to the precision of his work is reflected in the statements he makes about its uncertainties. It was apparent to the reporters of this discussion that, although fitting methods more informative than least squares are beginning to appear, even the information that can be extracted from a least squares fit is not being used as fully as it might be. Improved methods of data analysis call for improved interfacing between the statistician and the experimentalist. Increased perception of both faulty data and imperfect models will make it easier to communicate with precision not only what has been discovered from an experiment but also what still remains to be discovered.

II

Molecular Geometry of Free Molecules

Victor W. Laurie

Definitions and General Theory of Interatomic Distances

To characterize the geometry of molecules, a model of point nuclear masses with fixed internuclear distances has proved very useful in many instances. As long as internuclear distances are not required with an accuracy of better than a few hundredths of an angstrom, this model is quite adequate,* and it is possible to speak of "the" geometrical parameters of a molecule. If greater accuracy is required, however, it becomes necessary to consider the consequences of the nonrigidity of molecules.

Because of the differing ways in which effects of nonrigidity enter, different experimental techniques do not yield exactly the same values for structural parameters. Furthermore, more than one set of structural parameters may be derivable from the same experimental data, depending on the way in which corrections are made for nonrigidity. As a result, structural parameters reported in the literature have a variety of meanings. In this paper various types of structural parameters commonly reported for gas-phase molecules will be defined and the interrelations among them discussed. A similar discussion has been given by Kuchitsu and Cyvin.[1]

* In the present discussion we shall not consider the special problems introduced by large amplitude vibrations such as methyl torsions (see H. Dreizler, this volume) or ring puckering (see W. J. Lafferty, this volume). Also, it is assumed that perturbations, such as nuclear hyperfine structure, have been corrected.

SPECTROSCOPIC RESULTS

Spectroscopic methods constitute one of the two major sources of gas-phase structural information. Either from pure rotational spectroscopy or from vibration/rotation fine structure, the rotational energy levels in a particular vibrational state can be obtained.* For a rigid molecule the rotational energy levels are characterized by the reciprocals of the moments of inertia that, in turn, are functions of the atomic masses and Cartesian coordinates in the principal axis system. The rigid-rotor formalism fits the rotational energy levels of most molecules remarkably well. That this should be true is not immediately obvious as is demonstrated by the literature of the early 1930's.[2–4] What the rotational energy levels of a vibrating polyatomic molecule should be in general was not known until Wilson and Howard[5] carried out their treatment showing that, to second-order in perturbation theory and with allowance for centrifugal distortion, the rotational energy levels of a vibrating molecule could be fit with "effective" moments of inertia. The relation between these effective moments of inertia and physically meaningful parameters is complex, however.

The experimental parameters actually determined are the effective rotational constants B_v^g from which the effective moments of inertia are obtained by the relation:

$$I_g^v = \frac{\hbar^2}{2B_v^g} \tag{1}$$

where v indicates the vibrational states involved and $g = a,b,c$. The vibrational dependence of the effective moments of inertia may be written as:

$$I_g^v = I_g^e + \sum_s (v_s + 1/2\, d_s)\, \epsilon_s^g\,, \tag{2}$$

where the summation is over the vibrational states, each characterized by a quantum number v_s plus a degeneracy d_s. I_g^e is the moment of inertia of the equilibrium configuration representing the minimum in the Born–Oppenheimer potential surface.[6] ϵ_s is a complex function of both harmonic and anharmonic force constants. In general, it is not possible to calculate it and only for small molecules is it experimentally feasible to measure it. Consequently, relatively few polyatomic equilibrium

* For excited electronic states, see D. A. Ramsay, this volume.

structures have been determined.[7] Therefore, spectroscopic determinations of structure generally must make use of effective rotational constants. These rotational constants are commonly experimentally determined to one part in 10^6 or better. However, the contribution of vibrational effects even in the lowest vibrational state is of the order of one part in 10^3 to 10^2. To see how the vibrational effects enter into the determination of structure by spectroscopic means, we shall first look at diatomics where the effects are simplest and then consider polyatomics.

Diatomic Molecules

For a diatomic molecule the effective moment of inertia has a well-defined physical significance since the effective rotational constant is inversely proportional to the average value of the square of the bond length. Thus the effective moment of inertia for a diatomic molecule is given by:

$$I_v = \mu \langle r^{-2} \rangle^{-1} . \tag{3}$$

The bond length determined from the effective moment of inertia is therefore:

$$r_v = \langle r^{-2} \rangle^{-1/2} .$$

It should be emphasized that the physical significance attributable to the effective moment of inertia of a diatomic molecule does not extend to polyatomic molecules where the vibrational effects are more complicated. This point is sometimes overlooked in the literature.

For purposes of comparison it is convenient to define the quantity:

$$\xi = (r - r_e)/r_e . \tag{5}$$

In Table 1 a comparison of several different kinds of bond length for diatomic molecules is made. In the tabulated bond lengths quantities of interest are given by:

$$\langle \xi \rangle = -a_1 (3B_e/\omega_e)(v + 1/2) \tag{6}$$

$$\langle \xi^2 \rangle = (2B_e/\omega_e)(v + 1/2), \tag{7}$$

where ω_e is the harmonic vibrational frequency and a_1 is a dimension-

TABLE 1 Comparison of Bond Lengths for Diatomic Molecules[a]

Bond Length	r/r_e
Effective $r_v = \langle r^{-2} \rangle^{-1/2}$	$1 + \langle \xi \rangle - \frac{3}{2} \langle \xi^2 \rangle$
Average $\langle r \rangle$	$1 + \langle \xi \rangle$
rms $\langle r^2 \rangle^{1/2}$	$1 + \xi - \frac{1}{2} \langle \xi^2 \rangle$
Substitution r_s	$1 + f \langle \xi \rangle - \frac{3}{2} \langle \xi^2 \rangle$

[a] $\langle \xi \rangle$, $\langle \xi^2 \rangle$, and f are defined in the text.

less cubic anharmonic constant.[8] In these equations only terms linear in the vibrational quantum number have been retained.

As can be seen from Table 1, the difference of $\langle r \rangle$ from r_e is due entirely to anharmonicity. This is readily understood on physical grounds. For a harmonic potential, the vibrational amplitude is symmetric about the equilibrium value. Introduction of anharmonicity, however, skews the potential so that displacement for which $r > r_e$ have greater probability than those for which $r < r_e$. Consequently, the average value of r is greater than r_e. Note that $\langle r \rangle$ can be obtained from r_v by a correction involving the harmonic force constant only.

If an atom is isotopically substituted, the change in the effective moment can be used to calculate the coordinate of the atom.[9] From the coordinates, one then obtains the so-called substitution bond length, r_s.[10] From Table 1 it can be seen that r_s depends on both the harmonic and anharmonic force constants. The factor f in the table is given by:

$$f = (\mu/m_1)\langle 1 + (\mu_1/\mu)^{1/2} \rangle^{-1} + (\mu/m_2)\langle 1 + (\mu_2/\mu)^{1/2} \rangle^{-1} , \tag{8}$$

where m_1 and m_2 denote the masses of the atoms in the parent molecule and μ_1 and μ_2 are the reduced masses of the isotopically substituted molecules; f ranges from 0.414 for deuterium substitution in H_2 to an upper limit of one half when both μ_1 and μ_2 are nearly equal to μ. Thus, as pointed out by Costain,[10] r_s is approximately equal to the average of the effective and equilibrium bond lengths.

Polyatomic Molecules

For polyatomic molecules the physical significance of the effective moment of inertia is complicated by the presence of Coriolis terms. Thus one has the relation:

$$\mathbf{I}_v = \langle \mathbf{I}_g^{-1} \rangle^{-1} + \text{Coriolis terms.} \tag{9}$$

To obtain an expression for the components of the instantaneous moment of inertia tensor, we may write the expansion:

$$I_{\alpha\beta} = I_\alpha^e\, \delta_{\alpha\beta} + \sum_s a_s^{\alpha\beta}\, Q_s + \sum_s \sum_t A_{st}^{\alpha\beta}\, Q_s Q_t \tag{10}$$

where

$$a_s^{\alpha\beta} = \left(\frac{\partial I_{\alpha\beta}}{\partial Q_s}\right)_e \tag{11}$$

$$A_{st}^{\alpha\beta} = \left(\frac{\partial^2 I_{\alpha\beta}}{\partial Q_s \partial Q_t}\right)_e . \tag{12}$$

One obtains an expression for the inverse moment of inertia tensor:

$$(I^{-1})_{\alpha\beta} = \frac{\delta_{\alpha\beta}}{I_\alpha^e} - \sum_s a_s^{\alpha\beta}\, Q_s - \sum_s \sum_t (A_{st}^{\alpha\beta} - \sum_\gamma a_s^{\alpha\gamma}\, a_s^{\beta\gamma}/I_\gamma) Q_s Q_t . \tag{13}$$

Note that the term quadratic in the normal coordinates now contains first derivatives as well as second due to the effect of inverting the components of the moment of inertia tensor. To get an expression for the effective moments of inertia, Eq. (13) must first be vibrationally averaged and reinverted to give:

$$\langle (I^{-1})_{\alpha\beta}\rangle^{-1} = I_\alpha^e\, \delta_{\alpha\beta} + \sum_s a_s^{\alpha\beta}\, \langle Q_s\rangle + \sum_s \left(A_{ss}^{\alpha\beta} - \sum_\gamma a_s^{\alpha\gamma}\, a_s^{\beta\gamma}\big/ I_\gamma\right) \langle Q_s^2\rangle . \tag{14}$$

Finally, one must add the Coriolis contribution:

$$-4 \sum_s \sum_t \left[\zeta_{st}^\alpha\, \zeta_{st}^\beta\, \omega_s^2 / (\omega_s^2 - \omega_t^2)\right] \langle Q_s^2\rangle , \tag{15}$$

where the ζ_{st} have the usual significance.[11] Thus, it can be seen that there is no clear-cut relation between the effective moments of inertia and physically well-defined structural parameters.

However, there are two common operationally defined types of structure that are determined from effective moments of inertia. The more common, the so-called effective or r_0 structure, is somewhat loosely defined. In practice, any structural parameter that requires for its determination fitting one or more of the second moment relations is designated as r_0. r_0 structures are not uniquely defined since, for any overdetermined system, the value of structural parameters obtained depends somewhat on the manner in which the data are treated and the values are isotopically dependent. This problem is examined in more detail by Schwendeman (this volume).

The second common type of operationally defined structure is the so-called substitution or r_s structure.[10] The structural parameter is said to be an r_s parameter whenever it has been obtained from Cartesian coordinates calculated from changes in moments of inertia that occur on isotopic substitution at the atoms involved by using Kraitchman's equations.[9] In contrast to r_0 structures, r_s structures are very nearly isotopically consistent. Nonetheless, isotope effects can cause difficulties as discussed by Schwendeman. Watson[12] has recently shown that to first-order in perturbation theory a moment of inertia calculated entirely from substitution coordinates is approximately the average of the effective and equilibrium moments of inertia. However, this relation does not extend to the structural parameters themselves, except for a diatomic molecule or a very few special cases of polyatomics. In fact, one drawback of r_s structures is their lack of a well-defined relation to other types of structural parameters in spite of the well-defined way in which they are determined. It is occasionally stated in the literature that r_s parameters approximate r_e parameters, but this cannot be true in general. For example, for a linear molecule Watson[12] has shown that to first order:

$$[z_s(i)]^2 = \partial I_0/\partial m_i \, , \tag{16}$$

where $z_s(i)$ is the substitution coordinate for an atom of mass m_i.

Noting that $1/2\ (I_e + I_0) \approx I_s$, where I_s is calculated entirely from substitution coordinates, Watson[12] has suggested using:

$$I_m = 2I_s - I_0 \tag{17}$$

as an approximation to I_c. He has applied this method to several small molecules with good success. However, a very large amount of accurate isotopic data is required and this method is somewhat restricted in its applicability.

For diatomic molecules we have seen that the average bond length can be obtained from the effective rotational constant by a correction involving harmonic force constants only. For polyatomic molecules, removing the harmonic part of the vibrational dependence of the effective moments of inertia leads to the moments of inertia of the average configuration.[8,13] It is then possible to calculate structural parameters for the average nuclear positions. Structural parameters calculated in this way from ground-state rotational constants have been designated $\langle r \rangle$[8] or r_z.[13] This type of structure has a well-defined physical significance, and vibration/rotation effects such as the inertial defect for planar molecules are absent. Since they are well-defined physically, they can be compared with electron diffraction data as discussed below. Since they involve a vibrational average, these structures are isotopically dependent. If data from more than one isotopic species are used in an analysis, serious error may result unless the isotopic dependence is taken into account. Unfortunately, this requires knowledge of the anharmonic force constants and for most molecules can only be approximated.

Since anharmonic contributions to the effective moments of inertia tend to be larger than the harmonic contributions,[8] the isotope dependence of the average configuration probably contributes greatly to the difficulties encountered with r_0 structures. Similarly, very small changes in the average configuration on isotopic substitution can account for much of the trouble associated with small r_s coordinates.[14] Thus, future progress in our understanding of the errors associated with spectroscopic determinations of structure depends in large part on our ability to obtain increased knowledge of anharmonic parameters.

ELECTRON DIFFRACTION

Electron diffraction is the other of the two important sources of gas-phase structural data. As discussed by Hedberg (this volume), the intensity of electrons scattered by molecules is modulated by the interatomic distances, both bonded and nonbonded. Since interatomic distances enter explicitly into electron diffraction determinations, the method is in some ways more direct than spectroscopy. Moments of inertia are functions of Cartesian coordinates of individual atoms rather than distances between atoms. On the other hand, electron diffraction is much more susceptible to experimental error than spectroscopic techniques.[15] Problems with structural determinations by spectroscopic methods often stem almost entirely from model error, whereas in electron diffraction both experimental and model error are important. Experimental and model error in electron diffraction are discussed elsewhere in this volume by Hedberg, and we shall confine ourselves here to definitions of the various structural parameters that arise in electron diffraction studies and the relationships among them and spectroscopic quantities.

The contribution to the intensity of scattered electrons that depends on an interatomic distance r_{ij} is proportional to[1]:

$$\exp(-1/2\, \ell_{ij}^2 s^2) \sin s(r_{ij} - \kappa s^2), \tag{18}$$

where s, the scattering variable, $= 4\pi \sin(\theta/2)/\lambda$, θ is the scattering angle and λ the wavelength of the incident electrons, ℓ_{ij}^2 is the mean square amplitude of vibration along the internuclear distance, and κ is an asymmetry parameter arising from anharmonicity. In the argument of the sine function, r_{ij} is defined as r_a (a for argument), which is equal to the center of gravity of the radial distribution curve $P(r)/r$ and corresponds to Bartell's $r_g(1)$.[16] In general, Bartell defines $r_g(n)$ as:

$$r_g(n) = \int_0^\infty r^{-n+1} P(r)\,dr \bigg/ \int_0^\infty r^{-n} P(r)\,dr, \qquad (19)$$

where $r_g(n)$ represents the center of gravity of the $r^{-n}P(r)$ curve. Through appropriate Fourier transformation of the experimental intensity data, the curve $P(r)/r$ can be found. From this curve r_a for each interatomic distance can be extracted in principle. Having obtained r_a, one can then relate it to $r_g(0)$, the center of gravity of the radial distribution function $P(r)$, through the relation:

$$r_a = r_g - \frac{\ell^2}{r}. \qquad (20)$$

The exact specification of the mean square amplitude ℓ^2 and r have been discussed.[1] Often ℓ^2 is determined directly from the electron diffraction data themselves or from vibrational force constants.

$r_g(0)$ has an important physical significance. It is often referred to in the literature simply as r_g. It corresponds to the instantaneous interatomic distance averaged over all the vibrations of the molecule according to the Boltzmann distribution law. Thus it is a temperature-dependent parameter. If we define a local Cartesian coordinate system for two atoms i and j such that the z direction is along the equilibrium ij distance and x and y are perpendicular to z, we can[1] write:

$$r_g = \langle [\Delta x^2 + \Delta y^2 + (r_e + \Delta z)^2]^{1/2} \rangle_T, \qquad (21)$$

where the subscript T indicates a thermal average over the Boltzmann distribution of vibrational states. A small centrifugal stretching effect has been neglected. To a good approximation r_g may be rewritten:

$$r_g = r_e + \langle \Delta z \rangle_T + \frac{\langle \Delta x^2 \rangle_T + \langle \Delta y^2 \rangle_T}{2 r_e}. \qquad (22)$$

The term in Δz is primarily determined by anharmonic force constants, whereas the terms in Δx^2 and Δy^2 are primarily functions of the harmonic force constants. For diatomic molecules where the vibrational force constants are usually well characterized, it is possible to calculate the vibrational averages and thus obtain r_e from r_g.[1] For polyatomic molecules the situation is more complex since the anharmonic constants are usually not available. However, by treating bond-stretching as a diatomic system, approximate calculations have been carried out.[1]

Since r_g is an average over the instantaneous configuration of the molecule, relations between bonded and nonbonded distances are slightly different from those found in the equilibrium or average configuration. This gives rise to what has been called the "shrinkage" effect.[1] For example, consider the case of a linear XY_2 molecule. Since the molecule possesses a bending motion, at any instant in time the molecule is likely to be in a slightly bent configuration. Thus the instantaneous YY distance is frequently less than twice the instantaneous XY distance and this is reflected in r_g. Similarly, planar molecules can appear to be slightly nonplanar. One method of removing these effects is to convert r_g into a parameter representing the structure of the average configuration. Note that, while a linear XY_2 molecule may at any instant in time be bent, it is on the average linear. Bond distances in this case represent the projection of the instantaneous bond length of the molecular axis. Parameters associated with the average configuration are given by:

$$r_\alpha = r_e + \langle \Delta z \rangle_T + \frac{\langle \Delta x \rangle_T^2 + \langle \Delta y \rangle_T^2}{2 r_e}. \tag{23}$$

The terms in Δx and Δy are much smaller than the term in Δz, and one can write:

$$r_\alpha = r_e + \langle \Delta z \rangle_T . \tag{24}$$

If one extrapolates to $T = 0$, one obtains the equivalent of the spectroscopic r_z. Thus it is possible to make comparisons between electron diffraction and spectroscopic values if they are both converted into r_z parameters. The relation between r_g and r_α is:

$$r_g - r_\alpha = \frac{\langle \Delta x^2 \rangle_T + \langle \Delta y^2 \rangle_T}{2 r_e}. \tag{25}$$

Thus r_α can be obtained from r_g by a correction involving harmonic force constants only. If r_g is extrapolated to $T = 0$, it is possible to obtain r_z. This temperature extrapolation involves the anharmonic constants and is perhaps the most uncertain step in going from r_g to r_z in many cases. Determination of r_e from r_g is also limited by a lack of knowledge of anharmonic constants. Thus, as with spectroscopic determinations, anharmonicity of vibrations is a major limiting factor.

This work was supported in part by the National Science Foundation.

REFERENCES

1. K. Kuchitsu and S. J. Cyvin, *Molecular Structures and Vibrations,* S. J. Cyvin, Ed., Elsevier, Amsterdam (1972).
2. H. B. G. Casimir, "The Rotation of a Rigid Body in Quantum Mechanics," Leyden thesis, J. B. Wolters, the Hague, Netherlands (1931).
3. C. Eckart, *Phys. Rev.,* **46**, 383 (1934).
4. J. H. Van Vleck, *Phys. Rev.,* **47**, 487 (1935).
5. E. B. Wilson and J. B. Howard, *J. Chem. Phys.,* **4**, 260 (1936).
6. For discussion of the validity of the Born–Oppenheimer approximation, see, for example, P. R. Bunker, *J. Mol. Spectrosc.,* **42**, 478 (1972).
7. Y. Morino and E. Hirota, *Annu. Rev. Phys. Chem.,* **20**, 139 (1970).
8. D. R. Herschbach and V. W. Laurie, *J. Chem. Phys.,* **37**, 1668 (1962).
9. J. Kraitchman, *Am. J. Phys.,* **21**, 17 (1953).
10. C. C. Costain, *J. Chem. Phys.,* **29**, 864 (1958).
11. H. H. Nielsen, *Rev. Mod. Phys.,* **23**, 90 (1951).
12. J. K. G. Watson, *J. Mol. Spectrosc.* (in press).
13. T. Oka, *J. Phys. Soc. Japan,* **15**, 2274 (1960).
14. V. W. Laurie and D. R. Herschbach, *J. Chem. Phys.,* **37**, 1687 (1962).
15. K. Kuchitsu, *Molecular Structures and Vibrations,* S. J. Cyvin, Ed., Elsevier, Amsterdam (1972), Ch. 10.
16. L. S. Bartell, *J. Chem. Phys.,* **23**, 1219 (1955).

Kenneth Hedberg

Critical Evaluation of Structural Information from Gaseous Electron Diffraction

For the nonexpert to learn how to evaluate structural information from reports of gas electron-diffraction experiments is not an easy task, nor is it always easy for the expert either (as those who are called upon to referee manuscripts will know). In any structural analysis there are a large number of factors–some experimental, some related to the procedures used–that can affect the results. The reader must determine whether the author has recognized all these factors, many of which are different for different molecules. He must then determine whether the author has investigated or estimated their effects and appropriately taken them into account. In the event that the reader is unsatisfied with the treatment, he must attempt to determine the extent to which the structural claims are to be relied upon.

Because the factors mentioned above are closely connected to the nature of both the electron-diffraction experiment and the analysis of the data, the first three sections to follow are meant to provide a necessary background of knowledge about the method.[1–5] Nonessential details have been ignored in these sections, and to minimize confusion I have tried to avoid commonly used symbols for intensity functions unless these functions are indeed those usually identified with the symbol. Various aspects of the method and the limitations they impose on the structural results are next discussed.[6–8] Finally, I note some points that, it is hoped, will provide the basis for a critical evaluation of reported results.

EXPERIMENT

The important features of the geometry of the electron-diffraction experiment are shown in Figure 1. The gas to be studied is led into a highly evacuated chamber by means of a nozzle. A narrow beam of monochromatic electrons passes the nozzle tip at right angles in such a way as to intersect the gas jet. The diffraction pattern is generally recorded on photographic plates positioned perpendicular to the undiffracted beam at accurately known distances below the nozzle tip, i.e., distances from the center of the scattering region. The scattered intensity falls off very rapidly with increasing scattering angle 2θ [= arctan (d/h)]. To obtain photographic plates that are neither over- nor underexposed over a suitable range of scattering angle, the scattered intensity is modified by a mechanical device called a "rotating sector." This device is a spiral- or heart-shaped cam with an angular opening approximately proportional

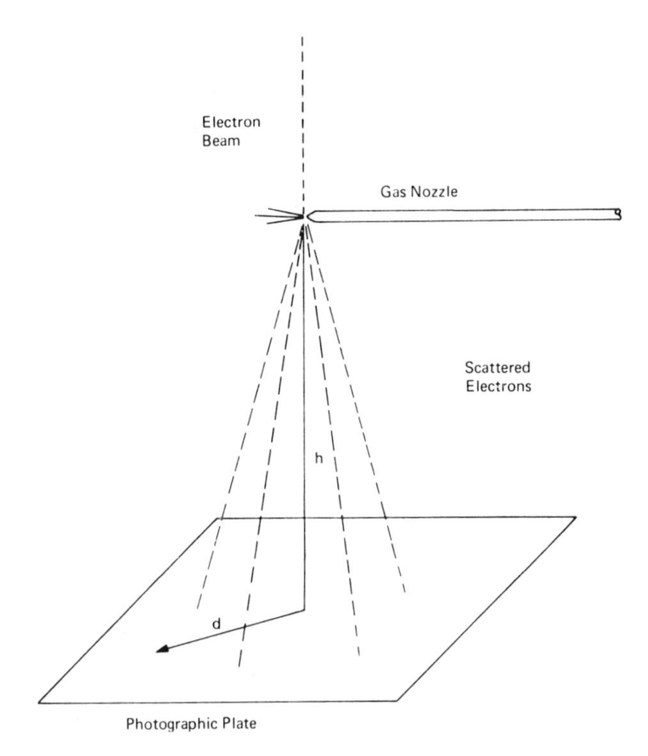

FIGURE 1 Geometry of the gas electron-diffraction experiment.

FIGURE 2 A typical electron-diffraction photograph. The substance was SiF_4.

to r^3, where r is the radial distance in the plane of the sector. It is mounted parallel to and a few millimeters above the photographic plate and is rotated rapidly in its own plane during exposures. The undiffracted beam is caught in a beam stop. Diffraction photographs are usually made at two or three different distances h to obtain data both at as small and at as large scattering angles as possible. Figure 2 shows a typical plate made at h = 30 cm.

RETRIEVAL OF THE DIFFRACTION DATA

Diffraction data are obtained in the form of a continuous optical-density distribution that, in an ideal experiment, is radially symmetric. The density distribution is measured with a microdensitometer, generally by making scans across the diameters of the plates. Because the emulsion grain contributes "noise," a given plate may be scanned many times along different diameters, or, alternatively, it may be scanned once or twice while being rapidly rotated about the axis of the diffraction rings. The results of the scans are usually digitized records in the form of punched cards or tape of the optical density (or transmission) at accurately known radial intervals; occasionally, strip chart records are obtained as an intermediate step. A typical result from a densitometric scan is shown in Figure 3. The relation between scattered electron intensity and, say, photographic density depends on the type of photographic emulsion and on the development methods. Accordingly, the conversion to scattered intensity is done using a calibration curve that may be obtained in a variety of ways.[9,10] The results of these steps may be symbolized by I_p, the scattered intensity striking the photographic plate. The data are in digital form, each item representing the intensity (on a relative, arbitrary scale) at a known radius from the center of the pattern.

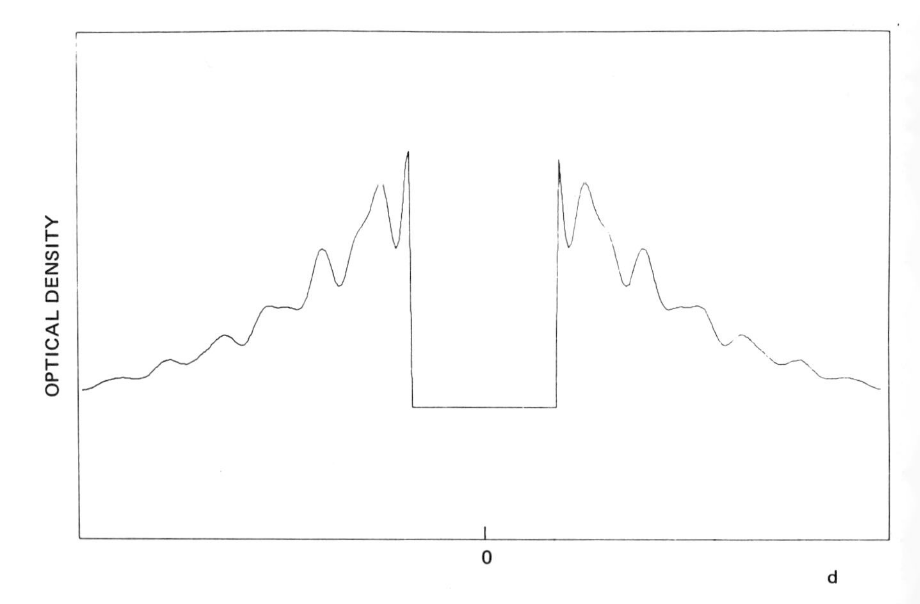

FIGURE 3 A typical densitometer record. The curve was made from the plate of Figure 2.

REDUCTION OF THE DIFFRACTION DATA

For the purpose of structure analysis, the total scattered intensity (I_{Tot}) may be regarded as comprising two components: I_{MS}, the molecular-structure-sensitive part, and I_{B}, a non-structure-sensitive part or background. The latter contains three components: inelastic scattering, "atomic" scattering, and unavoidable scattering from parts of the apparatus such as apertures. The following formulas show the relationships among these quantities and their theoretical forms:

$$I_{\mathrm{Tot}} = I_{\mathrm{MS}} + I_{\mathrm{B}} \tag{1}$$

$$I_{\mathrm{MS}} = \mathrm{const} \sum_{i>j} n_{ij} |f_i(s)||f_j(s)| (r_{ij}s)^{-1} \cos|\eta_i(s) - \eta_j(s)| \times \exp(-\ell_{ij}^2 s^2/2) \sin(r_{ij} - \kappa_{ij}s^2)s \tag{2}$$

$$I_{\mathrm{B}} = \mathrm{const} \sum_i |f_i|^2 + I_{\mathrm{Inel}} + I_{\mathrm{App}} \,. \tag{3}$$

In these formulas n_{ij} is the number of symmetry-equivalent distances of length r_{ij}, $|f_i|$ is the magnitude of the complex scattering amplitude for electrons, s is equal to $4\pi\lambda^{-1} \sin\theta$ (where 2θ is the scattering angle), η_i is a phase factor, ℓ_{ij}^2 is the mean square amplitude of vibration associated with distance r_{ij}, κ is an anharmonicity constant, I_{Inel} is the inelastic scattered intensity, and I_{App} the intensity scattered from the apparatus. Of these many quantities one usually attempts to measure only certain of the r's and ℓ's and, occasionally, a few of the n's; the remaining quantities are taken from tables or calculated from theory.*

The problem of reducing the diffraction data may be stated as that of extracting from I_{p} a set of numbers representing a form of the molecular-structure-sensitive scattering convenient for analysis for the r's and ℓ's. I_{MS} itself is not particularly convenient since its amplitude is a very rapidly varying function of s. The molecular intensity functions used are of three types arising from the preferences and practices of individual laboratories. Which is best is, of course, a matter of continuing discussion, but it is fair to say that in carefully done work the results are essentially independent of the intensity function used to obtain them.

The relation between I_{Tot} and I_{p} is given by

$$I_{\mathrm{p}} = I_{\mathrm{Tot}} \cdot \alpha(s) \cdot \cos^3 2\theta \,. \tag{4}$$

* For example, the important quantities $|f_i|$ and η_i have been tabulated by Schafer *et al.*[11]

The factor $\alpha(s)$ is a known function representing the modifying effect of the rotating sector, and $\cos^3 2\theta$ takes into account the fact that every point on the plane photographic plate is not equidistant from the scattering point. Now the form of I_B modified by the two factors just mentioned tends to be smooth so that the oscillations of the similarly modified I_{MS} can be easily recognized, as seen in Figure 4. Accordingly, the undesired background can be approximated (because of the unknown component from the apparatus it cannot be calculated from theory) by

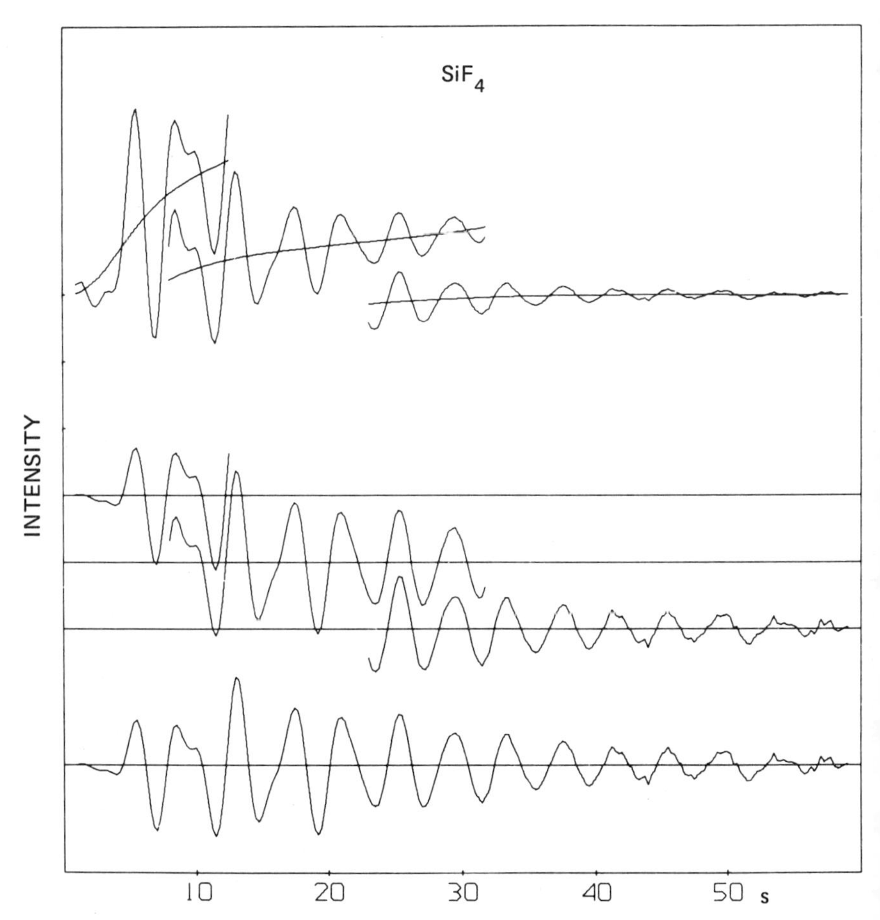

FIGURE 4 Electron-diffraction intensity curves. The three uppermost curves are in the form $s^4 I_{Tot}$ and show $s^4 I_{MS}$ superposed on $s^4 I_B$. They represent data from experiments with distances h equal to 75, 30, and 12 cm obtained from records such as that in Figure 3. The lowermost four curves are in the form sI_M. The last curve is a composite of the three immediately preceding. All curves are for SiF_4.

a "hand-drawn" or computer-constructed curve that passes through the estimated nodes of the oscillations. It is the subsequent handling of this background that constitutes the principal difference between the methods used in different laboratories. One of these may be termed the "background division" method and the other the "background subtraction" method. In the background division method one obtains from the measured values of I_p using Eq. (1) and (4) the function

$$sM(s) = s\,[(I_{\text{Tot}} - I_{\text{App}})/(I_{\text{B}} - I_{\text{App}}) - 1]$$

$$= \text{const} \sum_{i>j} \eta_{ij}|f_i(s)||f_j(s)|r_{ij}^{-1}\,(I_{\text{B}} - I_{\text{App}})^{-1} \cos|\eta_i(s) - \eta_j(s)| \times$$

$$\exp(-\ell_{ij}^2 s^2/2)\sin(r_{ij} - \kappa_{ij}s^2)s. \quad (5)$$

In the background subtraction method one calculates a function $sI_{\text{m}}(s)$ given by

$$sI_{\text{m}}(s) = s^5(I_{\text{Tot}} - I_{\text{B}})$$

$$= \text{const} \sum_{i>j} \eta_{ij}F_i(s)F_j(s)r_{ij}^{-1} \cos|\eta_i(s) - \eta_j(s)| \times$$

$$\exp(-\ell_{ij}^2 s^2/2)\sin(r_{ij} - \kappa s^2)s, \quad (6a)$$

where $F_i(s) = s^2|f_i(s)|$. In this method it is frequently convenient to calculate a slightly different function $I'(s)$:

$$I'(s) = sI_{\text{m}}(s)/[F_k(s)F_\ell(s)]. \quad (6b)$$

It is seen that $sM(s)$, $sI_{\text{m}}(s)$, and $I'(s)$ differ only in what may be termed the coefficients of the interference functions. For the purposes of this discussion the reduction of the diffraction data may be said to be complete when the treatment of I_p has led to a set of numbers expressible by Eq. (5), (6a), or (6b) over a range $s_{\text{min}} < s < s_{\text{max}}$ at intervals Δs. Figure 4 shows typical $sI_{\text{m}}(s)$ curves from plates made at different distances h.

TRIAL STRUCTURE

Because the intensity functions expressive of the experimental data are rather complicated, it is possible to deduce a trial structure from them by inspection only in cases of very simple molecules. However, these functions and what is called the radial distribution function $D(r) \sim P(r)/r$, where $P(r)$ is the probability of finding a pair of atoms between r and

$r + dr$, comprise a pair of Fourier transforms.[12] Accordingly, we may calculate

$$D(r) = \text{const} \sum_{s_{\min}}^{s_{\max}} sM(s) \exp(-Bs^2) \sin rs\, \Delta s \tag{7a}$$

or

$$D(r) = \text{const} \sum_{s_{\min}}^{s_{\max}} I'(s) \exp(-Bs^2) \sin rs\, \Delta s. \tag{7b}$$

The exponential convergence factor minimizes series termination errors. To the extent that $|f_i(s)||f_j(s)|/(I_B - I_{App})$ in the case of Eq. (7a) and F_iF_j/F_kF_ℓ in the case of Eq. (7b) are nearly constant, and $|\eta_i(s) - \eta_j(s)|$, $s_{\min}$ and κ are equal to zero, the individual peaks of the radial distribution curves will have nearly Gaussian shapes. In these cases $D(r)$ will be given to good approximation by

$$D(r) = \text{const} \sum Z_iZ_jr_{ij}^{-1} (4B + 2\ell_{ij}^2)^{-1/2} \exp[-(r - r_{ij})^2/(4B + 2\ell_{ij}^2)]^{1/2}. \tag{8}$$

Figure 5 shows the radial distribution curve calculated according to Eq. (7b) from the intensity curve of Figure 4 using data in the unobserved range $s < s_{\min}$ from a theoretical curve.

From the positions of the peaks of $D(r)$ one has trial values of r_{ij} and from their half widths at half height, say, one may calculate trial values of ℓ_{ij}. The geometry of a trial model may be constructed from the former and tested by calculation of theoretical intensity and radial distribution curves corresponding to it.

STRUCTURE REFINEMENT

Nowadays refinement of the trial structure is nearly always carried out by a least squares procedure based on intensity curves[13] in the forms defined by Eq. (5), (6a), or (6b). The geometry of the molecular model is usually expressed in some suitable set of internal coordinates—such as bond distances, bond angles, and torsion angles—and the dynamics of the model by a set of ℓ_{ij}. It is apparent from the forms of the intensity functions that the least squares procedure is nonlinear in these parameters. The usual linearizing methods based on Taylor's series expansions are used so that the parameters actually adjusted are the shifts in the internal coordinates and in the ℓ_{ij}. The successful conclusion of the re-

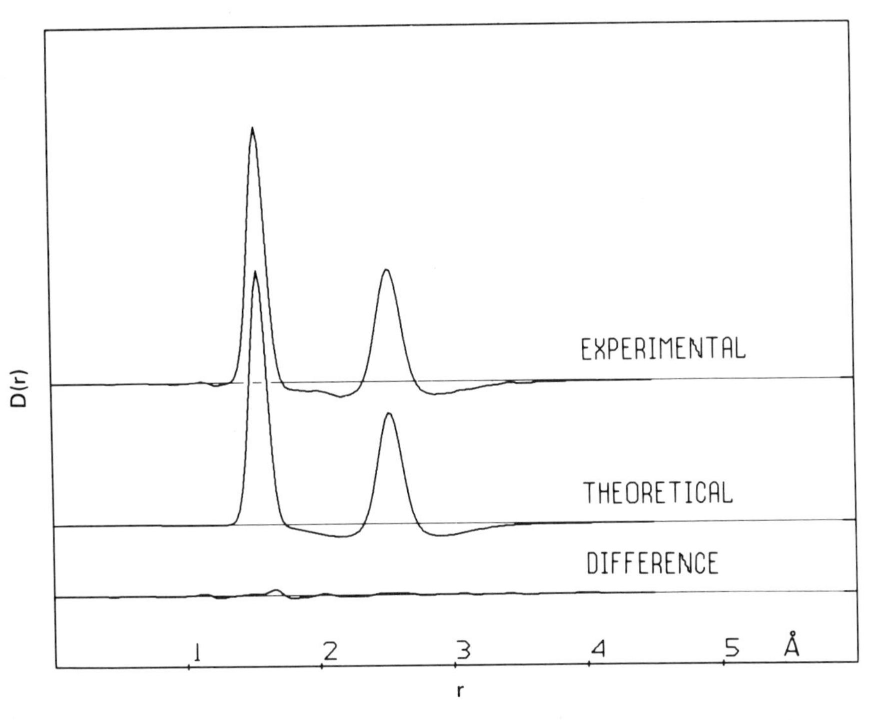

FIGURE 5 Radial distribution curves. The experimental curve is calculated using Eq. (7b) from the composite curve of Figure 4 with data from the unobserved region $0 < s < 2$ taken from a theoretical curve.

finement process leads to the best estimates of the values of the internal coordinates (often termed the "independent parameters"), other internal coordinates including distances calculated from them ("dependent parameters"), and the ℓ_{ij}; also obtained is the variance–covariance matrix or, alternatively, the correlation matrix.

FACTORS AFFECTING REPORTED STRUCTURES

From the foregoing it is apparent that the process of determining structures by electron diffraction involves many experimental, procedural, and interpretive matters that influence the reported results. A critical evaluation of the results requires an insight into the workings of these matters, the more important of which are summarized below. For ease of visualization the discussion is presented mostly in terms of the radial distribution curves.

1. *Definition of the structural parameters* As the accuracy of structural techniques has improved, one has been forced to consider the differences between the nominally similar quantities provided by these techniques. The interatomic distances provided by Eq. (6a) to (8) are termed r_a or r_g[1-5] and are represented by the centers of gravity of the $D(r)$ (or $P(r)/r$) peaks. They differ from the distances r_e, r_g, r_o, r_s, etc., by amounts in some cases substantial. Relations exist by which one type may be converted to another[14] of which Eq. (9) and (10) are examples pertinent to electron diffraction.

$$r_a = r_g - \ell^2/r_e. \tag{9}$$

$$r_g = r_e + \delta r + 3a\ell^2/2. \tag{10}$$

The term δr and $3a\ell^2/2$ are corrections for centrifugal distortion and anharmonicity, respectively. Table 1 shows the magnitudes of these types of distances calculated for CO_2 and SO_2. It will be noted that the differences between r_e and r_g or r_a are 0.004–0.007 Å, which is substantially greater than experimental error.

2. *Geometrical constraints* For r_a and r_g in the case of CO_2 the differences $2r(CO) - r(O\ldots O)$ are about 0.005 Å, greater than experimental error. This difference, at first sight apparently in conflict with the known $D_{\infty h}$ equilibrium symmetry of CO_2, is a consequence of molecular vibration, particularly bending, and is termed "shrinkage."[15] Shrinkage exists to some degree in most molecules but becomes an important phenomenon in linear molecules and in coplanar molecules having more than three atoms. To the extent that a structure determination assumes a geometrical constraint and shrinkage corrections are simul-

TABLE 1 Some Parameter Values for CO_2 and SO_2[a]

	r_e	r_g	r_a
CO_2			
C–O (Å)	1.1600	1.1652	1.1646
O. . .O (Å)	2.3200	2.3251	2.3244
∠OCO (deg)[b]	180.00	172.27	172.64
SO_2			
S–O (Å)	1.4308	1.4360	1.4351
O. . .O (Å)	2.4696	2.4771	2.4758
∠OSO (deg)[b]	119.31	119.20	119.22

[a] K. Kuchitsu, private communication.
[b] Apparent angles calculated from the distance values.

taneously ignored in these types of molecules, the parameter values will reflect a sort of compromise fit to the distance spectrum. In the case of CO_2, for example, the assumption of $D_{\infty h}$ symmetry without a simultaneous shrinkage correction will lead to a shorter CO and a longer O . . . O distance than corresponds to the center of gravity of the radial distribution peaks; the effects are similar in larger molecules. In general, one expects the differences to be considerably smaller for the bond and *gem* nonbond distances than for the longer distances.

3. *Geometry of the trial model* Because the structure-refinement procedures are based on functions that are nonlinear in the parameters, it is necessary that the trial model to be refined be similar to the final model. When this is not the case, the least squares procedure may lead to an incorrect structure. The existence of more than one structure in satisfactory agreement with observation is quite common in larger molecules, although it often happens that some may be ruled out on chemical or other grounds. The problem may be understood by a consideration of the form of the radial distribution curve. The peaks of this curve are often composites of peaks arising from two or more interatomic distances. When the weights of, say, two of these distances are similar, it may be readily seen that exchange of their positions might provide equally good fits to the composite. Figure 6 shows a remarkable example in which four structures for SOF_4 (three quite different but all with molecular symmetry C_{2v}), give equally good agreement with observation. Table 2 summarizes the geometries of the models.

4. *Correlated parameters* It was suggested earlier that the component peaks of the $D(r)$ curve were essentially Gaussian in shape. If two or more such peaks having about the same r_{ij} and ℓ_{ij} are summed, the result is approximately Gaussian also. Such a condition prevails very frequently, especially in organic molecules, and leads to high correlations among the several r's and ℓ's determining the shape of the composite peak. The situation is easily visualized in terms of two components. The fit to an essentially Gaussian composite peak provided by two components separated by Δr will tend to be maintained by decreasing the ℓ_{ij}'s as Δr increases. So far as the composite peak is concerned, the result is a manifold of models of approximately equal quality that is terminated only by Δr large enough to destroy the essentially Gaussian character of the composite. The practical consequence is the following more-or-less obvious rule: No more than the minimum number of parameters required by the model to reproduce the $D(r)$ peak may be independently refined.*

*As noted by Hilderbrandt and Bonham[4] this amounts to two parameters per Gaussian peak, i.e., a width parameter (ℓ) and a position parameter (r).

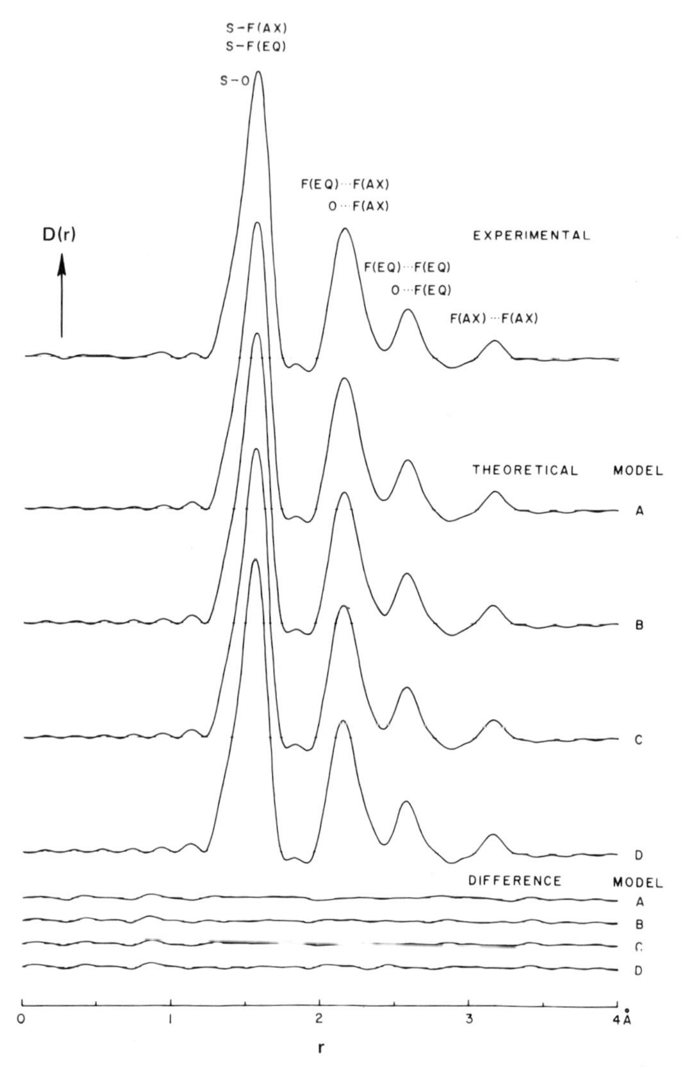

FIGURE 6 Radial distribution curves for SOF_4. The curves correspond to the models of Table 2.

5. *Assumed values for nonrefinable parameters* As may be deduced from the foregoing sections, the analysis of the structures of most molecules involves a number of assumptions about the values of certain parameters. The geometrical parameters are defined by the model and thus are usually many fewer than the number of different interatomic distances. However, a corresponding model of the molecular dynamics,

TABLE 2 Parameter Values for SOF_4[a]

	A	B	C	D
S = O (Å)	1.402 (3)	1.403 (3)	1.406 (4)	1.405 (4)
S – F(eq) (Å)	1.548 (4)	1.552 (4)	1.535 (4)	1.533 (5)
S – F(ax) (Å)	1.578 (4)	1.575 (4)	1.593 (4)	1.596 (4)
∠F(eq)SF(eq) (deg)	117.9 (2.3)	110.2 (1.8)	118.7 (3.3)	114.9 (3.3)
∠F(ax)SO (deg)	90.6 (0.4)	90.7 (0.4)	98.0 (0.3)	97.8 (0.3)
∠F(ax)SF(eq) (deg)	89.7 (0.2)	89.6 (0.2)	85.9 (0.2)	85.8 (0.2)
∠F(eq)SO (deg)	121.0 (1.1)	124.9 (0.9)	120.6 (1.7)	122.5 (1.7)

[a] Parenthesized values are errors estimated as 2σ. Data from Gundersen and Hedberg.[17]

which might serve to establish connections among the ℓ_{ij}, is much more difficult to formulate. Accordingly, the ℓ_{ij} are usually viewed as independent variables equal in number to the number of symmetrically non-equivalent r_{ij}. In most cases it is impossible to refine all geometrical and dynamical parameters, and assumed values must be assigned to some. Fortunately, experience suggests that many of the ℓ_{ij} values may be guessed or calculated from abbreviated potential functions with some confidence; others involve terms of such low weight that their values are, in any case, unimportant. When it is necessary to assign a value to one of the parameters involved in an important $D(r)$ peak, however, it is clear that the remaining parameter values will reflect that assumption.

6. *Sample purity* The matter of sample purity is somewhat more complex than might be thought at first. The effect of an impurity on the diffraction pattern depends on its mole fraction in the sample vapor and on its scattering power relative to that of the material of interest. (The structure-sensitive scattering from a molecule is approximately proportional to $\Sigma n_{ij} Z_i Z_j / r_{ij}$.) However, the effect of an impurity on the desired structure determination need not bear much relation to its effect on the diffraction pattern. If the impurity has a distance distribution sufficiently different from that of the substance of interest, the peaks of the $D(r)$ curve will be resolved; accordingly, the parameter values of the two molecules will be essentially uncorrelated and the desired results unaffected.* On the other hand, small amounts of an impurity with a distance spectrum similar to the substance of interest can seriously disturb the results. It is perhaps worth noting that the

*An amusing example is that of B_2Cl_4 in which $SiCl_4$ was unexpectedly found as an impurity.[16] The distances in the two molecules are different enough to have permitted the determination of the structures of both.

diffraction experiment itself may involve a stage of purification through the vaporization or liquid or solid samples.

7. *Amount and weighting of data* The amount of data used in a structure determination may at first appear to be large compared with the number of measured parameter values: One to five plates at two or three nozzle-to-plate distances are usual and may generate several hundred data points. Because data from plates made at the same distance are in a sense duplicate sets, they are often combined (averaged) leading to a single set of perhaps 100–200 values. These data, in fact, are correlated (consider, for example, that the positioning of the background affects sections of the I_{Tot} curve) so that the number of independent observations may be much smaller. Under these circumstances the quality and weighting of the data assume great importance.

The weighting of the data is done in a variety of ways and is at present such a confused subject that virtually no discussion of it can be fruitful. Diagonal unit and nonunit weight matrices are used as well as nondiagonal matrices that attempt to take into account the correlation mentioned above. The use of the different functions defined by Eq. (5), (6a), and (6b) each with a unit weight matrix is itself equivalent to different weighting. It is probably fair to state that the standard deviations obtained with any of the commonly used weighting schemes are much smaller than most investigators are prepared to accept at face value.

8. *Neglect of anharmonicity* From the form of Eq. (5) and (6a) one sees that anharmonicity manifests itself primarily in high angle data. The values of κ are of order 10^{-6} Å^3 for atoms other than H or D and thus the correction term κs^2 is greater than about 0.001 Å only for $s > 30$–35. Thus, in most cases the anharmonicity correction to distances will be less than experimental error on those distances. For bonds involving hydrogen or deuterium, however, the correction for anharmonicity may be several thousandths of an angstrom even with data in the range $s < 30$.

9. *Errors in apparatus constants, photographic calibrations, and theoretical factors* Every structure determination is based on measured values of λ (the electron wavelength), h, and d. The random errors in these quantities are usually small, perhaps considerably less than 0.1 percent, but an occasional large aberration in a given case is always a possibility. The effect of these errors is nearly entirely confined to the *size* of the molecule, i.e., neither the molecular shape nor the amplitudes of vibration are affected. Another apparatus constant involves that unfortunate but at present necessary device the rotating sector. The effect of errors in its calibration is probably strongly dpendent on the way in which individual laboratories handle their data. In most cases these errors are of minor concern only.

Errors in the correction for nonlinearity of photographic response and in the theoretical scattering amplitudes affect primarily the values of the amplitudes of vibration and, at least for the former, can be significant.

HELPFUL HINTS FOR CRITICAL EVALUATION OF STRUCTURAL INFORMATION

The factors discussed in the foregoing paragraphs are the more important ones from an even larger group that play a part in the structural results obtained from electron-diffraction investigations of gases. That the magnitude of their aggregate effect is often not easy to estimate is proved by the presence of occasional gross errors even in some carefully done studies. However, the existence of the factors themselves is not hard to establish in any given case, and one can at least be alert to possible error arising from them. The following is meant to be a guide:

1. Look for violations of the rule stated in item 4 above. If such violations occur, the values and error estimates for the individual parameters determining the shape of the $D(r)$ peak in question may be meaningless. Keep in mind, however, that parameter values for the rest of the structure may at the same time be correct: It is quite possible to determine, say, bond distances to a few thousandths of an angstrom and to know nothing of the longer range structure.
2. Learn to use the correlation matrix. The author's choice of parameters is made as a matter of convenience and interest to him. Often another set would have been better for the reader. It is possible to transform the set given into the set of interest by simple methods. As an example consider a single $D(r)$ peak arising from a pair of different type bond distances taken as parameters. The error on each will be relatively large and the correlation coefficient close to -1.0. Converted to the set $(r_1 + r_2)/2$ and $r_1 - r_2$, the error on the former will be relatively small and that on the latter large; that the correlation coefficient connecting them may be small is also important. In other words, the average distance may be well determined and relatively independent of the split.
3. Do not rely on the published differences between experimental and theoretical radial distribution and intensity curves as an index of correctness of the structure. These curves reflect, among other things, background removal methods that may suggest better agreement between model and experiment than in fact exists. Moreover, the quality of agreement between observed and calculated curves can only improve as more and more adjustable parameters are introduced, a circumstance that sometimes leads to excellent agreement in the cases of quite complicated molecules when the rule of item 4 above is violated.

4. Look carefully at the author's definition of his error estimates. There is no consistent convention. If the effects of systematic error and correlation among the observations have not been specifically included, 3σ or 4σ are often quoted where σ is obtained from the least squares result. When these effects are reflected in the σ values, 2σ or 3σ may be quoted. These quantities are arbitrary and perhaps often optimistic.

5. Remember that the results are no better than the model. What needs to be emphasized is that electron diffraction is quite incapable of distinguishing small differences between models: Benzene and a slight distortion of it having carbon–carbon bonds of alternate length would give essentially identical diffraction patterns. Since most structure determinations involve some arbitrary assumptions, one should attempt to ascertain if these are reasonable. Predictions of the effects of changes in these assumptions are difficult, but item 4 in the previous section suggests one type of behavior.

6. Determinations of geometry are relatively more accurate than determinations of root-mean-square amplitudes. This is a matter of experience. Aspects of the geometry giving rise to broad, low-weight peaks in the $D(r)$ curve may be poorly determined. Be cautious concerning geometrical deductions about the coplanarity or noncoplanarity of molecules when the possibility of large amplitude motions has not been considered. (Note the apparent angles calculated for CO_2 in Table 1.)

7. Convert r values to a common type before making detailed comparisons between results from different methods.

8. Do not allow any set of results to override an intuition that past experience has shown to be reliable, before those results have been thoroughly studied in connection with the preceding hints.

9. If hints 1–8 do not suggest answers to questions about a structure determination, ask the author for clarification.

10. If all else fails, offer to assist in a reinvestigation of the structure by another electron-diffraction laboratory.

I am grateful to Lise Hedberg for preparation of several of the figures.

REFERENCES

1. L. S. Bartell, in *Techniques of Chemistry,* Vol I: *Physical Methods of Chemistry,* A. Weissberger and B. W. Rossiter, Ed., Wiley-Interscience, New York (1972), Ch. II.
2. K. Kuchitsu, *MTP International Review of Science,* Physical Chemistry, Series I, Vol. 2 (1972).
3. J. Karle, in *Determination of Organic Structures by Physical Methods,* Vol. V, F. C. Nachod and J. J. Zuckerman, Ed., Academic Press, New York (1972).

4. R. L. Hilderbrandt and R. A. Bonham, *Ann. Rev. Phys. Chem.*, **22**, 279 (1971).
5. S. H. Bauer in *Physical Chemistry, An Advanced Treatise*, Vol. IV, D. Henderson, Ed., Academic Press, New York (1970), Ch. 14.
6. H. M. Seip in *Specialist Periodical Reports of the Chemical Society: Molecular Structure*, L. E. Sutton and G. A. Sim, Ed., Series 20, Vol. 1 (in press).
7. K. Kuchitsu in *Molecular Structures and Vibrations*, S. J. Cyvin, Ed., Elsevier Publishing Co., Amsterdam (1972), Ch. 10.
8. H. M. Seip and R. Stölevik in *Molecular Structures and Vibrations*, S. J. Cyvin, Ed., Elsevier Publishing Co., Amsterdam (1972), Ch. 11.
9. I. Karle and J. Karle, *J. Chem. Phys.*, **18**, 957 (1950).
10. L. S. Bartell and L. O. Brockway, *J. Appl. Phys.*, **24**, 656 (1953).
11. L. Schafer, A. C. Yates, and R. A. Bonham, *J. Chem. Phys.*, **55**, 3055 (1971).
12. J. Waser and V. Schomaker, *Rev. Mod. Phys.*, **25**, 671 (1953).
13. K. Hedberg and M. Iwasaki, *Acta Cryst.*, **17**, 529 (1964).
14. K. Kuchitsu and S. J. Cyvin in *Molecular Structures and Vibrations*, S. J. Cyvin, Ed., Elsevier Publishing Co., Amsterdam (1972), Ch. 12.
15. O. Bastiansen and M. Traetteberg, *Acta Cryst.*, **13**, 1108 (1960).
16. R. Ryan and K. Hedberg, *J. Chem. Phys.*, **50**, 4986 (1969).
17. G. Gundersen and K. Hedberg, *J. Chem. Phys.*, **51**, 2500 (1969).

R. H. Schwendeman

Structural Parameters from Rotational Spectra

Because of their relation to the molecular moments of inertia, spectroscopic rotational constants have been one of the two most important sources of information for the determination of bond distances and bond angles in free molecules. The purpose of this paper is to describe several computational methods by which bond distances and bond angles may be derived from rotational constants. The sources and magnitudes of the uncertainties in molecular parameters that result from the methods will also be discussed.

Derivation of molecular parameters from rotational constants is not a new subject, and many discussions of the various methods have previously been presented. One of the more complete of the recent accounts is that by Gordy and Cook.[1] Definitions of the different molecular parameters are given by Laurie elsewhere in this volume and have also been given recently by Kuchitsu and Cyvin.[2] The fundamental papers on this subject include the series by Herschbach and Laurie[3-5] and by Morino, Oka, and co-workers.[6-9] The present discussion will concentrate on a description of the computational strategies that may be employed and the problems that occur in practical cases. For the usual reasons of familiarity the examples will be dominated by work done at Michigan State University.

PRELIMINARY CONSIDERATIONS

Before rotational constants are used for the determination of molecular parameters, a number of experimental problems may have to be considered. For the purpose of this paper we will assume that the assignment of rotational transitions is correct and that the measurement uncertainties in the rotational constants have been properly assessed. Structure determinations begin with the rotational constants of molecules in the ground vibrational state. When equilibrium structural parameters are to be determined, the ground-state constants are adjusted to equilibrium values as described below. In most cases, however, the ground-state values—contaminated by vibrational effects—must be used. It is sometimes desirable to remove perturbing effects of one or more vibrational modes. The most common example of this involves internal rotation of a methyl group in a molecule with a low potential barrier to internal rotation ($V_3 \lesssim 1$ kcal/mol). For example, in methoxydifluorophosphine ($V_3 = 422$ cal/mol)[10] the interaction between internal and overall rotation alters the A rotational constant by 0.2 percent. This change will have a big effect on determining the structure and should be removed. If Herschbach's[11] perturbation treatment of internal rotation is employed in such a calculation, the proper denominator corrections described by Stelman[12] should be included.

If the molecule under study contains one or more nuclei with quadrupole coupling constants, the spectrum will show hyperfine structure, and the rotational constants for structure calculations must be obtained from hypothetical unsplit frequencies. For small quadrupole coupling constants, as with ^{14}N and ^{11}B, the hypothetical unsplit frequencies may be taken to be the simple or weighted average of the frequencies of the hyperfine components. For intermediate coupling constants, as with ^{35}Cl or ^{37}Cl, frequency shifts calculated from first-order theory should be subtracted from the measured frequencies of the hyperfine components of a transition and the results averaged to give the hypothetical unsplit frequency. For large coupling constants, as with ^{79}Br, ^{81}Br, or ^{127}I, second- or higher order theory should be employed.

The rotational constants are the coefficients of the squares of the components of the rotational angular momentum in the rotational Hamiltonian. For many molecules a simple Hamiltonian containing only squares of angular momenta is sufficient to express the frequencies of low J ($J \lesssim 5$) transitions to high accuracy (±0.5 MHz or better). In some cases, however, particularly with molecules containing less than three heavy (i.e., nonhydrogenic) atoms, a Hamiltonian with fourth and even higher powers of angular momenta may be necessary. When such centrifugal distortion terms must be included, it is much more difficult to obtain precise rotational constants. The problems involved have been discussed recently by Kirchhoff.[13] Even when centrifugal distortion terms do not seem to be needed, there may be an effect on the rotational constants. It turns out that the important structural information is in the differences between the moments of inertia of isotopically different species. For this reason it is most desirable to use the frequencies of the same set of rotational transitions to obtain rotational constants in all the isotopic species. The centrifugal distortion effects should then tend to cancel.

THEORY

The Hamiltonian most commonly used to interpret rotational spectra is as follows:

$$H = h[A_v L_a^2 + B_v L_b^2 + C_v L_c^2]. \tag{1}$$

In this equation A_v, B_v, and C_v are the rotational constants and L_a, L_b, and L_c are the projections of the rotational angular momentum on the principal inertial axes of the molecule. The Hamiltonian in Eq. (1) is often referred to as a "rigid-rotor" Hamiltonian, even though significant vibrational effects appear in the rotational constants. To good approximation

$$A_v = A_e - \sum_k \alpha_k^A \left(v_k + \frac{d_k}{2} \right), \tag{2}$$

where A_e is the rotational constant for the molecule at rest at equilibrium, α_k^A is a vibration/rotation interaction parameter, and v_k and d_k are the vibrational quantum number and degeneracy, respectively, of the kth vibrational mode. We use the single letter v to stand for the set $v_1, v_2, \ldots, v_n$. Equations similar to Eq. (2) may be written for B_v and C_v.

If the Hamiltonian in Eq. (1) is adequate to interpret the rotational spectrum of the molecule, the experimental frequencies may be used to obtain values of the rotational constants. In most cases rotational constants for the ground vibrational state (i.e., A_0, B_0, C_0) will be available. To obtain equilibrium rotational constants, rotational spectra in the first excited state of every vibrational mode must be analyzed.

The relations between the rotational constants A_e and A_v and the corresponding principal moments of inertia $I_a^{(e)}$ and $I_a^{(v)}$ are as follows:

$$I_a^{(e)} = K/A_e \tag{3a}$$

$$I_a^{(v)} = K/A_v, \tag{3b}$$

where $K = h/8\pi^2$. Similar equations hold for the other moments of inertia. Until 1969, the accepted value of K was 505531 a.m.u.·Å²·MHz, which was used with the atomic mass unit scale based on ^{16}O. The currently accepted value of K is 505376 u·Å²·MHz to be used with the unified atomic mass scale based on ^{12}C. In some of the earliest work K = 505548 a.m.u.·Å²·MHz was used. The differences in these values of K appear to lead to differences of less than 0.0005 Å in distances and 0.05° in angles in the derived structures.

Equation (3) may be combined with Eq. (2) to yield

$$I_a^{(v)} = I_a^{(e)} + \sum_k \epsilon_k \left(v_k + \frac{d_k}{2} \right), \tag{4}$$

where the ϵ_k are a new set of vibration/rotation constants related to the α_k. The ϵ_k have been shown[3-9] to depend about equally on both quadratic and cubic vibrational potential constants. For this reason the quadratic force constants that are available for many molecules do not provide sufficient information for conversion of effective moments of inertia into equilibrium values.

For many purposes it is useful to introduce planar second moments, $P_{\alpha\beta}$, which in an arbitrary Cartesian coordinate system fixed in the molecule are defined as

$$P_{\alpha\beta} = \sum_i m_i \alpha_i \beta_i - M\alpha_o\beta_o \qquad \alpha, \beta = x, y, z. \tag{5}$$

In Eq. (5) m_i is the mass of ith atom, M is the mass of the whole molecule, α_o and β_o are coordinates of the center of mass, and the sum is over all the atoms in the molecule. The $P_{\alpha\beta}$ define a 3 X 3 symmetric tensor as follows:

$$P = \begin{bmatrix} P_{xx} & P_{xy} & P_{xz} \\ P_{yx} & P_{yy} & P_{yz} \\ P_{zx} & P_{zy} & P_{zz} \end{bmatrix}. \tag{6}$$

The eigenvalues of the P tensor are the principal second moments, P_{aa},

P_{bb}, and P_{cc}, which are related to the principal moments of inertia by expressions like the following:

$$I_a = P_{bb} + P_{cc} \tag{7}$$

$$P_{aa} = \tfrac{1}{2}(I_b + I_c - I_a). \tag{8}$$

An important advance in structure determination was made by Kraitchman in 1953.[14] He showed that the coordinates of an atom that was isotopically substituted in a molecular species–referred to as the parent species–are related to the principal second moments by expressions like the following:

$$a_s{}^2 = \mu^{-1}(P'_{aa} - P_{aa})\left(1 + \frac{P'_{bb} - P_{bb}}{P_{bb} - P_{aa}}\right)\left(1 + \frac{P'_{cc} - P_{cc}}{P_{cc} - P_{aa}}\right). \tag{9}$$

In Eq. (9) a_s is the a coordinate of the substituted atom in the principal inertial axis system of the parent species; P'_{aa} and P_{aa} are the principal second moments of the substituted and parent species, respectively; and $\mu = M(M' - M)/M'$, where M' and M are the molecular weights of the two species. Similar equations for b_s and c_s may be obtained from Eq. (9) by cyclic permutation of the subscripts.

The principal application of the Kraitchman equations [Eq. (9)] is for the determination of the atomic coordinates, a_s, b_s, and c_s. From a study of the rotational spectrum of the parent and of a species with single isotopic substitution the coordinates of the substituted atom may be determined. These coordinates are referred to as "substitution" coordinates or r_s coordinates. Each new species yields new coordinates, and since all of the coordinates are in the same coordinate system, the calculation of substitution or r_s bond distances and bond angles is a simple process. Costain[15] demonstrated that there are definite advantages to the use of the Kraitchman equations to obtain molecular parameters. These advantages are sufficient to make the use of Kraitchman's equations the preferred method of structure determination from ground-state rotational constants.

It is clear from the form of Eq. (9) that the location of atoms is more sensitive to differences in moments of inertia than to the absolute magnitudes of the moments. Also, isotopic substitution at a molecular site provides information mainly about the location of that particular site and only slightly about the location of other atoms. Therefore, the best structures obtained from rotational constants are those for which data are available for species with isotopic substitution at every nonequivalent molecular site.

An approximate expression for the uncertainty in a coordinate may be obtained by noting that the last two factors in Eq. (9) are each approximately equal to 1 and are in most cases an order of magnitude less sensitive to uncertainties in the moments than is the second factor. The expression that results is

$$\delta a_s = \frac{\delta(P'_{aa} - P_{aa})}{2\mu a_s}, \tag{10}$$

where δa_s is the uncertainty in a_s and $\delta(P'_{aa} - P_{aa})$ is the uncertainty in the difference $P'_{aa} - P_{aa}$.

Now, we may rewrite Eq. (8) as

$$P_{aa} = P_{aa}^{(e)} - \frac{\Delta_{aa}}{2} = \tfrac{1}{2}(I_b + I_c - I_a). \tag{11}$$

In this equation, $P_{aa}^{(e)}$ is the appropriate planar second moment for the equilibrium configuration of the molecule, and Δ_{aa} may be referred to as a "pseudoinertia defect." Since

$$P'_{aa} - P_{aa} = P_{aa}^{(e)\prime} - P_{aa}^{(e)} - (\Delta'_{aa} - \Delta_{aa})/2, \tag{12}$$

one expression for $\delta(P'_{aa} - P_{aa})$ is

$$\delta(P'_{aa} - P_{aa}) = -(\Delta'_{aa} - \Delta_{aa})/2. \tag{13}$$

If one uses 0.006 $u \cdot Å^2$ as an estimate of the difference in pseudoinertia defect on substitution, the Costain rule[16] for uncertainty in a substitution coordinate is obtained as follows:

$$\delta a_s = \left| \frac{0.0015}{a_s} \right|. \tag{14}$$

It has been assumed that $\mu \sim 1$ u.

Equation (14) demonstrates that it is very difficult to locate atoms that are close to a principal inertial plane. However, if only one *a, b,* or *c* coordinate is close to an inertial plane, the appropriate first moment relation may be used to compute that coordinate. For example,

$$a_1 = -\left(\sum_{i=2}^{N} m_i a_i\right) / m_1. \tag{15}$$

Similarly, a product of inertia relation may be used, as for example,

$$a_1 = -\left(\sum_{i=2}^{N} m_i a_i b_i\right) \bigg/ m_1 b_1. \tag{16}$$

Since the uncertainties in the other coordinates contribute to the uncertainty in a_1 by either Eq. (15) or Eq. (16), a comparison of uncertainties is usually required to determine which is the best way to calculate a particular small coordinate. Of course, if isotopic data are not available, Eq. (15) or Eq. (16) may be used in lieu of Eq. (9) to compute a coordinate. Unfortunately, Eq. (15) and (16) are not satisfied exactly by substitution coordinates, and this must be taken into account.

Recently, Watson[17] has shown that a first-order treatment of isotope effects leads to the approximate equation

$$I_\alpha^{(e)} = 2I_\alpha^{(s)} - I_\alpha^{(o)}, \tag{17}$$

where $I_\alpha^{(s)}$ is the α moment of inertia ($\alpha = a, b, c$) evaluated from substitution coordinates. With this equation, ground-state moments of inertia for several isotopic species may be used to estimate equilibrium moments of inertia. Watson refers to structural parameters derived from this estimate of equilibrium moments as r_m parameters. He has tested the method with success for a number of diatomic and triatomic molecules, but has found that for others the experimental errors are magnified enough to make the structures useless. The advantage of the method is that it is not necessary to obtain spectra in excited vibrational states, so that the many problems with the detection of exceptionally weak spectra and the characterization of the perturbations and resonances that can affect excited states are avoided. The disadvantages are the requirements for complete r_s structures of high accuracy and the fact that the approximations are not good enough for hydrogen–deuterium substitution.

Several of the procedures for deriving structural parameters from moments of inertia make use of the method of least squares. Since the relation between moments of inertia and Cartesian coordinates or internal coordinates is nonlinear, an iterative least squares procedure must be used.[18] In this procedure an initial estimate of the structural parameters is made and derivatives of the n moments of inertia with respect to each of the k coordinates are calculated based on this estimate. These derivatives make up a matrix D with n rows and k columns. We then define a vector X to be the changes in the k coordinates and a vector B to be the differences between the experimental moments and the calculated moments. We also define a weight matrix W to be the inverse of the matrix of the variances and covariances of the observed moments. Then, the least squares solution that minimizes

$$\widetilde{(DX - B)}W(DX - B)$$

is

$$\tilde{D}WDX = \tilde{D}WB. \tag{18}$$

The matrix W contains information about the uncertainties in the moments of inertia and the correlations between them. Unfortunately, this information is nearly impossible to obtain or even estimate, because the vibration/rotation interactions lead to large model errors, which ordinarily cannot be characterized. That is, even if the moments contained no experimental error, they would not agree with moments computed with exact structural parameters. As a result of this fundamental lack of information, Eq. (18) has always been used with a diagonal weight matrix.

One effect of the use of a diagonal weight matrix is to make the solution to Eq. (18) depend on the particular linear combination of moments of inertia used. Thus, use of principal moments of inertia, or principal planar second moments, or differences in either of these all lead to different structural parameters even if the same set of rotational constants is used to derive the moments. By comparing structural parameters derived from each set of moments, some insight into the effect of the correlations between the moments and the magnitudes of the model errors may be obtained.

By contrast it may be shown that the solution of Eq. (18) is invariant to linear combination of the components of X. Therefore, determination of Cartesian coordinates or determination of internal coordinates should lead to the same result. It should be pointed out, however, that the principal axis conditions may be taken into account in different ways in Cartesian coordinate and internal coordinate calculations. We include the center-of-mass and product-of-inertia relations along with the moments when we fit Cartesian coordinates. On the other hand, internal coordinates are independent of the origin of the coordinate system. Thus, holding some Cartesian coordinates constant while others are varied is not the same as holding some internal coordinates constant while others are varied. This is true even if the same atomic positions are involved.

Equation (18) is easily solved for X if the matrix $\tilde{D}WD$ is nonsingular. If, however, the moments are not sensitive to one or more of the coordinates, or if two or more coordinates are nearly linearly dependent,

$\tilde{D}WD$ will be singular, or nearly so, and Eq. (18) cannot be solved. In such a case it is necessary to add more moments of inertia, hold some coordinates constant, or add restricting conditions in the form of constant bond lengths, constant bond angles, or symmetry restrictions (e.g., a symmetric methyl group). One method of checking for singularities or near singularities is to examine the eigenvalues of $\tilde{D}WD$; small or negative eigenvalues are evidence of singularities.

A final point in the use of iterative least squares concerns the possible interaction between two or more of the coordinates to be adjusted during the fitting. In iterative least squares the effect of second and higher derivatives of the moments with respect to the coordinates is ignored. If the adjustment in the coordinates is small, the effects of the second and higher derivatives are not important. However, under some circumstances the effect of the mixed second derivative may be crucial. It may prove necessary to uncouple certain sets of coordinates and alternate least squares adjustment of each set until the structure is determined as well as possible. This problem is particularly acute with hydrogen coordinates because of the low mass and, consequently, small contribution to the moments. We have found it necessary on occasion to alternate fitting the coordinates of hydrogen atoms and heavy atoms because of the effects of interaction. As a result of this kind of effect, it is necessary to iterate a nonlinear least squares problem essentially to convergence. It is sometimes possible to be misled by the results of early iterations.

STRUCTURE CALCULATIONS

The simplest, most direct, and most precise determination of bond distances and bond angles from rotational constants is from equilibrium values of these constants. Equilibrium parameters have a well-defined interpretation and are virtually invariant to isotopic substitution. Unfortunately, the required spectra in the first excited vibrational states are nearly always very difficult to obtain. In addition, the rotational constants must be free of the effects of perturbations and resonances. As a result, equilibrium structures have been obtained only for diatomic molecules and a few small polyatomic molecules. An example is the structure of SO_2 obtained by Morino *et al.*[19] (Table 1). Also shown in the table is the approximate r_e structure called the r_m structure by Watson.[17]

Ground-state rotational constants and rotational constants for the average molecular configuration are related by terms that depend on the quadratic force constants[3–9] and may be determined for many molecules. Therefore, bond distances and bond angles for the average

TABLE 1 Comparison of Equilibrium, Mass Dependence, Substitution, Effective, and Average Structures for SO_2[a]

Type of Structure	r(SO)	∠OSO
r_e	1.4308	119.32
r_m	1.4306	119.34
r_s	1.4328	119.36
r_o	1.4336	119.42
r_z	1.4349	119.21

[a] The SO bond distances are in Å, the OSO angles in degrees. The r_e and r_z structures are from Morino *et al.*,[19] the r_m structure from Watson.[17] The rotational constants for the r_s structure are from Steenbeckeliers[27] and Starck.[28]

configuration (i.e., r_z parameters), which have an appealing physical interpretation, may often be computed. Unfortunately, r_z parameters are not invariant to isotopic substitution and, thus, may be obtained only for molecules with three or less structural parameters. The structure of the average configuration of SO_2 is shown in Table 1.

For most molecules the data required to obtain an r_e structure or an r_z structure are not available. It is therefore necessary to compute some form of effective structural parameters from the ground-state rotational constants of one or more isotopic species. As indicated above, r_s parameters are preferred. The r_s parameters for SO_2, obtained by assuming $^{32}S^{16}O_2$ to be the parent molecule and $^{34}S^{16}O_2$ and $^{32}S^{16}O^{18}O$ to be the singly substituted species, are compared with the r_e, r_m, and r_z parameters in Table 1. The r_o parameters of SO_2, obtained by adjusting the SO distance and the OSO angle to provide a best fit to the ground-state rotational constants of $^{16}S^{32}O_2$, are also shown.

Sulfur dioxide is a very simple molecule for structure calculation in that it contains only two independent parameters. In Tables 2 and 3 the results of several different calculations of the structural parameters of 2-chloropropane (Figure 1) are compared.[20] Fifteen independent parameters are required to completely determine the structure of 2-chloropropane. The reason for presenting the comparison shown in Tables 2 and 3 is that 2-chloropropane is a somewhat unusual example in that rotational spectra have been analyzed and ground-state rotational constants obtained for a parent molecule and molecules with single isotopic substitution at every nonequivalent site, a total of eight isotopic species in all. The data are therefore sufficient for a complete r_s structure. Since for many molecules structures must be calculated with a more limited

TABLE 2 Comparison of Selected Bond Distances, Bond Angles, and First and Second Moments of 2-Chloropropane Obtained by Different Methods of Calculation[a]

	r_s		p-Kr		r_o		
Parameter	Complete	Partial	I's	P's	I's	P's	Experiment
CCl	1.797	1.798	1.800	1.803	1.807	1.805	
CC	1.520	1.523	1.517	1.521	1.522	1.523	
CH_s	1.091	1.091	1.099	1.091	1.096	1.094	
CH_2	1.099	1.091	1.094	1.091	1.092	1.092	
HCCl	105.2	105.4	104.9	105.0	105.0	105.0	
CCC	112.8	112.5	113.0	112.7	113.3	113.1	
CCH_1	111.0	110.6	110.9	110.7	111.1	110.9	
H_1CH_3	109.2	108.9	109.0	108.9	109.1	109.2	
$\Sigma m_i a_i$	0.1450	0.1450	0.0000	0.0000	0.0000	0.0000	
$\Sigma m_i c_i$	−0.3574	0.0000	0.0000	0.0000	0.0000	0.0000	
$\Sigma m_i a_i c_i$	0.0128	0.0000	0.0000	0.0000	0.0000	0.0000	
I_a	62.3375	62.4001	62.2086	62.3894	62.6358	62.6363	62.6389
I_b	110.1750	110.2376	110.2959	110.3654	110.5601	110.5612	110.5657
I_c	156.8886	156.8887	156.9402	157.0102	157.5482	157.5491	157.5573

[a] The bond distances are in Å, bond angles in degrees, first moments in u·Å, and the second moments in $u \cdot Å^2$. The rotational constants for these calculations are from Tobiason and Schwendeman.[20] The conversion factor used is 505 376 $MHz \cdot u \cdot Å^2$. The different methods of calculation are described in the text.

TABLE 3 Comparison of Selected Bond Distances and Bond Angles of 2-Chloropropane Calculated for Data from Eight Isotopic Species and for Data from Seven Isotopic Species[a]

	r_s		p-Kr, I's	
Parameters	8 Species	7 Species	8 Species	7 Species
CCl	1.797	1.805	1.800	1.817
CC	1.520	1.505	1.517	1.517
CH_s	1.091	1.117	1.099	1.095
HCCl	105.2	103.5	104.9	104.1
CCC	112.8	114.6	113.0	113.9
CCH_1	111.0	111.9	110.9	111.3

[a] See footnote *a*, Table 2. The species deleted for the seven-species calculation is that with ^{13}C substitution at the central carbon atom.

FIGURE 1 Model of 2-chloropropane showing the labeling of the hydrogen atoms.

set of data, the comparisons shown in Tables 2 and 3 demonstrate the magnitudes of the differences in bond distances and bond angles to be expected.

The 2-chloropropane molecule also provides an example of the effect of the calculation of small coordinates by the use of first-moment and product-of-inertia relations. The c coordinates of the Cl atom and the two equivalent out-of-plane C atoms are $\sim$0.1 Å or less, which is usually taken to be the region of difficulty for the Kraitchman equations. In this case the c coordinates of the Cl and out-of-plane C atoms were computed by simultaneous solution of the equations $\Sigma m_i c_i = 0$ and $\Sigma m_i a_i c_i = 0$ after the other coordinates had been computed by means of the Kraitchman equations. A comparison of the c coordinates computed by the two methods is given in Table 4 and the bond distances and bond angles resulting from the adjusted c coordinates are given in the column headed r_s (partial) in Table 2. It is seen that interatomic distance differences as large as 0.003 Å and angle differences as large as 0.4° result.

TABLE 4 Comparison of the c Coordinates of the Cl Atom and the Out-of-Plane C atom of 2-Chloropropane Determined by Kraitchman's Equations or by Assuming $\Sigma m_i c_i = 0$ and $\Sigma m_i a_i c_i = 0$[a]

Atom	Kraitchman's Equations	$\Sigma m_i c_i = 0$ $\Sigma m_i a_i c_i = 0$
Cl	0.0409	0.0462
C	0.1222	0.1308

[a] The coordinates are in Å. The data are taken from Tobiason and Schwendeman.[20]

Instead of fitting the moments of inertia by means of the Kraitchman equations, the moments of inertia or the planar second moments may be fit directly by adjusting the bond distances and bond angles to give the best fit in the least squares sense. This may be done in a variety of ways, as indicated above. To obtain the values in the column headed p-Kr (I's),* the parameters were adjusted to give a best fit to differences between the moments of inertia of the isotopically substituted species and the corresponding moments of the parent species. The structure labeled p-Kr (P's) is a best fit to differences in planar second moments. The numbers in the column labeled r_0 (I's) resulted from a direct fit of all

* We use the symbol p-Kr (i.e., pseudo-Kraitchman) for a structure obtained by least squares fitting to differences in moments of inertia.

the moments of inertia, and the ones in the column labeled r_0 (P's) are from a fit of all the planar second moments. For each of the least squares fits the values obtained were independent of whether Cartesian coordinates or internal coordinates were used as the fitting parameters.

The most important aspect of the comparison in Table 2 is that the bond distances vary by several thousandths of an angstrom and the bond angles vary by several tenths of a degree. We have found this to be typical. It is probably not surprising to find the greatest variation for parameters that depend strongly on the two small c coordinates. There is, however, a certain stability to the numbers in Table 2. That is, although the variations in parameters are significant, they are an order of magnitude smaller than variations which are indicative of difficulty in a structure determination. Unfortunately, it is only recently that general computer programs that make these calculations easy to do have become available. All the calculations described in this paper were performed or checked with one or both of two new computer programs STRFIT and STRFTQ.[21]

The moments of inertia of the parent species computed for the various structures are given at the bottom of the columns in Table 2. The values may be compared with the experimental values for the parent species, which are given in the last column. The agreement for the r_s and p-Kr structures is poor because these structural parameters have been computed from differences in moments of inertia rather than the moments themselves. By contrast, the moments for the r_0 structures are close to the experimental values.

Values of the nontrivial first moments and product of inertia for the various structures of 2-chloropropane are also given at the bottom of Table 2. The values are essentially zero for all but the substitution structures. The reasons for this difference were given above. The first moments for the r_s structure are somewhat larger than is typical for r_s structures, but the product of inertia is about average size.

In Table 3 the values of selected parameters of 2-chloropropane for the complete r_s calculation and the p-Kr (I's) calculation from Table 2 are compared with closely related calculations performed after omission of the experimental data for the species with ^{13}C substitution at the central carbon atom. The purpose of this calculation is to demonstrate how different the structure reported would have been if data for only seven isotopic species had been available. For the r_s calculation with seven species the coordinates of the central carbon atom were determined by adjusting them to give a best least squares fit to the principal axis conditions. For the p-Kr calculation with seven species the moments of inertia of the parent species were included in the fit along with the six remaining sets of differences in moments of inertia. The conclusion we draw from the comparison in Table 3 is that it is important to have data for all the isotopic species.

In spite of the comparison just made, the data for many molecules are insufficient in number or not of the right kind to compute a complete r_s structure. These molecules fall roughly into three groups. In the first group are the molecules for which the data are simply insufficient in kind; that is, molecules with isotopic substitution at every nonequivalent site are available, but some of the species are multiply substituted. For example, for vinyl chloride[22] the species available were $H_2CCH^{35}Cl$, $H_2CCH^{37}Cl$, $H_2C^{13}CH^{35}Cl$, $H_2{}^{13}CCH^{35}Cl$, $D_2CCD^{35}Cl$, $DHCCD^{35}Cl$ (*cis* and *trans*), $D_2CCD^{37}Cl$, and $DHCCD^{37}Cl$ (*trans*). Although all the H atoms have been deuterated, no singly deuterated species were available. Consequently, Kivelson *et al.*[22] computed r_s coordinates for the Cl atom and the two C atoms and then varied the H coordinates to give a best fit to the differences between the moments of inertia of the various deuterated species and the corresponding moments of the parent species. A structure obtained in this way is probably as reliable as a complete r_s structure.

In the second group of molecules for which a complete r_s structure cannot be obtained are those for which isotopic labeling at every nonequivalent site either has not been done or cannot be done. In Table 5 the mixed r_s, p-Kr structures (which we call simply Kr structures) and

TABLE 5 Comparison of Bond Lengths, Bond Angles, First and Second Moments for r_0 and Kr Structures of NF_2CN and PF_2CN[a]

	NF_2CN			PF_2CN		
Parameter[b]	r_0	Kr	Experiment	r_0	Kr	Experiment
X–F	1.398	1.403		1.567	1.566	
X–C	1.392	1.375		1.811	1.815	
C≡N	1.151	1.161		1.158	1.157	
∠F*X*F	102.9	102.2		99.1	99.2	
∠F*X*C	104.7	105.5		97.1	96.9	
∠*X*CN	169.7	173.6		171.5	171.2	
$\Sigma m_i a_i$	0.0000	0.0663		0.0000	0.0253	
$\Sigma m_i c_i$	0.0000	−0.4409		0.0000	0.0132	
$\Sigma m_i a_i c_i$	0.0000	−0.3719		0.0000	−0.0181	
I_a	49.8860	50.0284	49.8857	68.2603	68.2665	68.2606
I_b	113.2622	113.4070	113.2602	155.3559	155.3454	155.3565
I_c	154.2262	154.0847	154.2273	195.0911	195.0843	195.0905

[a] The bond distances are in Å, the angles in degrees. The first moments are in u·Å and the second moments in u·Å^2. The data in this table were taken from Lee *et al.*[23,24]

[b] X = N for NF_2CN and X = P for PF_2CN.

the r_0 structures of PF_2CN and NF_2CN are compared. For PF_2CN the ^{13}C- and ^{15}N-substituted species were prepared and analyzed[23]; thus, r_s coordinates of the C and N atoms could be calculated. The coordinates of the P and F atoms were then adjusted to give a best fit to the moments of inertia of the parent species, two first-moment conditions, and one product-of-inertia condition. In the table this Kr structure is compared with the usual r_0 structure in which all the coordinates are varied to give the best fit to the three sets of moments of inertia. For NF_2CN the situation was similar, except that the species with ^{15}N in the NF_2 group was also available[24] so that r_s coordinates could be obtained for the entire NCN geometry. The point to be made here is that, even though more data are available for NF_2CN and even though the structures of the molecules are very similar, the two structures for PF_2CN are more comparable than the two structures for NF_2CN. Also, the fit to the moments of inertia and first moments are better for PF_2CN. The reason for this, of course, is that as more atomic coordinates are determined by Kraitchman's equations, there are fewer degrees of freedom left, and also the difference between the experimental moments of inertia and the contributions from the r_s coordinates becomes less reliable. The results of these factors may be seen in Table 5. Incidentally, one of the most important features of the structures of these two molecules–namely, the fact that the PCN and NCN angles are less than 180°–is determined from very small c coordinates of the C atoms. We believe that the uncertainties in these angles given in the papers are properly assessed, but it should be realized that small coordinates always carry an extra degree of uncertainty in structural analysis from rotational constants, simply because the data are insensitive to the locations of atoms near a principal inertial plane.

In the third group of molecules for which complete r_s structures have not been obtained are those for which the number of independent rotational constants available is less than the number of independent parameters. In these cases some of the structural parameters must be obtained from other data or assumed. The most logical source of other data is an electron diffraction investigation. Unfortunately, the distance or angle parameters determined by electron diffraction studies are not the same as those obtained from rotational constants. (See Laurie, this volume, Kuchitsu and Cyvin,[2] and Kuchitsu.[25]) This lends an extra degree of uncertainty to this procedure.

In general, when structural parameters must be assumed, one should be very cautious about accepting the derived structural parameters. If the investigator has shown that the derived parameters depend only slightly on the assumed values, as may occur with certain hydrogen parameters, or when there is good chemical reason to believe that a structural parameter may be transferred from a similar molecule, it may be possible to properly assess the reliability of the derived parameters. Most investigators will admit to being overoptimistic on this point, however; it is therefore probably good policy to be especially wary of structures derived on the basis of assumptions.

ESTIMATION OF UNCERTAINTIES

The estimation of uncertainties in structural parameters derived from rotational constants is complex. The first question is whether to report uncertainties in the parameters as estimates of effective values of the parameters or as estimates of equilibrium values of the parameters. For example, r_s coordinates are defined operationally as functions of ground-state planar second moments. The only uncertainty in an r_s coordinate as an estimate of an r_s coordinate is from experimental uncertainty in the planar second moments, and this is easy to compute from Eq. (10). As estimates of equilibrium values of the coordinates, however, r_s coordinates contain an additional contribution to their uncertainty from the neglected pseudoinertial defects. This contribution may be estimated from the Costain rule [Eq. (14)].

The r_0 and p-Kr coordinates are determined by adjusting coordinates to fit moments of inertia and differences in moments of inertia, respectively. This is again an operational definition, so the only uncertainty is the result of experimental uncertainty in the rotational constants, and this may be treated by standard methods.[18] However, the standard deviation between observed rotational constants and those computed from the final structure should not be used in such a calculation. In the comparison shown above the structural parameters of NF_2CN are more reliable than those of PF_2CN, but the deviation between observed and calculated moments is greater for NF_2CN. The standard deviations of the rotational constants are best obtained from the fit of the rotational spectra. These values should be used to estimate the experimental uncertainties in r_0 and p-Kr coordinates. The uncertainties in r_0 or p-Kr coordinates as estimates of the equilibrium coordinates are very difficult to compute. The Costain rule is probably satisfactory for p-Kr coordinates, but less so for r_0 coordinates.

A second question concerning estimation of uncertainties is how to propagate the uncertainties in Cartesian coordinates through to bond distances or bond angles. The usual technique is to use the standard relation

$$\delta r = \left\{ \sum_{k,n} \left(\frac{\partial r}{\partial x_{nk}} \right)^2 (\partial x_{nk})^2 \right\}^{1/2} \tag{18}$$

in which x_{nk} is the kth coordinate of atom n with uncertainty δx_{nk} and r is a structural parameter with uncertainty δr. The sum extends over only those coordinates that contribute to r. The difficulty with the use of Eq. (18) is that positive or negative δx_{nk} may not be equally probable. It is known that the sign of that part of δx_{nk} obtained from the Costain rule is usually the same as the sign of x_{nk}. That is, the effect of neglect of the pseudoinertial defect is to give a coordinate that is smaller than the equilibrium value of the coordinate. By contrast, the experimental contribution to δx_{nk} is of either sign. For this reason a procedure of assessing δr, which takes these factors into account, was introduced some years ago.[20]

For atoms that lie close to a principal plane the Costain rule gives large uncertainties. If only one or two atoms lie close to a principal plane, a better estimate of the small coordinates may be obtained from Eq. (15) and/or Eq. (16) and correspondingly better estimates of the uncertainties in these coordinates may be obtained from derivatives of these equations and uncertainties in the other coordinates.

We recommend that experimental uncertainties in the coordinates be assessed by Eq. (10) and the vibration/rotation contributions be assessed by the Costain rule [Eq. (14)] or by first moment or product of inertia relations. Then, either the procedure introduced by Tobiason and Schwendeman[20] should be used to propagate the uncertainties into distances or angles, or the two contributions should be added together and used with Eq. (18) to generate distance and angle uncertainties. In either case the uncertainties in distances should range from a few thousandths to a few hundredths of an angstrom, whereas uncertainties in angles should range from tenths of a degree to a few degrees. Uncertainties as small or smaller than 0.001 Å or 0.1° should be viewed with skepticism with an extra requirement of proof by the investigator.

SPECIAL PROCEDURES

Two special procedures employed in structure calculations will be mentioned here. One is the double substitution method introduced by Pierce[26] and used for the determination of the coordinates of atoms near a principal inertial plane, which is the situation for which the Kraitchman equations become unreliable. In Pierce's method rotational constants must be available for four isotopic species, two of which are defined as parent molecules. Each of the remaining two species is related to one of the parent molecules by having single isotopic substitution at the near-plane atom site. In essence the method involves taking the difference between the two Kraitchman equations for the small coordinate. Since the Kraitchman equations involve differences in moments of inertia, the double substitution method involves second differences. The second differences lead to a further cancellation of the effects of the inertial defects. Since two parents are defined, a shift in coordinate systems is involved, but this may usually be accurately characterized. The principal drawbacks of the double substitution method are the extra isotopic substitution required and the highly accurate rotational constants required for the second differences. While Pierce's method has been very successful in some cases, it has not had widespread usage.

The second special procedure to be mentioned here is the method that takes into account what we refer to as the "Laurie correction."[4] This is a method for correcting for the fact that an average bond distance between two atoms shrinks when a heavier isotope is substituted for one of the atoms. Table 6 is a comparison of average distances in isotopic species of HCl and CO to demonstrate the effect.[4] When deuterium is substituted for hydrogen in HCl the average bond distance shrinks by 0.0046 Å as a result of the vibrational changes. However, when ^{37}Cl is substituted for ^{35}Cl in either HCl or DCl, the bond length change is less than 0.0001 Å. This is presumably a result of the fact that during vibration the lighter hydrogen atom does nearly all the moving. When ^{18}O is substituted for ^{16}O or ^{13}C is substituted for ^{12}C in CO, the average bond length decreases 0.0001 Å. Evidently, substitution of a heavy (i.e., nonhydrogenic) atom leads to small, but detectable changes in average bond length.

The shrinkage effect just described suggests a method for correcting

TABLE 6 Comparison of Average Bond Distances for Isotopic Species of HCl and CO[a]

Molecule	r_z
$H^{35}Cl$	1.2904
$H^{37}Cl$	1.2904
$D^{35}Cl$	1.2858
$D^{37}Cl$	1.2858
$^{12}C^{16}O$	1.1323
$^{13}C^{16}O$	1.1322
$^{12}C^{18}O$	1.1322

[a] Bond distances are in Å. The data are taken from Laurie and Herschbach.[4]

TABLE 7 Comparison of Selected Bond Distances and Bond Angles of 2-Chloropropane Calculated before and after the Laurie Correction to the Rotational Constants[a]

	r_s		p-Kr, I's	
Parameter	Before	After	Before	After
CCl	1.797	1.803	1.800	1.806
CC	1.520	1.519	1.517	1.518
CH_2	1.099	1.103	1.094	1.096
CCC	112.8	113.1	113.0	113.1
CCCl	109.5	109.3	109.5	109.2
CCH_3	109.6	109.4	109.6	109.2

[a] See footnote *a*, Table 2.

the moments of inertia of the various isotopic species before they are used in a structure calculation. One starts by calculating the moments of inertia for each isotopic species for some assumed structure. Then, for each isotopic species the bond distance to each deuterium atom is reduced by 0.002–0.005 Å, the deuterium isotope effect for average distances. Each distance to a heavy atom that has been substituted by a heavier isotope is reduced 0.00005–0.0001 Å, unless the distance is to a hydrogen atom, in which case it is not reduced at all. The moments of inertia are recalculated for each isotopic species with one or more adjusted distances. The difference between the moments of inertia of each species with and without the corrected distances is then applied as a correction to the experimental moments of inertia. The corrected moments of inertia are then used in the ordinary way in a structure calculation.

As an example of this method four sets of some selected bond distances and bond angles are given for 2-chloropropane in Table 7. The r_0 and complete r_s structures from Table 2 are repeated and compared with corresponding structures computed from moments of inertia after the Laurie correction has been made. For purposes of the calculation it was assumed that CH distances decreased 0.002 Å upon deuteration and that heavy atom distances decreased 0.00005 Å upon substitution of a heavier isotope. The result of these small changes is to alter bond distances by up to 0.006 Å and bond angles by up to 0.3°. Unfortunately, rules for the effect of isotopic substitution on bond angles have not been given and probably do not exist. In addition, the effect of using corrections for average bond distances has not been assessed. For these reasons the Laurie correction is most useful for assessing uncertainties and as an aid in determinng the magnitudes of the differences in pseudo-inertia defects.

SUMMARY

So many factors affect the accuracy of bond distances and bond angles derived from rotational constants that nearly every molecule is a unique case. Nevertheless, there are several key questions that should be answered by every report of structural analysis from rotational spectra. Some of these are as follows:

1. Have the frequencies of the rotational transitions been measured accurately, have any perturbing effects been correctly eliminated, and have the rotational constants and their uncertainties been evaluated properly?
2. How many isotopic species have been studied, and how many atoms have not been substituted?
3. How many atoms lie near principal inertial planes, and how have the small coordinates been treated?
4. Have several strategies for the calculation of distances and angles been employed and the results intercompared?
5. Have the experimental uncertainties in the rotational constants and the contributions from rotation/vibration been included in the assessment of uncertainties?

Discussion of the answers to questions like these are the mark of careful analysis. There is a wealth of data and a large number of excellent structure determinations from rotational constants in the literature.

The research in this paper was supported in part by grants from the National Science Foundation.

REFERENCES

1. W. Gordy and R. L. Cook, *Microwave Molecular Spectra,* Interscience, New York (1970), Ch. 13.
2. K. Kuchitsu and S. J. Cyvin, *Molecular Vibrations and Structure Studies,* S. J. Cyvin, Ed., Elsevier, Amsterdam (1972), Ch. 12.
3. D. R. Herschbach and V. W. Laurie, *J. Chem. Phys.,* **37**, 1668 (1962).
4. V. W. Laurie and D. R. Herschbach, *J. Chem. Phys.,* **37**, 1687 (1962).
5. D. R. Herschbach and V. W. Laurie, *J. Chem. Phys.,* **40**, 3142 (1964).
6. T. Oka, *J. Phys. Soc. Japan,* **15**, 2274 (1960).
7. T. Oka and Y. Morino, *J. Mol. Spectrosc.,* **6**, 472 (1961).
8. Y. Morino, K. Kuchitsu, and T. Oka, *J. Chem. Phys.,* **36**, 1108 (1962).
9. M. Toyama, T. Oka, and Y. Morino, *J. Mol. Spectrosc.,* **13**, 193 (1964).
10. E. G. Codding and R. H. Schwendeman, *Inorg. Chem.,* **13**, 178 (1974).
11. D. R. Herschbach, *J. Chem. Phys.,* **31**, 91 (1959).
12. D. Stelman, *J. Chem. Phys.,* **41**, 2111 (1964).

13. W. H. Kirchhoff, *J. Mol. Spectrosc.*, **41**, 333 (1972).
14. J. Kraitchman, *Am. J. Phys.*, **21**, 17 (1953).
15. C. C. Costain, *J. Chem. Phys.*, **29**, 864 (1958).
16. C. C. Costain, *Trans. Am. Cryst. Assoc.*, **2**, 157 (1966).
17. J. K. G. Watson, *J. Mol. Spectrosc.*, **48**, 479 (1973).
18. W. C. Hamilton, *Statistics in Physical Science, Estimation, Hypothesis Testing, and Least Squares,* Ronald Press, New York (1964).
19. Y. Morino, Y. Kikuchi, S. Saito, and E. Hirota, *J. Mol. Spectrosc.*, **13**, 95 (1964).
20. F. L. Tobiason and R. H. Schwendeman, *J. Chem. Phys.*, **40**, 1014 (1964).
21. R. H. Schwendeman, to be published.
22. D. Kivelson, E. B. Wilson, Jr., and D. R. Lide, Jr., *J. Chem. Phys.*, **32**, 205 (1960).
23. P. L. Lee, K. Cohn, and R. H. Schwendeman, *Inorg. Chem.*, **11**, 1917 (1972).
24. P. L. Lee, K. Cohn, and R. H. Schwendeman, *Inorg. Chem.*, **11**, 1920 (1972).
25. K. Kuchitsu, "Gas Electron Diffraction," Ch. 6 in *MTP International Review of Science,* Ser. 1, Vol. 2, *Physical Chemistry,* G. Allen, Ed., Butterworths, London (1972).
26. L. Pierce, *J. Mol. Spectrosc.*, **3**, 575 (1959).
27. G. Steenbeckeliers, *Ann. Soc. Sci. Bruxelles,* **82**, 331 (1968).
28. B. Starck, *Molecular Constants from Microwave Spectroscopy* (Landolt–Börnstein Tables, New Series, Group II, Vol. 4), Springer, Berlin (1967).

D. A. Ramsay

Critical Evaluation of Molecular Constants from Optical Spectra

The basic equation in experimental optical spectroscopy is

$$\nu_i = T_i' - T_i'' + \delta_i$$

where ν_i is the measured wavenumber of the ith line, T_i' and T_i'' are the term values of the corresponding upper and lower states, and δ_i is the experimental error in the measurement. These quantities are usually expressed in cm^{-1}, and it is commonly assumed that the δ_i are uncorrelated, though in actual practice the systematic errors in a set of measurements frequently exceed the random errors.

In atomic spectroscopy the term values depend primarily on electronic quantum numbers and the process of analysis consists of reducing a number of measurements to a term scheme. The confidence in an analysis increases as the system becomes more overdetermined, and the process becomes more definite as the accuracy of the measurements improves. Other information is also used to facilitate the assignments of the lines, e.g., relative intensities, the observation of certain lines in absorption, the splittings of lines by magnetic fields; theoretical calculations of terms and multiplet splittings may sometimes be helpful.

In molecular spectroscopy the term values depend on vibrational and rotational quantum numbers, as well as on the electronic quantum numbers.* The intensity of each electronic transition is distributed over a number of vibrational bands and the intensity within each band is distributed over a number of rotational transitions. Since the rotational transitions obey simple selection rules (e.g., $\Delta J = 0, \pm 1$), the lines form branches in which the spacings usually increase or decrease in a regular manner. The process of rotational analysis consists in recognizing these branches and assigning rotational quantum numbers to the lines. For a $^1\Sigma$–$^1\Sigma$ band of a diatomic molecule this process is relatively straightforward, but for a band of a polyatomic molecule, which is also an asymmetric top, analysis might be quite difficult due to the overlapping of a large number of different branches. For heavier molecules, e.g., substituted benzenes, the number of rotational transitions in a given band may be so large that it is no longer possible to resolve all the individual lines; thus the problem is one of attempting to analyze a rotational contour. Most rotational analyses are based on models for the two combining states and are frequently assisted by computer techniques. The success in analyzing complex spectra, however, still depends largely on the skill and experience of the spectroscopist. In the following discussion it will be assumed that correct rotational assignments have been made, and the problems that will be investigated are the accuracy with which measurements can be made and the methods by which molecular constants are obtained from rotational assignments.

LINE WIDTHS AND ACCURACY OF MEASUREMENTS

The accuracy with which the wavenumber of the center of a symmetrical isolated line can be measured is $\sim \frac{1}{20}$ of the line width. For a gas at a sufficiently low pressure so that pressure broadening is unimportant, the width of the line is determined by the Doppler effect and is

$$\Delta\nu = \frac{2\nu}{c}\sqrt{\frac{2RT\ln 2}{M}} = 7.162 \times 10^{-7}\,\nu\sqrt{\frac{T}{M}},$$

where $\Delta\nu$ is the width at half maximum intensity, R is the gas constant, M the molecular weight, and T the thermodynamic temperature. If $M = 50$ and $T = 300$ K, then $\nu/\Delta\nu \sim 6 \times 10^5$. Some typical Doppler widths for different regions of the optical spectrum are given in Table 1.

In the visible and near ultraviolet regions of the spectrum, large Ebert or Czerny–Turner spectrographs (3.4–10 m) with resolving power of

*Hyperfine effects have been resolved in the optical spectra of a few diatomic molecules.

TABLE 1 Doppler Widths of Lines (assuming $T = 300$ K and $M = 50$) and Typical Accuracies of Measurements for Different Spectral Regions

λ	$\nu(\text{cm}^{-1})$	Doppler Width $\Delta\nu(\text{cm}^{-1})$	Typical Measurement Accuracy (cm^{-1})
1200 Å	83 333	0.15	0.04
2000 Å	50 000	0.09	0.006
3000 Å	33 333	0.06	0.003
5000 Å	20 000	0.035	0.002
10000 Å	10 000	0.018	0.001
3 μm	3 333	0.006	0.001
5 μm	2 000	0.004	0.001
10 μm	1 000	0.002	0.002

∿600000 are in current use; the resolving powers of these instruments are comparable with those needed to resolve the Doppler widths of the lines. Furthermore, the dispersions of these instruments are matched to the plate grain characteristics of photographic plates. For example, the 7.3 m Ebert spectrograph at Ottawa is used in the twentieth order for work at 3000 Å and has a reciprocal dispersion of 0.10 Å/mm. With a resolving power of 600 000, two lines 0.005 Å apart are just resolved and are separated by 0.05 mm in the focal plane. Such lines are easily photographed with fast plates that permit the resolution of 40 lines per millimeter. Furthermore, to locate the center of a line to $\frac{1}{20}$ of its line width, it is necessary to measure the plate with an accuracy of 0.0025 mm, i.e., 2.5 μm. With comparators equipped with photoelectric scanning devices, it is possible to achieve a reproducibility, though not necessarily an accuracy, in the measurement of spectral features to about 1 μm. In the vacuum ultraviolet the practical resolving power is less, although resolving powers of the order of 300 000 have been achieved in the region of 1500 Å. The Doppler widths of the lines are also greater, so the wave-number accuracy of measurements is appreciably less than in the visible region of the spectrum. In the infrared the line widths are much smaller, but the resolving powers of the grating instruments in use rarely exceed 200 000. Nevertheless, the measurements in the near infrared are more accurate than those that can be made in the visible. Estimates of typical measurement accuracies in different regions of the optical spectrum are given in the last column of Table 1. It should be stressed that these accuracies apply only to the measurement of single lines and that the accuracies with which blended and overlapped lines can be measured are less.

Although lines in the visible region of the spectrum can be measured with an accuracy of a few thousandths of a cm^{-1}, it is frequently found that different sets of measurements of the same band differ by 0.01 cm^{-1} or more. Since the systematic errors of measurement are considerably larger than the random errors, it is important to measure a band completely, rather than in sections as is the practice in some laboratories. Furthermore, it is important that a sufficient number of calibration lines be measured and that these should extend beyond the limits of the band. One source of systematic error involves the alignment of the slit of the comparator parallel to the spectral lines; others involve errors in the comparator screw and variation in the thickness of the oil film on the screw during measurement. There may be other sources of systematic error, not all of which may be known. For example, it is usually assumed that the spectral lines on a photographic plate are straight and parallel, whereas it is known that, in fact, the lines are slightly curved, and careful measurement reveals that there is a very slight fanning of the lines along the photographic plate. Another source of error may involve nonuniform shrinking or swelling of the gelatin film during the photographic process. Finally, although spectroscopists apply vacuum corrections to measurements of plates taken in air, the large majority adopt the vacuum corrections appropriate to a pressure of 760 torr. The sensitivity of these corrections to small changes in pressure is quite appreciable; for example, a change in pressure of 1 torr changes the vacuum correction by 0.013 cm^{-1} at 3000 Å. Such errors usually appear in the calculated band origins and, hence, in the vibrational term values of the molecule. It is of interest to enquire also what is the dispersion of this error across a band. For a band at 3000 Å with a rotational width of 10 Å photographed in air at a pressure of 700 torr, the difference in the vacuum correction at each end of the band introduces an error of 0.003 cm^{-1} if corrections appropriate to atmospheric pressure are applied. Such a systematic error is comparable with the accuracy of measurement. It should be stressed that none of the above systematic errors is eliminated by statistical treatment of the data.

METHODS FOR EVALUATING MOLECULAR CONSTANTS

Combination Differences

One of the traditional ways for determining rotational constants involves the method of combination differences.[1] If two lines have one state in common, then the difference between the wave numbers of the two lines gives an interval between two of the term values for the other state.

Thus for a simple $^1\Sigma$–$^1\Sigma$ band we have

$$\Delta_2 F''(J) = F''(J+1) - F''(J-1) = R(J-1) - P(J+1)$$

and

$$\Delta_2 F'(J) = F'(J+1) - F'(J-1) = R(J) - P(J).$$

It is normally thought that the procedure of taking combination differences separates the effects of the two states, and this indeed would be true for perfect measurements. Since, however, measurement errors cannot be assigned exclusively to either state, the effect of these errors is to produce a small correlation between the two states.

It is usual to assume a model for the rotational energy levels, for example,

$$F(J) = B\,J(J+1) - D\,J^2(J+1)^2$$

in which case

$$\Delta_2 F(J) = (4B - 6D)(J + \tfrac{1}{2}) - 8D(J + \tfrac{1}{2})^3.$$

Values for B and D are obtained graphically by plotting $\Delta_2 F(J)/(J + \frac{1}{2})$ against $(J + \frac{1}{2})^2$ and determining the intercept and slope of the best straight line; alternatively, the data can be fitted by computer methods. The former method is more subjective, and sometimes the quoted error limits are overoptimistic as has been discussed by Albritton *et al.*[2] in the case of the atmospheric oxygen bands.

One disadvantage of the combination difference method is that, in general, it utilizes only a part of the data. Furthermore, if a line is overlapped then the combination differences involving this line are less accurate, even though the other lines involved in the combination differences may be perfectly good.

An interesting example of the effect of systematic errors was recently[3] found in a re-examination of the $\tilde{A}^2\Sigma^+ - X^2\Pi_i$ band system of NCO using higher resolving power than in earlier work.[4] An energy level diagram for the 000–000 band is shown in Figure 1. Even with the higher resolving power, no splittings were observed between the principal branches and the satellite branches, indicating that the spin splittings in the $^2\Sigma^+$ state are very small indeed. If a graph* of $[R_1(J) - P_1(J)/4J$ is plotted against J^2, then a good straight line is obtained (Figure 2), and

*A combination difference plot for the upper state is shown here rather than for the lower state, since the energy levels of the latter follow a more complicated formula.

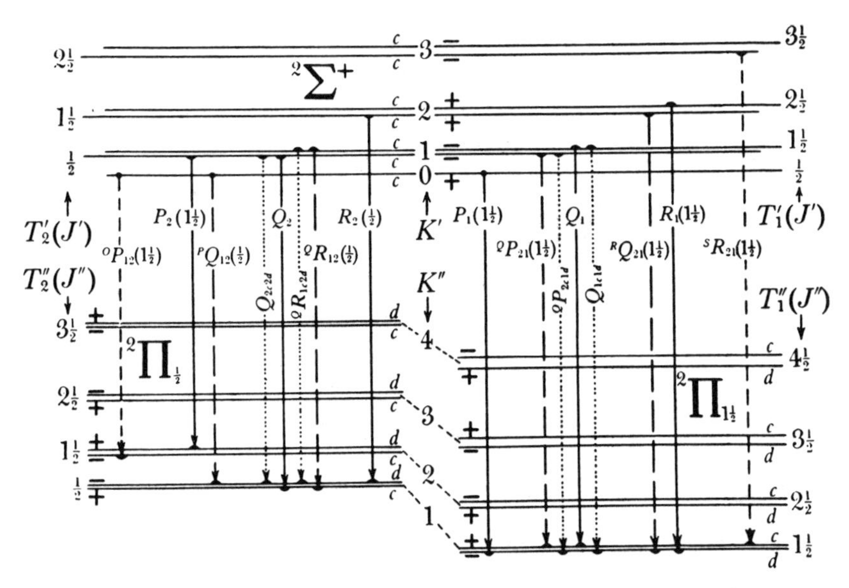

FIGURE 1 Energy level diagram for a $^2\Sigma^+ - {}^2\Pi_i$ transition. From Jevons.[13]

from the intercept we obtain $B' = 0.40203$ cm^{-1}. Now[5] a microwave B-value is known for the ground state, viz., $B'' = 0.389516$ cm^{-1} and the value of $B'–B''$ (= 0.01264 ± 0.00001 cm^{-1}) is well determined from the optical spectrum. Combining these values we obtain $B' = 0.40216$ cm^{-1}, which lies considerably outside the scatter of the points plotted in Figure 2. To investigate this apparent dilemma, the *ground*-state combination differences, $R_1(J-1) - P_1(J+1)$, were obtained from the optical data and were also calculated using the microwave constants (Table 2). It is seen that there is a small, but systematic, discrepancy between the two sets of numbers. The reason for this discrepancy is that while the P_1 lines are single lines, the "R_1 lines" are unresolved blends of an R_1 branch and its $^RQ_{21}$ satellite branch. For low values of J, the two branches have comparable intensities, but the spin splittings are too small to be resolved or even to produce a noticeable increase in the widths of the lines; the measured frequencies, however, are systematically displaced from the frequencies of the individual components. Combination differences using unresolved branches may therefore give misleading results.

Two-State Fit

In this method it is assumed that accurate formulas are available for the rotational energy levels in the two combining states. Every line is

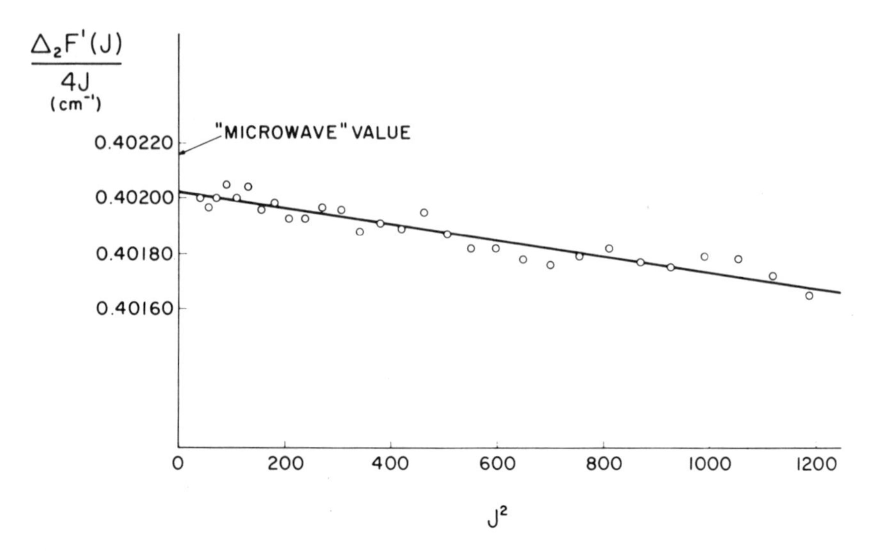

FIGURE 2 Combination difference plot, $[R_1(J) - P_1(J)]/4J$ against J^2, for the 000-000 band of the $\tilde{A}^2\Sigma^+ - \tilde{X}^2\Pi_i$ transition of NCO. Data from Bolman *et al.*[4]

TABLE 2 Ground-State Combination Differences for NCO

J	$\Delta_2''F_1(J)^a$ Optical	$\Delta_2''F_1(J)^b$ Microwave	Difference
5.5	9.310	9.310	0.000
6.5	10.861	10.862	−0.001
7.5	12.411	12.413	−0.002
8.5	13.963	13.964	−0.001
9.5	15.510	15.516	−0.006
10.5	17.064	17.067	−0.003
11.5	18.612	18.618	−0.006
12.5	20.163	20.169	−0.006
13.5	21.712	21.720	−0.008
14.5	23.265	23.270	−0.005
15.5	24.816	24.820	−0.004
16.5	26.364	26.372	−0.008
17.5	27.915	27.922	−0.007
18.5	29.465	29.471	−0.006
19.5	31.012	31.021	−0.009
20.5	32.566	32.571	−0.005

[a] $R_1(J-1) - P_1(J+1)$; data from Bolman *et al.*[3]
[b] Calculated using $B'' = 0.389516$ and $D'' = 2.3 \times 10^{-7}$ cm^{-1}; data from Saito and Amano.[5]

used in the determination of molecular constants, and suitable weights can be given to the individual measurements if so desired. A disadvantage of the method is that errors in the Hamiltonian for one state cause errors in the molecular constants for the other state. Examples of this behavior are provided by molecules where one state is perturbed, or where the rotational energy levels of one state can be fitted much more accurately than the energy levels of the other state, e.g., NH_2 and PH_2.

The application of the method to the spectra of diatomic molecules has been discussed by Zare *et al.*[6] who call it the "method of direct approach." The method has been applied more extensively to the spectra of polyatomic molecules where it is sometimes called the "whole line fit method." Some salient points will now be illustrated by reference to the analysis of many bands of glyoxal by the present author and his colleagues.[7,8] In a typical analysis 1000–2000 rotational transitions are assigned, and the standard deviation of the fit lies between 0.010 and 0.020 cm^{-1}. Three rotational constants and three centrifugal distortion constants* are determined for each state in addition to the band origin. The standard deviation of the fit is much larger than the accuracy of measurement of a single isolated line ($\sim$0.002 cm^{-1}) and is caused partly by the extensive overlapping of the lines in the spectrum.

One estimate for the accuracies of the molecular constants can be obtained from the statistical standard deviations. However, experience shows that these estimates are invariably too low, owing to the neglect of systematic errors. More reliable estimates can be found by two methods. In the first, several different bands with one state in common are analyzed independently and the results compared. Thus values for the ground-state rotational constants of glyoxal-d_2 obtained from the independent analyses of five bands are reproduced in Table 3. It is seen that the values for A'', B'', and C'' are consistent to ±0.00025, ±0.00004, and ±0.00004 cm^{-1}, respectively, whereas the standard deviations of these constants obtained from the analysis of the 0–0 band are ±0.00004, ±0.000008, and ±0.000009 cm^{-1}, respectively. In this example a realistic estimate for the accuracy is roughly five times the standard deviation.

In the second method, the ground-state rotational constants obtained from the analysis of an optical spectrum are compared with the constants obtained from a microwave investigation. Some values for *cis*-glyoxal are given in Table 4. Again it is found that it is necessary to multiply the statistical standard deviations of the constants by factors from 3 to 6 to obtain realistic estimates of the accuracy. The correlation matrix obtained from the analysis of the optical spectrum is given in Table 5.

*The molecule is very close to a prolate symmetric top ($\kappa \sim -0.99$), and other centrifugal distortion constants do not appear to be significant.

TABLE 3 Rotational Constants for Glyoxal-d_2 (in cm^{-1})[a]

Band	A''	B''	C''
0–0	1.2354	0.15924	0.14115
2^1_0	1.2354	0.15921	0.14110
4^1_0	1.2353	0.15916	0.14109
5^1_0	1.2352	0.15916	0.14107
7^2_0	1.2357	0.15923	0.14114

[a] Data from Agar *et al.*[8]

It is important to realize that this matrix depends only on the nature of the rotational assignments and not on the actual measurements. It is seen that there is very strong correlation between corresponding constants in the two states, indicating that the differences in the rotational constants are much more accurately determined than the absolute values. To demonstrate this point, a further computer run was carried out with A'', B'', and C'' fixed at their final values. The statistical standard deviations for A', B', and C' decreased from 0.000067, 0.000024, and 0.000031 cm^{-1} to 0.000018, 0.0000029, and 0.0000025 cm^{-1}, i.e., by factors of 4, 8, and 12, respectively. Next, the microwave values for A'', B'', and C'' were used and the run repeated. The changes in A', B', and C' matched the changes in A'', B'', and C'' to the fifth decimal place, and the standard deviation of the fit increased only marginally, i.e., from 0.012 to 0.013 cm^{-1}. It appears therefore that for bands analyzed by the two-state fit method, the constants may vary in a correlated manner by an order of magnitude times the statistical standard deviations without an appreciable decrease in the quality of the fit.

TABLE 4 Rotational Constants for *cis*-Glyoxal (in cm^{-1})

	Microwave[a]	Optical[b]
A''	0.89141	0.8910 ± 0.0002[c]
B''	0.20684	0.2066 ± 0.0001
C''	0.16821	0.1681 ± 0.0001

[a] From Durig *et al.*[14]
[b] From Currie and Ramsay.[15]
[c] Three times the standard deviations.

TABLE 5 Correlation Matrix Obtained from the Analysis of the 0–0 Band of *cis*-Glyoxal[a]

	ν_0	A'	B'	C'	D_K'	D_{JK}'	D_J'
ν_0	1.0000	−0.6598	−0.7439	−0.5911	+0.4404	+0.6193	+0.4792
A''	+0.6607	−0.9913	−0.4084	−0.2113	+0.9149	+0.8584	+0.1882
B''	+0.7467	−0.4228	−0.9957	−0.6164	+0.2245	+0.6121	+0.6122
C''	+0.5896	−0.2161	−0.5944	−0.9993	+0.1080	+0.3566	+0.9494
D_K''	−0.4360	+0.9081	+0.2054	+0.0999	−0.9797	−0.7545	−0.0964
D_{JK}''	−0.6384	+0.9049	+0.6096	+0.3687	−0.8167	−0.9890	−0.8710
D_J	−0.4696	+0.1854	+0.5773	+0.9507	−0.0982	−0.3486	−0.9994

[a] To save space the correlation coefficients between rotational constants in the same state have been omitted.

One-State Fit

In this method it is assumed that the molecular constants of one state are known and can be fixed while the molecular constants of the other state and the band origin are varied. The method has the advantage that all the data can be used; unfortunately, however, the Hamiltonians used may not reproduce the energy levels exactly. As indicated above, the standard deviations of the constants are much smaller than for a two-state fit. The method can be used with either the upper or lower state constants fixed and is particularly useful if the ground-state rotational constants are known from microwave spectroscopy. Unfortunately, there is no case known where the excited state constants are also determined with microwave accuracy; otherwise it would be possible to check the accuracy of the one-state fit method and the significance of the statistical standard errors of the rotational constants.

Term Value

This method was introduced in 1965 by Åslund[9] and has been extensively used by the Swedish workers and others in analyzing the spectra of diatomic molecules. The initial objective is to reduce a system of assignments and measurements to a set of term values by least squares methods. The term values for the upper and lower states are then fitted independently to rotational energy level expressions. It has always been assumed that upper and lower state energy levels could be separated in this way; recently, however, Albritton and co-workers[10] have shown that the upper and lower state term values are correlated. A correlation matrix is given by these authors for two upper and one lower state of

H_2. Indeed in fitting the term values to energy level expressions, it is necessary to take into account this correlation matrix; otherwise, incorrect values for the standard deviations are obtained and the rotational constants may be in error by the order of one (correct) standard deviation.

As an example, the table of molecular constants given by Albritton *et al.*[10] for ThO is reproduced in Table 6. The measurements and assignments are taken from a paper by Edvinsson, Selin, and Åslund[11] and are reduced to molecular constants in a variety of ways. The first column shows the values obtained using the "method of direct approach," i.e., the two-state fit method. The corresponding values obtained by the original term value method (i.e., neglecting correlation) are shown in column two. The constants agree with those in column 1 to within one standard deviation of the latter. However, the accuracies of the rotational constants in column 2 are overestimated by a factor of $\sim$5, and the accuracy of the band origin is underestimated by a similar factor. In column 3 are given the corresponding numbers obtained by the term value method in which the correlation between the upper and lower state levels is taken into account; the results agree with those given in column 1. In the last column, the molecular constants obtained by combination differences are given. These constants differ from those given in column 1 by 1.7–3.0 times the standard deviations of the latter; furthermore, the standard deviations in column 4 are from 1.2 to 1.7 times larger than those in column 1.

One of the disadvantages of the term value method is that the term values are separated into disconnected blocks by parity and sometimes by "missing" lines. It is necessary to connect these blocks by a molecular model or to insert "dummy" lines. These disadvantages are overcome in the two-state fit method. One of the important applications of the term value method is in the treatment of many bands with states in common. In this way the degree of overdetermination of the term values is enhanced. However, to obtain significant values for the molecular constants and their standard deviations, it is necessary to include the correlation that complicates the computational procedures considerably. It would appear to be more expedient to attempt to fit all the bands simultaneously by an extension of the two-state fit method, producing a procedure that might be called the "many-state fit method."

One of the limitations of the two-state fit method is that it presupposes that the energy levels of both states can be fitted with comparable accuracy. To obviate this difficulty, Åslund[12] has recently modified his term value approach to permit the determination of the molecular

TABLE 6 Molecular Constants Determined by Different Methods of Reduction from the (0,0) Band of the ThO $G^1\Delta - H^1\Phi$ System[a,b]

	Direct Approach	Traditional Term Value Approach	Improved Term Value Approach	$\Delta_2 F$ Combination Differences
$\nu_0(0,0)$	12 691.686(2)	12 691.685(10)[c]	12 691.686(2)	
B_0'	0.317586(20)	0.317571(3)	0.317586(21)	0.317552(25)
D_0'	$1.974(14) \times 10^{-7}$	$1.959(3) \times 10^{-7}$	$1.974(15) \times 10^{-7}$	$1.936(23) \times 10^{-7}$
B_0''	0.325827(20)	0.325811(3)	0.325827(21)	0.325789(27)
D_0''	$1.902(15) \times 10^{-7}$	$1.885(3) \times 10^{-7}$	$1.902(16) \times 10^{-7}$	$1.858(25) \times 10^{-7}$
degrees of freedom	293	79[d] 107[e] 106[f]	79[d] 214[g]	77[h] 77[i]
$\hat{\sigma}$	0.014	0.011[d] 0.038[e] 0.037[f]	0.011[d] 0.015[g]	0.020[h] 0.021[i]

[a] Reproduced from Albritton, Harrop, Schmeltekopf, Zare, and Crow[10] by kind permission of the journal and of the authors.
[b] The number(s) in parentheses are the uncertainty in the last digit(s) that corresponds to one standard deviation. All units are reciprocal centimeters.
[c] ν_0 (0,0) is the difference $F_0' - F_0''$, where $F_0' = 12\,690.698(7)$ and $F_0'' = -0.987(7)$.
[d] Step one.
[e] Step two, upper state.
[f] Step two, lower state.
[g] Step two.
[h] Upper state.
[i] Lower state.

constants of one state and the term values for the other state. It will be interesting to follow the development of this method.*

CONCLUSIONS

One important conclusion of this paper is that the errors in molecular constants produced by systematic errors in measurements frequently exceed the statistical errors. For this reason it is important for authors to try to assess realistic error limits and to state how these are related to the statistical errors, e.g., "the error limits quoted are three times the standard deviations."

The effects of systematic errors are of two kinds. A constant shift in the wave numbers of the lines of a band–produced, for example, by an alignment error in the measuring process–can be absorbed into the band origin and, hence, into the vibrational intervals of the molecule. The effect produces no serious problems unless a series of bands is being fitted simultaneously and the vibrational scheme is overdetermined. The second kind of systematic error lies within a given band, and the effects are more difficult to assess. Suppose that one is trying to fit a single branch of a diatomic molecule by the one-state fit method and that the systematic error in measurement between low and high J lines is δ. The errors introduced into the rotational constants B and D will then be of order of magnitude $\delta/J(J+1)$ and $\delta/J^2(J+1)^2$, where J refers to the high J value. If $\delta = 0.01\ \mathrm{cm}^{-1}$ and $J = 30$, then the errors introduced into B and D are $\sim 10^{-5}$ and $10^{-8}\ \mathrm{cm}^{-1}$, respectively.

Another important source of error in molecular constants is the error introduced by fitting a given model to a set of energy levels. It is frequently found that the molecular constants differ slightly from model to model, by amounts larger than the so-called standard deviations. This fact must be borne in mind when attempting to deduce other parameters (e.g., molecular geometries) from a set of constants.

Of the four methods discussed above, the two-state fit method is the most accurate if satisfactory expressions are available for the energy levels of the two states and there are no earlier precise constants available for either state (e.g., from microwave studies). This method is equivalent to the term value method with correlation included. The rotational constants in the two states are highly correlated, and it is frequently possible to change the molecular constants in a correlated manner by an order of magnitude times the standard deviations without seriously impairing the quality of the fit. If the molecular constants of one state

* The new method also makes possible the determination of the molecular constants of one (or several) states and the term values of the rest of the system.

are known very accurately, then the constants of the other state are best determined by the one-state fit method, and the standard deviations are usually smaller by nearly an order of magnitude than those obtained from the two-state fit method. Finally, combination differences are useful in assigning spectra and for obtaining moderately accurate molecular constants. Higher accuracy, however, can be obtained by using one of the other methods.

The author wishes to acknowledge helpful discussions with Drs. D. L. Albritton, F. W. Birss, and N. Åslund.

REFERENCES

1. G. Herzberg, *Spectra of Diatomic Molecules,* D. Van Nostrand Co., Inc., Princeton, N.J. (1950).
2. D. L. Albritton, W. J. Harrop, A. L. Schmeltekopf, and R. N. Zare, *J. Mol. Spectrosc.,* **46**, 103 (1973).
3. P. S. H. Bolman, J. M. Brown, A. Carrington, I. Kopp, and D. A. Ramsay, to be published.
4. R. N. Dixon, *Philos. Trans. R. Soc. (London),* **A252**, 165 (1960).
5. S. Saito and T. Amano, *J. Mol. Spectrosc.,* **34**, 383 (1970).
6. R. N. Zare, A. L. Schmeltekopf, W. J. Harrop, and D. L. Albritton, *J. Mol. Spectrosc.,* **46**, 37 (1973).
7. F. W. Birss, J. M. Brown, A. R. H. Cole, A. Lofthus, S. L. N. G. Krishnamachari, G. A. Osborne, J. Paldus, D. A. Ramsay, and L. Watmann, *Can. J. Phys.,* **48**, 1230 (1970).
8. D. M. Agar, E. J. Bair, F. W. Birss, P. Borrell, P. C. Chen, G. N. Currie, A. J. McHugh, B. J. Orr, D. A. Ramsay, and J.-Y. Roncin, *Can. J. Phys.,* **49**, 323 (1971).
9. N. Åslund, *Ark. Fys.,* **30**, 377 (1965).
10. D. L. Albritton, W. J. Harrop, A. L. Schmeltekopf, R. N. Zare, and E. L. Crow, *J. Mol. Spectrosc.,* **46**, 67 (1973).
11. G. Edvinsson, L.-E. Selin, and N. Åslund, *Ark. Fys.,* **30**, 283 (1965).
12. N. Åslund, in press (1973).
13. W. Jevons, *Report on Band-Spectra of Diatomic Molecules,* The Physical Society, London, 1932.
14. J. R. Durig, C. C. Tong, and Y. S. Li, *J. Chem. Phys.,* **57**, 4425 (1972).
15. G. N. Currie and D. A. Ramsay, *Can. J. Phys.,* **49**, 317 (1971).

Discussion Leader: KOZO KUCHITSU

Reporters: RICHARD L. HILDERBRANDT
LISE NYGAARD

Discussion

POLO commented that progress in determining anharmonic force constants depended on further refinements in the theory describing the vibrational dependence of the effective rotational constants. Because of this interrelation, he suggested, further developments in this area may be deadlocked. BUCKINGHAM agreed that the importance of anharmonic contributions to interatomic distances in diatomic and triatomic molecules does not augur well for our ability to obtain refined values in larger molecules, since the number of independent cubic anharmonic constants is proportional to N^3, where N is the number of atoms, compared with a number of harmonic constants proportional to N. However, LAURIE noted that present investigations indicate many of the anharmonic force constants are of lesser consequence and that reasonable estimates of the corrections needed can be made using bond stretching anharmonicity only.

In reply to a question by PLÍVA, HEDBERG explained that the parameter κ used to correct electron diffraction data for anharmonicity was an effective correction involving the contribution of several anharmonic force constants, rather than a true anharmonic constant. IBERS remarked that the neglect of the anharmonicity parameter κ in the case of methane would certainly influence the deduced C–H bond length by more than the 0.001 Å figure suggested by Hedberg in the presentation. HEDBERG agreed that in the particular case of methane studied by Bartell and Kuchitsu the effect was more like 0.007 Å.

HOPE observed that the intensity data for small scattering angles would be quite sensitive to bonding effects produced by valence electrons. He inquired whether systematic calculations had ever been made using different values of $s_{\min}$ to determine the influence on deduced values of r and l. HEDBERG replied that the effect would depend on the choice of the intensity function used in the analysis and that this varied from laboratory to laboratory. The effect on r appears to be less than the effect on l, but the errors in both r and l are generally greater than the effect in any event. SCHOMAKER asked whether weighting schemes based on errors in the data were used in electron diffraction least squares refinement. HEDBERG replied that mathematical weighting functions based on estimated experimental errors were used for this purpose.

MEIBOOM inquired whether the damping in the intensity curve observed for cyclooctatetraene was possibly the result of the peculiar bond shift inversion process that has been observed in this molecule or whether it was a more general effect. HEDBERG replied that such damping is generally observed and is always larger for nonbonded distances since the amplitudes of vibration are usually 30–40 percent larger than for bonded distances.

NELSON inquired about the difficulty of reporting the sensitivity of desired parameters to the choice of model. (Model in this context refers to simplifying assumptions regarding the geometry.) HEDBERG replied that the problem was indeed difficult and, in fact, intractable in the case of larger molecules. EVANS urged that Tukey's suggestion be adopted by electron diffractionists; that is, difference curves should be multiplied by a substantial factor so as to provide more meaningful information.

In the discussion of Schwendeman's paper LAURIE re-emphasized the point that the Costain uncertainties should always be added to the Kraitchman coordinate in propagating the uncertainty to an error estimate for a derived parameter. SCHWENDEMAN agreed that the Costain uncertainty refers to the absolute value of the Kraitchman coordinate, which is known to err by being too small. NELSON observed that as a result these uncertainties should be classified as systematic errors. SCHWENDEMAN agreed and pointed out that in his own method of analysis the Costain uncertainties were propagated in such a manner as to produce asymmetric uncertainties that, in turn, reflect systematic errors.

SNYDER asked whether the motion of the two methyl groups in 2-chloropropane was coupled or if they rotated independently. SCHWENDEMAN replied that in this case the torsional barriers were evidently so high that the torsional motion was simply treated as one of the vibrations. No coupling is detectable by this treatment. In molecules with lower barriers a more elaborate treatment of the data is needed.

KIRCHHOFF initiated a lengthy discussion clarifying the analytic approach employed by Ramsay. The discussion revolved about Ramsay's comment regarding the apparent insensitivity of the overall fit to deliberate variations in the ground-state rotational constants. MILLS questioned why the five independent studies indicated an overall variation in A, "which was 3 to 5 times that of the standard deviation in A." CRAWFORD suggested that this discrepancy might be due to a deficient model, citing the well-known effects described by those famous physicists: Fermi, Coriolis, and Centrifugal. TUKEY remarked that the overall insensitivity to such a variation was to be expected by the very strong linear dependence among the parameters. He pointed out that in this case the linear combination $A' - A''$

TABLE 1 Connections between Electron Diffraction and Spectroscopic Measurements

	Molecular Geometry	Motion of Atoms	Electronic Properties	
ED	Diffraction intensity	Intensity damping	Small angle intensity	Inelastic scattering
	↓	↓	↓	↓
	Internuclear distances	Mean square amplitudes	Electron distribution	Electronic excitations
	↕	↕	↘	↙
	Rotational constants	Normal vibrations	Electronic energy Electronic wavefunction	
	↑	↑	↑	
SP	Rotational spectra Vibration/rotation spectra	Vibrational spectra	Electronic spectra	

could be determined precisely whereas the linear combination $A' + A''$ carried an extremely large uncertainty. As a result A' and A'' are strongly correlated.

KUCHITSU contributed the following comments on the combination of spectroscopic and electron-diffraction methods for determining accurate molecular geometry:

"As mentioned in the previous papers, spectroscopy (SP) and electron diffraction (ED) have their own merits and demerits and, in many cases, a combined use of them should result in the most accurate structure for free molecules.* The connections among the structural parameters determined by the ED and SP methods are shown in Table 1. Information about geometry comes from the bonded and nonbonded internuclear distances determined by ED and also from the rotational constants determined by SP. One has to test in the first place whether these values are consistent with each other. One of the experimental examinations is shown in Figure 1.† For BF_3, the discrepancies between the measured ED intensities (dots) and those calculated from the rotational constants B_0 (measured by Ginn *et al.*‡ by high-resolution infrared spectroscopy), with corrections for vibrational effects, are well within the estimated limits of experimental error. The r_z (B–F) distances determined by ED and SP independently are $1.311_1 \pm 0.001_2$ Å and $1.311_2 \pm 0.0008$ Å, respectively.

"Figure 2 shows a correlation diagram among the rotational constants and various distance parameters,§ where the symbol H indicates corrections dependent on

* K. Kuchitsu, *MTP International Review of Science,* Series 1, Volume 2, *Physical Chemistry,* G. Allen, Ed., Butterworths, London (1972), Ch. 6.

† K. Kuchitsu and S. Konaka, *J. Chem. Phys.,* **45,** 4342 (1966).

‡ S. G. W. Ginn, J. K. Kenny, and J. Overend, *J. Chem. Phys.,* **48,** 1571 (1968).

§ K. Kuchitsu and S. J. Cyvin, *Molecular Structures and Vibrations,* S. J. Cyvin, Ed., Elsevier, Amsterdam (1972), Ch. 12.

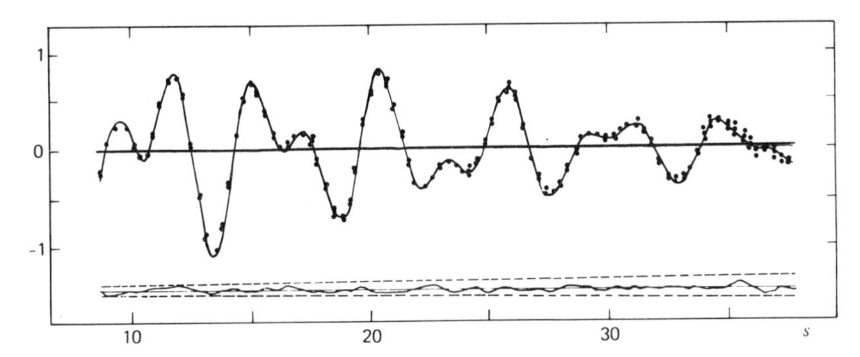

FIGURE 1 Reduced molecular intensity for BF_3 (Kutchitsu and Konaka, 1966). Values observed from three photographic plates are shown in dots. The solid curve is calculated from the rotational constants B_0 measured by high-resolution spectroscopy with corrections for vibrational effects. The lower curve is the difference between the observed and theoretical curves for one of the photographic plates showing the magnitude of random experimental error, which is within the estimated limits of experimental error shown by the broken lines.

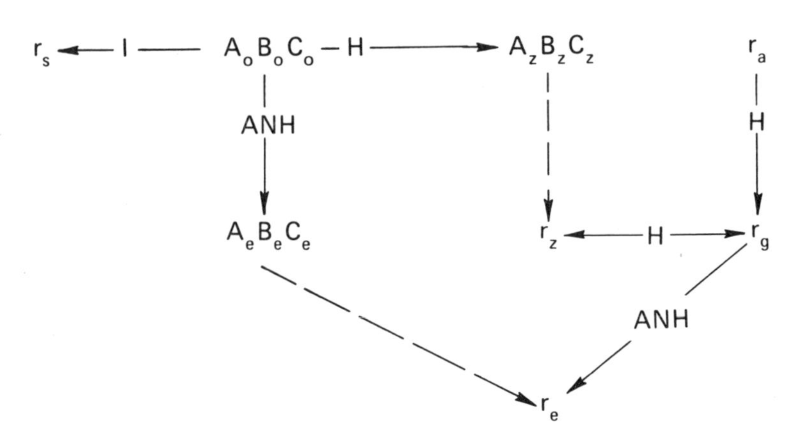

FIGURE 2 Diagram showing the relationship among the rotational constants and distance parameters determined by spectroscopy and gas electron diffraction. Symbols H and ANH indicate harmonic and anharmonic corrections for vibrational effects, respectively, and I stands for isotopic substitution.

the harmonic force constants only, whereas ANH indicates those dependent on the anharmonic constants. The equilibrium structure, r_e, is unquestionably the best defined structure since it has neither vibrational nor isotopic effects. However, it is only rarely determined because of the difficulty in estimating ANH corrections and for most molecules the average structures (r_g, r_z, r_α, etc.) seem to be the second best because of their clear physical significance. One can make a reasonable estimate of random and systematic uncertainties in the average structures.

"For comparison of bond lengths in related molecules, r_g is a convenient representation since differences in the r_g bond distances should be very nearly equal to the differences in the corresponding r_e bond distances. This statement needs a brief explanation: r_g is defined by

$$r_g = r_e + \langle \Delta r \rangle \approx r_e + \tfrac{3}{2} a \langle \Delta r^2 \rangle,$$

where Δr is an instantaneous displacement of the bond length and a is a Morse-like parameter representing the stretching anharmonicity of the bond. For similar bonds, e.g., C–C single bonds, a and $\langle \Delta r^2 \rangle$ are nearly constant, so that

$$r_g(\text{C–C, molecule 1}) - r_g(\text{C–C, molecule 2}) \approx r_e(\text{C–C, molecule 1}) - r_e(\text{C–C, molecule 2}).$$

On the other hand, the r_z bond distances correspond to $r_e + \langle \Delta z \rangle$, where Δz is the projection of Δr onto the equilibrium bond direction, and are more likely influenced by molecular environment, particularly by bending vibrations.

"For representation of angles, however, r_g is not suitable because nonbonded r_g distances contain so-called shrinkage effects. The r_z notation is more suitable since the r_z angle is defined with respect to the well-defined average positions of atoms.

"Though the operational structures r_s and r_0 provide very useful experimental information on geometry, a precise comparison of r_s or r_0 bond lengths in related

TABLE 2 C–C Single Bond Lengths (Å)[a]

Molecule	Length	Molecule	Length
$r_s(r_0)$ Distances			
$(CH_3)_2C{=}O$	1.507 ± 0.003	$CH_3(H)C{=}O$	1.501 ± 0.005
$(CH_3)_2C{=}CH_2$	1.507 ± 0.003	$CH_3(H)C{=}CH_2$	1.501 ± 0.004
r_g Distances			
$(CH_3)_2C{=}O$	1.520 ± 0.003	$CH_3(H)C{=}O$	1.514 ± 0.004
$(CH_3)_2C{=}CH_2$	1.508 ± 0.002	$CH_3(H)C{=}CH_2$	1.506 ± 0.003

[a]Data from Kuchitsu (1972) and Kuchitsu and Cyvin (1972).

TABLE 3 Differences between r_g and r_s Distances[a] (Å)[a]

Molecule	Bond	r_g	$r_g - r_s$
Acetone	C=O	1.212	–0.010
	C–C	1.518	0.011
Acetyl chloride	C=O	1.187	–0.005
	C–C	1.508	0.009
	C–Cl	1.798	0.009
Acrolein	C=O	1.217	–0.002
	C=C	1.345	0.000
	C–C	1.484	0.014
Acrylonitrile	C≡N	1.167	0.003
	C=C	1.343	0.004
	C–C	1.438	0.012
t-Butylchloride	C–C	1.528	–0.002
	C–Cl	1.828	0.025
Isobutene	C=C	1.342	0.012
	C–C	1.508	0.001
Propane	C–C	1.532	0.006
Propylene	C=C	1.342	0.006
	C–C	1.506	0.005
Propynal	C≡C	1.211	0.002
	C=O	1.214	–0.001
	C–C	1.453	0.008

[a] Uncertainties in r_g and $r_g - r_s$ range from 0.002 to 0.005 Å and from 0.004 to 0.010 Å, respectively.

molecules requires caution, since r_s and r_0 may contain appreciable systematic errors. For example, the C–C bond lengths in four simple molecules are listed in Table 2. By use of SP data only, one is led to the conclusion that the C–C bonds in propene and acetaldehyde are equal and are only slightly shorter than those in isobutene and acetone. In fact, the r_g distances from ED, which are consistent with the rotational constants, clearly show that there are significant variations in the C–C bond lengths, acetone> acetaldehyde> isobutene$\gtrsim$ propene. The differences in the r_g and r_s distances selected from recent accurate experiments (Table 3) range mainly from 0.014 Å to –0.002 Å, but differences as large as 0.025 Å (C–Cl in *t*-butyl chloride) and –0.010 Å (C=O in acetone) may not be rare exceptions.

"The use of SP data is often imperative for an accurate ED analysis. Symmetry, force constants, and rotational constants determined (or estimated) by SP are helpful in the ED analysis (Figure 3), and closely spaced distances and hydrogen parameters, which are often difficult to determine by ED data alone, may sometimes be determined with reasonable accuracy. For example, the multiple possibilities of the SeO_2F_2 structures mentioned by Hedberg may be solved if rotational constants can be taken into the analysis.

"Figures 4, 5, and 6 illustrate systematic comparisons of the average structures

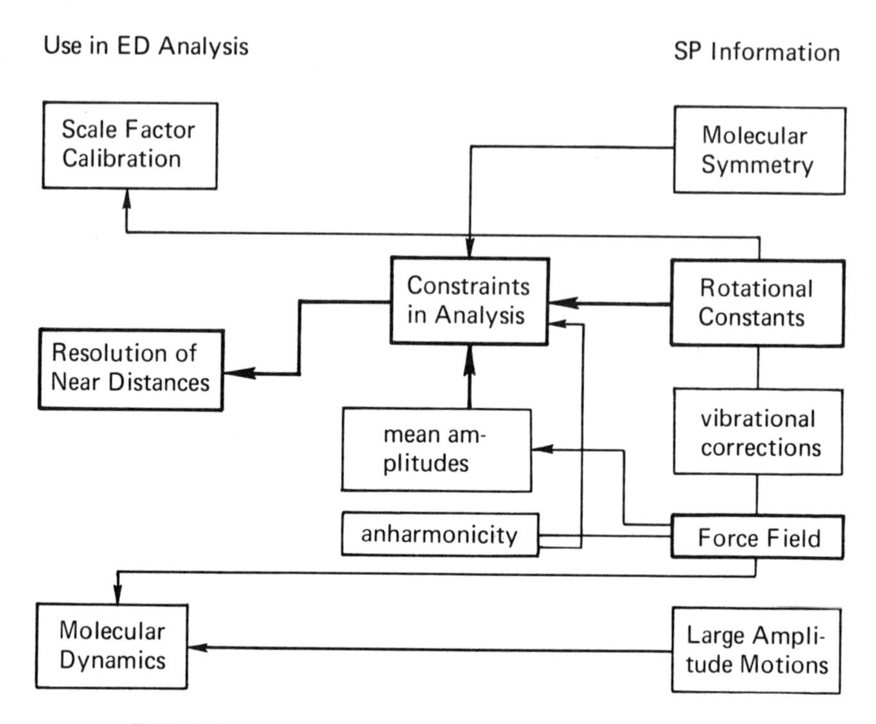

FIGURE 3 Use of spectroscopic data in the electron-diffraction analysis.

determined by the present method showing regularities in the r_g(C–C) bond lengths in conjugated aliphatic molecules* (Figure 4), the carbon valence angles adjacent to a single bond and a double bond (Figure 5), and the r_g(C=O) and r_g(C–N) bond lengths in simple amides in the gas and crystal phases (Figure 6)."

LIDE observed that perhaps Kutchitsu's presentation places electron diffraction in too favorable a light compared with microwave studies. He inquired as to how one could account for the serious discrepancies that have occasionally occurred when the same structure has been studied in different electron diffraction laboratories. KUCHITSU replied that the intensities must be measured very carefully and precisely to obtain accurate results and that this is not always done. HEDBERG commented that some authors are perhaps too optimistic in quoting resolution of closely spaced distances in the radial distribution curve. He pointed out that it would be a better practice to quote average r's and probable differences than to quote individual r's. HEDBERG and KUCHITSU pointed out in addition that failures to correct the model for certain vibrational effects (shrinkage effects) followed by analysis based on a geometrical model may often lead to erroneous conclusions regarding even the fundamental symmetry of the molecule. The reader should be

* Kuchitsu, *ibid.*

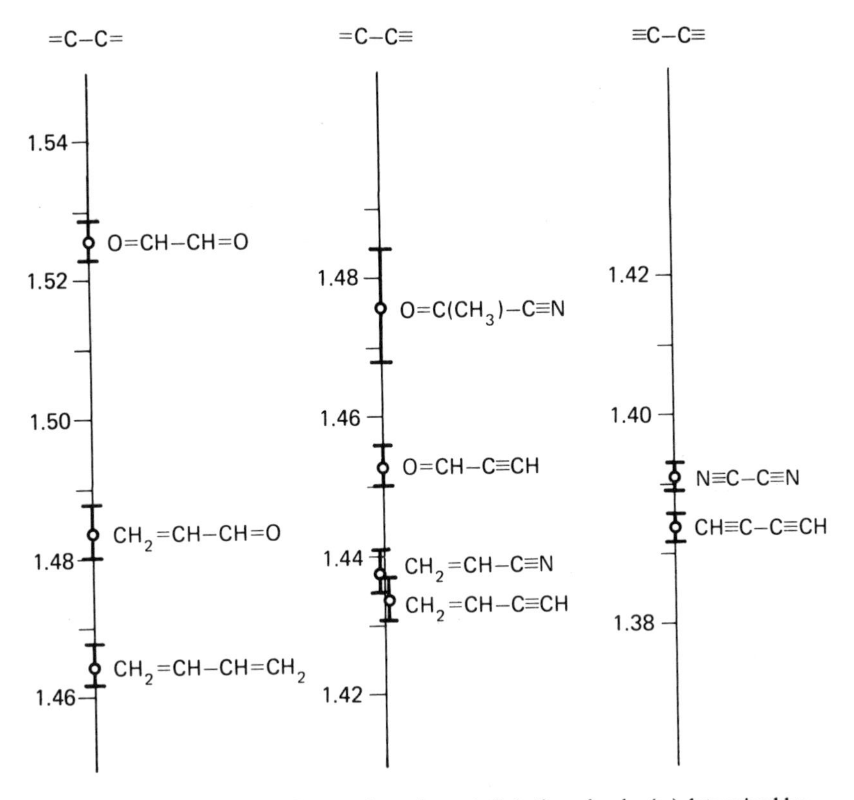

FIGURE 4 C–C single bond distances in conjugated aliphatic molecules (r_g) determined by gas electron diffraction. The structures are consistent with the rotational constants determined by spectroscopy. Vertical bars represent limits of experimental error. [M. Sugié and K. Kuchitsu, *J. Mol. Struct.* (in press).]

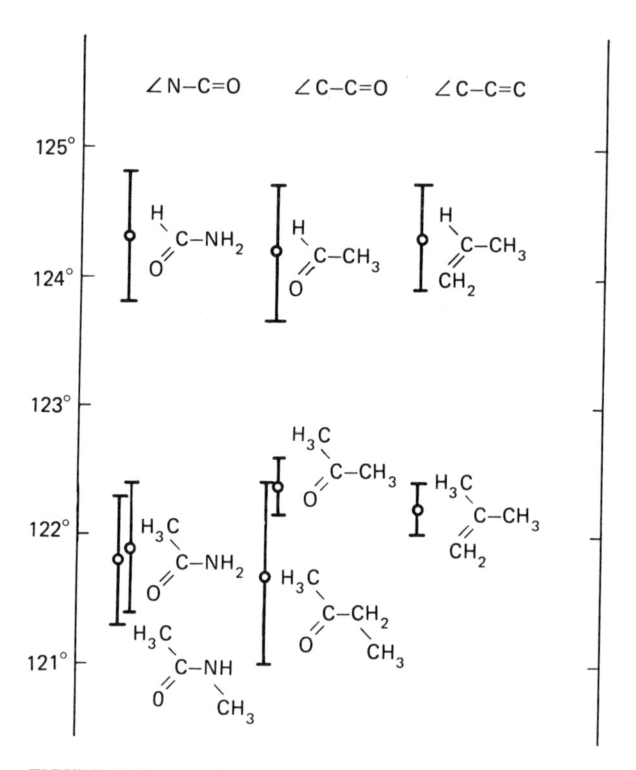

FIGURE 5 Systematic comparisons of the carbon valence angles adjacent to a single bond and a double bond, ∠N–C=O, ∠C–C=O, and ∠C–C=CH_2, determined by gas electron diffraction [M. Kitano and K. Kuchitsu, *Bull. Chem. Soc. Jap.*, **47**, 67 (1974)]. Vertical bars represent estimated limits of experimental error. The effect of methyl substitution on the carbon valence angles is clearly shown.

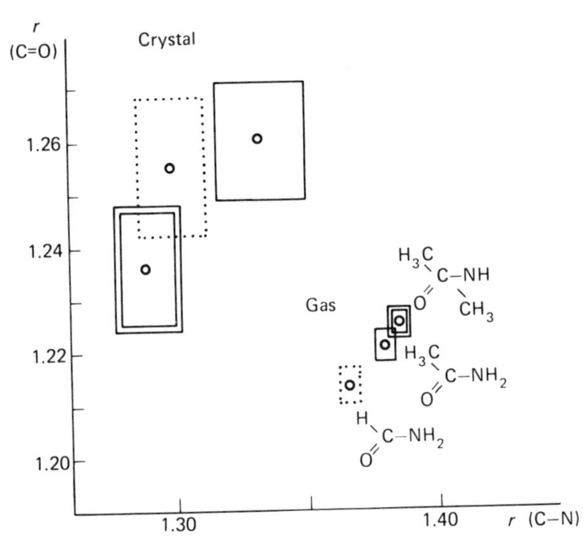

FIGURE 6 The C=O and C—N bond lengths in simple amides in the gas phase (determined by electron diffraction, r_g) [M. Kitano and K. Kuchitsu, *Bull. Chem. Soc. Jap.*, **46**, 384, 3048 (1973); **47**, 67 (1974)] and in the crystal phase (determined by x-ray diffraction).

cautioned regarding slight deviations from planarity or slight twists in the molecular framework. Such effects can in fact be difficult to separate from vibrational effects.

A discussion involving LIDE, NELSON, ABRAHAMS, and SCHOMAKER led to the consensus that a careful study on a standard sample in several electron diffraction laboratories would provide a useful evaluation of experimental and analytical techniques. KUCHITSU remarked that steps have been taken in this direction and that CO_2 and benzene were now being used by many laboratories for calibration purposes.

SCHOMAKER pointed out that electron diffractionists were not the only scientists subject to erroneous interpretation of data, citing the case of a misinterpretation regarding the bond angle in methylene by Herzberg *et al.* RAMSAY replied that in the original publication two possible interpretations were indicated although the incorrect linear model was favored. BUCKINGHAM remarked that Boys in some early *ab initio* calculations predicted that the bent model was to be expected.

III

Molecular Geometry in the Solid and Liquid States

Lawrence C. Snyder and Saul Meiboom

Molecular Structure from NMR in Liquid Crystalline Solvents

Nuclear magnetic resonance (NMR) in liquid crystalline solvents is a relatively recent method for the determination of molecular geometry. Just a decade ago Saupe and Englert[1] showed that solute molecules give highly resolved NMR spectra, the structure of which is largely due to intramolecular magnetic dipole–dipole interactions, and that suitable spectrum interpretation can yield quantitative information on molecular geometry. In the subsequent years both the experimental techniques and the supporting theoretical understanding required for this structure tool have developed rapidly.[2] Compounds that have been studied include benzene, ethylene, acetonitrile, cyclobutane,[3] cyclooctatetraene, cyclohexane,[2] and numerous other organic molecules.[3] It is characteristic of this method that geometry can only be determined up to a scale factor. However, NMR in liquid crystalline solvents should provide information on ratios of distances to a few parts in a thousand. It is perhaps the only method for studying the quantitative structure of molecules in liquid solutions. The main factors that limit the accuracy of the interpretation of NMR in liquid crystals in terms of molecular geometry are the effects of vibrations,[4] the possibility of pseudodipolar interactions,[4] and possible deformation of the molecular geometry by the anisotropic solvent.[5]

A nematic liquid crystal is most commonly employed as the NMR solvent for structure determination. The nematic phase is a fluid that exhibits a long-range orientational order of the molecules. A typical example is *p, p'*-di-*n*-hexyloxyazoxybenzene, which on heating changes from a crystalline solid into the nematic phase at 84 °C and from the nematic phase to a normal isotropic liquid at 127 °C. The magnetic field of the NMR spectrometer aligns the nematic phase, which in turn provides an anisotropic environment that partially orients the solute molecules. For an oriented solute molecule, the intramolecular nuclear dipole–dipole interactions do not average to zero over the anisotropic molecular tumbling. These direct magnetic dipolar interactions are typically as large as 1000 Hz and almost completely determine the structure of the NMR spectra of solute molecules. A set of solute proton NMR spectra are shown in Figure 1. They are typically several thousand hertz wide with a line width of a few hertz.

For ease of reference we summarize here the principal equations. For a more detailed discussion, see Meiboom and Snyder,[2] Diehl and Khetrapal,[3] and Snyder.[4]

The spin Hamiltonian required to interpret and simulate the NMR spectra of a solute molecule has three terms:

$$\mathcal{H} = -(2\pi)^{-1} \sum_i \gamma_i (1 - \bar{\sigma}_{izz}) H_z I_{zi} + \sum_{i>j} J_{ij} [I_{zi} I_{zj} + \tfrac{1}{2}(I_{+i} I_{-j} + I_{-i} I_{+j})] + \sum_{i>j} D_{ij} [I_{zi} I_{zj} - \tfrac{1}{4}(I_{+i} I_{-j} + I_{-i} I_{+j})]. \quad (1)$$

The first term is the Zeeman interaction with the applied field with a mean shielding $\bar{\sigma}_i$ for nucleus i, which may differ somewhat from the value in isotropic solvents due to the anisotropy of the shielding tensor. The second term contains the well-known isotropic part J_{ij} of the indirect spin–spin coupling of nuclei i and j. For pairs of protons the J_{ij} are a few hertz. The major contribution to the parameters D_{ij} of the last term derives from the direct (dir) magnetic dipole interactions of pairs of nuclei, which will be denoted by D_{ij}^{dir} . The contribution from the indirect interactions is often negligible and will be discussed below.

It is the interpretation of the parameters D_{ij} that yields information on molecular geometry. The direct dipolar coupling parameter D_{ij}^{dir} of nuclei i and j of a molecule can be written as an average over all molecular motions, including the vibrations and tumbling of the molecule

$$D_{ij}^{\text{dir}} = (2\pi)^{-1} \gamma_i \gamma_j \hbar \langle (1 - 3\cos^2\Theta_{ij}) r_{ij}^{-3} \rangle_{\text{av}}, \quad (2)$$

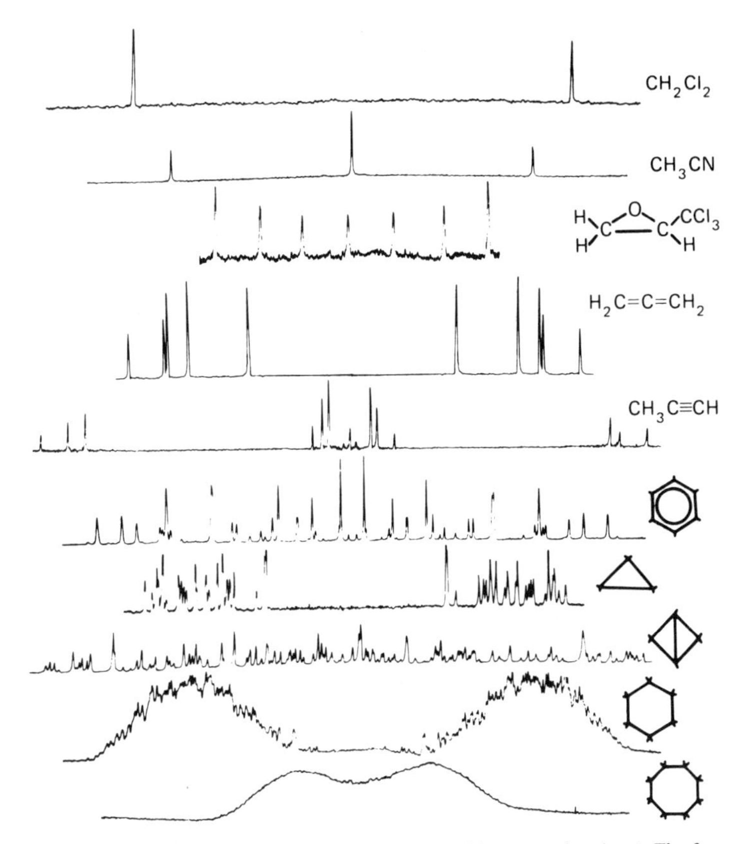

FIGURE 1 Proton NMR spectra of compounds dissolved in a nematic solvent. The frequency scale of the different spectra varies, but the overall width is of the order of 2 kHz. The compounds for the five bottom traces are benzene, cyclopropane, bicyclobutane, cyclohexane, and cyclooctane.

where γ_i and γ_j are the gyromagnetic ratios of nuclei i and j, the distance between the nuclei is r_{ij}, and Θ_{ij} is the angle between the internuclear vector and the applied magnetic field direction. If the molecular geometry and vibrations are assumed to be independent of the orientation of the molecule with respect to the magnetic field, the averaging process over the anisotropic motion of the molecule can be characterized by one to five motional constants (the actual number depending on the symmetry of the system). Then, too, D_{ij}^{dir} can be written as the sum of at most five products, each consisting of a motional constant multiplying a term depending only on the vibrationally averaged geometry of the molecule.

To relate D_{ij}^{dir} to the anisotropic motion of a molecule in a liquid crystalline solvent, we employ the function $P(\theta, \Phi)$, defined as the probability per unit solid angle of a molecular orientation specified by the angles θ and Φ, the polar coordinates of the applied magnetic field direction relative to a molecule-fixed Cartesian coordinate system. We expand $P(\theta, \Phi)$ in real spherical harmonics:

$$P(\theta, \Phi) = (1/4\pi) + c_x P_x + c_y P_y + c_z P_z + c_{3z^2-r^2} D_{3z^2-r^2} + c_{x^2-y^2} D_{x^2-y^2} + c_{xz} D_{xz} + c_{yz} D_{yz} + c_{xy} D_{xy} \ldots . \tag{3}$$

The P spherical harmonics have been included for generality. In the absence of an electric field, they will vanish in liquid crystal solvents.

Written explicitly, the average of Eq. (2) becomes

$$\langle (1 - 3\cos^2\Theta_{ij}) r_{ij}^{-3} \rangle_{\text{av}} = \iint P(\theta, \Phi)(1 - 3\cos^2\Theta_{ij}) \times r_{ij}^{-3} P(q_1 \ldots q_n) d\theta d\Phi dq_1 \ldots dq_n . \tag{4}$$

Here the coordinates q_k describe the internal molecular motions and displacements from the equilibrium geometry, and $P(q_1 \ldots q_n)$ is the configurational probability density.

If we disregard the vibrations by setting the configurational probability density equal to a δ function, the integration of Eq. (4) gives

$$D_{ij}^{\text{dir}} = -K_{ij} 5^{-1/2} [c_{3z^2-r^2} (2z_{ij}^2 - x_{ij}^2 - y_{ij}^2) r_{ij}^{-5} + c_{x^2-y^2} \sqrt{3}(x_{ij}^2 - y_{ij}^2) r_{ij}^{-5} + c_{xz} 2\sqrt{3}(x_{ij} z_{ij}) r_{ij}^{-5} + c_{yz} 2\sqrt{3}(y_{ij} z_{ij}) r_{ij}^{-5} + c_{xy} 2\sqrt{3}(x_{ij} y_{ij}) r_{ij}^{-5}] . \tag{5}$$

Here $z_{ij} \equiv z_i - z_j$, $x_{ij} \equiv x_i - x_{ij}$, and $y_{ij} \equiv y_i - y_j$, where x_i, y_i, and z_i are the coordinates of the ith nucleus in the molecule-fixed system, and K_{ij} is a constant.

DETERMINATION OF A MOLECULAR GEOMETRY

The process of obtaining a molecular geometry is conveniently divided into three steps: the measurement of the experimental spectrum, the analysis of the experimental spectrum in terms of the D_{ij}'s of Eq. (1), and the interpretation of the D_{ij} in terms of molecular geometry through Eq. (2) and Eq. (5).

Measurement of the Spectrum

The solute molecule is dissolved in the liquid crystal solvent at low concentration. A variety of nematic solvents are available, some of which are nematic at room temperature. Representative high-resolution proton NMR spectra are given in Figure 1. Because the solvent order depends on composition and temperature, it is important that temperature and composition gradients at the NMR probe be minimized if the narrow line widths of a few hertz are to be obtained. The spectra of Figure 1 show the rapid increase of spectral complexity with the number of nuclei. The spectra become almost continuous and uninterpretable at about 10 spins. Simplified proton NMR spectra can be obtained by partial deuterium substitution and decoupling.[6] This has been described for cyclohexane, but has not been used extensively. Proton double resonance is also a useful experimental technique for the identification of spectral lines.[6]

Analysis of the Spectrum

Analysis of the spectrum means finding the set of spin Hamiltonian parameters $\overline{\sigma}_i$, J_{ij}, and D_{ij} of Eq. (1), which permit an accurate computer simulation of the observed spectrum. We assume that the number and kind of magnetic nuclei in the molecule to have been established by chemical analysis. The determination of the magnitude of the parameters is of necessity a trial and error process, starting with a reasonable guess and refining the values till a good fit is obtained. Values for the indirect couplings J_{ij} can often be obtained from NMR in isotropic phases, as can approximate values for $\overline{\sigma}_i$.[7] A set of starting values for the D_{ij} is usually obtained from Eq. (5) by assuming a molecular geometry and approximate bond lengths and angles and guessing the motional constants. At this stage some of the difficulties in using Eq. (5) are ignored. Motional and geometric parameters are varied to minimize the root-mean-square (rms) deviation of experimental and calculated peak positions. However, as a last step it may prove necessary to vary the D_{ij} independently of the assumed geometry in order to obtain the virtually perfect fit of the experimental and calculated peak positions that should be possible in terms of the Hamiltonian, Eq. (1).

Interpretation of the D_{ij} for Molecular Geometry

The geometry referred to in the preceding paragraph is the simplest approximation to the molecular geometry, the one most easily deduced from the spectrum, and the one most often reported. In this interpretation, the D_{ij} of Eq. (1) are equated to the D_{ij}^{dir} of Eq. (5), and the requirement of averaging over molecular vibrations is disregarded. An assumption of molecular rigidity is also implicit in Eq. (5). As all basic uncertainties and inaccuracies in a determination of a molecular structure are due to the above approximations and assumptions, most of the remainder of this paper will be devoted to their discussion.

APPROXIMATIONS

The above interpretation of the spectrum and D_{ij}'s in terms of geometry is not rigorous. Reasons follow.

Pseudodipolar Interactions

The anisotropy of the indirect (ind) spin–spin coupling of nuclei can lead to a term D_{ij}^{ind} called a pseudodipolar interaction, which enters the spin Hamiltonian of Eq. (1) in the same way as the direct dipolar interactions of Eq. (2).[4] Thus we may write

$$D_{ij} = D_{ij}^{\text{dir}} + D_{ij}^{\text{ind}}. \tag{6}$$

Since only the D_{ij}^{dir} component can be interpreted for geometry, a knowledge of the relative magnitude of D_{ij}^{ind} is required. In some cases D_{ij}^{ind} can be estimated experimentally by running spectra in solvents that give different orientational parameters. This was done by Gerritsen and MacLean[8] for 1,1-difluoroethylene, for which they found $D_{\text{FF}}^{\text{ind}} \cong \frac{1}{4} D_{\text{FF}}^{\text{dir}}$. So far, major pseudodipolar interactions have only been observed between fluorine atoms. Fortunately, theoretical estimates of D_{ij}^{ind} find it to be negligible relative to D_{ij}^{dir} for H–H interaction and also probably for C–H and F–H interactions.[9]

Effect of Vibrations

To take account of vibrations, the configurational probability $P(q_1 \ldots q_n)$ in Eq. (4) has to be introduced. If one makes the additional assumption that this function is independent of molecular orientation, then the integration in Eq. (4) can be factorized, and the vibrations are accounted for by introducing average D_{ij}.

$$D_{ij}^{\text{dir}} = \int D_{ij}^{\text{dir}} (q_1 \ldots q_n) P(q_1 \ldots q_n) dq_1 \ldots dq_n. \tag{7}$$

Here $D_{ij}^{\text{dir}}(q_1 \ldots q_n)$ is the direct dipolar interaction for the molecular configuration specified by $q_1 \ldots q_n$.

To evaluate Eq. (7), one can expand $D_{ij}(q_1 \ldots q_n)$ in a Taylor series about the equilibrium geometry, neglecting terms higher than quadratic. Assuming that motions in the coordinates adopted are independent of each other, cross terms can be neglected and one obtains Eq. (8)

$$D_{ij} = D_{ij}^{\text{e}} + \sum_k \left(\frac{\partial D_{ij}}{\partial q_k}\right)^{\text{e}} \overline{q_k} + \frac{1}{2} \sum_k \left(\frac{\partial^2 D_{ij}}{\partial q_k^2}\right)^{\text{e}} \overline{q_k^2}. \tag{8}$$

The derivatives of the direct coupling parameters at the equilibrium configuration can easily be evaluated from Eq. (5). The superscript "e" indicates quantities evaluated at the equilibrium geometry, and the bars indicate averages over the configurational probability function. The expansion Eq. (8) opens the way for using normal coordinates and their mean square amplitudes to estimate the effect of the terms quadratic in the displacements on the observed dipolar couplings. In a later section we review recent work in which vibrational corrections were applied to find an average geometry for benzene.[10, 11]

Dependence of Molecular Geometry upon Orientation

The derivation of Eq. (5) assumed that the molecular geometry is independent of orientation with respect to the nematic optic axis. If that is not true, the expectation value of Eq. (2) may be non-zero even if the motional constants are zero. Evidence for some distortion has been found in the spectra of tetramethylsilane and neopentane,[5] tetrahedral molecules for which zero motional constants are expected. The observed D_{CH} of about -7 Hz and D_{HH} of 2 Hz can be interpreted in terms of a variation of the Si–C–H or C–C–H bond angles of about 0.1° during tumbling. Since the H–H interactions within methyl groups are of order 1000 Hz for oriented molecules, this effect is apparently relatively unimportant for geometry determinations.[5]

Rapid Interconversion between Conformers

If rapid interconversion occurs, the observed D_{ij} are the weighted averages of the corresponding quantities of the conformers.[12] This can complicate the interpretation, because the contributing conformers may have lower symmetry and require a larger number of motional and geometric parameters.

SPECIAL FEATURES OF LIQUID CRYSTAL NMR FOR STRUCTURAL STUDIES

Structural studies by NMR in liquid crystal solvents are distinguished by several special features: the matter of time scales, spin Hamiltonian symmetry and connectivity, and the influence of the liquid solvent phase.

NMR Time Scale

There are two important time scales. The one we shall call the NMR time scale; it has a characteristic time that is the reciprocal of the spectrum width in hertz, typically about 10^{-3} s. The other is the time scale of molecular tumbling, with a characteristic time of the order of 10^{-11} s. The rate of molecular motions relative to these time scales determines major features of the symmetry and structure of the observed spectra.

The movement of a flexible molecule between structures in times much longer than the characteristic time of the NMR scale gives an NMR spectrum that is a superposition of those of the separate conformers: The conformers give spectra as if they were separate molecules. Molecular motions that are fast on the NMR time scale but slow relative to molecular tumbling are considered to be between conformers. The observed NMR spectra correspond to a spin Hamiltonian that is a weighted mean of those of the individual conformers. In general, each such conformer will be characterized by its own structure, molecule-fixed coordinate system, and motional constants.

We shall refer to molecular motions that are fast on the tumbling time scale as "vibrations" of a conformer.

Spin Hamiltonian Symmetry and Connectivity

The analysis of an NMR spectrum produces a spin Hamiltonian that is characterized by a set of parameters. There are parameters associated with individual nuclei: These are chemical shielding constants σ_i (and nuclear quadrupole coupling constants). There are also parameters associated with pairs of nuclei: these are the indirect spin–spin couplings J_{ij} and the direct dipolar couplings D_{ij}. Spectrum analysis places indices on these parameters, even though their assignment to molecular positions has yet to be made. Often subsets of the D_{ij} are found to be equal and can be interpreted to give the symmetry of the spin Hamiltonian. The connectivity of the indices of indirect and direct interactions gives additional information on the molecular structure.

The molecular point group symmetry on the time scale of molecular tumbling determines the number of motional constants that must be employed. In other words, the symmetry of each conformer must be considered. On the other hand, the symmetry of the spin Hamiltonian is characterized by permutation groups. It is based only on the magnetic nuclei, averaged over the NMR time scale. Any operation of the molecular point group expressed as a permutation of the magnetic nuclei is a member of the spin Hamiltonian permutation group. The group of the spin Hamiltonian often has a larger number of symmetry elements than that of the molecule because it excludes the nonmagnetic nuclei and is an average on the NMR time scale.[13]

Influence of the Liquid Solvent Phase

At this time little is known on whether the geometry of a molecule in the liquid phase is generally significantly different from that of the free molecule. The geometries determined so far are quite similar to those determined in the gas phase. Also, the measurements on tetramethylsilane and neopentane,[5] referred to above, indicate that deformations due to the anisotropy of the solvent are quite small.

AVERAGE MOLECULAR STRUCTURES FROM NMR

It is convenient to think in terms of four different structures:

1. The "NMR structure" (r_n), which is defined as the set of internuclear distances calculated directly from the observed dipolar interactions.
2. The "dipolar structure" (r_d), which is defined as the set of internuclear distances computed from the dipolar interactions without correction for vibrations, but in which slower motions, if present (e.g., the rotation of methyl groups and the interconversion of conformers such as ring inversion or pseudorotation), have been accounted for.

3. The average structure (r_α), which is deduced from observed dipolar interactions corrected to remove the effects of vibrations proportional to their mean square amplitudes in Eq. (8).

4. The equilibrium structure (r_e), which is defined in the usual way as the configuration of the nuclei for which the electronic energy is a minimum; a calculation of this structure from NMR data would require accounting for all motions, including anharmonicity.

Because of the additional averaging involved, the NMR structure will have a symmetry that is equal to or higher than that of the dipolar structure, which in turn is equal to or higher than that of the equilibrium structure. A corollary of this fact is that any actual molecular symmetry properties will always appear in the NMR structure: i.e., in setting up the Hamiltonian Eq. (1), the number of different D_{ij} and $\bar{\sigma}_{ij}$ values that can be assumed is reduced by any *a priori* symmetry properties of the molecule one may know or assume, independent of any vibrations or other molecular motions.

BENZENE

We choose benzene as an example to illustrate the structures that can be obtained from proton NMR in nematic solvents, with and without corrections for the effect of the vibrations. The experimental proton NMR spectra were carefully measured by Englert, Diehl, and Niederberger[10] for ordinary benzene and for benzene containing one ^{13}C nucleus. Their nematic solvent was *N*-[*p*-ethoxybenzylidene]-*p*-*n*-butylaniline, which is nematic between 38 and 80 °C. These spectra were interpreted for structure by Diehl and Niederberger,[11] whose main results we shall summarizc here.

If one assumes that benzene is a planar molecule with a sixfold axis of symmetry, which we take to be the *z* axis of the molecule-fixed coordinate system, then only the motional constant $c_{3z^2-r^2}$ is nonzero. By making use of Eq. (8) the dipolar interaction for any pair of nuclei may be written as:

$$D_{ij} = D_{ij}^{e}\left(1 - 3\frac{\overline{\Delta Y}}{r_e} + 6\,\frac{\overline{(\Delta Y)^2}}{r_e^2} - 3\,\frac{\overline{(\Delta X)^2}}{2r_e^2} - 9\,\frac{\overline{(\Delta Z)^2}}{2r_e^2}\right), \tag{9}$$

where r_e is the equilibrium separation of the nuclei, ΔY is their displacement from equilibrium along their equilibrium internuclear vector, and ΔX their displacement in the plane and perpendicular to ΔY. Their separation perpendicular to the plane is ΔZ. In this discussion the labeling of Z and Y has been interchanged from that given by Diehl and Niederberger.[11]

The simplest interpretation of the D_{ij} values obtained from the spectrum analysis is for r_d:

$$r_{ij\mathrm{d}} = (k_{ij} D_{ij})^{-\frac{1}{3}}, \tag{10}$$

where $k_{ij} = (2\pi)^{-1}\gamma_i\gamma_j\hbar\langle\langle 1 - 3\cos^2\theta_{ij}\rangle\rangle_{av}$, and the average in braces is taken to be the same for all bonds. It may be shown using Eq. (9) that:

$$r_d = r_e + \overline{\Delta Y} + \frac{\overline{(\Delta X)^2} + 3\overline{(\Delta Z)^2} - 4\overline{(\Delta Y)^2}}{2r_e}. \tag{11}$$

It is notable that r_d is not equal to $\overline{(r^{-3})}^{-\frac{1}{3}}$:

$$\overline{(r^{-3})}^{-\frac{1}{3}} = r_e + \overline{\Delta Y} + \frac{\overline{(\Delta X)^2} + \overline{(\Delta Z)^2} - 4\overline{(\Delta Y)^2}}{2r_e}. \tag{12}$$

The difference occurs because out-of-plane motions change the orientation of a bond as well as its length.

Estimates of the mean square displacements in benzene have been made from normal coordinate calculations by Brooks, Cyvin, and Kvande.[14] These have been used by Diehl and Niederberger to estimate the last term of Eq. (9) for each dipolar interaction of benzene and then estimate the average structure r_α:

$$r_\alpha = r_e + \overline{\Delta Y}. \tag{13}$$

In estimating the r_d structure for benzene Diehl and Niederberger assumed r_d(C–C) = 1.398 Å. They assumed D_{6h} symmetry and varied r_{CH}/r_{CC} and $\langle\langle 1 - 3\cos^2\theta\rangle\rangle_{avg}$ to give a minimum rms difference of the calculated and experimental values for seven observed couplings: three between protons and four between protons and carbon-13. Their r_d structure from NMR is presented in Table 1 together with r_g structures from electron diffraction[15,16] and an r_o structure from Raman spectroscopy.[17] They have estimated the r_α structure in a similar way, taking r_α(C–C) = 1.398 Å and reducing the r_d distances by the estimated contribution of the mean square vibrational amplitudes. The NMR r_α structure is given in the lower part of Table 1 along with their estimate of the r_α structures for electron diffraction.

In the r_d structure from NMR, the C–H bond length is seen to be too

TABLE 1 Comparison of Benzene Structures[a, b]

	NMR	Electron Diffraction[c]	Electron Diffraction[d]	Raman[e]
	r_d	r_g	r_g	r_o
CH	1.142 ± 0.001[f]	1.090 ± 0.02	1.116 ± 0.009	1.084 ± 0.005
CC	(1.398)[g]	1.400 ± 0.005	1.401 ± 0.002	1.397 ± 0.001
CH/CC	0.817 ± 0.001	0.778 ± 0.015	0.796 ± 0.007	0.776 ± 0.004
	r_α	r_α	r_α	
CH	1.101 ± 0.001	1.073 ± 0.02	1.099 ± 0.009	
CC	(1.398)[g]	1.397 ± 0.005	1.398 ± 0.002	
CH/CC	0.788 ± 0.001	0.768 ± 0.015	0.786 ± 0.007	

[a] From Diehl and Niederberger.[11]
[b] All lengths in angstroms.
[c] Data from Almenningen *et al.*[15]
[d] Data from Kimura.[16]
[e] Data from Langseth and Stoicheff.[17]
[f] Errors in NMR results quoted as including 2.5 times standard deviation and uncertainties in scale factor.
[g] Assumed.

long relative to the C–C bond length as determined by other methods. This has been observed for cyclopropane, cyclobutane, and bicyclobutane. It is due to the large bending motions of bonds to hydrogen, relative to bonds bearing only first row atoms.[2] Bending motions tend to reduce the angular average of Eq. (2) and thus the dipolar coupling. This is interpreted as a relative lengthening of the bonds to hydrogen in an r_d structure. The effect of bending on couplings of protons to each other is much smaller.

Our general conclusion is that it is probably possible to determine r_α structures from NMR that are of comparable accuracy to those obtained from other methods. However, the required fractional corrections of observed interactions for vibrational effects will be larger.

REMARKS

A number of criteria by which to judge the reliability of an analysis can be formulated:

1. Experimental peak frequencies should preferably be accurate to within a few hertz. Thus one should have a well-resolved spectrum, reasonably narrow peaks, and a reliable frequency scale. In particular, the linearity of the frequency scale should be established.

2. The interpretation of the spectrum in terms of D_{ij}, J_{ij}, and $\overline{\sigma}_j$ should be checked by a complete spectrum simulation. An almost perfect fit, both as to peak frequencies and peak intensities, should be obtainable. The number of peaks should, of course, be larger than the number of adjustable parameters. This is generally the case if the molecule contains more than three or four magnetic nuclei.

3. In an interpretation in terms of a molecular geometry, the number of motional parameters plus the number of independent coordinates describing the geometry should be less than the number of independent D_{ij} measured in order to have a check on the internal consistency. A good additional check on the interpretation is to run a spectrum in a different nematic solvent. A fit should be possible with only the motional parameters changed (and possibly some changes in the $\overline{\sigma}_j$).

If we restrict ourselves to H–H and C–H interactions, for which the pseudodipolar terms can be neglected, then the major systematic errors are due to molecular vibrations. As discussed above, they are estimated to amount to about 3 percent of the directly bonded C–H distance relative to more remote nuclear distances. In general, accurate corrections have not yet been undertaken.

A few final remarks on the scope of the method. Proton (or fluorine) NMR on normal (that is without isotopic substitution) compounds is generally restricted to molecules containing up to 10 magnetic nuclei. For a larger number the spectral complexity and peak overlap become excessive. More complex molecules can be tackled by partial deuterium substitution; the usefulness of this technique will depend to a large extent on the availability of molecules with deuterium substitution at specified sites. The positions of the carbon atoms can in principle be determined in two ways: by proton NMR, if peaks due to species containing a ^{13}C nucleus can be detected, and by ^{13}C NMR. Both methods are practical on material with natural abundance of ^{13}C, though in many cases low peak intensity will present a problem. Carbon-13 NMR on partially deuterated compounds may be an attractive possibility, but to our knowledge no such work has as yet been undertaken. How far, in respect to the complexity of the molecules studied, the NMR method will develop seems hard to predict at this time.

REFERENCES

1. A. Saupe and G. Englert, *Phys. Rev. Lett.*, **11**, 462 (1963).
2. S. Meiboom and L. C. Snyder, *Acc. Chem. Res.*, **4**, 81 (1971).
3. P. Diehl and C. L. Khetrapal, in *NMR, Basic Principles and Progress*, P. Diehl, E. Fluck, and R. Kosfeld, ed., Springer, New York (1969).
4. L. C. Snyder, *J. Chem. Phys.*, **43**, 4041 (1965).

5. L. C. Snyder and S. Meiboom, *J. Chem. Phys.*, **44**, 4057 (1966).
6. S. Meiboom, R. C. Hewitt, and L. C. Snyder, *Pure Appl. Chem.*, **32**, 251 (1972); R. C. Hewitt, S. Meiboom, and L. C. Snyder, *J. Chem. Phys.*, **58**, 5089 (1973); L. C. Snyder and S. Meiboom, *J. Chem. Phys.*, **58**, 5096 (1973).
7. K. Wuthrich, S. Meiboom, and L. C. Snyder, *J. Chem. Phys.*, **52**, 230 (1970).
8. J. Gerritsen and C. MacLean, *J. Mag. Res.*, **5**, 44 (1971).
9. R. Ditchfield and L. C. Snyder, *J. Chem. Phys.*, **56**, 5823 (1972).
10. G. Englert, P. Diehl, and W. Niederberger, *Z. Naturforsch.*, **26a**, 1829 (1971).
11. P. Diehl and W. Niederberger, *J. Mag. Res.*, **9**, 419 (1973).
12. S. Meiboom and L. C. Snyder, *J. Am. Chem. Soc.*, **89**, 1038 (1967).
13. H. C. Longuet Higgins, *Mol. Phys.*, **6**, 445 (1963).
14. W. V. F. Brooks, S. J. Cyvin, and P. C. Kvande, *J. Phys. Chem.*, **69**, 1489 (1965).
15. A. Almenningen, O. Bastiansen, and L. Fernholt, *K. Nor. Vidensk. Selsk. Skr.*, **3**, 1 (1958).
16. Quoted in Ref. 11 as private communication of M. Kimura, to be published by K. Tamagawa and T. Iijiwa.
17. A. Langseth and B. P. Stoicheff, *Can. J. Phys.*, **34**, 350 (1956).

David P. Shoemaker

Well-Behaved Crystal Structure Determinations

This paper was to have been prepared by Walter C. Hamilton. It is a matter of great sadness to us all that he has been taken from us. Aside from our personal loss, his absence has been a substantial scientific loss to this conference. I cannot feel that I am in any real sense a replacement for him. My paper by no means covers all that Walter Hamilton would have included, in view of his unique position at the forefront of many areas of structural crystallography as well as my own relative detachment from direct technical involvement in crystal-structure determination over the past few years.

The title, "Well-Behaved Crystal Structure Determinations," itself indicates a significant restriction on the subject matter, since not many such determinations are well behaved in all respects. The misbehaviors, the subject of following papers, truly test the mettle of the crystallographer. I shall here lay some groundwork for those papers and attendant discussions by describing the model of crystal structure that ordinarily forms the basis of maximizing the fit with diffraction data, the experimental conditions that are favorable to the determination of "well-behaved" crystal structures, and the ways in which results obtained under favorable conditions should be interpreted. In short, I shall discuss the aspects of a structure determination that give the reader of a crystal structure paper that "warm happy feeling" of confidence in the validity of the scientific work and of the results presented. The coverage of topics reflects to some extent my point of view as a former co-editor of *Acta Crystallographica* (succeeded in that position, in fact, by Walter Hamilton), rather than as an expert in every topic covered. The coverage is not encyclopedic; for a more complete summary of the kinds of information that should appear in a crystal structure paper, the reader of a paper (and, it should not be necessary to add, the writer also) is referred to "Notes for Authors" in *Acta Crystallographica.*[1]

The crystal structure model implicit in the usual refinement methods may be summarized as follows. The structure consists of atoms and/or ions with assumed spherically symmetric electron distributions, independently undergoing thermal vibration around individual rest positions. The thermal vibration in this model is assumed to be harmonic but not necessarily isotropic. The time-average electron density is infinitely periodic in three dimensions. Kinds of information presumed known before structure determination and refinement and incorporated into the model are dimensions of the unit cell (lattice constants), symmetry of the time-average structure (space group), numbers and kinds of various atoms and/or ions in the unit cell, and the quantum-mechanically calculated electron densities of the spherical atoms and ions (as reflected in the tabulated atomic form factors). In cases where the phase of the wave scattered by an atom differs from that of the incident wave by other than $-\pi$ (i.e., in cases of anomalous dispersion), real and imaginary corrections to the atomic form factor, also calculated quantum mechanically, must be included. With this model, the parameters to be adjusted by use of the diffraction data are the positional parameters of the rest positions of the atoms or ions and the root-mean-square (rms) atomic vibrational amplitudes in the principal directions. Interatomic distances, bond angles, dihedral angles, deviations of atoms from the mean plane of a ring or other group, etc., are calculated in this model by elementary geometry from the atomic parameters.

The simple model implies an infinite perfect crystal. The crystal specimen studied is necessarily finite (rarely more than 0.5 mm in size) and, if "perfect" by conventional criteria, would be utterly unsuitable for collection of intensity data of the kind required for ordinary structure determination and refinement. What is needed is an "ideally imperfect" crystal shot through with dislocations and intergrain boundaries so that it behaves, so far as diffraction is concerned, like a "mosaic" of perfect crystal blocks of the order of micrometers or tenths of micrometers in size, tilted with respect to one another by angles of the order of a few seconds of arc and scattering independently (incoherently) with respect to one another. The assumption of an "ideally imperfect" crystal permits use of the so-called "kinematic" diffraction theory in which the (properly corrected) intensity from a diffraction plane is proportional to the square of the magnitude of the structure factor (Fourier transform of the time or lattice average electron density), rather than the more complicated "dynamical" theory that must be applied for perfect crystals.

As a preface to a simple mathematical description of the model, let us review with a few equations the relation between the crystal structure and the diffraction experiment. The crystal structure is defined in terms of *direct space,* while the diffraction experiment is basically concerned with *reciprocal space.* In the equations below, applicable to a single tiny perfect crystallite (i.e., one mosaic block isolated from the real crystal), the equations concerned with direct space (for which the positional vector is $\mathbf{r}$, expressed in Eq. (1a) with base vectors $\mathbf{a}$, $\mathbf{b}$, $\mathbf{c}$ and dimensionless coordinates x, y, z) are given on the left, while those concerned with reciprocal space (for which the positional vector is $\mathbf{h}$, expressed in Eq. (1b) with base vectors $\mathbf{a}^*$, $\mathbf{b}^*$, $\mathbf{c}^*$ and dimensionless coordinates η, κ, λ) are given on the right:

$$\mathbf{r} = x\mathbf{a} + y\mathbf{b} + z\mathbf{c} \tag{1a}$$

$$\mathbf{h} = \eta\mathbf{a}^* + \kappa\mathbf{b}^* + \lambda\mathbf{c}^* \tag{1b}$$

$$\mathbf{r}_{mnp} = m\mathbf{a} + n\mathbf{b} + p\mathbf{c} \tag{2a}$$

$$\mathbf{h}_{hkl} = h\mathbf{a}^* + k\mathbf{b}^* + l\mathbf{c}^* \tag{2b}$$

$$\rho_{xtl}(\mathbf{r}) = \rho_\infty(\mathbf{r}) \cdot S(\mathbf{r}) \tag{3a}$$

$$F_{xtl}(\mathbf{h}) = F_\infty(\mathbf{h}) * T(\mathbf{h}) \tag{3b}$$

$$\begin{aligned}\rho_\infty(\mathbf{r}) &= \rho_c(\mathbf{r}) * L(\mathbf{r}) \\ &= \int \rho_c(\mathbf{r}')L(\mathbf{r}-\mathbf{r}')d\mathbf{r}'\end{aligned} \tag{4a}$$

$$\begin{aligned}F_\infty(\mathbf{h}) &= F(\mathbf{h}) \cdot \Lambda(\mathbf{h}) \\ &= F_{hkl} \cdot \Lambda(\mathbf{h})\end{aligned} \tag{4b}$$

$$L(\mathbf{r}) = \frac{1}{V}\sum_m \sum_n \sum_p \delta(x-m)\delta(y-n)\delta(z-p) \tag{5a}$$

$$\Lambda(\mathbf{h}) = \sum_h \sum_k \sum_l \delta(\eta-h)\delta(\kappa-k)\delta(\lambda-l) \tag{5b}$$

When the direct space coordinates are allowed to take on all integral values $m, n, p,$ the vector termini $\mathbf{r}_{mnp}$ define the direct lattice [Eq. (2a)]; similarly, when the reciprocal space cordinates take on all integral values $h, k, l,$ the vector termini $\mathbf{h}_{hkl}$ define the reciprocal lattice [Eq. (2b)]. In Eqs. (3), (4), and (5) the quantities on the left are Fourier integral transforms of the respective ones on the right and vice versa. By the Fourier folding theorem, wherever a product (indicated by a dot) occurs on one side, a convolution (folding or Faltung, indicated by an asterisk) must occur in the other and vice versa. Equations (5a) and (5b) define direct and reciprocal lattice functions; the necessity that they constitute a Fourier transform pair defines the relations (not given here) between direct and reciprocal basis vectors. Equation (3a) defines the electron density in the case of x-ray diffraction, or what we may call more generally the "scattering density" in the case of either x-ray or neutron diffraction; scattering density is the product of the density ρ_∞ for an infinite crystal times a shape factor S, which has the value unity inside the crystal and the value zero outside. **The function ρ_∞ constitutes the description of the crystal structure to which the concepts of lattice periodicity and crystal symmetry as expressed by the space group apply.** Equation (4a) describes the density ρ_∞ as a convolution of the density function ρ_c for *one unit cell* (the contents of a parallelepiped with lattice points at the corners but none on the edges or faces or in the interior) with the direct lattice function. Equation (4b) contains the quantity F_{hkl}, the Fourier transform of ρ_c, known as the "structure factor." These equations are schematically represented for the one-dimensional case by Figure 1. The reciprocal lattice function Λ effectively "samples" the structure factor at reciprocal lattice points hkl; the value of F_c between lattice points is immaterial. The evaluation of the Fourier integral transform of $F_\infty(\mathbf{h})$, to obtain $\rho_\infty(\mathbf{r})$, amounts to a Fourier series summation of F_{hkl} over the reciprocal lattice points hkl.

The intensity of diffraction by the single tiny crystallite depends on the square of the magnitude of F_{xtl}, the Fourier transform of the crystallite density function ρ_{xtl}. This is shown schematically in Figure 2. The "peak shape" at each reciprocal lattice point depends on the square of the magnitude of T, the Fourier transform of the crystal shape function S. The width of the peak decreases, and the ratios of the heights of the ripples to that of the main peak also decrease, as the crystallite size increases. In the mosaic model of the real crystal the individual perfect crystallites scatter incoherently (i.e., with random phase) with respect to each other, and their intensities are additive; owing to the tilts of the mosaic blocks with respect to one another, however, the maxima of the various blocks may appear at slightly different places so that the

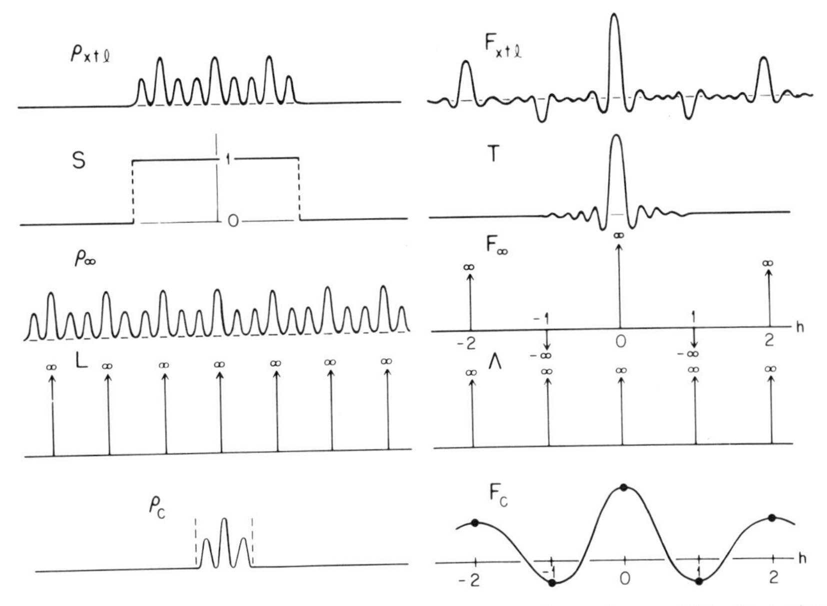

FIGURE 1 A one-dimensional crystal structure model for a tiny perfect crystallite, illustrating, on the left, the scattering density function for the crystal, the shape function, the infinite structure model, the lattice, and the contents of a single unit cell; the Fourier transforms of these functions are given on the right.

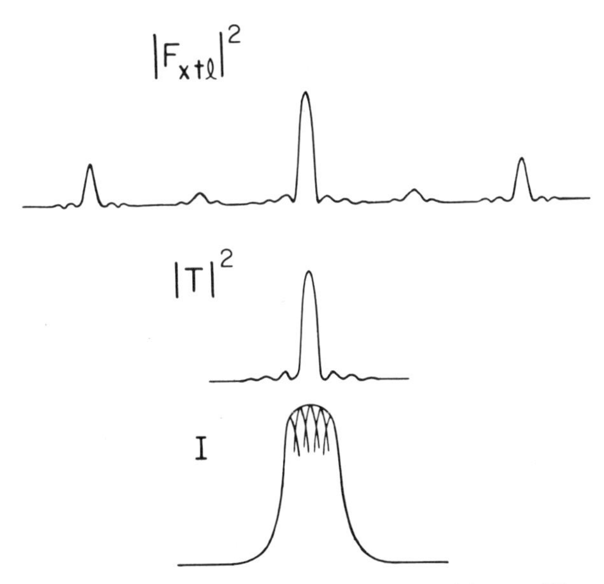

FIGURE 2 The square of the Fourier transform of the crystallite scattering density function and of the shape function (Figure 1) and a schematic intensity function resulting from mosaic spread.

total intensity function at a reciprocal lattice point may be as schematically indicated by curve I in Figure 2.

Our brief summary of the conventional model is completed with the following equations, which also mainly constitute Fourier transform pairs:

$$\rho_c(\mathbf{r}) = \sum_{\text{cell}} \rho_i(\mathbf{r}-\mathbf{r}_i) \tag{6a}$$

$$F(\mathbf{h}) = \sum_{\text{cell}} F_i(\mathbf{h}) \tag{6b}$$

$$F_i(\mathbf{h}) = f_i(\mathbf{h}) \exp(2\pi i \mathbf{h}\cdot\mathbf{r}_i) \tag{6c}$$

$$\rho_i(\mathbf{r}) = \rho_i^{\text{o}}(\mathbf{r}) * u_i(\mathbf{r}) \tag{7a}$$

$$f_i(\mathbf{h}) = f_i^{\text{o}}(\mathbf{h})\cdot t_i(\mathbf{h}) \tag{7b}$$

$$u_i(\mathbf{r}) = (|\mathbf{A}|/\pi^3)^{1/2} \exp(-\mathbf{r}\cdot\mathbf{A}_i\cdot\mathbf{r}) \tag{8a}$$

$$t_i(\mathbf{h}) = \exp(-\mathbf{h}\cdot\mathbf{B}_i\cdot\mathbf{h}) \tag{8b}$$

$$B_i = \pi^2 \mathbf{A}_i^{-1} \tag{9}$$

$$\rho_i^{\text{o}}(\mathbf{r}) = \rho_i^{\text{o}}(|\mathbf{r}|) \tag{10a}$$

$$f_i^{\text{o}}(\mathbf{h}) = f_i^{\text{o}}(|\mathbf{h}|) \tag{10b}$$

Equation (6a) describes the density in one unit cell simply as the sum of the densities ρ_i due to independent atoms in the unit cell. The latter are assumed in Eq. (7a) to be the convolution of ρ_i^{o} (the densities of the atoms at rest) with probability distribution functions, which Eq. (8a) describes as three-dimensional gaussian functions characterized by the symmetric tensors A_i, each with six independent components. It is directly implied by Eq. (7a) and (8a) that all atoms are vibrating independently and harmonically. For x-ray diffraction, ρ_i^{o} is spherical and very roughly gaussian with a half width of the order of 1 Å. For neutron diffraction–since the nucleus is extremely small–ρ_i^{o} may be taken as proportional to a three-dimensional delta function (except in the case of magnetic scattering involving unpaired electron spins and orbital moments). The values of the "atomic form factors" f_i^{o} for electrons, and the scattering lengths (square roots of cross sections) for neutrons, are tabulated for the various kinds of atoms and ions or nuclei in *International Tables,* Volumes III and IV.[2]

The principal deficiencies in this model are the assumed sphericity of atomic electron distributions (deviations from which are discussed in more detail in a following paper by Stewart) and the implicit independence of thermal motion of different atoms. The former ignores the electrons forming bonds between atoms, and the latter ignores collective motion of groups of atoms in semirigid moieties or molecules. From the standpoint of electron density the model is particularly deficient in regard to hydrogen atoms bonded to other atoms; the electron density of the bonded hydrogen atom is highly aspherical and is shifted toward the atom to which the hydrogen is bonded so that the x-ray method may be in error by nearly 0.1 Å in the relative positions of the two atoms. This particular deficiency is not possessed by the neutron method. Molecular librations have the effect of seriously distorting the distribution function for the atomic position away from the gaussian function expected for harmonic motion and shifting the apparent center of gravity of the atom, so that the interatomic distances calculated on the simple model can be short by as much as hundredths (more often, thousandths) of an angstrom.

For structural work x-ray and neutron crystallographic diffraction methods are in some ways much more powerful than gas-phase electron diffraction or molecular spectroscopy in that molecular structures of considerable size can be determined. On the other hand, for determining interatomic distances in small molecules, the crystallographic method is at a disadvantage with respect to the others mentioned in that there is no clear-cut, unambiguous way to correct for thermal motion so as to obtain interatomic distances rigorously comparable with those of various kinds derivable by the other methods (see Hedberg and Laurie, this volume). This is because nothing precisely analogous to the radial distribution function as described for electron diffraction of gases by Hedberg is available in crystal structure work. The Patterson function of crystal structure analysis, which is the Fourier transform of the *square* of the magnitude of F_∞ (i.e., the Fourier sum of $|F_{hkl}|^2$) is only approximately analogous to the radial distribution function (apart from being a three-dimensional rather than a one-dimensional vector function). Being strictly lattice periodic, it cannot take account of the non-lattice-periodic variation of rms relative displacement of atom pairs with distance apart. Thus, ideally, the "self-interaction" peak at the origin of Patterson space should not contain atomic vibrational effects at all, while the peaks at other lattice points should be affected by the relative thermal motions of atoms whose rest positions are related by lattice translations; in fact, in the calculated Patterson function the peak at the origin is no different from peaks at other lattice points. The information needed to complete the analogy is, in principle, available in the form of thermal diffuse scattering, which may be observed in reciprocal space between lattice points; its accurate measurement is so difficult, however, that it has only been attempted in special cases. This thermal diffuse scattering often tends to "peak" at reciprocal lattice points, sharply enough that it is not altogether subtracted off with background. Integrated intensity measurements and structure factor values derived from them may thereby be subject to appreciable error. The problem has been discussed by Young.[31]

Accordingly, correction of interatomic distances for thermal motion generally requires certain assumptions. Schomaker and Trueblood[3] have shown how corrections can be made on the assumption that the molecule in the crystal (undergoing translational, librational, and "screw" vibrational motions) is *rigid,* an assumption that corresponds to the zero-order treatment in gas electron diffraction and molecular spectroscopy. For some cases where the rigid body assumption is inapplicable, Busing and Levy[4] have given a treatment based on a "riding model" in which atom B (a hydrogen atom, for example) is said to "ride" on atom A if atom B has all of the translational motion of atom A plus an additional motion uncorrelated with the instantaneous position of atom A; this treatment may yield a valid correction to the apparent A–B distance provided there is no significant contribution from molecular libration. An excellent review of these treatments and other related considerations has been given by Johnson and Levy.[5] The data required for such treatments include, besides the values of positional parameters of the atoms in the molecule, values of the six thermal parameters for each atom (constituting the six independent components of the symmetric tensor $\boldsymbol{B}_i$); these are parameters of which the claimed accuracy is often subject to considerable suspicion, especially when photographic data have been used. Neutron diffraction data are particularly favorable to such treatments, being less subject to systematic error due to absorption than are x-ray data and being unaffected by vagaries of electron density. The efficacy of the rigid-body treatment in a favorable case is exemplified by the work of Hamilton's group[6] on the structures of L-serine and D,L-serine as determined by neutron diffraction. Here the geometry of the same serine molecule in two different crystalline environments may be compared. Although certain differences between corresponding distances in the two environments appeared significant in the absence of thermal correction, these differences became insignificant (except for those involving atoms bonded to hydrogen) after rigid-body corrections were applied.

We now turn our attention to the experimental side of crystal structure determination. The geometrical requirements for x-ray reflection from a set of Bragg lattice planes are summarized in the diagram of the Ewald construction (Figure 3). Here s_o and s are unit vectors in the direction of the incident and reflected rays, respectively, and the requirement for a particular set of planes with Miller indices h, k, l to diffract is that the corresponding reciprocal lattice point lie on the surface of the "Ewald sphere" (a sphere of radius $1/\lambda$ with the tail of the s_o/λ vector as center). In the figure the necessary condition is fulfilled for $h, k, l = 2, 1, 0$ (also, by accident, for 0, 4, 0). The intensity measured in practice is an "integrated intensity," or total number of photons or neutrons collected from the diffracted beam while the reciprocal lattice point passes from one side of the surface of the Ewald sphere to another as the crystal specimen rotates or precesses around its center in the x-ray beam. This is shown schematically in Figure 4. The intensity function, I_{crystal}, is an integration over a probability distribution of mosaic tilt vectors ϵ as indicated by the equation

$$I_{\text{crystal}}(\mathbf{h}) = \frac{V_{\text{crystal}}}{V_{\text{mosaic}} V_c^2} |F_{hkl}|^2 \int d\epsilon \, p(\epsilon) |T(\mathbf{h} - \mathbf{h}_{hkl} - \mathbf{h}_{hkl} \times \epsilon)|^2 . \tag{11}$$

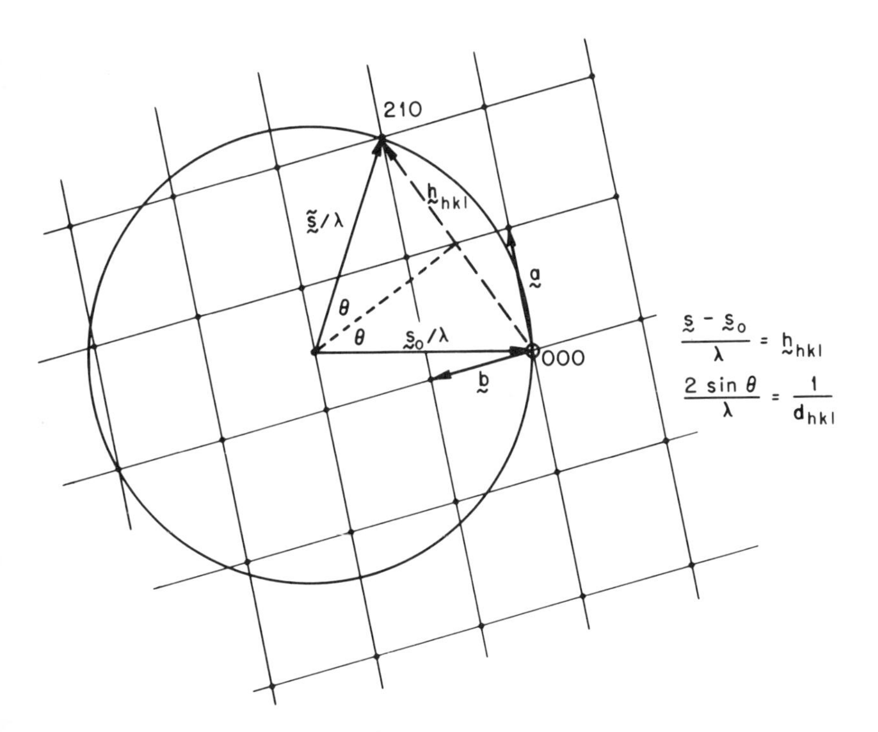

FIGURE 3 Ewald construction.

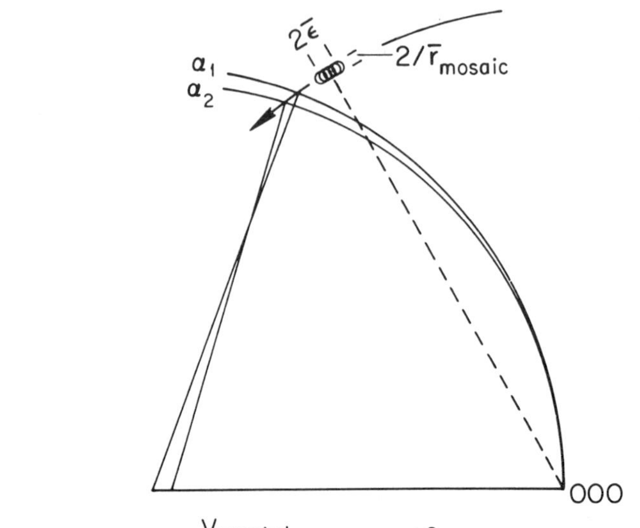

$$I_{\text{crystal}}(\mathbf{h}) = \frac{V_{\text{crystal}}}{V_{\text{mosaic}} V_c^2} \left| F_{hkl} \right|^2 \int d\boldsymbol{\epsilon}\, p(\boldsymbol{\epsilon}) \left| T(\mathbf{h} - \mathbf{h}_{hkl} - \mathbf{h}_{hkl} \times \boldsymbol{\epsilon}) \right|^2$$

FIGURE 4 Passage of a reciprocal lattice point through the Ewald sphere of reflection to yield an integrated intensity.

In this schematic representation the intensity function is spread transversely because the effect of mosaic tilts exceeds the "natural width" of the reflection (half width of $T(\mathbf{h})$), as in what Zachariasen[7–10] refers to as a "type I" crystal. (In "type II" crystals the effect of mosaic tilts is less than that of the natural width.) Not one but two Ewald spheres are involved because the incident radiation has two components, differing in wavelength by a few parts per thousand.

Figure 5 shows in rough schematic form a conventional four-circle x-ray diffractometer, in which the four angles 2θ, ϕ, χ, and ω can be independently set. The "integration," to obtain the integrated intensity, may be obtained by scanning over 2θ, with ω set equal to θ (the "θ–2θ scan"), or by scanning over ω with 2θ fixed (the "ω scan"), or in other ways. At one end or the other of the scan, more often both, the counter is held steady to obtain background counts. The integrated intensity is given by

$$I_{hkl}^{\text{integrated}} = k[C_S - (t_S/t_B)C_B], \tag{12}$$

where C stands for counts (number of x-ray photons or neutrons detected) t for time spent counting, S for scan, and B for background. It is important that not only the intensity be measured but also that the uncertainty in the intensity be estimated. An important, but by no means the only, contribution to the uncertainty is the statistical uncertainty in the numbers of counts; other random errors arise from such sources as fluctuations in the incident beam intensity. Systematic errors that may in many respects behave like random errors may arise from crystal centering errors, nonuniformity of the beam, inaccurately corrected absorption effects, etc. Many crystallographers use an expression such as

$$\sigma(I) = k[C_S + (t_S/t_B)^2 C_B + K^2 I^2]^{1/2} \tag{13}$$

for the estimated uncertainty. Here k is the same scale factor as in Eq. (12) and K is an empirical constant that may be arrived at from systematic comparisons of measurements made on different crystals of the same substance, or on symmetry-equivalent reflections with the same crystal, and/or from general experience.

The reduction of the intensity values to observed structure factors is illustrated by the following equations:

$$(\text{Energy}) = I_{hkl}^{\text{integrated}} = \frac{I_0}{\omega} V_{\text{crystal}} Q_{hkl} = K' Q_{hkl} \tag{14}$$

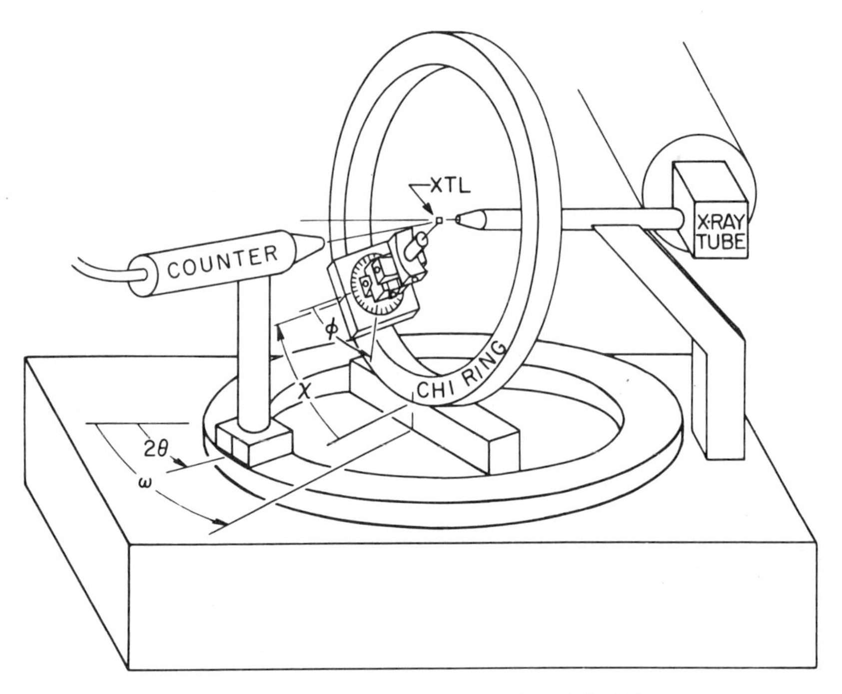

FIGURE 5 Schematic diagram of a conventional four-circle diffractometer.

$$Q_{hkl} = K'' \cdot Lp \cdot A \cdot E \cdot |F_{hkl}|^2 \tag{15}$$

$$Lp = (1 + \cos^2 2\theta)/\sin 2\theta \tag{16}$$
(unpolarized, equatorial rotation)

$$A = \int_{\text{crystal}} \exp \left\{ -\mu_{\text{lin}} \left[t_1(\mathbf{r},\mathbf{s}_0) + t_2(\mathbf{r},\mathbf{s}) \right] \right\} d\mathbf{r} \tag{17}$$

$$E = \exp \left[-gQ_{hkl} \right] \quad \text{(Darwin)} \tag{18}$$

$$F_{hkl} = |F_{hkl}| \exp[i\alpha_{hkl}] = A_{hkl} + iB_{hkl} \tag{19}$$

Here I_0 is the incident beam intensity, ω is the angular speed of rotation of the crystal during the integration process, K' and K'' are empirical scale factors, Q_{hkl} is the "diffracting power" of the crystal, Lp is the Lorentz polarization factor, A and E are absorption and extinction factors, and $|F_{hkl}|$ the absolute magnitude of the structure factor. The sign or phase of the structure factor, α_{hkl}, is not directly determinable from the diffraction experiment.

"Absorption" means diminution of coherent x-ray intensity in the crystal through inelastic processes such as atomic absorption and fluorescence, photoelectron emission, and Compton effect; "extinction" means intensity diminution due to loss through diffraction by fortuitously oriented mosaic blocks. The simple extinction expression due to Darwin, given in Eq. (18), is only a rough approximation; more accurate treatments will be mentioned in what follows. In Eq. (17) the absorption factor is expressed in terms of the linear absorption coefficient μ_{lin} (calculated from tabulated values of the elemental atomic or mass absorption coefficients, updated values of which will appear in Vol. IV of *International Tables,*[2] the path length t_1 of the incident ray from the crystal surface to the point of diffraction **r**, and the path length t_2 of the diffracted ray from that point to the crystal surface. The calculation of the absorption factor in practice demands use of a computer with a sophisticated program and requires also accurate knowledge of the shape and dimensions of the crystal specimen. The problem of measuring the shape and dimensions of an object as small as a diffraction specimen with sufficient accuracy is by no means trivial, and some workers prefer to retain the simple parallelepipedal habit possessed by some specimens naturally, rather than to grind the crystal to a somewhat rough sphere with a somewhat ill-defined radius. The problem of measurement is especially severe for chips or fragments with conchoidal fracture surfaces and re-entrant angles, such as diffraction specimens of intermetallic compounds are often compelled to be.

The problem of estimating loss of intensity by secondary extinction is usually, from a practical point of view, confined to a few (rarely more than a few dozen) reflections of high intensity and small reflecting angle. In the final stages of refinement a relative correction proportional to the measured intensity ($F^2/\sin 2\theta$) will often suffice, the proportionality factor (which in theory is related to the "mosaic spread") being empirically adjusted to give the best fit between calculated and observed structure factors for the reflections affected. For cases of severe extinction it may be advisable to use the 1963 theory of Zachariasen[11] or even to go beyond it to the more general theory given by Zachariasen in 1967–1969[7–10] for isotropic mosaic spread distributions and isotropic mosaic blocks, or the theory of the anisotropic case given by Coppens and Hamilton in 1970.[12] The later Zachariasen and the Coppens and Hamilton papers contain both general equations and equations specialized on the assumption that primary extinction (due to dynamical effects) is negligible, but contain a warning that, if this is not so, the complications of dynamical theory—including the Borrmann effect (anomalous transmission)—must be explicitly considered.

The simultaneous diffraction of x rays by two sets of planes, or (what is substantially equivalent geometrically) the rediffraction of an already diffracted beam by a second set of planes so that it appears to arise from a third set of planes (the "Renninger" effect) has been known for many years and has disturbed crystallographers by apparently contributing intensity to reflections that should be extinguished by symmetry. In 1965 Zachariasen[13] pointed out that the conditions under which most x-ray diffraction work was being done give rise to the possibility or certainty of double or multiple diffraction, which is capable of altering the intensities of reflections; Burbank[14] in the same year further elaborated these conditions for intrinsic and systematic multiple diffraction, which arise from the symmetry of the experiment in relation to that of the reciprocal lattice. A typical situation where systematic multiple diffraction is encountered is illustrated by Figure 6a. The tetragonal or cubic crystal is being rotated around the normal to the equatorial plane, the trace of which is shown on the basal reciprocal lattice layer, and the circles are intersections of the Ewald sphere of reflection on that layer for different angles of rotation. When a reflection in the low-angle region where secondary extinction is significant is examined with rotation of the crystal around the diffraction vector (which is possible with a four-circle diffractometer), a strong reflection may show pronounced dips in the otherwise constant intensity profile

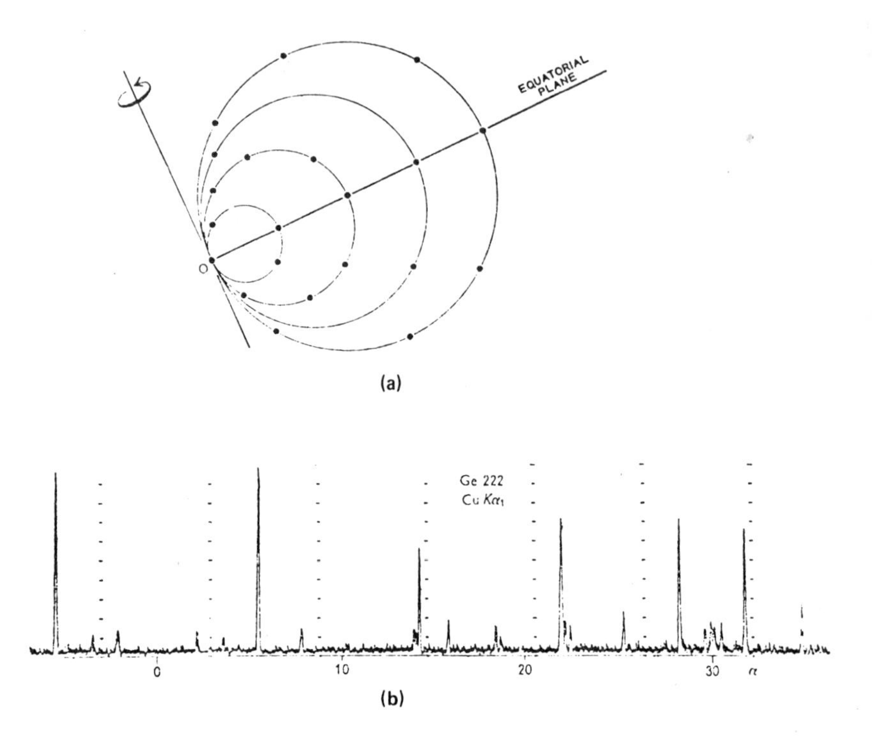

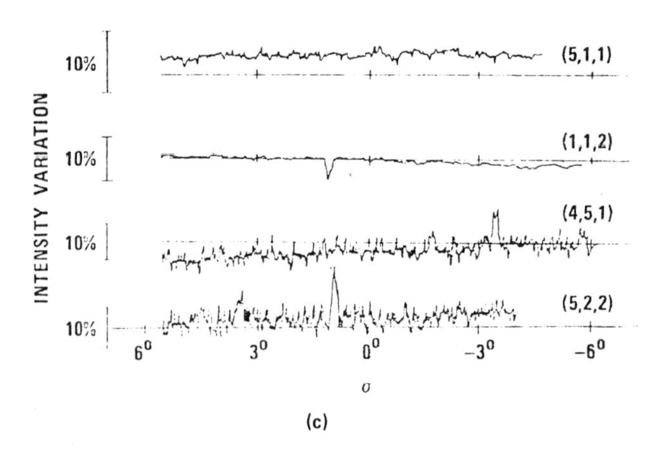

FIGURE 6 (a) Graphical construction for multiple diffraction on a square planar reciprocal lattice net.[14] (b) Dependence, on rotation angle around the diffraction vector, of multiple reflection contribution to the apparent reflection intensity of 222, a forbidden reflection, in germanium.[30] (c) Multiple reflection effects for several unrelated reflections, in rotation around the diffraction vector, showing peaks and dips.[31]

when conditions for double diffraction are realized, while a weak or extinguished reflection may show peaks, and an intermediate reflection may show both (see Figure 6b, c). Possibly the best readily available protection from multiple diffraction effects is to use the full power of the computer-controlled diffractometer, by mounting the crystal in an arbitrary orientation, or at least by "missetting" it by an arbitrary angle, then letting the computer use the actual orientation matrix to calculate the diffractometer settings.

A crystal structure determination begins by determination of the unit cell, lattice type, and space group. The "Laue symmetry," or point symmetry of the intensity distribution in reciprocal space, is usually easy to determine from diffraction photographs, and in the absence of measurable deviations from Friedel's law is always centrosymmetric. Consistent with a given lattice type and Laue symmetry there may be several possible space groups, choice among which may be reduced through use of "systematic extinctions" indicating glide planes or screw axes. In some cases, as in $P2_1/c$ (centrosymmetric) and $P2_1 2_1 2_1$ (noncentrosymmetric), the choice given by systematic extinctions is unique (although there are cases, such as 2_1 along a very short axis or a situation with many general extinctions due to a pseudostructure, where the validity of extinctions as due to space group is subject to some statistical uncertainty). In others, the choice may narrow down to one centrosymmetric space group (e.g, P2/m) plus one or more noncentrosymmetric space groups (e.g., P2 and Pm). Without atoms present that would produce anomalous dispersion yielding significant deviations from Friedel's law, there is no way of making the choice with x-ray data alone short of determining and refining the structure successfully with a trial space group. Where the resulting structure is clearly noncentrosymmetric and and far from being pseudocentrosymmetric, this is unobjectionable (though possibly expensive); when the resulting structure appears to be centrosymmetric, however, it may be hard to rule out some degree of deviation from true centrosymmetry (e.g., small deviation of atomic positions from apparent mirror planes). To avoid this difficulty and to achieve the greatest economy of effort and expense, it is advisable to make use of additional (i.e., nondiffraction) information. An L-amino acid (or any substance showing an optical rotation in a nonrotating solvent) cannot crystallize in a centrosymmetric space group. However, a centrosymmetric molecule or a racemic mixture *can* crystallize in a noncentrosymmetric space group. The crystal morphology (face development, etch patterns) may give the answer. Physical measurements that may indicate absence of a center of symmetry in certain circumstances are piezoelectricity (Giebe–Scheibe test), pyroelectricity, and second-harmonic generation in a laser beam.[15] None of these effects exists in all noncentrosymmetric crystal classes; piezoelectricity is absent in 432 and second-harmonic generation in 422 and 622 also, while pyroelectricity shows only the presence of a polar axis.

The accurate determination of the lattice constants is important for at least two reasons: (1) From the standpoint of interpretation of results, interatomic distances depend on products of parameter differences with values of lattice constants and cannot be more relatively accurate or precise than the latter. If interatomic distances of about 1.5 Å are to be given with an estimated standard deviation of 0.003 Å, a lattice constant of 15 Å must be determined to much better than 0.03 Å, preferably to 0.010 Å or less. (2) If a diffractometer is being used, accurate lattice constants are essential to calculating reciprocal lattice coordinates and diffractometer settings for the reflections for which intensities are to be measured. Preferably, the lattice constants to be reported should be measured with the same crystal specimen and the same diffractometer that are to be used for intensity data collection and should be the result of least squares refinement based on careful measurements of reciprocal lattice coordinates for a dozen or more high angle reflections. The diffractometer (and the procedure for using it) should be accurately aligned, calibrated with standards, and demonstrably capable of yielding the accuracy required; otherwise it is not much good for intensity work. Users of photographic techniques for intensity collection sometimes employ powder photography for obtaining precise lattice constants. This is nonobjectionable, provided it is reasonably clear that the powder and single-crystal specimen are equivalently representative of the substance being investigated in composition and in crystal structure. The powder method is capable of very high precision, often more than usable. Many substances have linear thermal expansion coefficients of $3 \times 10^{-5}\ K^{-1}$ or more (often very much more for molecular crystals). It is of no use to quote a lattice constant with a relative estimated standard deviation of, say, 3×10^{-5} unless the temperature is specified. Similarly, it is important that diffractometer data be taken at a reasonably uniform temperature.

The "warm happy feeling" can be fostered if the density of the crystal specimen (or a reasonable proxy) has been determined accurately (rather better than 1 percent) and if the number of formula units (Z) per unit cell calculated from the density, formula weight, and lattice constants comes out an integral value consistent with the space group, within an uncertainty range consistent with the data used. Deviations of Z from an integral value often evoke a variety of excuses (voids, air bubbles, partial solvation or desolvation, etc.) but are sometimes symptomatic of an error in chemical composition or of a crystal specimen not representative of the bulk material. A good value of Z, a set of lattice constants for the crystal specimen that agree with those of a powder, and (if necessary) a powder intensity profile that agrees with one calculated from single crystal intensities, together support the assumption that the crystal specimen and the bulk material are equivalent and of the assumed composition. However, an elemental analysis of the material ahead of time is often warranted as insurance against using the diffractometer and computer themselves as extremely expensive analytical instruments.

The experimental details of data collection are too many to discuss in depth here. The crystal specimen should be small enough that the longest path of an incident and reflected ray does not exceed three or four times the reciprocal of the linear absorption coefficient. The specimen should be of roughly uniform proportions if possible. As already stated, it should be of such a shape that the dimensions are easy to measure accurately for computing absorption corrections. The crystal must be mounted on a fiber sufficiently stiff that it does not move out of the center of the beam due to gravity or air currents and with a cement that does not turn the apparatus into a very expensive hygrometer. These conditions are more stringent than with photographic methods since the Weissenberg and precession cameras are somewhat forgiving as long as the crystal stays in the beam and the spot is not clipped by the layer-line screen. The diffractometer itself should be aligned so that all axes of rotation intersect at all times at the intended center of the crystal to an accuracy of 0.001 cm. A full-circle diffractometer is preferable to a quarter-circle instrument since it permits collecting data on Friedel pairs even for triclinic crystals, allows random orientation of the crystal for avoidance of systematic multiple scattering, and permits intensity measurements in nearly the full 4π steradians of reciprocal space. This last feature is of value for obtaining data from which estimated standard deviations of measured intensities can be determined and for providing indications of crystal or diffractometer misalignment if present. In view of the possibilities of intensity changes due to crystal decomposition, crystal movement, or fluctuations in incident beam intensity, a set of a half dozen or more "standard" reflections, chosen to represent various intensity ranges and regions in the reciprocal lattice, should be repeatedly measured at intervals throughout the data collection. For the same reasons it is advisable to "jump around" to some extent in the reciprocal lattice rather than to collect all data in a strict digital sequence. An even better procedure, if sufficient diffractometer time were available, would be to measure the entire data set three or more times on the same crystal and extrapolate every reflection back to zero time. Since radiation damage may be accompanied by structural changes that affect different reflections by different amounts, the reliability of the determination—at least as it concerns radiation-sensitive moieties (multiple bonds, strained rings, etc.)—may be suspect if the decreases in check-reflection intensity much exceeds 20 percent or if the resulting structure shows anomalously high temperature factors for certain atoms or groups.

The radiation chosen is usually CuK_α or MoK_α, for which $\lambda = 1.5418$ and 0.7107 Å, respectively. The latter has advantages of lower absorption and larger accessible reciprocal lattice volume but can give rise to difficulties in separating adjacent reflections in structures with large unit cells. The choice of radiation may be made to enhance anomalous dispersion effects for determining an absolute configuration or to avoid fluorescence (e.g., from a cobalt-containing compound with copper radiation). Radiation monochromatized by crystal reflection is often useful for reducing radiation damage in the crystal due largely to the white radiation component; more, it can increase accuracy by greatly reducing the background, especially for weak reflections. Use of a monochromator, however, may give rise to a narrowing of the incident beam or uneven intensity distribution if the monochromator crystal has insufficient mosaic spread. Many other details of data collection are beyond the scope of this paper: e.g., whether to use θ–2θ scan, or ω-scan, how wide the scan range should be, how long to count background on each side, or whether to use balanced filters.

It is rare indeed that all conditions of gathering intensity data are optimized in all of the ways indicated above. Where they are not, possibilities of systematic error in the results may creep in. Failure adequately to determine the parameters for absorption correction throws into serious question the value of anisotropic temperature factors for analyzing the thermal motion of molecules and of atoms in molecules so as to make bond-length corrections. The same effect can result from loss of intensity due to crystal decomposition, while data collection is following a digital sequence along a particular direction, if not compensated with the aid of repeated measurements on standard reflections.

Many determinations are still being made with data obtained from Weissenberg and precession photographs; indeed, photographic methods have in recent years received a strong boost through the development of computer-assisted automatic film scanners. At its best, photographic work may approach diffractometer work in accuracy, but there are inherent limitations due to emulsion nonuniformity and geometric restrictions. An effect of geometric restrictions may be, in certain cases, unavoidable systematic multiple diffraction. Film also lacks the capability of rejecting fluorescence radiation, which a diffractometer may handle with a pulseheight discriminator.

Having collected all needed intensity data under the most favorable conditions possible, the crystallographer processes the data, applying absorption corrections and, if necessary, corrections for decomposition of the crystal, and arrives at his "data set," consisting of the values of $|F_{hkl}^{\mathrm{obs}}|$ or $|F_{hkl}^{\mathrm{obs}}|^2$, either unscaled or with a rough scale factor calculated by statistical methods. Each datum should be accompanied by a standard deviation σ that represents random error (and possible random effects of systematic errors) as derived, for example, with Eq. (13).

We pass over the elucidation of the trial structure by Patterson methods or statistically based "direct methods"; the pitfalls awaiting the unwary in this part of the job are the subject of Donohue's paper (this volume). We arrive at a trial structure for which the calculated values of the structure factor (or its square), F_{hkl}^{cal} (or $|F_{hkl}^{\mathrm{cal}}|^2$) are in rough agreement in magnitude with $|F_{hkl}^{\mathrm{obs}}|$ (or $|F_{hkl}^{\mathrm{obs}}|^2$) (after scaling the latter to the former), and the sign or phase of the former can be provisionally assigned to the latter. The structure is now ready to be refined by least squares to determine the value of the positional and thermal parameters that minimize some function of the residuals between calculated and observed structure factors, such as

$$R_1 = \sum_{hkl} w_{hkl}^{(1)} (|F_{hkl}^{\mathrm{obs}}| - |F_{hkl}^{\mathrm{cal}}|)^2 \tag{20}$$

or

$$R_2 = \sum_{hkl} w_{hkl}^{(2)} (|F_{hkl}^{\mathrm{obs}}|^2 - |F_{hkl}^{\mathrm{cal}}|^2)^2 \tag{21}$$

Much has been said (see chapters by Tukey, Andrews, and Nicholson, this volume) concerning the Gaussian method of least squares (characterized repeatedly during discussion as the "method of ill repute") and possible modifications or replacements of it. Nevertheless, it is by far the most frequently used method for crystal structure refinement and will continue to be for some time to come.

Refinement requires an iterative approach owing to the nonlinearity of the problem. In principle, refinement with either R_1 or R_2 should converge to the same result (if weights $w^{(1)}$ and $w^{(2)}$ are consistent), but the latter, besides being possibly more convenient for noncentrosymmetric structures, obviates a certain problem about "unobserved" reflections (the intensities of which are not significantly above background). The problem of what to do about unobserved reflections has nettled crystallographers for a long time. A widespread school of thought says: Leave them out of refinement if the ratio of the F or F^2 to its σ does not exceed a certain value such as three. However, it can be argued that these null observations are among the most powerful observations we have. Moreover, it can be argued that omitting them below a certain arbitrary value actually introduces a statistical bias that may affect the result. (Imagine a hypothetical structure representing a small distortion from an ideal structure for which a certain class of reflections vanish; by virtue of the distortion, the reflections of this class are on the edge of observability by the criterion we use. Random error will cause about half of these to increase above this level; they are included in the refinement; the other half, being "unobserved," will be neglected. Thus the distortion will be exaggerated. This is admittedly an artificially contrived situation but a little of it may lurk in many structures.) Therefore, it appears, we should include all reflections, observed and unobserved, in the refinement. What about those for which the observed intensity is *below* background, so that $|F_{hkl}^{obs}|^2$ is negative, and $|F_{hkl}^{obs}|$ evidently imaginary? Rejecting those, which many would do, can similarly introduce a bias, as pointed out recently by Hirshfeld and Rabinovich[32]; thus they should be included. An argument in favor of refining with R_2 is that this can be done more naturally than with R_1. Inclusion of "unobserved" reflections, on the other hand, raises the possibility of introducing a bias, owing to the strong nonlinearity of $|F|^2$ near its zero value if refining with R_2, or to the possibility of including F with the wrong sign if refining with R_1, especially when there are significant model errors (e.g., erroneous atomic form factors). A cue to the possibility of such model errors may be provided by a goodness-of-fit parameter differing significantly from unity (see below). The problem is more complicated when visually estimated photographic data are used, in which case an "unobserved reflection" is one having an intensity less than the minimum value for which a diffraction spot is visible to the eye. Generally, such "unobserved reflections" are left out of refinement. Alternatively, they may be put in when $|F^{cal}|$ exceeds the limiting value $|F^{lim}|$, or put in with $|F^{obs}|$ set at the appropriate mean value of the implied distribution with a weight derived from the second moment of that distribution. The last procedure is frequently objected to on the (not entirely correct) assumption that gaussian least squares requires that all observations belong to normally distributed populations. The question of what to do with "unobserved reflections" remains one on which the last word has not been said.

The assignment of weights, $w^{(1)}$ or $w^{(2)}$, ought to be a very simple matter if one has values of σ available: $w = 1/\sigma^2$. "Unit weights" have been much used in the past, but happily their use seems to be declining as more and more workers realize that the computational convenience that unit weights give is trivial and that their use invalidates any serious estimation of the validity of the refinement results based on the final residuals.

Choice of which parameters to refine is a matter of some judgment. In the model as previously described, the parameters are an overall scale factor (since intensities are virtually never measured on an absolute scale, although this is possible in principle), three positional parameters for each atom unless some are eliminated by symmetry, and six tensor components for anisotropic vibration again subject to reduction on account of symmetry. Other parameters may be added to take explicit account of occupancy factors (e.g., in alloys), disorder, extinction (isotropic or anisotropic), and unresolved experimental problems such as crystal decomposition. In principle, one can also introduce parameters to take account of distortions of atomic electron distributions (bonding electrons, unshared pairs) and effects of molecular motion. One can go the other way and refine on a simpler model—one in which the six anisotropic thermal parameters for an atom are replaced by a single isotropic temperature factor, as is often done with hydrogen atoms when heavier atoms are refined anisotropically. The point to be chosen on the scale between sub- and superrefinement must be chosen with two points in mind: (1) the importance of "overdetermination" as a check on the appropriateness of the model, since increasing the number of parameters decreases the overdetermination factor, and (2) the probable significance of any additional parameters, the addition of which to the model is being considered, in relation to statistical tests such as the 1965 $\mathcal{R}$ ratio test of Hamilton.[16] Actual settlement of a question such as whether one should refine with isotropic or anisotropic thermal parameters for certain atoms or all atoms may require that complete refinement to convergence be carried out with two (or more) sets of parameters and that the Hamilton $\mathcal{R}$ ratio test be applied to determine whether the refinement with the enlarged parameter set has produced any significant change. The Hamilton test can also be used to determine whether the diffraction experiment can make a significant choice between models with, for example, different space groups, or different absolute configurations, etc. It is worth mention here that the validity of significance testing of this kind can be vitiated by systematic error, or by strong deviations from normal error distributions.

The progress of refinement is followed by the magnitudes of the parameter shifts and should not be terminated until the shifts are all much smaller than the estimated standard deviations. It is also followed by the decline in the conventional R index

$$R = \sum_{hkl} ||F_{hkl}^{obs}| - |F_{hkl}^{cal}|| \Big/ \sum_{hkl} |F_{hkl}^{obs}| \tag{22}$$

and/or the weighted R index

$$R_w = \sum_{hkl} w_{hkl}||F_{hkl}^{obs}| - |F_{hkl}^{cal}|| \Big/ \sum_{hkl} w_{hkl}|F_{hkl}^{obs}| . \tag{23}$$

It has been common practice in the past to quote the values of these "compressed indicators" (normally ranging between 0.03 and 0.10 for a good determination) as an index of the quality of the determination and do nothing more in addition than quote the values of the estimated standard deviations of the parameters, which are given by

$$\sigma_i = B_{ii}^{1/2} \sigma_1 , \tag{24}$$

where B_{ii} is the ith diagonal element of the reciprocal normal equations matrix and σ_1 is the standard deviation of an observation of unit weight:

$$\sigma_1 = \left[\sum_{hkl} w_{hkl}||F_{hkl}^{obs}| - |F_{hkl}^{ref}||^2 / (n-m) \right]^{1/2} , \tag{25}$$

where ref = "refined", i.e., the calculated F_{hkl} value at convergence of the refinement. This quantity, also called the "goodness of fit," is often not quoted, although it is usually given routinely by the standard least squares computer programs. In principle, it is a much better guide to the quality of the determination than the conventional R factors or the quoted parameter standard deviations themselves. If observational standard deviations have been correctly assessed, if errors in the model are insignificant in comparison to random errors in the data, and if there are no significant systematic errors, σ_1 should be close to 1.0. Although significant model errors may bring the value to 3 or 4 or even more, a value in excess of unity can result from underassessment of experimental errors, or a value close to unity may result from errors in the model being offset by overassessment of experimental errors. In principle, assuming that random-error uncertainties in the data have been assessed correctly, a goodness of fit near unity should permit the conclusion that the structure model is correct and that one is entitled to place reasonable confidence in the validity of the estimated standard deviations in the parameters. This, however, is not altogether free of the risk that concealed systematic errors may affect the parameter values, some perhaps more than others; uncorrected or poorly corrected absorption may have very little effect on positional parameters but considerably greater effect on thermal parameters. Moreover, such a systematic error, having resulted in erroneously shifted thermal parameters, may not significantly affect the apparent goodness of fit. Generally, if the value of σ_1 is much in excess of unity, there may be defects in the model (which may be remedied in certain cases by additional refinement with a larger set of parameters) or systematic errors, and it would be prudent to multiply the quoted estimated standard deviations by a factor of two or more before making any judgments based on values of structure parameters or differences between them.

A value of σ_1 close to 1.0 can certainly enhance the "warm happy feeling," while one in excess of 1.0 does not necessarily condemn the determination.

Ideally, at the end of a determination that has made use of hundreds or thousands of observations and is thus amenable to statistical analysis at a reasonably high level, one should not merely take comfort in satisfactory values of parameters such as R and σ_1 but rather undertake an overall statistical analysis of the residuals to ascertain whether the uncertainties in the data were properly assessed and whether there is an opportunity to detect errors in the model or systematic errors in the data. In the past this has rarely been done, in part because it is a lot of work and in part because most crystallographers do not really know how to go about it. Recently, Abrahams and Keve[17] (later also Hamilton and Abrahams[18]) showed that the objectives sought can be largely achieved by use of an "extended indicator"–the normal probability plot. In the application of this technique to the problem at hand, the normalized residual

$$\delta R_{hkl} = \left(|F^{obs}_{hkl}| - |F^{ref}_{hkl}|\right) \Big/ \sigma \; |F^{obs}_{hkl}| \qquad (26)$$

obtained at the end of least squares refinement is ordered and plotted against the quantiles expected for a normal distribution. If the δR_{hkl} belong to a single statistical population and are normally distributed and the σ's are correctly assessed and scaled, the resulting plot should closely approach a straight line of unit slope and zero intercept, although perhaps with some scatter at both ends. A non-zero intercept reveals incorrect scaling of F^{obs} to F^{cal}, a slope different from unity reveals under or overestimation of the σ's, and significant deviations from linearity may reveal defects in the model, systematic error in the data, or changes in the conditions of the experiment that may have led to variations in the random error (e.g., instabilities in the x-ray generator, electrical noise adding to scan and background counts, losses of counts due to a faulty component, and movement of the crystal in the beam). In an example given by Abrahams[19] (Figures 7 and 8), a δR plot of 310 pairs obtained for $CuInS_2$ ($R = 0.047$, $\sigma_1 = 1.08$) shows a large and systematic departure from linearity over the entire central portion of the plot. Examination of the experimental conditions revealed that one complete layer of data was affected by equipment malfunction and was out of scale with the remainder of the data. Rescaling of this layer and further refinement (to $R = 0.038$, $\sigma_1 = 0.99$) gave a respectably linear δR plot except at the ends where the few aberrant points correspond to reflections of low intensity. It is worth noting that both the R index and the goodness of fit were entirely respectable before this pathological condition was detected and corrected; without a detailed examination of the residuals such as that afforded by the δR plot, the condition would most probably have escaped notice, and the results previously obtained would have been accepted. Surely, as much as almost any other one thing, a proper δR normal probability plot could contribute to the "warm happy feeling."

A contemporary account of the state of the art of assessment of reliability of crystallographic structural information would be incomplete without mention of modern methods of comparing data sets and parameters sets from two or more independent determinations, although this is too large a subject to be covered more than briefly here. Hamilton[20] proposed testing the sum of the squares of normalized parameter differ-

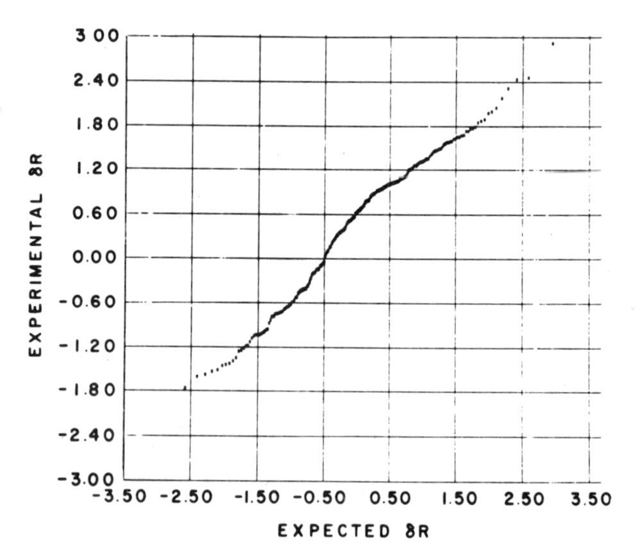

FIGURE 7 Normal probability plot for $CuInS_2$, *before* rescaling of faulty data.[19]

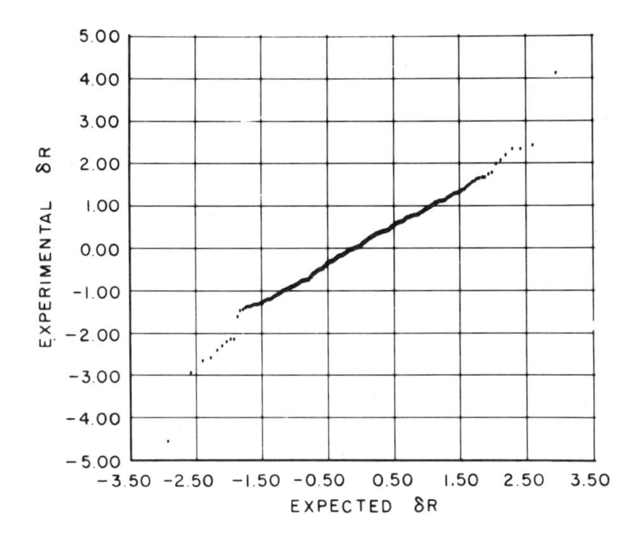

FIGURE 8 Normal probability plot for $CuInS_2$, *after* rescaling of faulty data.[19]

ences against χ^2 to test the hypothesis that the parameter differences are normally distributed in a single population and showed that such testing could effectively reveal which parameters were and which were not significantly different in the results of two separate investigations. The normalized moduli of parameter differences can also be ordered and plotted against moduli of expected quantiles in a "half-normal probability plot." Such plots are particularly effective in assessing the claimed estimated standard deviations in the parameters. An excellent example has been provided by studies of the amino acid L-asparagine monohydrate by Hamilton's group[21] and others. Figure 9 shows half-normal plots of the ordered normalized statistic

$$\delta p_i = ||p_i^{(1)}| - |p_i^{(2)}|| / [\sigma_i^{(1)^2} + \sigma_i^{(2)^2}]^{1/2} \tag{27}$$

vs. expected quantiles for half-normal order statistics. In the two plots on the left, thermal parameters are compared and on the right, positional parameters. The two top plots result from the independent neutron study of Ramanadham, Sikka, and Chidambaram,[22] and the two bottom ones are the x-ray results of Kartha and de Vries[23] as further refined by G. Kartha (private communcation to W. C. Hamilton). The approximate linearity of plots (a), (b), and (c) indicates that the differences δp_i for these cases are approximately normally distributed, while in (d) the large deviations from linearity suggest the presence of systematic

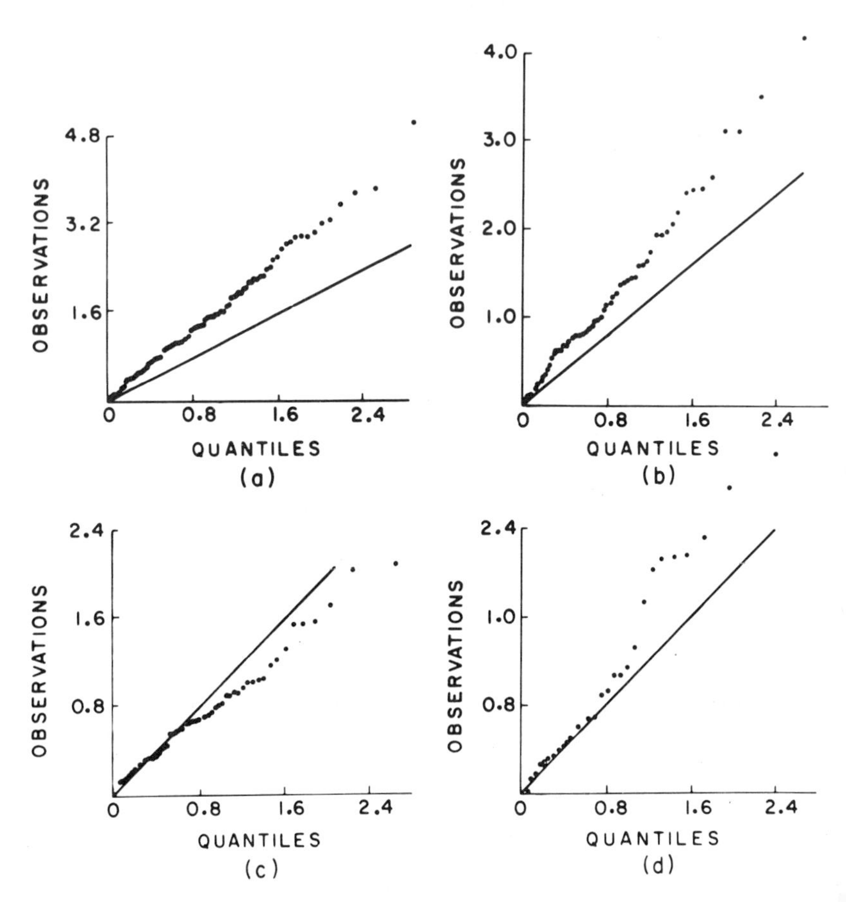

FIGURE 9 Half-normal probability plots for L-asparagine monohydrate (data from Verbist *et al.*,[21] figure from Abrahams[19]; see text).

error in the x-ray parameters. In all cases except (c), the pooled standard deviations [the denominator in Eq. (27)] appear to have been underestimated.

In refining the positional and thermal parameters corresponding to a given "trial structure" that may have some atoms missing (e.g., solvent molecules), some atoms misidentified or badly misplaced, or other structural defects, it may be possible to obtain convergence and even an acceptable $\mathcal{R}$ index with no glaringly obvious clue (except perhaps a less than encouraging goodness of fit) that anything is wrong. The "warm happy feeling" is strongly encouraged by the calculation of a "difference Fourier" density map containing no pathological features. Such a difference map may also, in favorable cases, indicate subtle deficiencies in the model, such as failure to allow for concentration of electrons in bonds and in nonbonding pairs (see Stewart, this volume).

Finally, the "warm happy feeling" benefits strongly from indications of internal consistency and plausibility of the structural results themselves: rational values for ordinary distances and angles, benzene rings that are regular and planar, etc. It does no harm to have more than one molecule per asymmetric unit and to show (by chi square testing or half-normal probability plots where appropriate) that the structural results are mutally consistent.

In the foregoing, which describes crystal structure determination as currently practiced, scans and counts are taken for fixed times, and the experimental σ's and corresponding least squares weights are thereby established. In 1967 and 1968, independently of each other, Hamilton[24] and Shoemaker[25] considered the proposition that the least squares weights might be chosen to optimize the precision of a given parameter or set of parameters, and the counting times and scan rates varied accordingly, subject to the constraint of a fixed total counting time for the determination. (Subsequently, it turned out that V. Schomaker in an unpublished work had independently investigated this problem some years earlier.) Equations were presented with which counting times and scan rates can be estimated based on the choice of parameter or set of parameters to be optimized. The original equation of Hamilton,[24] as modified by Shoemaker and Hamilton,[26] is

$$t_j + t_j^\circ = \text{const}\ (Lp)^{-1/2} \left|\frac{\partial |F_j|^2}{\partial \xi}\right| \cdot \frac{1}{|F_j|^3}, \tag{28}$$

where $j = hkl$, $t_j^\circ = \lambda_j/\kappa_j^2$, $\lambda_j = \text{const}/Lp$, κ_j^2 = variance in F from sources other than counting statistics. This equation gives the time t_j for counting reflection j under conditions where background is negligible and κ^2 can be assumed to be proportional to F^2, in order to maximize the weight of parameter ξ; the constants must be determined by normalization. More general equations are given by Shoemaker[25] and Shoemaker and Hamilton.[26] The equations should be most powerful if at least a rough structure is known ahead of time, but can be used with expectation values of structure factor derivatives. An interesting result of the studies is the conclusion that some reflections should not be counted at all, because the time spent on them is needed for other reflections on which the parameters to be optimized more sensitively depend. The proposed optimization procedures have not yet been put to the experimental test. It is argued by some that they are unlikely to have much effect in the case of x-ray diffraction, where, if higher precision is needed, the total counting time can be increased without exorbitant cost until the variance due to counting statistics is only a small part of the total variance, under which condition the optimization procedure becomes ineffectual. There is also some question concerning the limitations that may be placed on optimization by errors in the model. It seems likely that the proposed procedure may be more useful in neutron diffraction, where counting rates are generally much smaller, other random errors are better controlled, and the model is somewhat better (being independent of the vagaries of atomic electron distribution).

Notwithstanding the inherent possibilities of optimization, it is to be expected that for some time to come structural information from x-ray and neutron diffraction will be obtained more or less as it is today. There may be some improvements in instrumental capabilities, but perhaps the greatest improvements will come with more general and widespread appreciation among crystal structure workers of the experimental and statistical realities to which their results are subject, and with a correspondingly improved capability of assessing the reliability of the structural information obtained from their determinations.

In this brief survey many of the experimental and interpretive aspects of structural crystallography have been covered very lightly or not at all. For further details, the reader is referred to standard texts such as the one by Stout and Jensen.[27] A useful guide to the applications of statistics to the interpretation of physical measurements, with particular emphasis on structural crystallography, is provided by Hamilton's excellent book.[28] A useful survey of structural crystallography, which parallels this one in some aspects but provides more examples of the interpretation of structural results for organic molecules, has been given by Stewart and Hall.[29]

I am pleased to acknowledge valuable discussions with Sidney C. Abrahams, Carroll K. Johnson, James A. Ibers, and Verner Schomaker during the preparation of this paper. In particular, Dr. Abrahams has critically read it in its first version and has provided me with Figures 7, 8, and 9, which are from his Walter C. Hamilton Memorial Symposium paper,[19] and Dr. Johnson has made available technical assistance for preparing some of the figures. I dedicate this paper to the memory of the late Walter C. Hamilton to whom I am indebted for many valuable and educational discussions over recent years.

REFERENCES

1. Notes for Authors, *Acta Cryst.*, **B29**, 155 (1973).
2. International Union of Crystallography, *International Tables for X-Ray Crystallography,* Kynoch Press, Birmingham. Vols. I (1952), II, III, IV (in press, 1974).
3. V. Schomaker and K. N. Trueblood, *Acta Cryst.*, **B24**, 63 (1968).

4. W. R. Busing and H. A. Levy, *Acta Cryst.,* **17**, 142 (1964).
5. C. K. Johnson and H. A. Levy, in *International Tables for X-Ray Crystallography,* Vol. IV (2), Sec. 5., Kynoch Press, Birmingham (in press).
6. M. N. Frey, M. S. Lehmann, T. F. Koetzle, and W. C. Hamilton, *Acta Cryst.,* (in press).
7. W. H. Zachariasen, *Acta Cryst.,* **23**, 558 (1967).
8. W. H. Zachariasen, *Acta Cryst.,* **A24**, 212 (1968).
9. W. H. Zachariasen, *Acta Cryst.,* **A24**, 421 (1968).
10. W. H. Zachariasen, *Acta Cryst.,* **A25**, 102 (1969).
11. W. H. Zachariasen, *Acta Cryst.,* **16**, 1139 (1963).
12. P. Coppens and W. C. Hamilton, *Acta Cryst.,* **A26**, 71 (1970).
13. W. H. Zachariasen, *Acta Cryst.,* **18**, 705 (1965).
14. R. D. Burbank, *Acta Cryst.,* **19**, 957 (1965).
15. S. C. Abrahams, *J. Appl. Cryst.,* **5**, 143 (1972).
16. W. C. Hamilton, *Acta Cryst.,* **18**, 502 (1965).
17. S. C. Abrahams and E. T. Keve, *Acta Cryst.,* **A27**, 157 (1970).
18. W. C. Hamilton and S. C. Abrahams, *Acta Cryst.,* **A28**, 215 (1972).
19. S. C. Abrahams, Paper presented at Walter C. Hamilton Memorial Symposium, Brookhaven National Laboratory, June 15, 1973 (submitted to *Acta Cryst.,* in press).
20. W. C. Hamilton, *Acta Cryst.,* **A25**, 194 (1969).
21. J. L. Verbist, M. S. Lehmann, T. F. Koetzle, and W. C. Hamilton, *Acta Cryst.,* **B28**, 3006 (1972).
22. M. Ramanadham, S. K. Sikka, and R. Chidambaram, *Acta Cryst.,* **B28**, 3000 (1972).
23. G. Kartha and A. de Vries, *Nature* (*Lond.*), **192**, 862 (1961).
24. W. C. Hamilton, Abstract E6, ACA Summer Meeting, Minneapolis, Minn., August 20–25, 1967.
25. D. P. Shoemaker, *Acta Cryst.,* **A24**, 136 (1968).
26. D. P. Shoemaker and W. C. Hamilton, *Acta Cryst.,* **A28**, 402 (1972).
27. G. H. Stout and L. H. Jensen, *X-Ray Structure Determination, A Practical Guide,* The Macmillan Company, New York (1968).
28. W. C. Hamilton, *Statistics in Physical Science,* Ronald Press, New York (1964).
29. R. F. Stewart and S. R. Hall, in *Determination of Organic Structures by Physical Methods,* Vol. 3, p. 74, Academic Press, New York (1971).
30. H. Cole, F. W. Chambers, and H. M. Dunn, *Acta Cryst.,* **15**, 138 (1962).
31. R. A. Young, *Acta Cryst.,* **A25**, 55 (1969).
32. F. L. Hirshfeld and D. Rabinovich, *Acta Cryst.,* **A29**, 510 (1973).

James A. Ibers

Problem Crystal Structures

As a lover of cats I assert that I have never seen an ugly cat. In the same spirit, as a practical structural chemist I assert that I have never been involved in a crystallographic study that did not present some problems. By necessity, then, I limit my remarks on the topic of "problem crystal structures" to selected examples that hopefully will illustrate a few, but surely far from all, of the problems that can befall both the aware and the unaware.

Problems in crystallographic studies may be conveniently classified according to the time of their occurrence or recognition. Problems at the beginning of the study are usually rather obvious. These include the inability to produce suitable crystals or the ability only to produce crystals that are unstable in the x-ray beam. The collection of intensity data is therefore impossible, and the investigator seeks another problem. In the middle of a crystallographic study the most embarrassing problem is the inability to find a suitable trial structure, a procedure that is necessary before refinement can be effected and chemical information derived. Since by this time there has been a considerable investment of time and money in the study, the problem is often a desperate one. It is generally true that the longer one seeks to find a satisfactory trial structure, the more computing time is wasted. Inability to solve the structure may result from ill-advised experimental procedures, including determination of an incorrect space group, or it may result from limitations on the methods employed by the investigator or on his experience and imagination. Problems that occur near or at the end of the structure investigation include refined solutions that do not give one that "warm, happy feeling." It will mainly be these problems that I will consider here. To limit my discussion I will focus generally on specific problems that we have encountered over the past five years or so. It may be more instructive to others, and it is certainly kinder, to drag one's own skeletons out of the closet.

It is appropriate at this point to indicate our personal motivation for carrying out structural studies, the types of compounds we study, and the experimental conditions we employ. In a very general sense we are interested in the bonding of small molecules and ions, e.g., O_2, N_2, NO, N_2R^+, olefins, and acetylenes, to transition-metal complexes. Because of our interest in bonding, we seek the best solutions we can attain. Rapid, qualitative answers to conformational problems are not our interest. Since those transition-metal systems that bind small molecules generally have the metal in a low oxidation state, and since a low oxidation state is usually stabilized by ligands of the type PR_3 (R = alkyl or aryl), solution to our problems involves typically the determination of a large number of structural parameters. With only a few exceptions the intensity data are obtained at room temperature on a Picker FACS-1 computer-controlled diffractometer. Usually the ratio of observations to variables is at least 10, and it is often 20 to 30.

Table 1 provides a summary of the types of problems we have tackled in the last three years and the various problems we have encountered. We have had considerable difficulty in finding a solution in about 20 percent of the cases, although we have been lucky in that none remains unsolved. We have obtained results grossly different from expectation about 10 percent of the time, and we have often found unexpected sol-

TABLE 1 Some Statistics on Problems Attacked since November 1970

Light atom structures	2	Published or completed	40
Heavy atom, first row	21	Unsolved	0
Heavy atom, second row	13	Unsatisfactory solution	2
Heavy atom, third row	10	In progress	4
	46		46
Problems with data collection	1	Disorder of one sort or another	7
Problems with solution	8	Unexpected solvent molecule	3
Crystals very hard to grow or compound difficult to make	11	Resultant formulation different from that expected	5

vent molecules in the structures. More importantly, we have had great difficulty in obtaining suitable crystals in almost 25 percent of the cases. This fact should be kept in mind by those who advocate x-ray structure analysis not only as the ultimate analytical method but also as one of the cheapest. Clearly, the man-hours involved in preparation, crystallization, and recrystallization can be very expensive.

All of the structures tackled have been refined by least squares methods and, with few exceptions, by full-matrix least squares methods. The total computing investment over the past three years amounts to approximately 300 hours of central-processor time on a CDC-6400. When the necessary calculations have exceeded the capacity of the CDC-6400 (about 250 variables), we have turned to a remote hookup with the CDC-7600 at Lawrence Berkeley Laboratory.

EXAMPLES OF PROBLEMS ENCOUNTERED

The purpose of this paper is to inform nonspecialists about those features of a completed structure determination that lead one to view the answers with suspicion. While many criteria may be employed in such an evaluation, some of which have been discussed in preceding papers, the ultimate criterion is one's own "chemical intuition." Such intuition must be used with caution, however, or one may overlook subtle, genuine chemical effects that might be of importance. With this in mind, let us pass along through a "Rogue's Gallery of Problem Crystal Structures."

Problem 1

Compound: $[RhH(NH_3)_5](ClO_4)_2$.[1]

Space Group Determination: Initially, the space group was determined from normally exposed photographic films. Extinctions noted were hkl, none; $hk0$, h odd; $0kl$, $k + l$ odd; $h0l$, none. These are indicative of space groups Pnma or $Pn2_1a$, since the Laue symmetry is *mmm*. There are four formula units in the cell so that no symmetry need be imposed on the cation in $Pn2_1a$, whereas in Pnma either a mirror plane or a center of symmetry is imposed.

Solution and Results: A Patterson function suggested that the Rh atom was at $x\frac{1}{4}z$. Initial solution and refinement was carried out in Pnma. The cation had an imposed mirror plane, which would not necessarily require disorder of the H and *trans*-NH_3 ligands, but it was necessary to postulate such disorder in order to achieve concordant thermal parameters among the NH_3 groups. Both of the Cl atoms were found to lie on the mirror plane, necessitating two sets of O atoms also on this plane. In further calculations the space group $Pn2_1a$ was considered, but refinement of neither enantiomer in this space group led to significantly better results. The $RhH(NH_3)_5^{2+}$ ion was found to be essentially octahedral.

Detection of Troubles: There were several features of the refinement and of the resultant structure that made us suspicious: (1) The R index of about 11 percent was too high when compared with results we had obtained on comparable structures; (2) the O atoms of the ClO_4^- groups would not refine anisotropically (i.e., the thermal parameters became non-positive definite); (3) analysis of $|F_o|$ vs $|F_c|$ as a function of Miller indices revealed that agreement was particularly poor for indices with large values of k. This strongly suggested that the y coordinate of the Rh atom was incorrect; (4) most importantly, the Rh–N distances *cis* to the disordered H ligand ranged from 1.93(2) to 2.16(1) Å, where the estimated standard deviations are given in parentheses.

Resolution of the Difficulties: The problem was dropped for a period of almost a year. When it was resumed a series of very long exposures with Cu radiation was taken. These exposures clearly revealed a few "extra" reflections that we had previously believed were extinguished. The correct space group was thus found to be $P2_12_12_1$. Refinement proceeded smoothly in this space group to a final R index of 4.0 percent. The resultant cation is not disordered, the H ligand is readily positioned, and the Rh–N(*cis*) distances are self-consistent.[2] A final structure factor calculation indicates that approximately 30 "missing" reflections would have been observable had we not specifically programmed the diffractometer to omit reflections of this type.

Morals: (1) Collect all unique data in primitive space groups. (2) Take very long film exposures. (3) Do a total analysis of agreement of $|F_o|$ vs $|F_c|$ as a function of $|F_o|$, setting angles, and Miller indices. (4) Have faith in your chemical intuition.

Problem 2

Compound: $Os(CO)_3[P(C_6H_5)_3]_2$.

Space Group Determination: Normal film methods led to space group P3ml with three formula units in the cell.

Solution and Results: The problem was solved by standard methods and was found to refine satisfactorily. The resultant structure consisted of three independent molecules in the cell, each with imposed symmetry $3m$.

Detection of Troubles: A calculation of all intermolecular contacts, including those based on idealized positions for the H atoms of the phenyl rings, revealed some impossibly short H· · · H interactions.

Resolution of the Difficulty: Exceedingly long x-ray photographs indicated that the c axis had to be doubled. As a result, the space group became $P\bar{3}c1$ and refinement led to essentially the same structure as before. There are now two independent molecules in the cell, one with imposed symmetry 32, the other with imposed symmetry 3. The molecular structure did not change, save for rotations of the phenyl rings and these rotations give rise to the very weak l-odd reflections.[3]

Morals: (1) Take very long film exposures. (2) Calculate and examine all intermolecular contacts at the end of a structure refinement.

Problem 3

Compound: *cis*-$PtCl_2[P(CH_3)_3]_2$.

Space Group: $B2_1$ with four formula units in the cell.

Solution and Results: Solution and refinement were carred out in a normal manner (for those days) with the neglect of anomalous scattering of the Pt atom. The result was the expected square planar configuration about the Pt atom.

Detection of Troubles: Each Cl atom is *trans* to P, and hence it was most surprising that the two Pt–Cl distances differed significantly, as did the two Pt–P distances.

Resolution of the Difficulty: Some years earlier we[4] had added anomalous scattering terms to the usual least squares program. In 1966 the "polar dispersion error" was discovered.[5] When the above problem came to our attention, we felt that the difficulty might be in an incorrect space group determination. Ensuing calculations using our modified least squares program quickly revealed that the difficulty arose from the neglect of anomalous dispersion. On the basis of the usual criteria of equivalent bond distances and of the R-index test,[6] we were able to determine the correct enantiomer and to obtain a satisfactory structure.[7]

Morals: (1) When there is the possibility that the space group can be acentric and, especially, if it can be polar (i.e., origin of space group not fixed by symmetry elements), collect intensities for $\overline{hkl}$ as well as hkl (Friedel pairs). (2) Believe your chemical intuition.

Additional Remark: All of the structures done before 1965, and many done since, that contain a heavy atom on a polar axis will be incorrect by an amount that is large compared with the standard deviations given. The magnitude of the error resulting from the neglect of anomalous scattering can be estimated.[8]

Problem 4

Compound: $[Co(NO)(NH_3)_5]Cl_2$.

Space Group and Crystal Data: Space group Cmcm, a = 10.459(7), b = 8.753(6), c = 10.459(7) Å, but Laue symmetry $4mm$. Previous investigators[9,10] had explained the fact that the observed systematic absence does not correspond to any tetragonal space group. They postulated that the observed diffraction pattern is from a macroscopically twinned crystal, in which the basic unit is orthorhombic, space group Cmcm, with two of the dimensions equal. One individual of the twin is then derived from the other by interchange of the equal axes. Each observed reflection is a superposition of the hkl reflection from one indi-dividual and the $\bar{l}kh$ reflection from the other. The observed reflections fall into two categories: Those for which either hkl or $\bar{l}kh$ is absent in Cmcm and those for which both are present.

Solution and Refinement: The former set of reflections may be indexed for each individual, and it was used in the form of a partial data set to solve the structure. Refinement was carried out by Fourier methods, admittedly somewhat biased by the incomplete nature of the data set. The resultant solution consisted of essentially octahedral $Co(NO)(NH_3)_5^{2+}$ ions.

Detection of Troubles: The Co–N–O linkage was linear, which, although not surprising when the structure determinations were carried out, was surprising after the discovery of bent M–N–O linkages.[11] Moreever, in two previous studies the N–O distance was found to be 1.26 Å[9] or 1.43 Å,[10] and both values for this distance seemed impossibly long.

Resolution of the Difficulty: Data were recollected and refinement was effected through the use of all of the data. For those reflections to which both individuals of the twin contribute, the composite reflection "*hkl*" was taken to be

$$F^2(\text{“}hkl\text{”}) = aF^2(\text{“}hkl\text{”}) + (1-a)\,F^2(\text{“}\bar{l}kh\text{”}),$$

where a describes the volume fraction of the "*hkl*" component lattice. The resultant structure[12] is entirely satisfactory, with a Co–N–O angle of 119.0(9)°, an N–O bond length of 1.154(7) Å, and an obvious *trans* effect.

Morals: (1) Macroscopically twinned crystals can still lead to successful structures. (2) Partial data sets can introduce bias into the results.

Problem 5

Compound: $Pd_2(DBA)_3 \cdot CHCl_3$ (DBA = dibenzylideneacetone, PhHC=CHCOCH=CHPh, where Ph = phenyl).

Space Group: $P2_1/c$ with four formula units in the cell. There is no imposed symmetry.

Solution and Refinement: These were carried out by standard methods. The problem is a large one, involving the determination of approximately 290 variables based on around 3,800 observations. The final R index is 6.7 percent. The resultant structure is a novel one in which the C=C bonds of a given DBA ligand coordinate separately to two Pd atoms to yield a binuclear complex in which each Pd atom exhibits trigonal coordination.[13] The Pd–Pd distance is 3.25 Å.

Detection of Difficulties: The structure is a reasonably symmetric one with each of the DBA ligands in the extended *s-cis, s-trans* conformation. It is therefore disconcerting to find seemingly equivalent distances not only deviating significantly from one another but from the values expected. Thus we find C=C distances ranging from 1.03 to 1.33 Å; C–C distances ranging from 1.47 to 1.63 Å, and Pd · · · C distances ranging from 2.18 to 2.32 Å.

Resolution of the Difficulties: None has been found. On the assumption that the data are reliable, and we have internal checks that suggest that they are, the fault must be with the model. Perhaps a more sophisticated model for thermal vibrations of the atoms is needed, but the number of variables is already very large. Data collection at low temperatures is probably in order.

Moral: You can't win them all.

Problem 6

Compound: Co(OEP)(1-MeIm) (OEP = octaethylporphyrin dianion, 1-MeIm = 1-methylimidazole).[14]

Space Group: $P2_1/n$ with $Z = 4$ and no imposed symmetry.

Solution and Refinement: These proceeded by the usual methods. The problem is a large one, with 337 variables; because of the poor quality of the crystal, only 2019 reflections could be observed above background.

Detection of Difficulties: Difference maps in the course of the refinement provided ample indication that at least four of the eight ethyl groups are disordered.

Resolution of the Difficulties: None was found. The basic strategy of refinement of the disordered C(β) atoms of the ethyl groups involved the use of multiple positions and variable occupancies. Nevertheless, some of the C(α)–C(β) distances refined to 1.1 Å or shorter. With this model the R index is 9.1 percent. In a final model the C(β) atoms were fixed at chemically reasonable positions, while the positions of the other atoms in the structure were refined. The resulting R index is 10.4 percent. The structural parameters of interest, namely the distance of the Co atom from the mean plane of the porphyrin ring and the Co–N(Im) distance, did not vary significantly with change of refinement model. Of course, because of the generally poor agreement engendered by our inability to treat the disorder properly, the resultant standard deviations are larger than normal. Yet these are sufficiently small to enable us to reach our objective, an estimate of the movement of the proximal histidine group in cobalt-substituted hemoglobin upon oxygenation.[15]

Moral: It is sometimes possible to salvage the pertinent information of interest in a structure, despite disorder in other parts of the structure. You can't lose them all.

Problem 7

Compound: Ru(CO)(EtOH)(TPP) (EtOH = ethanol; TPP = tetraphenylporphyrin dianion).[16]

Space Group: $P\bar{1}$ with one formula unit in the unit cell; hence, the CO/EtOH portion of the molecule is disordered in view of the imposed crystallographic center of symmetry.

Solution and Refinement: Previously,[17] the structure had been solved and refined by standard methods on the assumption that the compound was $Ru(CO)_2$(TPP). This formulation was apparently supported by mass spectral results.

Detection of Difficulties: The resultant structure showed an unprecedented nonlinear Ru–C–O linkage with an angle of 153° and a very long Ru–C distance of 2.03 Å.

Resolution of the Difficulties: Both chemical and structural reinvestigations were undertaken.[16] They are in agreement that the compound is correctly formulated as Ru(CO)(EtOH)(TPP). A difference Fourier map in the late stages of refinement clearly reveals the C(β) atom of the ethanol group. It is readily shown that a superposition of a linear CO and a coordinated EtOH group will result in the apparent bent Ru–C–O linkage if the C(β) atom is undetected and the O–C(α) portion of the ethanol is interpreted as a CO.

Morals: (1) Examine difference maps carefully in the late stages of refinement. (2) Do not accept spectral information if the resultant formulation leads to a structure that contains unprecedented features.

Problem 8

Compound: $IrCl_2(NO)[P(C_6H_5)_3]_2$.[18]

Space Group: I2/a with four formula units in the cell and hence the imposition of twofold rotational symmetry on the molecule.

Solution and Refinement: These proceeded normally. The Ir atom was placed on the twofold axis, as was the N atom of the NO group. Refinement proceeded smoothly to an R index of 3.2 percent.

Detection of Difficulties: The N–O bond length was 0.90(2) Å and the Ir–N–O angle was 138(2)°.

Resolution of the Difficulties: The assumption that the N atom is on the twofold axis was abandoned. It is then necessary to postulate disorder of the N atom in addition to that assumed for the O atom. The resultant refinement, carried out by a group procedure,[19] led to the chemically more reasonable parameters N–O = 1.03(2) Å and Ir–N–O angle = 123.4(2.0)° and to the identical R index of 3.2 percent.

Moral: In the present instance either of the above models, as well as all intermediate ones, will yield identical agreement with the data. This is an instance where certain chemical features cannot be determined from the diffraction experiment. Similar problems will arise whenever a crucial atom is near to or on a symmetry element.[20]

Problem 9

Problem: Assessment of the geometry of the Me_5C_5 ring (Me = methyl).

Data Available: Three crystal structures have been determined to high accuracy in which a Me_5C_5 ring is attached to a transition metal. These include $(Me_5C_5)[Fe(CO)_2SO_2CH_2CH{=}CH(C_6H_5)]$[21] (I), $[(Me_5C_5)RhCl]_2HCl$[22] (II), and $(Me_5C_5)Rh(DBA)$[23] (III). Some of the results of these studies are given in Table 2.

Deductions from These Data: Generally speaking, the standard deviation of the C–Me distance in each of the structures, as estimated from

TABLE 2 Assessment of Trends in the Geometry of Me_5C_5 (distances in Å)

Quantity	Compound I[21]	Compound II[22]	Compound III[23]
C–C range	1.382–1.460(10)	1.403–1.440(8)	1.404–1.440(9)
C–C average	1.431(30)[a]	1.425(14)	1.418(16)
C–Me range	1.492–1.530(11)	1.479–1.503(11)	1.482–1.511(9)
C–Me average	1.509(16)	1.492(9)	1.502(11)

[a] This is the standard deviation of a single observation calculated on the assumption that the five distances averaged are from the same population.

the individual distances, is in sufficiently good agreement with that estimated from the least squares procedure to give credence to the hypothesis that these latter standard deviations are reliable. Accordingly, comparison of the standard deviation of an individual C–C distance, as obtained from the inverse matrix, with that calculated on the assumption that these distances are from the same population suggests that there are indeed significant differences among these distances. Such a conclusion would be tenuous if only one such structure determination were available. But consistent evidence from three such determinations makes the conclusion more probable. Naturally, the problem can be examined in greater detail by reference to deviations in metal-C distances and to deviations among C–C–C angles and by application of more rigorous statistical tests.

Moral: Small differences from expected chemical symmetry are difficult to assess in a given crystal structure and recourse should be made to several crystal structures before such differences are used as justification for a given theoretical argument.

Problem 10

Problem: Assessment of the symmetry of the M_4Y_4 core (cubane-like structure where M is a metal and Y is a ligand) in various crystal structures.

Data Available: The geometry of the M_4Y_4 core has been determined to varying accuracy in at least 16 crystal structures.[24] It is of theoretical interest to assess, in a given case, whether the approximate symmetry of this core is T_d, D_{2d}, or lower.

Procedure: Briefly, the procedure developed consists of the transformation of the positional parameters of the M_4Y_4 atoms, together with their standard deviations, to an orthonormal coordinate system. One then performs a least squares analysis in order to adjust rotations about the orthonormal axes to obtain the best fit to an assumed symmetry. One can then test the variance of an observation of unit weight by an appropriate chi square test. Further details are given by Averill *et al.*[24]

Results: In general, it is found that the M_4Y_4 core deviates significantly from T_d or D_{2d} symmetry. However, if one assumes that the standard deviations, as derived from the x-ray experiment, are underestimated by a factor of 3 to 4,[25] then of the 16 structures examined the symmetry of the M_4Y_4 core is found to be T_d in seven cases, D_{2d} in seven cases, and lower than D_{2d} in two cases.

Moral: The assessment of minor apparent variations in symmetry between related molecules in different crystal structures is risky at best and cannot be used objectively to support theoretical calculations.

CONCLUSION

This has been an abbreviated tour through a "Rogue's Gallery of Problem Crystal Structures." The overall morals include the following: (1) If you don't believe the results, know enough to detect the false steps in experiment or calculation that are involved; (2) whereas problems in diffraction studies are minimized by a healthy observation-to-parameter ratio, it is still possible to make mistakes; (3) when all else fails, believe your chemical intuition and that "warm, happy feeling" and don't be impressed by elegant models, or low *R* indices.

On the other hand, one can take the position of Pareto[26]: "Give me a fruitful error any time, full of seeds, bursting with its own corrections. You can keep your sterile truth for yourself."

It is a pleasure to acknowledge the fruitful, happy, occasionally panic-filled associations I have had with the many colleagues whose results I have used freely here. The work reported has been supported by the National Science Foundation, the National Institutes of Health, the Donors of the Petroleum Research Fund administered by the American Chemical Society, and by the Advanced Research Projects Agency and the National Science Foundation through the Northwestern University Materials Research Center.

REFERENCES

1. B. A. Coyle, "The structures of three inorganic complexes." Ph.D. thesis, Northwestern University (1969).
2. B. A. Coyle and J. A. Ibers, *Inorg. Chem.*, **11**, 1105 (1972).
3. J. K. Stalick and J. A. Ibers, *Inorg. Chem.*, **8**, 419 (1969).
4. J. A. Ibers and W. C. Hamilton, *Acta Cryst.*, **17**, 781 (1964).
5. T. Ueki, A. Zalkin, and D. H. Templeton, *Acta Cryst.*, **20**, 836 (1966).
6. W. C. Hamilton, *Acta Cryst.*, **18**, 502 (1965).
7. G. G. Messmer, E. L. Amma, and J. A. Ibers, *Inorg. Chem.*, **6**, 725 (1967).
8. D. W. J. Cruickshank and W. S. McDonald, *Acta Cryst.*, **23**, 9 (1967).
9. D. Hall and A. A. Taggart, *J. Chem. Soc.*, 1359 (1965).
10. D. Dale and D. C. Hodgkin, *J. Chem. Soc.*, 1364 (1965).
11. D. J. Hodgson, N. C. Payne, J. A. McGinnety, R. G. Pearson, and J. A. Ibers, *J. Am. Chem. Soc.*, **90**, 4486 (1968).
12. C. S. Pratt, B. A. Coyle, and J. A. Ibers, *J. Chem. Soc.*, (A) 2146 (1971).

13. T. Ukai, H. Kawazura, Y. Ishii, J. J. Bonnet, and J. A. Ibers, *J. Organometal. Chem.* (in press).
14. R. G. Little and J. A. Ibers, *J. Am. Chem. Soc.* (in press).
15. J. A. Ibers, J. W. Lauher, and R. G. Little, *Acta Cryst.* (in press).
16. J. J. Bonnet, S. S. Eaton, G. R. Eaton, R. H. Holm, and J. A. Ibers, *J. Am. Chem. Soc.*, **95**, 2141 (1973).
17. D. Cullen, E. F. Meyer, Jr., T. S. Srivastava, and M. Tsutsui, *J. Chem. Soc., Chem. Commun.*, 584 (1972).
18. D. M. P. Mingos and J. A. Ibers, *Inorg. Chem.*, **10**, 1035 (1971).
19. J. A. Ibers, *Acta Cryst.*, **B27**, 250 (1971).
20. See, for example, B. L. McGaw and J. A. Ibers, *J. Chem. Phys.*, **39**, 2677 (1963).
21. M. R. Churchill and J. Wormald, *Inorg. Chem.*, **10**, 572 (1971).
22. M. R. Churchill and S. W.-Y. Ni, *J. Am. Chem. Soc.*, **95**, 2150 (1973).
23. J. A. Ibers, *J. Organometal. Chem.* (in press).
24. B. A. Averill, T. Herskovitz, R. H. Holm, and J. A. Ibers, *J. Amer. Chem. Soc.*, **95**, 3523 (1973).
25. S. C. Abrahams, *Acta Cryst.*, **B28**, 2886 (1972).
26. V. Pareto, "Comments on Kepler," in *Familiar Quotations*, John Bartlett, 12th ed., Little Brown & Co., Boston (1949), p. 1198.

Jerry Donohue

Incorrect Crystal Structures: Can They Be Avoided?

This paper is largely historical in nature and may therefore be expected to overlap somewhat with others in this symposium, although my emphasis is probably rather more historical than most. Older crystallographers may find this survey interesting if only for nostalgic reasons, whereas structurally oriented chemists who are not diffraction analysts and who, accordingly, must believe what they are told may be interested for other reasons. Perhaps practicing crystallographers of any generation will be able to profit from the lessons of history, but, on the other hand, it may be true that Henry Ford was right.

TWO EARLY EXAMPLES

It was only 2 years after the discovery that x rays are diffracted by crystals that Mr. W. L. Bragg (as he was then) published[1] a table giving the lattices of rock salt, NaCl, zincblende, ZnS, and fluorspar, CaF_2, as face-centered cubic, but that of sylvite (KCl) as simple cubic. It is obvious today how this could have come about*—indeed, Bragg himself realized the true state of affairs. Nevertheless, there may be other points,

*Because K^+ and Cl^- are isoelectronic, their scattering power to x-rays is very nearly the same, and the face-centered unit cell thus appears to be simple cubic, one-eighth as large.

perhaps more subtle, that are not apparent in 1973 but may well be so 60 years hence.

The classic example of an incorrect structure and the damage it can create is that of pentaerythritol, $C(CH_2OH)_4$. The results of the first investigation[2] of this substance are crystal symmetry, C_{4v}; unit cell a = 6.16 Å, c = 8.76 Å, with two molecules; lattice, body centered; space group, C_{4v}^9, as all other body centered C_{4v} space groups can accommodate only 4, 8, or 16 molecules per cell; site symmetry of the molecule, and thus of the central carbon atom, C_{4v}. This means that the coordination at this atom must be either square planar or tetragonal bipyramidal and *not* tetrahedral. This rather startling result was soon verified in another laboratory.[3] A controversy then raged over the space group determination that was not settled to everyone's satisfaction until 14 years later[4] when a structure determination based on three-dimensional diffraction data was published. The true space group is S_4^2, and previous authors had been led into error by the unusual face development of the crystals. In the interim some 25 papers had been published on this crystal, some of which are concerned with the theory of the nontetrahedral carbon atom; they make very interesting reading today. The site symmetry of the central carbon atom is now S_4, but the controversy continues, having shifted to the point of how great the distortions from perfect tetrahedral the coordination there is. In a more recent thorough x-ray study[5] the C–C–C bond angles differ from 109° 28′ by nearly 3σ, and one of the positional parameters differs from that of the earlier study by 5σ.

ARE YOU CERTAIN WHAT'S IN THE X-RAY BEAM?

Polonium

Electron diffraction photographs of pure polonium powder, prepared by the method of M. Curie, were interpreted[6] in 1936 on the basis of a monoclinic unit cell having a = 7.43 Å, b = 4.30 Å, c = 14.13 Å, and β = 92°. The agreement between observed and calculated spacings was satisfactory. A structure based on space group C_2^3, with 12 atoms per unit cell, in three sets of the general position (calculated density 9.32 g/cm^3) gave moderately good agreement between observed and calculated intensities after trial-and-error adjustment of the nine positional parameters. This structure, as Strukturbericht type A19, was the accepted one for polonium for 10 years, until the discovery[7] that at room temperature polonium is actually a mixture of a low temperature (α) form and a high temperature (β) form. Later work[8] gives 54 °C for the transition

$\alpha \rightarrow \beta$ and 18 °C for the reverse transformation, with the two forms co-existing between these temperatures. The structure of α-Po is simple cubic, one atom per unit cell, and that of β-Po is simple rhombohedral, one atom per unit cell.[7] The preferred values of the lattice constants are, respectively, 3.366 Å and 3.373 Å,[9] and it is an interesting numerical coincidence that the average of the calculated densities is exactly that calculated for the spurious structure. The agreement between observed and calculated spacings for the α and β forms is also included in Table 1, where it may be seen that the 21-line pattern consists of six α lines, seven β lines, and eight coincident lines.

Fibrous Sulfur

By a very ingenious arrangement in which stretched filaments of fibrous sulfur could be continuously generated from liquid sulfur and exposed to an x-ray beam, the fiber diagram was recorded.[10] The procedure was designed to obviate the difficulty caused by crystallization of the sample on aging. The fiber diagram was indexed on a monoclinic unit cell having $a = 26.4$ Å, $b = 9.26$ Å, $c = 12.32$ Å, and $\beta = 79°15'$. The observed density then corresponds to 112 atoms per unit cell, and a structure consisting of ribbons of sulfur atoms was proposed.

Somewhat later[11] this flat chain structure was rejected on the grounds that the known geometrical properties of –S–S–S–S– systems should lead to a helical structure, as observed in two congeners of sulfur, selenium,

TABLE 1 Values of $1/d^2$ ($Å^{-2}$) for Polonium[a]

Observed	Calculated[b]	α[c]	β[d]	Observed	Calculated[b]	α[c]	β[d]
0.092	0.092	0.090	0.093	0.531	0.534, 0.542	0.536	0.527
0.158	0.152		0.155	0.717	0.722	0.718	0.716
0.180	0.181	0.179		0.800	0.794	0.806	
0.218	0.221		0.217	0.847	0.844		0.840
0.248	0.242		0.249	0.895	0.844, 0.904	0.898	0.871
0.264	0.252	0.269		0.992	0.964, 1.01	0.986	0.995
0.363	0.373	0.356		1.15	1.14, 1.18	1.16	1.15
0.370			0.374	1.25	1.26	1.25	
0.407	0.393, 0.402		0.405	1.47	1.46, 1.48, 1.49	1.52	1.46, 1.49
0.448		0.449		1.63	1.62, 1.63	1.61	1.61, 1.64
0.460	0.473		0.467				

[a] Data from Beamer and Maxwell.[7]
[b] For the monoclinic unit cell, see text.
[c] For simple cubic Po.
[d] For simple rhombohedral Po.

and tellurium, which consist of three-atom, one-turn helices. In sulfur, however, there should be seven-atom, two-turn spirals, and this would account for the unusual result that the number of atoms in the unit cell is divisible by seven.

However, it was later discovered[12] that, when aged, stretched fibers of sulfur were extracted with carbon disulfide, some of the reflections no longer appeared in the diffraction pattern. This result was explained[13] on the basis that the original diffraction pattern was the superposition of two patterns, that of true fibrous sulfur, plus oriented crystals of one of the allotropes of S_8. The earlier unit cell is thus clearly invalidated.

The correct structure of fibrous sulfur has been the subject of much investigation, and there is no reason to recount the whole story here, but only to remark that recent work[14,15] agrees that the substance consists of ten-atom, three-turn helices packed together in a complicated fashion.

Mercury

In 1922 a hexagonal structure was proposed for mercury,[16] based on powder data obtained at –78 °C, dry ice temperature. The 16-line pattern was indexed on a unit cell having a = 3.85 Å and c = 7.25 Å, and containing four atoms (in position $4f$ of space group D_{3d}^{2}-$P\bar{3}m$).

At the same time[17] a different structure was reported for mercury at –115 °C. This structure is rhombohedral, with the unit cell having $a = 3.031$ Å, $\alpha = \cos^{-1} \frac{1}{3}$, and one atom (in position $1a$ of space group D_{3d}^{5}-$R\bar{3}m$). In several other investigations this rhombohedral structure was also observed in temperature ranges between –150 and –80 °C. This led Wyckoff[18] to remark that "the data upon which the truly hexagonal structure was based thus remain entirely unexplained."

The hexagonal structure was considered by the editors of Strukturbericht to be very plausible, and they erected structure type A10 for it. An allotropic change could not be ruled out; furthermore, the hexagonal structure was in better agreement with the observed density than the rhombohedral structure.

This muddle was not cleared up until 1933,[72] when it was shown that five of the lines of the "hexagonal" powder pattern were actually from solid carbon dioxide; when these were ignored, the remainder of the diagram agreed with the rhombohedral structure and the powder data of the other (and subsequent) investigations.

HAS THE CORRECT UNIT CELL BEEN FOUND?

Orthotelluric Acid

In 1926 a cubic unit cell having a = 15.48 Å was assigned to orthotelluric acid[19] on the basis of Laue photographs. This lattice constant corresponds to 32 molecules per unit cell. Eight years later this result was criticized[20] as being unnecessarily large, and the true value was said to be a = 7.83 Å, with four molecules per cell. Reflections corresponding to the large cell were not observed. The tellurium atoms were placed in position 4a of space group O_h^5, and the oxygen atoms in position 24e, with 0.5 given as the approximate value of the x parameter.

Almost at once[21] one of the authors of the first paper countered with the presentation of a heavily exposed rotation photograph, which showed very weak odd layer lines that confirmed the larger unit cell. It is interesting that, while the larger cell is unequivocally (and easily) determined with Laue photographs, more dramatic (and more convincing?) evidence was proferred with a rotation photograph.

Chalcopyrite

In a very early contribution from the Throop College of Technology[22] the structure of chalcopyrite ($CuFeS_2$) was given as tetragonal (a = 5.24 Å, c = 5.15 Å) and the atoms positioned as shown in Figure 1a. This was the accepted structure for this substance for 15 years, when later work[23] in the same place, now named the California Institute of Technology, showed that the c-axis must be doubled, a situation leading to a somewhat different distribution of the iron and copper atoms (Figure 1b). This decision was also made on the basis of Laue photographs. It is probable that most of the present-day structure analysts would not know how to index such photographs. Also worth pointing out is the fact that all of lines listed for chalcopyrite in the ASTM file have $l = 2n$.

White (β) Tin

The first data obtained[24] from β-tin consisted of a powder pattern of some 20 lines, which was indexed on a tetragonal cell having a = 5.84 Å, c = 2.37 Å, and containing three atoms, at (000), ($\frac{1}{2}0\frac{1}{2}$), ($0\frac{1}{2}\frac{1}{2}$). The calculated density is 7.27 g/cm^3, compared with the observed value of 7.29.

This structure lasted for 5 years when it was shown[25] that in the true structure the c-axis is one-third longer, the unit cell having a = 5.84,

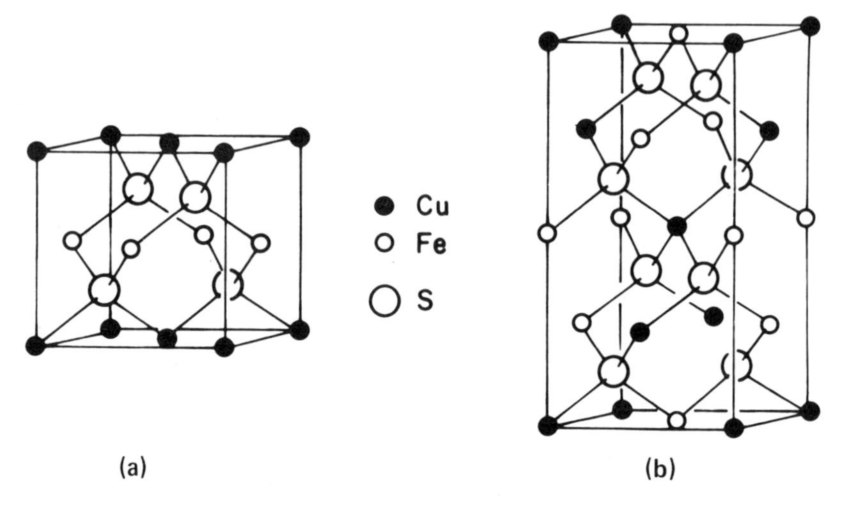

FIGURE 1 The two structures for chalcopyrite, $CuFeS_2$: (a) first structure[22]; (b) revised structure.[23]

$c = 3.15$ Å, and atoms at (000), $(\frac{1}{2}\frac{1}{2}\frac{1}{2})$, $(0\frac{1}{2}\frac{1}{4})$, $(\frac{1}{2}0\frac{3}{4})$. The calculated density is 7.24 g/cm^3. This new structure was based on more extensive data and is still the accepted one for β-tin.

Lead Chloride

Recently[26] powder photographs, 38 lines, of an HCl extract of an Sb–Pb alloy were indexed on an orthorhombic unit cell with the equation:

$$\sin^2\theta = Ah^2 + Bk^2 + Cl^2,$$

where $A = 0.00702$, $B = 0.00990$, $C = 0.00600$. Although the general agreement between observed and calculated values of $\sin^2\theta$ was "quite good," the authors were suspicious because for one line a discrepancy of roughly twice their estimated limit of error occurred. This suspicion was confirmed by the "figure of merit"[27] criterion, a very low value of 3 being obtained.

It was later possible to extract single crystals of the specimen, and from Weissenberg and oscillation photographs they were found to be $PbCl_2$, which has in the equation for $\sin^2\theta$ the constants $A = 0.0073$, $B = 0.0102$, and $C = 0.0290$.

It is seen that the spurious result gave close approximations to two of the constants, but not the third, and the authors concluded that it is dangerous to accept interpretations of powder photographs that are not backed by additional evidence.

L-Tyrosine

Also quite recently,[28] powder and fiber photographs of L-tyrosine were interpreted on an orthorhombic unit cell having $a = 13.89$ Å, $b = 21.08$ Å, $c = 5.842$ Å, and containing eight molecules. The space group Pnam was assigned to this crystal. After it had been pointed out[29] that a centric space group is not possible for L-tyrosine, the space group was implied to be P222. (Why this impossibility—which, by the way, is true not only for centric space groups but for those containing any symmetry elements of the second kind—was not noted by the authors, referees, and editors of the journal remains entirely unexplained.)

Single crystal data[30] soon revealed that the a-axis must be halved and that the correct space group is $P2_1 2_1 2_1$, with the unit cell containing four molecules—a favorite one for chiral organic molecules. Complete structure determinations have since been carried out[31] in this space group.

WHAT IS THE SPACE GROUP?

Gallium

In the first investigation of the structure of gallium[32] the substance was reported as tetragonal, eight atoms per unit cell, space group D_{4h}^{16}, and approximate values were given for the two positional parameters. This analysis was based on 13 powder lines, and the results were described as structure type A11 in Strukturbericht. This structure lasted for 6 years when it was discovered, on the basis of single crystal work,[33] that the true symmetry is orthorhombic, with two of the axes of equal length. The correct space group is D_{2h}^{18}, and subsequent work[34] has shown that the two axes, although not quite equal in length, differ by 0.15 percent at room temperature. Extrapolation shows that at some temperature near 140 °C the two axes would be *exactly* equal, but this is over 100 °C above the melting point.

C_s^4-*Cc or* C_{2h}^6-*C2/c?*

As is well known, the above two space groups have the same characteristic extinctions in an x-ray diffraction experiment, and other methods must be used to distinguish between them. If the true space group is the one of higher symmetry, then, when the structure is expressed in terms of the subgroup, there will be functional relationships between certain of the positional parameters. If the structure is refined in the subgroup, these relationships will not be exact. For the special case of molecular crystals with four molecules per unit cell, no molecular symmetry is required in the subgroup, but either a twofold axis or a center of symmetry in the higher space group.

Strontium Chloride Dihydrate The original description of this structure[35] is in terms of space group Ia (Ia is an alternate setting of Cc), $4SrCl_2 \cdot 2H_2O$ per unit cell, and it was stated that no reasonable structure could be found in the higher space group. However, it was later stated[36] that the final structure, as given in C_s^4, can be described in C_{2h}^6, and the positional parameters were tabulated on that basis.

Iron Pentacarbonyl The structure of $Fe(CO)_5$ was first described[37] in space group Cc, with four molecules per unit cell. The higher space group was rejected on the grounds that the molecule could not be centric. It was soon thereafter noticed[38] that, within experimental error, the molecule does contain a twofold axis, and the structure was satisfactorily refined in C2/c.

Dibenzyldisulfide This is another example of a four-molecule unit cell in either Cc or C2/c. The first description[39] was in terms of Cc, but an approximate functional relationship between parameters, which corresponds to a molecular twofold axis, was soon noted[40,41]: In C2/c the atoms come in pairs, and for every atom at xyz there is a second at $\bar{x}y\bar{z}$. Re-refinement of the published data[40] in C2/c resulted in average changes in the positional parameters of only 0.0003, with a maximum change of 0.0008 (0.007 Å); E statistics were said[41] to favor clearly the centric space group. Other statistical evidence, the $N(z)$ test, on the other hand, presented[42] by one of the authors of the original publication, was said to favor Cc. This question was then resolved by the observation[43] that no Bijvoet differences in intensity between hkl and $\bar{h}\bar{k}\bar{l}$ were detected, some of which for the noncentric structure were calculated to be very large. The correct space group is thus C2/c.

Sulfuric Acid The structure of H_2SO_4 was also described in terms of a four-molecule unit cell in Cc.[44] However, examination of the published parameters reveals that the sulfur atom could lie on a twofold axis and that the oxygen atoms come in pairs, very nearly related by a twofold axis. Unfortunately, the paper includes no structure factor (F) data, so the conjecture that the correct space group is C2/c cannot be tested further.

$Pn2_1a$ or Pnma?

These two space groups also have the same extinctions, and the situation is similar to the Cc-C2/c ambiguity. Although more examples might be found by a literature search based on the tables of Nowacki,[45] only three are given below.

Phosphorus Oxybromide In the first determination[46] of the crystal structure of $POBr_3$ the least squares refinement in Pnma "proved to be disastrous," whereas convergence in $Pn2_1a$ was attained after 30 (sic) cycles. The unit cell contains four molecules, and the parameters closely approximate a molecule lying on a plane of symmetry, as required by Pnma. Subsequent refinement of the data in that space group[47] led the authors to conclude that there is no reason to reject Pnma. One of the original authors countered by stating[48] that there is no reason to *accept* Pnma unless better evidence to the contrary is found. It seems to me that this is the wrong way around: The simpler assumption is the one to be chosen.[49] Meanwhile, someone should anomalously disperse this crystal, because, as pointed out,[47] statistical tests are inappropriate here.

Phosphorus Oxychloride This crystal is isotypic with $POBr_3$, and its structure was reported 2 years following the earlier one[50]; again, the parameters approximate a molecule with a symmetry plane. The structure was refined by least squares in $Pn2_1a$, and the paper includes the astonishing statement, "If the wrong space group is used in refining the structure, no problem arises because the final structure will contain the mirror plane, necessary for space group Pnma." The re-refinement of this structure has not been published. A statistical test is said to reject the centric space group at the 5 percent level,[48] but the previously mentioned objection[47] applies here also.

Basic Copper Molybdate The space group $Pna2_1$ has been assigned to $Cu_3Mo_2O_9$ after Pnam was rejected on the grounds that it led to improbably short metal–metal distances.[51] However, it has been pointed out (D. H. Templeton, private communication) that the published positional parameters correspond very closely to symmetry Pnam; no parameter has to be shifted more than 10 standard deviations and only one by more than 5 standard deviations to make the correspondence exact.

Detailed considerations of the higher space group have not yet been published.

L-Tryptophane

Powder photographs of L-tryptophane have been indexed, and the space group Pmmm was tentatively assigned to this crystal.[28] When it was pointed out that this is an impossible space group for this compound,[29] space group P222 was assigned instead. Although this space group is not impossible, it is a highly improbable one for a molecular crystal. Only three such crystals are listed as having this space group by Nowacki[45] and, of these, one assignment is in error and the other two are dubious.

LIGHT ATOMS-HEAVY ATOMS

It has been rather fashionable in recent years to locate hydrogen atoms in organic crystals, usually by the use of difference maps. Save for a few notable exceptions, the locations of the hydrogen atoms are usually known, so that no new information of significance is obtained, only a further reduction of the *R* value. Because of the relatively low x-ray scattering power of hydrogen atoms the errors in their positions are always relatively large.

Cyclopentadiene

During a perfectly straightforward least squares refinement of the structure of C_5H_6 one of the hydrogen atoms moved from its initial position to somewhere near the center of the ring.[52] The authors, not unnaturally, considered this ridiculous and decided "to return it forcibly to its logical position and to remove it from the refinement." No error was thus introduced into the literature.

Dihydridodi-π-cyclopentadienylmolybdenum

The structure of the above compound, $(C_5H_5)_2MoH_2$, was refined in the usual way.[53] Peaks in the difference maps near the molybdenum atom were assigned to hydrogen atoms, with Mo–H bond lengths of 1.2 ± 0.3 Å. The discussion of the bonding in this molecule was based in part on the position of these hydrogen atoms.

Reassessment of these data[54] included a least squares refinement in which the crucial hydrogen moved to a site 6.0 Å from the metal atom to which it was supposedly bonded. On the basis of this and other refinements it was concluded that there is "no significant evidence that the scattering data either contain a component due to H(6) or can distinguish between H(6) at widely different positions in the unit cell."

Another part of the discussion of the bonding[53] was based on differences in the observed C–C bond lengths. However, the re-assessment[54] led to the conclusion that these variations were not significant.

Decammine-μ-peroxo-dicobalt Pentanitrate

This compound contains the cation $(H_3N)_5CoO_2Co(NH_3)_5$, and in the first structure determination it was reported that the arrangement of the bridging group was symmetrical:

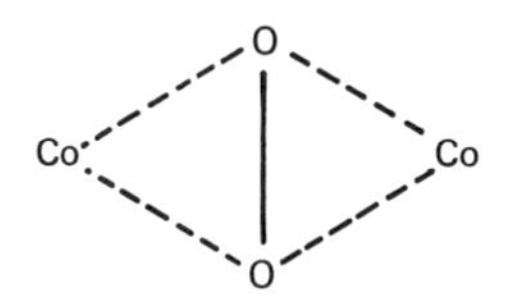

This arrangement was not found in the corresponding sulfate tris (bisulfate) salt, where a skewed arrangement was observed[56]:

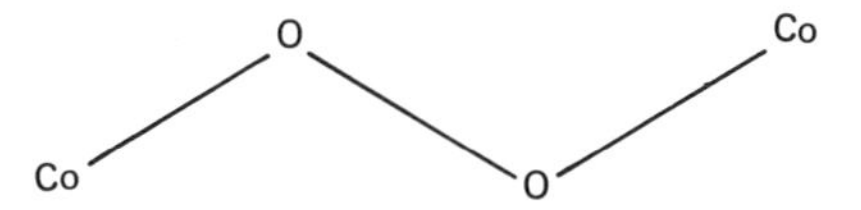

A new set of data was collected and refined.[57] Although the crystals suffer from severe disorder, it is clear that the skewed arrangement is correct, with the interatomic distances showing that each cobalt atom is bonded to only one of the oxygen atoms and that the bridging group is a superoxide radical, rather than a peroxide ion.

Potassium Fluorobromite

Powder data from $KBrF_4$ were interpreted[58] to show that the structure is based on space group D_{4h}^{18}-I4/mcm, the unit cell containing 4K in position 4*c*, 4 Br in position 4*b*, and 16 F in position 16*l*, leading to tetrahedral BrF_4^- ions. Because structural theory, involving use of octahedral sp^3d^2 orbitals by the bromine atom, predicts a square planar configuration for this ion, the data were re-examined, and an alternate structure with the potassium atoms in position 4*a* and the bromine atoms in position 4*d* was proposed.[59] Unfortunately, this drastic change in structure affects only those reflections with *l* odd, and but six of these were reported, of which two are obscured by impurities, two are poorly resolved, and one only approximately measured. The revised structure is thus essentially based on one line. A new structure determination was said to have been initiated[60] in 1957, but, so far as I am aware, no additional publication has appeared.

Silver Chlorite

A structure for $AgClO_2$ included the unusual feature of Ag–Cl bonds of length 2.20 Å, so that the crystals were said to consist of $AgClO_2$ molecules rather than Ag^+ and ClO_2^- ions.[61] The space group was given as Cmma, with four molecules in the unit cell, as determined from Weissenberg oscillation photographs.

The unusual feature, plus several other structural peculiarities, led to the collection of a new set of single crystal data.[62] Numerous faint reflections showed that the cell is not centered and that the true space group is Pcca. The new structure confirms the expected ionic character of this salt and contains no surprising features.

ELECTRON DENSITIES, HANDSOME AND OTHERWISE

In some quarters the refinement of structures has been accomplished solely by least squares, with the electron density functions sometimes being calculated only for decorative purposes. Alternately, it has not been unusual for a refinement to be carried out solely by the calculation of electron densities, with the phases having been determined by one or more of the standard methods. Discussed below are some examples of what can go wrong.

D (–)-*Isoleucine Hydrochloride Monohydrate*

```
        NH2  CH3
        |    |
HOOC–CH–CH–CH2–CH3
```

The amino acid isoleucine was investigated at a time when[63] the relative configuration at the two asymmetric carbon atoms was not known. Phase determination was facilitated by a parallel study of the isotypic hydrobromide. The *a*-axis projection of the first trial structure is shown in Figure 2a. The visual data used to calculate this projection refined to an R of 23 percent. A difference map calculated at this point revealed that one of the methyl groups must be displaced, and least squares re-

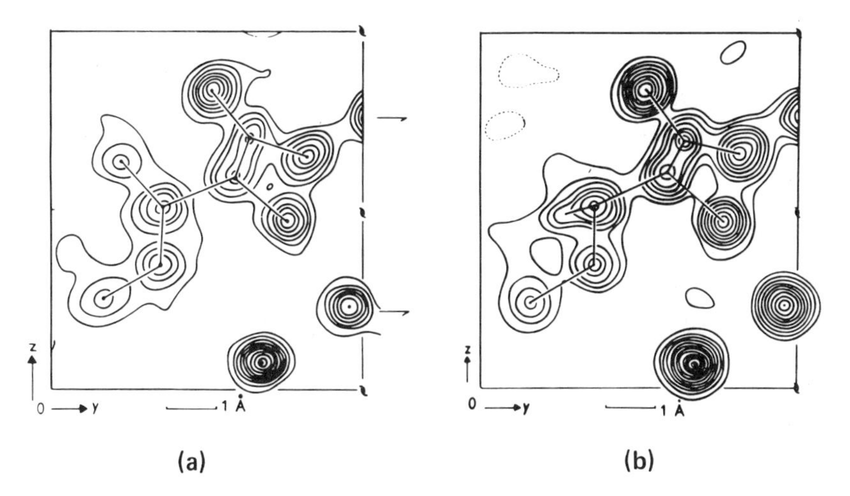

FIGURE 2 Electron density of D-isoleucine hydrochloride monohydrate projected down the *a*-axis. (a) Phases calculated with the incorrect structure; (b) phases calculated with the correct structure. From Trommel and Bijvoet.[63]

finement was continued, the value of *R* dropping to 18 percent. The new *a*-axis projection (Figure 2b) shows an inverted configuration at the β-carbon atom, and this was confirmed by the *c*-axis projection. A total of 120 *F* values was used in preparing the two projections of Figure 2; in going from the incorrect to the correct structure, only eight of them changed sign, and of these only one is of medium intensity and the rest are medium-weak (J. Trommel, personal communication).

Iso-*iridomyrmecin*

The cyclopentanoid lactone ($C_{10}H_{16}O_2$) was first investigated with the crystal at room temperature.[64] The crystal has a relatively low melting point of 58 °C, and data could be collected only out as far as 1.1 Å. A promising trial structure in projection down the *b*-axis was obtained and refined to an *R* of 20 percent by successive electron density projections and difference maps. The final projection is shown in Figure 3a. This projection, however, appears to be in accord with a chair conformation of the six-membered ring, a result that was found difficult to reconcile with the expected planarity of the lactone group. Instead of publishing a note announcing the discovery of an unusual lactone structure, the authors then collected data from a crystal at −130 °C. A total of 192 *F* values was observed out to spacings of about 0.8 Å, compared with the 72 observed previously. It was found that the structure would not refine to an *R* below 29 percent. Gross errors in the

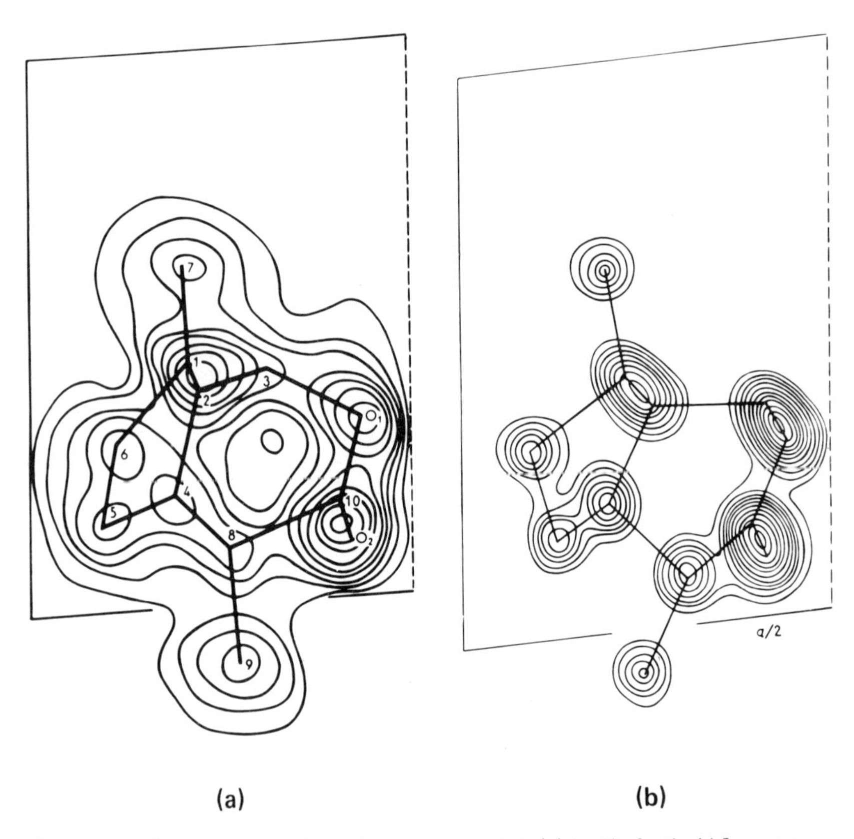

FIGURE 3 Electron density of *iso*-iridomyrmecin projected down the *b*-axis. (a) Incorrect structure based on low-resolution room temperature data; (b) correct structure, based on high-resolution, low-temperature data. From Schoenborn and McConnell.[64]

positions of some of the atoms were discovered by use of a difference map, and after corrections for these the structure refined readily to an R of 18 percent. The correctness of this structure was verified by refinement of three-dimensional data to an R of 14 percent. The final *b*-axis projection is shown in Figure 3b. Some of the atoms moved by as much as 1 Å, and it is interesting that the difference maps prepared with the low resolution room temperature data did not reveal the errors in the trial structure.

Bi-dioxacyclopentyl

On chemical grounds a certain substance having empirical formula $C_6H_{10}O_4$ was thought to be a naphthadioxane and a trial structure was

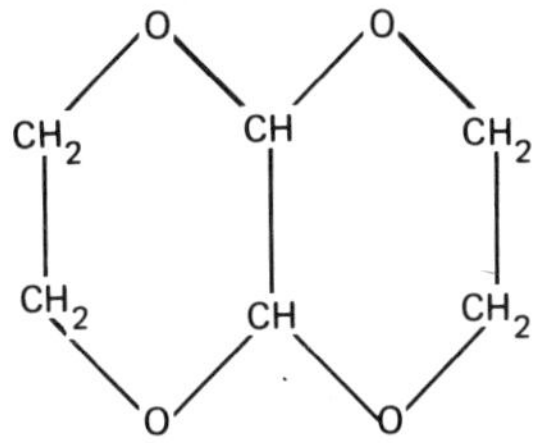

proposed on that basis.[65] The corresponding electron density map, however, has several objectionable features, even though, according to the authors, the value of R of 38 percent is low enough to indicate that the structure might be "essentially correct." These features (Figure 4a) are small spurious peaks and distortions and projected C–O bond lengths of 1.80 Å. The bi-dioxacyclopentyl structure was then tried. This gave

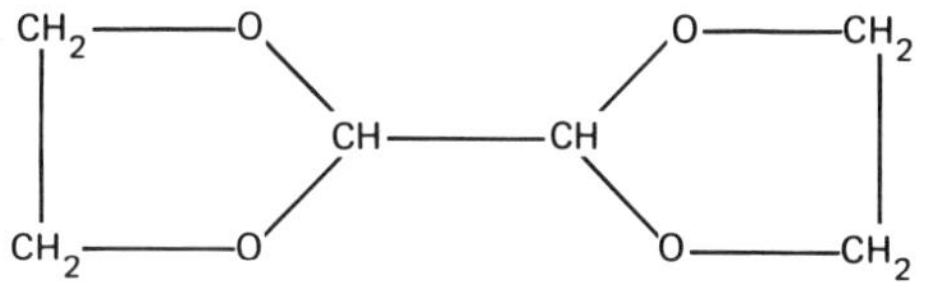

a much cleaner electron density projection (Figure 4b) and eventually refined to an R of 13 percent.

Azulene

Two independent investigations[66,67] agreed that azulene has an ordered structure, two molecules per unit cell in space group Pa. Although the diffraction data are also consistent with space group $P2_1/a$ this would require either a center of symmetry for the molecules (impossible) or

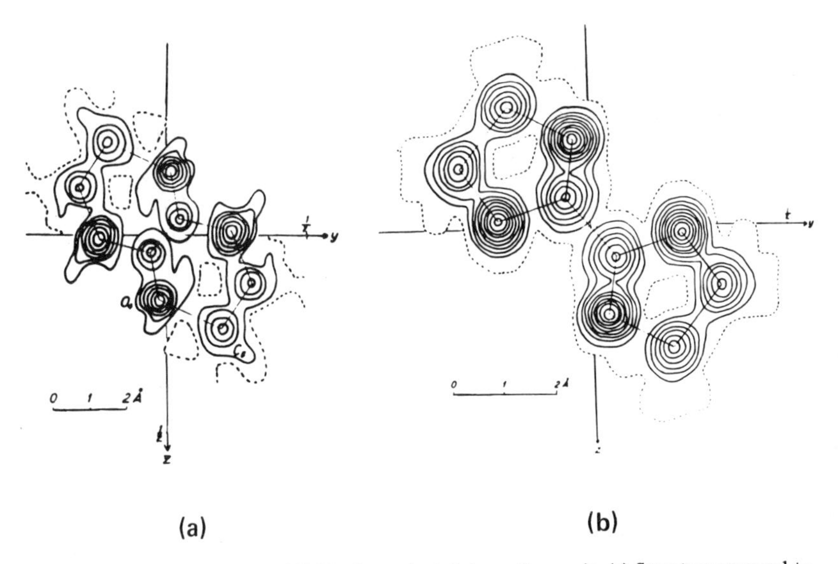

FIGURE 4 Electron density of $C_6H_{10}O_4$ projected down the *a*-axis. (a) Structure assumed to be naphthadioxane (incorrect); (b) structure assumed to be bi-1, 3-dioxacyclopentyl (correct). From Furberg and Hassel.[65]

disorder (said to be ruled out by intensity statistics). The ordered structure, in projection down the *b*-axis, is shown in Figure 5a and the corresponding electron density function in Figure 6a. This structure refined to an *R* of 11 percent.

Following a report[68] that the entropy provided support for the disordered structure, together with the inability to refine the *R* value for a full set of three-dimensional data below 22 percent, the structure was successfully refined with the disordered structure to an *R* of 6.5 percent.[69] The disordered structure is shown in Figure 5b and the corresponding electron density function in Figure 6b.

2-Phenylazulene

The situation with regard to 2-phenylazulene resembles somewhat that of azulene itself. The structure was first thought to be disordered,[70] as had been shown for azulene, and refined in projection to an *R* of 17 percent; however, later calculations based on the ordered model[71] reduced the *R* value to 14 percent. The two structures and the corresponding electron density projections are shown in Figure 7. At present, it is not known which model is correct, but it is certain that at least one of these electron density maps is incorrect.

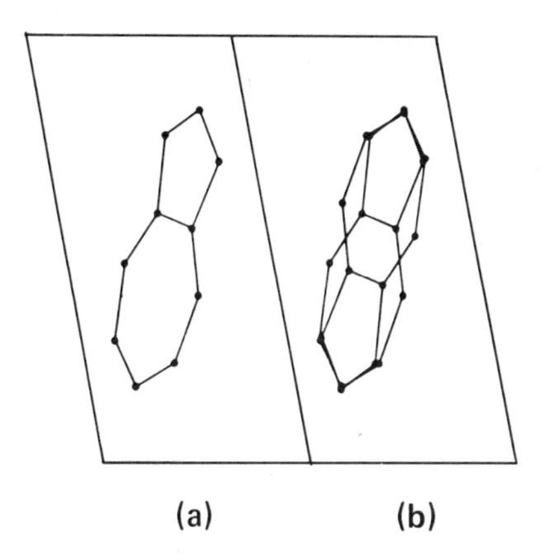

FIGURE 5 The ordered (a) and disordered (b) structures for azulene, viewed along the *b*-axis.

CONCLUSION

Following the presentation of this paper, I had numerous conversations with other participants in the conference. Various criteria, listed below, were suggested as indicating that a crystal structure, as reported in the literature, may be suspect. (These are independent, any one of them being sufficient.) No attempt is made to separate possible errors regarding the gross structure from possible errors in details of an essentially correct structure.

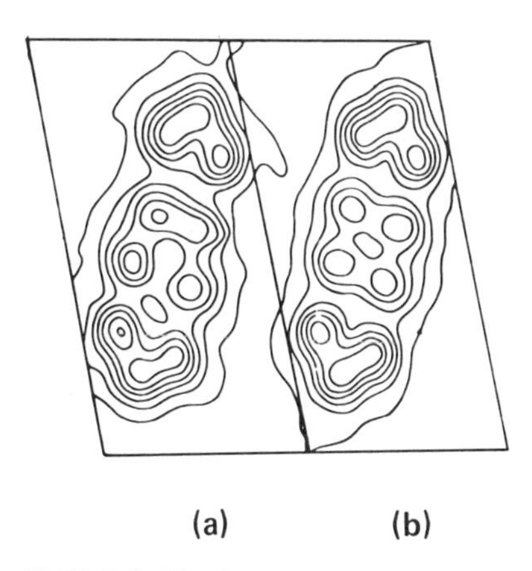

FIGURE 6 The electron density projections corresponding to Figure 5: (a) ordered, (b) disordered. From Robertson *et al.*[69]

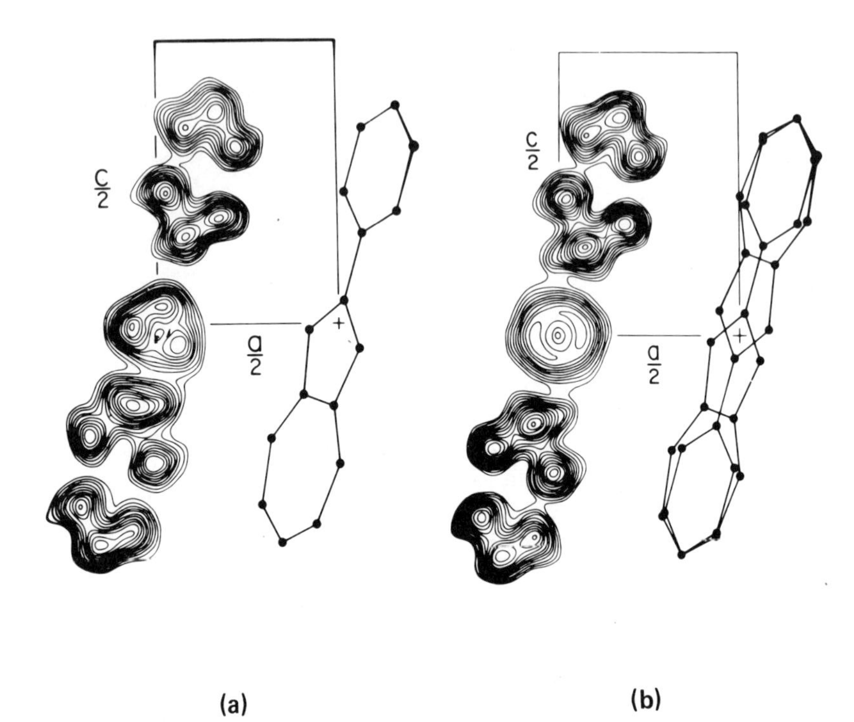

FIGURE 7 Electron density functions and structures of 2-phenylazulene viewed along the *b*-axis: (a) ordered, (b) disordered. From Donohue and Sharma.[71]

1. The work was done more than 10 years ago.

2. The data were obtained from powders or fibers instead of single crystals.

3. Only a partial data set, for example, the three prism zones, was collected, as opposed to a complete three-dimensional set.

4. The data were not collected to the limit of CuK_α radiation.

5. Systematic errors are present in the data, but not allowed for in the refinement.

6. The average percent difference between the observed and calculated structure factors (the R value) is greater than 10 percent.

7. There are several, say a dozen, individual major discrepancies between observed and calculated structure factors.

8. Values of the structure factors were not calculated for the accidentally absent reflections.

9. The observed and calculated structure factors, if not published, are not made available in accord with instructions in a footnote.

10. A final three-dimensional electron density function was not examined for inconsistent or suspicious features.

11. Assurance is not made that a least squares refinement is not hung up in a false minimum.
12. A space group ambiguity is resolved by intensity statistics alone.

Needless to say, if the above were strictly applied, a large number of interesting and important structures would be thrown open to suspicion. I might add that I do not, personally, agree with all of them, as some are certainly unnecessarily strict. However, it has also occurred to me that a view different from that of Henry Ford has been expressed by George Santayana: "Those who do not remember the past are condemned to relive it."

REFERENCES

1. W. L. Bragg, *Proc. R. Soc.*, **89A**, 248 (1914).
2. H. Mark and K. Weissberg, *Z. Phys.*, **17**, 301 (1923).
3. M. L. Huggins and S. B. Hendricks, *J. Am. Chem. Soc.*, **48**, 164 (1926).
4. F. J. Llewellyn, E. G. Cox, and T. H. Goodwin, *J. Chem. Soc.*, 883 (1937).
5. R. Shiono, D. W. J. Cruickshank, and E. G. Cox, *Acta Cryst.*, **11**, 389 (1958).
6. M. A. Rollier, S. B. Hendricks, and L. R. Maxwell, *J. Chem. Phys.*, **4**, 648 (1936).
7. W. H. Beamer and C. R. Maxwell, *J. Chem. Phys.*, **14**, 569 (1946).
8. J. M. Goode, *J. Chem. Phys.*, **27**, 1269 (1957).
9. J. M. Goode, "A redetermination of the lattice parameters of polonium." U.S. Atomic Energy Commission Report MLM-808 (1953).
10. K. H. Meyer and Y. Go, *Helv. Chim. Acta*, **17**, 1081 (1934).
11. L. Pauling, *Proc. Natl. Acad. Sci., U.S.A.*, **34**, 495 (1949).
12. J. A. Prins, J. Schenk, and P. A. M. Hospel, *Physica*, **22**, 770 (1956).
13. J. A. Prins, J. Schenk, and L. H. J. Wachters, *Physica*, **23**, 746 (1957).
14. F. Tuinstra, *Physica*, **34**, 113 (1967).
15. M. D. Lind and S. Geller, *J. Chem. Phys.*, **51**, 348 (1969).
16. N. Alsen and G. Aminoff, *Geol. Foren. Stockh. Forh.*, **41**, 124 (1922).
17. L. W. McKeehan and P. P. Cioffi, *Phys. Rev.*, **19**, 444 (1922).
18. R. W. G. Wyckoff, *The Structures of Crystals*, 2nd ed., p. 209. Chemical Catalog Co. Inc., New York (1931). This book also contains references to the other investigations mentioned in the previous sentence.
19. L. M. Kirkpatrick and L. Pauling, *Z. Krist.*, **63**, 502 (1926).
20. B. Gossner and O. Kraus, *Z. Krist.*, **88**, 298 (1934).
21. L. Pauling, *Z. Krist.*, **91**, 367 (1935).
22. C. L. Burdick and J. H. Ellis, *Proc. Natl. Acad. Sci., U.S.A.*, **3**, 644 (1917).
23. L. Pauling and L. O. Brockway, *Z. Krist.*, **82**, 188 (1932).
24. A. J. Bijl and N. H. Kolkmeijer, *Chem. Weekbl.*, **15**, 1077, 1264 (1918); *idem*, *Proc. Acad. Amst.*, **21**, 494 (1919).
25. H. Mark and M. Polanyi, *Z. Phys.*, **18**, 75 (1923); H. Mark, M. Polanyi, and E. Schmid, *Naturwiss.*, **11**, 256 (1923).
26. K. M. Lester and H. Lipson, *J. Appl. Cryst.*, **3**, 92 (1970).
27. P. M. de Wolff, *J. Appl. Cryst.*, **1**, 108 (1968).
28. B. Khawas and G. S. R. Krishna Murti, *Acta Cryst.*, **B25**, 1006 (1969).

29. B. Khawas and G. S. R. Krishna Murti, *Acta Cryst.*, **B25**, 2663 (1969).
30. R. Boggs and J. Donohue, *Acta Cryst.*, **B27**, 247 (1971).
31. A. Mostad, H. M. Nissen, and C. Rømming, *Tetrahedron Lett.* No. 24, 2131 (1971); M. N. Frey, T. F. Koetzle, M. S. Lehmann, and W. C. Hamilton, *J. Chem. Phys.*, **58**, 2547 (1973).
32. F. M. Jaeger, P. Terpstra, and H. G. Westenbrink, *Proc. Acad. Sci. Amst.*, **29**, 1193 (1926).
33. F. Laves, *Naturwiss.*, **20**, 472 (1932).
34. C. S. Barrett and F. J. Spooner, *Nature*, **207**, 1382 (1965).
35. A. T. Jensen, *K. Danske Vidensk. Sel'sk.*, **20**, No. 5 (1942).
36. A. J. C. Wilson, ed., *Struct. Rep.*, **9**, 161 (1955).
37. A. W. Hanson, *Acta Cryst.*, **15**, 930 (1962).
38. J. Donohue and A. Caron, *Acta Cryst.*, **17**, 663 (1964).
39. J. D. Lee and M. W. R. Bryant, *Acta Cryst.*, **B25**, 2497 (1969).
40. B. van Dijk and G. J. Visser, *Acta Cryst.*, **B27**, 846 (1971).
41. H. Einspahr and J. Donohue, *Acta Cryst.*, **B27**, 846 (1971).
42. J. D. Lee, *Acta Cryst.*, **B27**, 847 (1971).
43. R. Srinivasan and B. K. Vijayalakshmi, *Acta Cryst.*, **B28**, 2615 (1972).
44. R. Pascard, *C.R. Acad. Sci. Paris*, **240**, 2162 (1955).
45. W. Nowacki, *Crystal Data*, Williams and Heintz Map Corp., Washington, D.C. (1967).
46. K. Olie and F. C. Mijlhoff, *Acta Cryst.*, **B25**, 974 (1969).
47. L. K. Templeton and D. Templeton, *Acta Cryst.*, **B27**, 1678 (1971).
48. K. Olie, *Acta Cryst.*, **B27**, 1679 (1971).
49. J. Donohue, *Acta Cryst.*, **B27**, 1071 (1971).
50. K. Olie, *Acta Cryst.*, **B27**, 1459 (1971).
51. L. Kihlborg, R. Norrestam, and B. Olivecrona, *Acta Cryst.*, **B27**, 2066 (1971).
52. L. Liebling and R. E. Marsh, *Acta Cryst.*, **19**, 202 (1965).
53. M. Gerloch and R. Mason, *J. Chem. Soc.*, 296 (1965).
54. S. C. Abrahams and A. P. Ginsberg, *Inorg. Chem.* **5**, 500 (1966).
55. N.-G. Vannerberg and C. Brosset, *Acta Cryst.*, **16**, 247 (1963).
56. W. P. Schaefer and R. E. Marsh, *Acta Cryst.*, **21**, 735 (1966).
57. R. E. Marsh and W. P. Schaefer, *Acta Cryst.*, **B24**, 246 (1968).
58. S. Siegel, *Acta Cryst.*, **9**, 493 (1956).
59. W. G. Sly and R. E. Marsh, *Acta Cryst.*, **10**, 378 (1957).
60. S. Siegel, *Acta Cryst.*, **10**, 380 (1957).
61. R. Curti, V. Riganti, and S. Locchi, *Acta Cryst.*, **10**, 687 (1957).
62. J. Cooper and R. E. Marsh, *Acta Cryst.*, **14**, 202 (1961).
63. J. Trommel and J. M. Bijvoet, *Acta Cryst.*, **7**, 703 (1954).
64. B. P. Schoenborn and J. F. McConnell, *Acta Cryst.*, **15**, 779 (1962).
65. S. Furberg and O. Hassel, *Acta Chem. Scand.*, **4**, 1584 (1950).
66. J. M. Robertson and H. M. M. Shearer, *Nature*, **177**, 885 (1956).
67. Y. Takeuchi and R. Pepinsky, *Science*, **124**, 126 (1956).
68. H. H. Günthard, *Helv. Chim. Acta*, **38**, 1918 (1955).
69. J. M. Robertson, H. M. M. Shearer, G. A. Sim, and D. G. Watson, *Nature*, **182**, 177 (1958); *idem, Acta Cryst.*, **15**, 1 (1962).
70. B. D. Sharma and J. Donohue, *Nature*, **192**, 863 (1961).
71. J. Donohue and B. D. Sharma, *Nature*, **198**, 878 (1963).
72. M. C. Neuberger, *Z. Anorg. Allg. Chem.*, **212**, 40 (1933).

Discussion Leader: JACK M. WILLIAMS

Reporters: DAVID J. DUCHAMP
ROBERT K. BOHN

Discussion

DISCUSSION OF PAPER BY SNYDER AND MEIBOOM

WILLIAMS pointed out that in NMR vibrational effects cause a lengthening of C-H bonds, whereas in diffraction methods C-H bonds appear to be shortened. SNYDER agreed and pointed out that this was due to the dependence of D_{ij} on θ during a bending that involves a C-H bond. He noted that C-H bonds appear about 0.04 Å too long, relative to C-C distances, but that the former can be corrected to a precision of 0.01 Å in the relative distance. However, an increase in precision beyond this is at present relatively difficult.

MULLER asked if the assignment of the signs of the D_{ij}'s has caused problems. SNYDER responded that this had not been a problem because the sign depended on the orientation of the molecule with respect to the magnetic field. The D_{ij} signs are negative when the coupled pair is aligned with the magnetic field and positive when perpendicular to it.

HERBSTEIN asked if the method would work if a benzene molecule, for example, were contained in a clathrate crystal. SNYDER replied that, if the molecule migrates through the clathrate relatively rapidly so that it sees an average magnetic environment, then one might indeed be able to obtain a spectrum with very large dipole interactions and quite a bit of structure. But if it tends to stay put in the clathrate site for 0.001 s or so, one would observe the typical broad-band spectrum characteristic of a solid.

LIDE asked if elongated C-F bonds were also observed. SNYDER pointed out that in CH_3F an apparently normal C-F bond length, relative to C-H and H · · · H distances, was observed. Experimental results, as well as theoretical calculations, suggest that pseudodipolar interactions between protons and first-row elements are small relative to dipole–dipole couplings and they do not interfere with determinations of structure. However, pseudodipolar couplings are somewhat larger between first-row atoms, especially F · · · F and perhaps C · · · F, and could cause errors of 1–3 percent in bond lengths.

BUCKINGHAM pointed out that the separation of the dipolar coupling constants and the motional constants depends on the assumption that the molecule orients as a rigid body. In reality, there is coupling between the internal vibrations and the molecule's orientation in the anisotropic environment. He asked if this is a serious difficulty. SNYDER mentioned his experience with tetramethylsilane where couplings up to 7 Hz could be interpreted in terms of H–C–H angle changes of the order of 0.1°. He indicated that such couplings so far have corresponded to very small distortions because the coupled nuclei are bound to the same atom and are thus very close to one another. In summary, the coupling has not presented a problem in structure determination so far. However, it is something to be kept in mind, particularly for less rigid molecules.

DISCUSSION OF PAPERS BY SHOEMAKER, IBERS, AND DONOHUE

HOPE asked how the azulene trial structures were obtained, after stating that direct methods might have led to the disordered structure automatically. DONOHUE replied that the trial structures were obtained from packing considerations and a known center of symmetry. Since there were only 70 reflections for phenylazulene, direct methods were not feasible.

LIDE asked about cases where it was not possible to distinguish among a range of structures or space groups. Do authors make this clear in their papers? DONOHUE estimated that 1 percent of authors make this clear. IBERS added a standard deviation of 2 percent to Donohue's estimate. ABRAHAMS also responded that the problem of choosing between centro- and noncentrosymmetric point groups continues, somewhat unnecessarily, to plague current crystal structure determination. Although it is well known that x-ray intensity statistics, the full Patterson function, and pyro- and piezoelectric tests are often hard to use unambiguously, it does not seem to be widely recognized that detection of second harmonic generation in a laser beam is a very sensitive and definitive technique.* Furthermore, a commercial device suitable for routine use with powder specimens may reach the market soon.

ABRAHAMS added that if the atomic arrangement in a crystal is disordered, the crystal will give a diffractometer pattern that most likely contains diffuse scattering. He felt that, when an author postulates disorder, the burden of proof should rest with the author to demonstrate by additional experiment, such as the observation of diffuse scattering, that his model does indeed represent physical reality. DONOHUE pointed out that microscopic twinning might cause apparent disorder without producing diffuse scattering.

RUSH asked how seriously noncrystallographers should take hydrogen atom

* See S. C. Abrahams, *J. Appl. Cryst.*, **5**, 143 (1972).

positions as determined by the x-ray method. IBERS noted that in light-atom structures, and even in some first- and second-row transition-metal compounds, hydrogen atom positions can often be determined reasonably. DONOHUE was less optimistic. IBERS felt that, even though standard deviations are often unusually high (up to 0.1 Å), this information is superior to knowing nothing about hydrogen atom location.*

WILLIAMS pointed out that even in the case of neutron diffraction, hydrogen atom positions may not be well determined because of large thermal motion. In p-$CH_3C_6H_4NH_3^+$ $(HF_2)^-$, the apparent methyl C–H distances range from 0.92 Å to 0.96 Å; these short values are artifacts caused by large-amplitude torsional motion of the methyl group. Furthermore, it is very difficult to correct reliably for such effects.

The statement was made in an earlier session that the thermal parameters determined by crystal structure diffraction analysis should be considered virtually meaningless, in contrast to positional parameters for which estimated standard errors should be multiplied by a factor of five. LEVY commented as follows:

"This may have been the situation 20 years ago. Today it is pessimistic to a totally unwarranted degree. When based on carefully collected data and careful analysis, it frequently appears that the derived thermal parameters make good sense, in that the ellipsoids correlate well in orientation and magnitude with the chemical structure. Furthermore, there are cases in which x-ray and neutron data yield virtually identical thermal parameters. Some degree of caution in quantitative assessment of accuracy is nevertheless advisable. The reason is that the thermal parameters are sensitive to any systematic errors correlated with scattering angle, and such errors are quite difficult to assess.

"In evaluating the meaning of the measures of error given by crystallographers for interatomic distances, it is important to bear in mind the physical meaning of the quantity that is estimated.

"The positional parameters derived in the usual crystal structure determination represent the coordinates of the first moments of the distribution of instantaneous atomic centers as produced by zero point and thermal displacements from an equilibrium configuration. The distance between pairs of first-moment positions, together with a measure of precision derived in a straightforward way from the estimated standard errors of the parameters, constitutes a "raw" distance, and its precision measure is as reliable as are those of the positional parameters. This is the quantity usually quoted by crystallographers as raw or uncorrected interatomic distance.

"The raw distance, however, is never of rigorous significance outside the crystal in which it is determined. For comparisons between crystals, or between crystal and free molecule, one needs to estimate preferably the equilibrium distance, or failing

* Discussion group note: X-ray determinations of hydrogen-atom positions yield distances that are short (even in absence of large thermal motion) because of the spherical approximation of the hydrogen scattering factor. In general, x-ray results yield the correct interatomic vector between a hydrogen atom and the atom to which it is bonded, but the interatomic magnitude is usually short by ~10–20 percent. However, it is well known that omission of hydrogen atoms in the calculation of structure factors may lead to errors in the positions of nearby atoms.

that, the mean separation. These quantities, it can be shown, are never less than the raw distance and usually exceed it by increments of up to several times the precision measures for carefully determined structures.

"To estimate the magnitude of a correction for thermal motion, one needs—in addition to the mean square displacements of the pair of atoms, which are provided by crystal structure determination—a description of the manner in which the displacements are correlated with each other. The correlation is not available from crystal structive analysis, but may sometimes be deduced in reasonable approximation on physical grounds.

"Probably the most satisfactory model for estimation of bond distance corrections, when applicable, is the rigid body model. In this instance, the observed thermal displacements are interpreted as arising from the translations and librations of a rigid group, and the rotational amplitudes are then used to estimate the corrections. Elaborations of the method may include corrections for internal motion (estimated from spectroscopic frequencies on free molecules), and segmented body motions in more flexible molecules. A simple special case is the "riding model." In this instance one atom is considered to ride on another; it is a useful approximation when a light atom is bonded only to a considerably heavier one, as, for example, in –X–H bonds.

"In any case, the principal uncertainty lies in the incompleteness of the model used to describe the thermal motion. Consequently, the accuracy of a corrected distance is always less than the precision of the raw distance. The best accuracy is achieved when the crystal is hard and the temperature is low, so that only zero point motion obtains and the thermal displacements are minimal."

SUTTON inquired about how far it is possible *at present* to make significant comparisons between interatomic distances observed in the gas phase and in the crystal phase. Chemists need to do this to find out whether environment changes the structure of a molecule. "Significant" is an arbitrary word, but may be defined as a difference of more than 0.005 Å. HEDBERG seemed to best summarize the discussion that followed. He stated that there had been no concerted attempt known to him to compare solid-state and gas-phase results, but that there are a number of examples where differences are large. It was agreed that such comparative studies should certainly be made. TUKEY's feeling was that it was inappropriate until we first examined the reliability of the results obtained by each of the various methods. Once we can estimate standard deviations that include the systematic errors properly, then we can have a basis for informed discussion.

Concerning the "warm happy feeling," with respect to space group selection, MOSER noted that it is easy to account for extra reflections in retrospect. He asked how such reflections are dealt with when encountered, i.e., wrong space group or multiple reflection. SHOEMAKER suggested that, if only one such reflection is encountered, it could probably be ignored; however, the presence of more than two or three should be taken seriously. IBERS further suggested that, if symmetry absent reflections are observed, then they should be remeasured and their presence verified. TUKEY suggested that *all* reflections should be included in least squares refinement, regardless of their measured magnitude. It is wrong to omit any observation, even if that observation must be treated in a special way.

TUKEY further questioned why ellipsoids were used to represent thermal motion when chemical intuition suggested other shapes, e.g., "kidneys." JOHNSON noted that higher order terms are available, for example, the cumulant model.

CAUGHLAN asked whether obtaining good probability plots ensured that systematic errors were absent. TUKEY responded that it did not, but that the plots were considerably better than no test at all.

In discussion about significance testing, in particular Hamilton's $\mathcal{R}$ test, TUKEY said it was helpful especially in cases where many parameters were involved. SCHOMAKER expressed the following reservations:

"I do not believe that the $\mathcal{R}$ test is appropriate for many or most of the situations to which it is applied. This is not to say anything at all about the particular choice of the function $\mathcal{R}$ to be tested or to deprecate the use of a relevant (but not necessarily accurate) comprehensive statistic. Rather, I would argue that the language ('reject,' 'accept,' 'significant') and the statistical scheme used in the $\mathcal{R}$ test, viz., Hypothesis Testing (HT), have the connotations that are apt to mislead by giving the impression that the judgment reached in the test is a final judgment about the physical situation in question, even in the face of compelling reasons for suspending or even reversing the judgment.

"Consider first a case for which the HT scheme seems appropriate, although the language may very well mislead: You have an optically active natural product, newly isolated and chemically characterized but apparently involved in many a well-known biochemical mechanism. Your Friedel pairs suggest that the optical configuration is the opposite of the configuration that has been convincingly established by the chemists as agreeing with all biochemical expectation. The null hypothesis (H_0)—that the chemists are right—has high *a priori* probability ($\frac{1}{2}$?, 0.999?) and merits serious consideration in every respect. You want to know whether you have rendered it highly improbable or whether, instead, you need to get more evidence, besides checking every little detail, before you start talking too much. [Bijvoet at first privately communicated a similarly startling result about his wonderful pioneer tartaric acid determination; a little later, describing how well the work had gone to completion, he again wrote (to Hughes) that a 'trivial error of sign' had been discovered in the original treatment of the phase angles.] If you choose to use the $\mathcal{R}$ test, you may find a situation like the one delineated by Hamilton,* such that by careful choice of a critical value of $\mathcal{R}$ the chances of falsely rejecting H_0 (type I error) and of falsely rejecting the alternative (type II error) can both be made small. If so and if your $\mathcal{R}$ is greater than that critical value, you may feel statistically justified in publishing your unexpected configuration result.

"The situation is different, however, for the many applications similar to the question of deviation of the 'thermal' ellipsoids from simple spheres that was described briefly here by SHOEMAKER. In the absence of cubic local symmetry, there is now no physical basis for a presumption in favor of the null hypothesis (that the ellipsoids are precisely spherical), and to pose that it is, I contend, is inappropri-

* W. C. Hamilton, *Statistics in Physical Science,* Ronald Press, New York (1964), p. 49.

ate, frivolous, even downright wrong. We already *know* that H_0 is false, but if I proceed to test it in the usual way I engage in an exercise apt to lead me to a formal 'rejection' of the alternative, despite its overwhelmingly greater *a priori* probability. How? Because now the distribution of $\mathcal{R}$ values to be expected on the basis of H_0 (the *B*'s are anisotropic) will not be much different (my data do not lead to very precise estimates of the anisotropic components, and for the type of crystal in question the range of values to be expected is not very great), so that when, with great (ironically false) conservatism, I choose the critical $\mathcal{R}$ value so as to have a conventional low probability (say 0.05) of erroneously rejecting H_0, I automatically establish a high probability (0.95?, 0.9?, 0.4?) of erroneously rejecting the alternative."

IV

Protein Structure

L. H. Jensen

Critical Evaluation of Protein Crystallographic Methods

The methods of protein crystallography are subject to the same limitations and errors as are encountered with small structures. Because other factors may be dominant when dealing with proteins, some of these errors become inconsequential. However, there are errors associated with the special techniques that have been developed in protein crystallography, and these must be considered as well as the special properties of the crystals. To evaluate the methods and gain some idea of how errors affect the confidence that can be attached to the results, I will first treat briefly the nature of protein crystals and diffraction from them and then consider some of the special experimental and computational techniques.

INTENSITY DATA

Proteins are relatively large molecules with minimum molecular weights of about 6000. They crystallize with a large amount of solvent, often 40 percent or more, which is necessary to maintain the integrity of the crystals. If the solvent evaporates the crystals lose their order.

Because of the size of protein molecules and the solvent that crystallizes with them, the unit cells of protein crystals are large, ranging in volume from about 25 000 Å^3 to values greater than 10^6 Å^3. It can be shown that for structures with similar atom types (e.g., proteins with mostly first-row atoms) intensities from crystals of a given size are approximately proportional to $1/V$ where V is the unit cell volume. Thus for crystals with large unit cells, intensities are relatively less than for smaller structures and the proportion of weak reflections with poor signal-to-noise ratios is greater. In addition, intensities of the reflections decrease rapidly with increasing diffraction angle.

To maximize intensities, quite large crystals are desirable, usually with dimensions in the range 0.3–1.0 mm. For crystals in this size range and with the relatively long wavelength radiation that must be used, usually CuK_α, absorption can be serious. Because of the way protein crystals are mounted in capillaries (Figure 1), it is not possible to calculate an absorption correction with adequate accuracy. For data collected on a four-circle diffractometer, however, an approximate correction for absorption can readily be made by measuring the intensity of a particular reflection at $\chi = 90°$ as a function of ϕ. Figure 2 is an example of such an absorption curve for crystals of the protein rubredoxin. For the rhombus-shaped crystals of the native protein with dimensions approximately $0.3 \times 0.6 \times 0.6$ mm^3, differences in transmission were as large as a factor of 1.5, and for the somewhat larger HgI_4^{2-} derivative crystals,

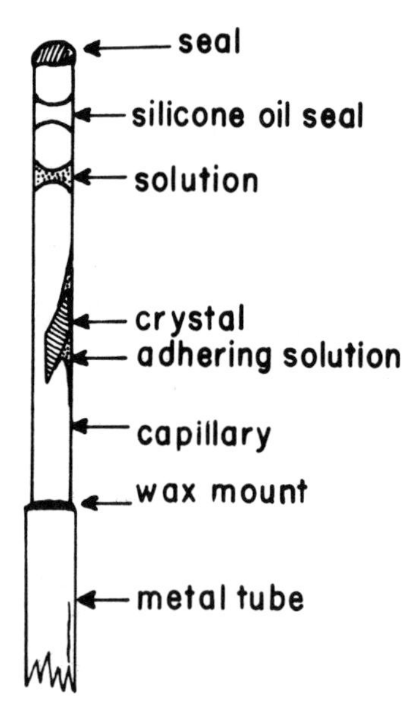

FIGURE 1 Mounting of protein crystal in capillary.

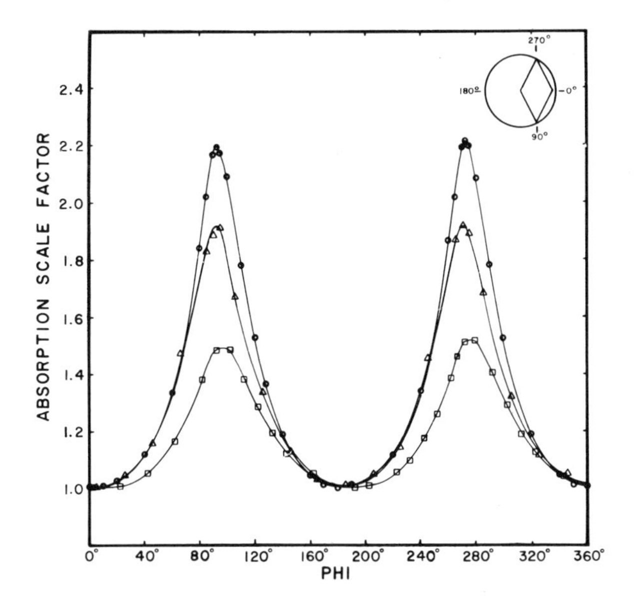

FIGURE 2 Plot of relative value of $1/T$ (absorption scale factor) as a function of ϕ at $\chi = 90°$ for native rubredoxin crystal and two derivative crystals. From Herriott *et al.*[7]

as large as a factor of four. Clearly, for absorption effects as large as these, accurate corrections are needed; an error of only 10 percent in the correction for the derivative crystals can lead to an error of approximately 20 percent in the structure amplitude. Fortunately, absorption is seldom as serious as in this example, but it is, nevertheless, a problem that must be dealt with carefully.

Protein crystals invariably are sensitive to x rays, and the usual practice is to correct for deterioration on the basis of a number of standard reflections, which are assumed to be representative of the data as a whole. Although the average intensity of the standard reflections decreases with increasing x-ray exposure, the effect is not solely a matter of deterioration because the intensities of different reflections change at different rates. Evidently, structural change is superimposed on the deterioration. It is necessary, therefore, to terminate data collection from any given crystal at some arbitrary point–10 percent deterioration has been used–and to continue with a fresh crystal. Even this procedure has its difficulties in that data from different crystals must be scaled together, introducing additional error, and different crystals, even from the same batch, may not be identical.

Each of the effects described here–the relatively weak diffraction from structures with large unit cells, the rapid decrease of intensities with increasing diffraction angle, possible large absorption effects, and deterioration and change of the structure in the x-ray beam—serves to impair the accuracy of diffraction data (and the derived structure factor amplitudes) that can be collected from protein crystals.

MULTIPLE ISOMORPHOUS REPLACEMENT PHASES

The expression for calculating the electron density is a three-dimensional Fourier series,

$$\rho(x,y,z) = (1/V)\sum_h \sum_k \sum_l |F_{hkl}| e^{i\alpha_{hkl}} e^{-2\pi i(hx + ky + lz)}, \tag{1}$$

where V is the volume of the unit cell, hkl are the indices of the reflections, $|F_{hkl}|$ are the amplitudes of the reflections, α_{hkl} are the phases of the reflections, and x,y,z are the coordinates at which the electron density is evaluated. When a data set has been collected and reduced to $|F_{hkl}|$, all quantities in Eq. (1) are known except α_{hkl}. This is the phase problem of x-ray crystallography.

The most general method of solving the phase problem for protein crystals is that of multiple isomorphous replacement in which two or more isomorphous heavy-atom derivatives are used.[1] The principle of the method is shown in Figure 3. In Figure 3a a circle with radius $|F_P|$, the amplitude of a reflection from the native protein, is shown with center at the origin, O. It is assumed that the heavy atoms in at least two derivatives have been located and referred to the same unit cell origin. This can be a difficult problem and mistakes can be made, but

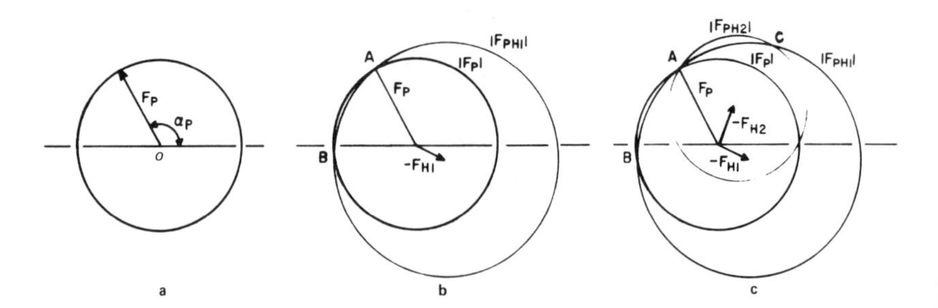

FIGURE 3 (a) Phase circle for a reflection with phase angle α_P. (b) Phase circle for derivative 1 intersects native circle at A and B. (c) Phase circle for derivative 2 intersects native circle at A and C.

for our purpose here we shall assume good derivatives, i.e., those for which this can be done. In Figure 3b, $-F_{H1}$, the heavy atom resultant vector for derivative 1 has been added, and with its terminus as center, a circle of radius $|F_{PH1}|$ is drawn. In general, the derivative circle will intersect the native circle in two points, corresponding to two possible phases for this reflection. The ambiguity can be resolved with a second derivative as shown in Figure 3c where $-F_{H2}$ is added for derivative 2, and with the terminus of this vector as center, a circle with radius $|F_{PH2}|$ is drawn. The common intersection corresponds to the phase α for this reflection.[2]

Because of experimental error, the three circles will not intersect at a point as in Figure 3c. In practice, the situation is as shown in Figure 4 where the derivative circles intersect the native one at points A_1 and A_2. The problem has been treated on the basis of error theory by Blow and Crick.[3] With an estimate of the total error E (in f_H from whatever cause and in the measured amplitudes), they calculate the probability of a phase angle $[P(\alpha)]$ according to the equation,

$$P(\alpha) = \exp(-x^2/2E^2), \tag{2}$$

where x is the lack of closure, i.e., the quantity to be added to $|F_{PH1}|$ to close the triangle in Figure 5.

In Figure 6a, $P(\alpha)$ is plotted as a function of α for the case shown in Figure 3b with the two peaks in the probability corresponding to the

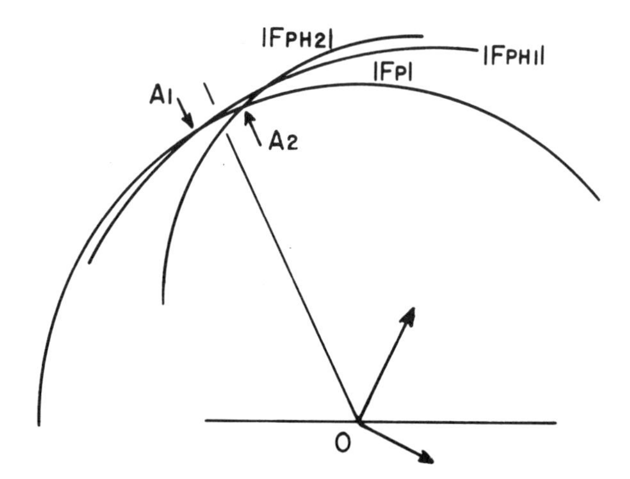

FIGURE 4 Experimental error causes derivative circles to intersect native circle at A_1 and A_2 near A.

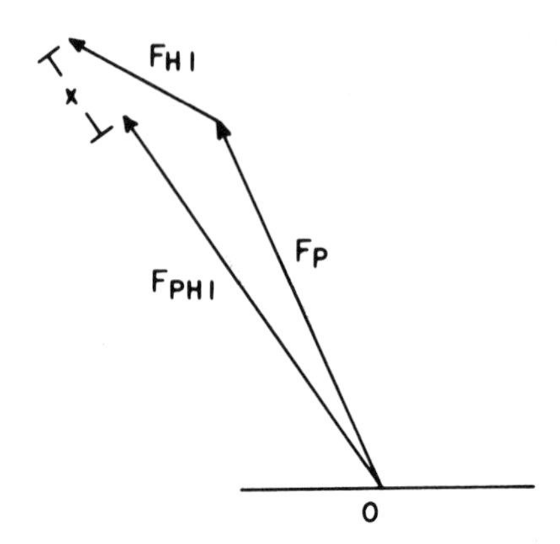

FIGURE 5 Triangle showing lack of closure error x.

intersections of the circles, A and B. A similar plot for the second derivative (circle $|F_{PH\,2}|$ in Figure 3c) is shown in Figure 6b. The combined probability is proportional to the product of the separate probabilities and is shown in Figure 6c.

The most probable phase corresponding to the greatest value of $P(\alpha)$ in Figure 6c may be used in calculating an electron density map. Blow and Crick show, however, that the most probable phases do not lead to an electron density map with minimum deviations from the true values in the least squares sense. They show, instead, that such a "best" map is obtained by use of *centroid* phases with amplitudes weighted by the *figure of merit.* What these terms mean can be seen by weighting the circumference of the phase circle in Figure 3a proportional to the probabilities in Figure 6c.[3,4] This is shown in Figure 7 where $P(\alpha)$ has

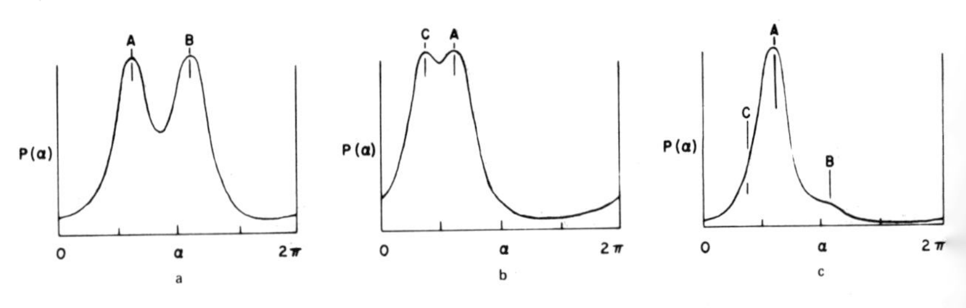

FIGURE 6 (a) Plot of relative probabilities corresponding to Figure 3b. (b) Plot of relative probabilities corresponding to second derivative in Figure 3c. (c) Combined probability, relative value $P(\alpha)_{(a)} \times P(\alpha)_{(b)}$.

been plotted around the circle of radius $|F_p|$ and the centroid of the whole distribution is indicated at *C*. The centroid phase is the angle *XOC* (Figure 7) and the ratio $OC/|F_p|$ is the figure of merit (m). The latter can be shown to be equal to the value of cos ϵ, where ϵ is the error in phase angle.

USE OF ANOMALOUS SCATTERING

It has been shown (Figure 3b, c) that a second derivative will resolve the phase ambiguity. It is possible to do this, however, not by a second derivative, but by measuring both reflections of each Friedel pair, I_{hkl} and $I_{\bar{h}\bar{k}\bar{l}}$ for a single derivative.[5,6] By considering the imaginary component of the scattering from the heavy atoms, it can readily be shown by a diagram similar to Figure 3c how the phase is determined.

Although the imaginary part of the scattering is much smaller than the real part, nevertheless, it is sufficiently large that it can be used in determining phases with a single derivative only.[7,8] If multiple derivatives are used, the additional data from measuring both reflections of each Friedel pair can substantially improve the phase angles.[9]

ERRORS IN PHASE

If the total error *E* has been estimated properly, the figure of merit (m) will be a measure of the error in phase. For reflections in the low resolution region (large d) the average figure of merit may be 0.9 or better, indicating standard errors of 26° or less, but for high resolution reflections (small d) the average figure of merit may drop to 0.7 or less cor-

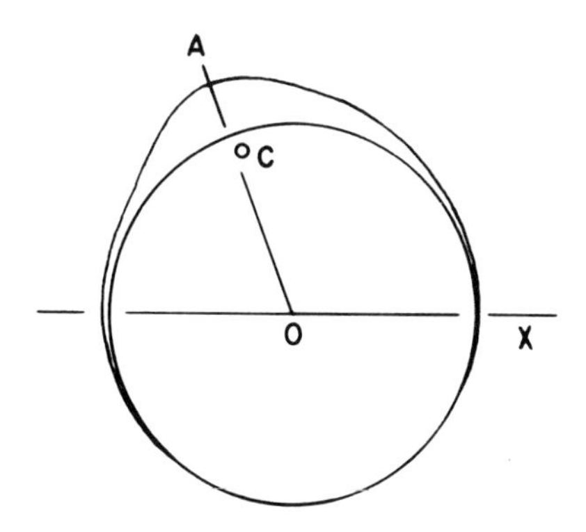

FIGURE 7 Combined probability in Figure 6c plotted around phase circle.

responding to standard errors of 45° or more. Consequently, the error in the experimental phases must be regarded as large.

ELECTRON DENSITY MAPS

The resolution that can be achieved in an electron density map based on x-ray diffraction data is related to the minimum spacing of the diffracting planes used to calculate the Fourier series. The limit of resolution is given by the equation

$$\text{L.R.} = 0.72 d_{\min} \tag{3}$$

and is often spoken of as the resolution.[10] Thus for CuK_α radiation, $d_{\min}$ is 0.77 Å, and if data to this limit were used, the theoretical resolution would be 0.52 Å. This will not be reached in practice, and in protein crystallography it is more nearly equal to $d_{\min}$. In this field it has thus become common usage to give $d_{\min}$ as the resolution.

As noted earlier, the intensities from protein crystals decrease rapidly with increasing diffraction angle, so rapidly in fact, that the diffraction pattern fades out long before the limit of CuK_α radiation is reached. Data for a few proteins have been collected to d values less than 2 Å, but it is fair to say that, even in the best maps available at the present time, we do not completely resolve covalently bonded light atoms.

It should be noted that the amount of data on which a map is based increases rapidly with increasing resolution (i.e., decreasing $d_{\min}$) and is proportional to $(1/d_{\min})^3$. Thus there are eight times as many data for 3 Å resolution, for example, as for 6 Å and for 2 Å resolution, 27 times as many.

Low resolution electron density maps are often calculated and may provide valuable information. For example, the first three-dimensional map of myoglobin was at 6 Å resolution, and it revealed the general features of the molecule largely because the α-helices, which make up about 75 percent of the molecule, were resolved.[11] Even if little helix is present, low resolution maps are useful in providing an overall view of the molecule. For example, Lipscomb and his colleagues[12] have calculated a 5.5-Å map of aspartate transcarbamylase, which has enabled them to describe the shape of and draw important inferences about this large enzyme molecule.

In a 4-Å resolution map, much of the polypeptide chain can be traced; but where different sections of the chain approach each other, they may not be resolved, and it is difficult or impossible to trace their course correctly. A good map at 3.0 Å resolution will usually enable one to trace the polypeptide chain throughout the molecule. Many of the side chains will be evident and a tentative identification of a few of the large, bulky ones may be made. The carbonyl groups that are important in fixing the plane of the peptide residues are usually not prominent in a 3-Å resolution map. Although a skeleton model with ideal bond lengths and angles can be fit in an approximate way to the map, the main chain torsional angles may be quite unreliable.

A 2.5-Å resolution map will be based on 73 percent more data than a 3-Å map and will show much more detail. In fact, for good maps, one is struck by the improved detail visible at 2.5 Å; qualitatively, at least, it appears more nearly proportional to the added data than to the improved resolution. Usually, the carbonyl oxygen atoms are prominent, and a model can be fit to the map with reasonable certainty. In a good map many of the amino acids can be identified, and one can be more nearly certain of the amino acid count if it is not already known. At 2 Å resolution almost twice as many data are potentially observable as at 2.5 Å; therefore, a 2-Å map shows much more detail. Unless there is disorder or other adverse circumstances, the amino acid count should be virtually certain, and half or more of the amino acids should be identifiable.

Difference electron density maps are often calculated to show changes between the native protein and some altered state where this includes deletion from or addition to the molecule such as binding an inhibitor to it.[13] Such maps are calculated by use of the *phases for the native protein* and differences in the amplitudes between the native and altered state as coefficients in the Fourier series. Such difference density maps are interpreted in the same way as electron density maps and with the same qualifications. They are subject to rather more uncertainty, however, since the phases are for the native protein rather than for the difference vectors between the native and altered state reflections.

At any resolution the hazard is in trying to interpret the electron or difference electron density maps beyond what is warranted. It should be noted, however, that it is possible to correctly deduce structural features beyond the limit of resolution, and there is an outstanding example of this in the annals of optical microscopy. It was his understanding of the principles of image formation that enabled the great nineteenth-century histologist Köllicker to infer cellular detail beyond the limit of resolution of the optical microscope, detail that has been confirmed only in more recent times by the electron microscope. In the case of proteins we have a great deal of additional information such as knowledge of the bond lengths and angles of the amino acids, of α-helices and pleated sheets, and sequence and other chemical information; and this is invaluable in extending the interpretation of the electron density maps beyond the nominal limit of the data on which they are based.

MODEL BUILDING AND COORDINATES

It has been common practice to fit a skeletal model to the electron density map, usually by means of a Richards comparator, a device with a half-silvered mirror that enables an observer to see a reflected image of a wire model projected within the three-dimensional map.[14] The model is adjusted by varying the torsional angles and in some cases distorting the interbond angles to allow for departures from ideal values, until the atoms are in positions that appear best to satisfy the electron density. It must be emphasized that the fit is a subjective one and that it is impossible to make an objective assessment. Errors in coordinates from models fit to maps at resolutions of 2–2.5 Å are usually estimated to be in the 0.2- to 0.5-Å range,[15] although errors much in excess of this may be made.

It should be recognized that coordinates for atoms within the molecule are likely to be much closer to their true values than are coordinates for atoms on the outside. Indeed, the electron densities for groups on the surface, and these are often ones of prime interest, may be so diffuse that even a sizable one may appear to be missing. Figure 8 shows the difference in quality that was observed for two tyrosine groups in the 2-Å electron density map for the protein rubredoxin.[16] Figure 8a shows a section of the map through Tyr 13, that is inside the molecule, and its OH is hydrogen bonded to the main chain carbonyl oxygen of residue 28. In contrast, Figure 8b shown Tyr 11 that is on the surface of the molecule. Clearly, there is much more uncertainty in fitting Tyr 11 than Tyr 13 and, accordingly, the coordinates must be much less precise.

In contrast to models fit to electron density maps as described above, they are sometimes constructed in a different way in protein studies. For example, a model of an enzyme substrate may be made and fit to the active site to test ideas concerning the mechanism of enzyme action. It should be clearly understood that such models cannot be accorded the same confidence that is placed in models fit directly to electron density maps.[17] Although the interpretation of electron density maps is subjective and may even be distasteful to those with a more objective bent, it must be emphasized that the contribution of such maps to our understanding of the structure and function of protein molecules is so great that it is essential to extract all possible information from them.

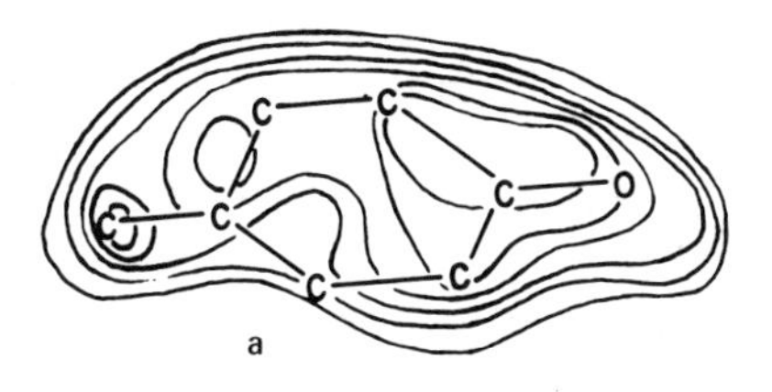

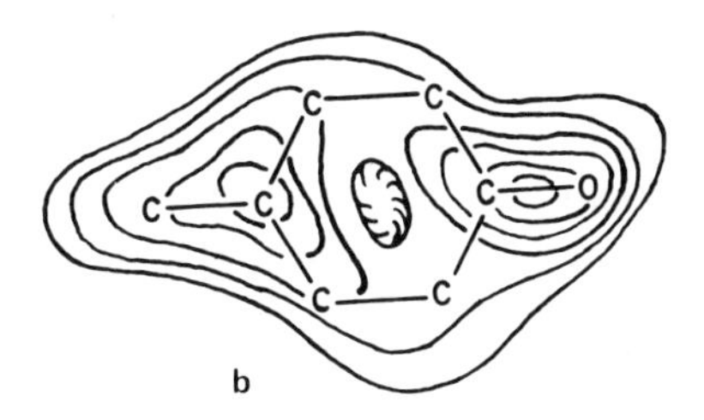

FIGURE 8 (a) Section through Tyr 13 in 2-Å resolution map of rubredoxin. (b) Section through Tyr 11 in 2-Å resolution map of rubredoxin. Both (a) and (b) from map based on experimental phases.

IMPROVING THE MODEL

We may now ask if the electron density maps can be improved or if it is possible to improve the model based on them. The answer is provided by crystallographers who do small structures and routinely refine their models to the limit imposed by the data. They use a variety of indices to measure the agreement between the observed amplitudes, $|F_o|$, and those calculated for the model, $|F_c|$. The most frequently used one (R), is defined by Eq. (4).

$$R = \sum ||F_o| - |F_c|| / \sum |F_o| \tag{4}$$

This is just the mean fractional error in the structure amplitudes.

When an approximate model has been found for a small structure, the initial value of R will often be in the range 0.3–0.4. If the structure is essentially correct, i.e., most atoms are within about 0.3 Å of their true positions, it can usually be improved to 0.1 or less by various refinement processes that minimize the differences between $|F_o|$ and $|F_c|$.

For those protein models for which structure amplitudes have been calculated, R values are usually in the range 0.4–0.5 even for the limited data available for these noncentrosymmetric structures.[15,18] There would appear to be no "in principle" reason why a protein model should not refine within the limits imposed by the extent and accuracy of the data; and, in fact, efforts were made to refine myoglobin, the first protein to be solved. Considerable effort was invested and progress was made, but the work has never been fully reported. It is clear from what has been reported, however, that the effort was an extensive one but that the methods used were not sufficiently effective.

Attempts to refine a much smaller protein, rubredoxin, have led to a decrease in R from 0.372 to 0.126 in eight cycles for 5,005 reflections to a resolution of 1.54 Å.[16] In this case the first four cycles were ΔF syntheses (difference maps) and R decreased to 0.224. The next four cycles were by the method of least squares. There can be little question that the refinement is genuine and that the phases are much improved. In support of this view, Figure 9a shows a section through Tyr 11 from the 2-Å resolution electron density map based on phases after eight refinement cycles. (This figure is to be compared with Figure 8b.) Figure 9b is a similar section through Tyr 11, but at 1.5 Å resolution. The improved definition of the atoms, compared with the 2 Å resolution sec-

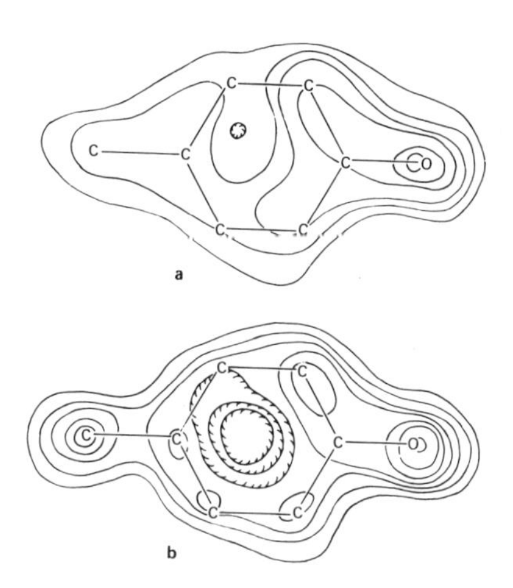

FIGURE 9 (a) Section through Tyr 11 in 2-Å resolution map of rubredoxin. (b) Section through Tyr 11 in 1.5 Å resolution map of rubredoxin. Both (a) and (b) based on calculated phases after eight refinement cycles. From Watenpaugh *et al.*[16]

tion in Figure 9a, is striking. Nevertheless, the refinement has not yet converged, and some atoms are still in impossible positions.

In considering the number of parameters in an unconstrained refinement in which x,y,z and B (the so-called thermal parameter) are varied for each atom, it is clear that the number of observations will exceed the number of parameters by a factor of no more than two for 2-Å data and four for 1.5-Å data, even if all possible reflections are considered and the atoms in the solvent are neglected. In practice, the degree of overdetermination may be less than this because many weak reflections can only be measured with very low precision.

By introducing constraints involving bond lengths and angles and other known features of protein molecules, the ratio of the number of observations to the number of parameters can be substantially improved. Moreover, the refinement should behave better and its cost should be much reduced.

Diamond[19,20] has written a pair of least squares programs with constraints that can be used to build a model and to refine it in real space. The first one smooths the errors in any protein coordinates supplied to it by fitting an ideal model in which the torsion angles are varied. These idealized coordinates serve as input to the refinement program in which both torsion and interbond angles may be varied to minimize the differences between the observed and calculated electron densities. It should be noted, however, that when used in this way, the process is not refinement in the usual crystallographic sense since the phases are not varied. There is no reason, however, why the electron densities should not be recalculated with α_{calc} after each cycle. With the phases allowed to vary, this would constitute refinement in the usual sense. A modification of this in which the program minimizes $(2|F_o| - |F_c|)\exp i\alpha_{calc}$ has been used successfully by R. Huber (personal communication) and colleagues[21] in refining bovine trypsin inhibitor to an R of 0.23.

Diamond's real space refinement program has been used on several proteins, and there can be little doubt that it improves the positions of most atoms in the model. When used with electron densities based on α_{iso}, however, the program may give poor positions for atoms in side chains where the electron density is diffuse, sometimes moving them to invade the electron density belonging to nearby atoms.[22] This adverse behavior can be avoided by introducing potential functions and minimizing the interaction energies among the atoms. Although this should place the atoms correctly, it is uncertain to what extent this is achieved at present, and the final check on atomic positions for molecules in the crystal must still come from a comparison of $|F_o|$ and $|F_c|$.

Extensive tests of phase refinement by the tangent formula of Hauptman and Karle have been made, and Sayre[23] has recently developed a method of phase refinement in which he minimized

$$\sum_h |a_h F_h - \sum_k F_k F_{h-k}|^2$$

as a function of the phases. Starting with the experimental phases for the 2.5-Å rubredoxin data, he has extended them to 1.5 Å by the tangent formula and refined them for the whole data set. In this way he has arrived at a set of phases that gives an electron density map with the atomic positions essentially indistinguishable from those derived from the conventional refinement.

REFERENCES

1. D. W. Green, V. M. Ingram, and M. F. Perutz, *Proc. R. Soc. Ser. A*, **225**, 287 (1954).
2. D. Harker, *Acta Cryst.*, **9**, 1 (1956).
3. D. M. Blow and F. H. C. Crick, *Acta Cryst.*, **12**, 794 (1959).
4. R. E. Dickerson, J. C. Kendrew, and B. E. Strandberg, *Acta Cryst.*, **14**, 1188 (1961).
5. J. M. Bijvoet, *Nature*, **173**, 888 (1954).
6. D. M. Blow and M. G. Rossmann, *Acta Cryst.*, **14**, 1195 (1961).
7. J. R. Herriott, L. C. Sieker, L. H. Jensen, and W. Lovenberg, *J. Mol. Biol.*, **50**, 391 (1970).
8. K. D. Watenpaugh, L. C. Sieker, L. H. Jensen, J. LeGall, and M. Dubourdieu, *Proc. Natl. Acad. Sci., U.S.A.*, **69**, 3185 (1972).
9. A. Arnone, C. J. Bier, F. A. Cotton, V. W. Day, E. E. Hazen, D. C. Richardson, J. S. Richardson, and A. Yonath, *J. Biol. Chem.*, **246**, 2302 (1971).
10. R. W. James, *Acta Cryst.*, **1**, 132 (1948).
11. J. C. Kendrew, G. Bodo, H. M. Dintzis, R. G. Parrish, H. W. Wyckoff, and D. C. Phillips, *Nature*, **181**, 662 (1958).
12. D. R. Evans, S. G. Warren, B. F. P. Edwards, C. H. McMurray, P. H. Bethge, D. C. Wiley, and W. N. Lipscomb, *Science*, **179**, 683 (1973).
13. L. N. Johnson and D. C. Phillips, *Nature*, **206**, 761 (1965).
14. F. M. Richards, *J. Mol. Biol.*, **37**, 225 (1968).
15. J. J. Birktoft and D. M. Blow, *J. Mol. Biol.*, **68**, 187 (1972).
16. K. D. Watenpaugh, L. C. Sieker, J. R. Herriott, and L. H. Jensen, *Acta Cryst.*, **B29**, 943 (1973).
17. W. N. Lipscomb, in *Bio-Organic Chemistry and Mechanisms*, p. 131, Proceedings of the Welch Foundation Conference on Chemical Research, Ed. by W. O. Milligan, Robert A. Welch Foundation, Houston, Texas (1972).
18. W. N. Lipscomb, G. N. Reeke, J. A. Hartsuck, F. A. Quiocho, and P. H. Bethge, *Philos. Trans. R. Soc. Lond.*, **B257**, 177 (1970); also, J. Kraut, personal communication.
19. R. Diamond, *Acta Cryst.*, **21**, 253 (1966).
20. R. Diamond, *Acta Cryst.*, **A27**, 436 (1971).

21. J. Deisenhoeffer and W. Steigmann, in *Abstracts, Stockholm Symposium on the Structure of Biological Molecules,* p. 98. Swedish National Committee for Cystallography, University of Stockholm (1973).
22. R. Diamond, *J. Mol. Biol.* (in press).
23. D. Sayre, *Acta Cryst.* (in press).

Carroll K. Johnson

Critical Evaluation of Protein Crystallographic Results

We wish to pursue the following questions: How reliable are protein crystal structure results under favorable conditions? How do the more commonly encountered departures from the ideal case influence the results? Before attempting to answer these questions, a brief synopsis of the field is presented.

Protein crystallography is a young, dynamic discipline that is attracting a rapidly growing number of crystallographers and physically oriented biochemists. At least 20 protein crystal structures have been reported[1] and there are over 30 protein crystallographic groups in the United States alone. The reason for this interest in protein crystallography is that the results obtainable can yield significant information on subjects such as enzyme catalysis mechanisms and molecular genetics. In addition, a protein structure analysis provides a stimulating crystallographic challenge in contrast to many smaller structures that are now solved quite routinely. Until recently, protein crystallographic research was done by those who pioneered in the field, and most of the current groups, with a few notable exceptions, have descended directly or indirectly from the laboratories of John C. Kendrew and Max F. Perutz of Cambridge University.

The author recently visited 10 protein crystallography laboratories in the United States as an outside crystallographer trying to understand the motivations, the problems, and the results in the field. The groups visited were all enthusiastic about discussing their results and were extremely hospitable and helpful. Protein structure groups are interdisciplinary and function either by gathering specialists in protein chemistry and crystallography in one laboratory or by forming interlaboratory collaborations.

Protein crystallography rests on a foundation with three cornerstones. The first is the isomorphous derivative technique[2] wherein heavy-atom modifications of the native protein crystal allow the classical x-ray phase problem to be solved completely objectively. The second cornerstone is the polymeric nature of the protein molecule that imposes stringent stereochemical constraints on its conformational freedom. As a consequence, stereochemical reasoning based on structural relations found in smaller molecules can be used legitimately to extend conclusions past the limits of the completely objective x-ray diffraction results. The third cornerstone is the existence of chemical methods for establishing the amino acid sequence of the protein. If the chemical and diffraction studies are carried out concurrently, they can complement each other to make the sequencing task somewhat easier. The complete chemical sequencing of a protein is quite time-consuming but progress toward automation of the process is being made.[3]

Technical papers describing protein crystal structure results, like all other scientific publications, are rich in the technical jargon of the field and thus difficult to read and evaluate by those outside the field. The following are some points that seem relatively important.

1. The chemical preparation step in which the protein is isolated, purified, and crystallized is critical in that the protein preparation must be chemically homogeneous; otherwise, the resulting disorder will muddle the electron-density map. The preparation of isomorphous derivatives by soaking native protein crystals in various mercury, platinum, lead, uranium, etc., solutions also is critical since several crystals of each derivative are required for x-ray data collection (because of irradiation damage) and all the crystals should have the same heavy-atom distribution and concentration. The protein structure documentation should provide evidence that the preparative protein chemistry is sound.

2. X-ray data collection from proteins is usually plagued by radiation damage to the specimen crystals, and the experimenter may tend to hurry the data collection excessively. The intensity statistics on equivalent reflections are useful indicators of the diffractionist experimental technique. The root-mean-square (rms) deviation from the mean should not

exceed 10 percent if an absorption correction and a radiation-damage decay correction are applied.

3. The location of the heavy-atom sites and the refinement of positions, occupancies, and temperature factors are critical steps in the crystallographic analysis. Although numerous R factors are given, they are of little value to the general reader and the only trustworthy indicator that all is well is a reasonably flat "residual" Fourier map for each heavy-atom derivative.[4] A useful empirical indicator of the quality of phase determination provided by a particular heavy-atom derivative is the ratio of the rms "lack of closure" error over the mean heavy-atom scattering. The phase information is of marginal value if the ratio exceeds one.[5] In favorable cases the ratio is around 1/3 to 1/2. These statistics are usually provided in the protein structure paper.

4. The figure of merit m for a given x-ray reflection is the distance from the center of a unit circle to the center of gravity for the experimental phase-angle, probability-density function plotted on the circle. This indicator is used to express the overall reliability of the isomorphous replacement phasing of the x-ray reflection. The figure of merit is a valid statistic only if the rms errors are assigned correctly.[6] The mean figure of merit $\overline{m}$ decreases with increasing $\sin\theta/\lambda$ and some protein crystallographers use a rule of thumb that, when $\overline{m}$ for a shell of data drops below 0.55, the isomorphous phasing process should be questioned. At small values of $\sin\theta/\lambda$, $\overline{m}$ is often 0.85 or better.

The phases can be improved from their experimental values and additional phases generated for high resolution data through various mathematical methods but the value of these techniques is not widely acknowledged. The simplest and least expensive of the methods involve successive numerical Fourier transforms to minimize the negative regions of the electron-density map (G. Kartha, personal communication, and Barrett and Zwick[7]). In this reviewer's opinion, this is a valid procedure that should receive more attention. Other techniques such as tangent formula refinement[8] and a new least squares direct methods approach[9] also seem promising but require appreciable amounts of computing time.

5. The overall mean figure of merit $\overline{m}$ is used in the following equation for the rms error[6] of the commonly used "best Fourier synthesis" electron-density map[10]:

$$\langle\Delta\rho\rangle = \frac{1}{V}\left[\sum(1 - \overline{m}^2)F^2\right]^{1/2}.$$

The sum is over the entire sphere of reflections, and V is the unit cell volume. The lowest contour in the electron-density map should be drawn at the density level of the rms error, which is often about 0.25 electron per cubic angstrom in a 2-Å resolution map.

6. The major features of a "high" resolution (2-Å) density map usually can be interpreted in terms of (a) the outline of a single molecule or structural unit, (b) the path for the folding of the backbone within the molecule except for loops or ends outside the nominal surface of the molecule, (c) the position and approximate orientation of distinctive structural elements, such as planar heme groups and certain planar aromatic side chains, inside the nominal surface, and (d) the positions of bonded heavy atoms such as iron and sulfur. At this point the known amino acid sequence is shifted along the internal backbone region to find the positions where bulky side chains fall in large density areas. The model building process then starts in earnest, and individual amino acid residues along the known sequence are added to the model while observing the optically superimposed electron density map through a half silvered mirror.[11] The builder positions the model to adopt a conformation as consistent as possible with (a) the density map, (b) the mechanical constraints of the wire model, and (c) his concepts of the most likely stereochemical principles for protein molecules. The interior of the molecule is usually reasonably well defined by the density and the packing, but the residues in the surface region may be poorly defined or disordered. Furthermore, small molecule substrates, inhibitors, and coenzymes, which are not part of the protein polymer, sometimes must be positioned quite subjectively in that the densities for such regions are low, and there are few reliable invariant stereochemical constraints to guide the model builder. This is unfortunate since some of the more biologically interesting aspects of a protein structure involve the nature of the small molecule interactions with the protein.

An interesting and objective insight into the accuracy of the protein crystallographer's interpretation of the side-chain regions of the density map is given by his score on the identification of side-chain residues before the amino acid sequence is available. This score seems to be between 50 and 75 percent with good quality density maps.

7. After the model has been assembled, positional coordinates are measured, and a "model-building" computer program is used to adjust the coordinates to minimize the strain energy within the model molecule[12] or the departure from idealized interatomic distances and angles.[13] In either case, the stereochemical preconceptions incorporated should be reported by the protein crystallographer so that an unsuspecting user of the "model-building coordinates" will not make the logical error of quoting the output results to prove the input assumptions.

8. The absence of adequate mathematical procedures for refinement and statistical testing of trial structures derived by model building is undoubtedly the weakest link in contemporary protein crystallography. The reasons for this are the mammoth size of the numerical calculations involved[14] and the basic fact that, in general, the protein diffraction data are inadequate to define atomic detail. However, one small protein, rubredoxin, has been refined successfully by traditional crystallographic least squares adjustment of unconstrained atomic parameters (see L. H. Jensen, this volume). The only general protein-refinement algorithm available at present is the "real-space refinement" program of Diamond,[15] which adjusts atomic parameters in a sliding "molten zone" to minimize the integral $\int (\rho_o - \rho_c)^2 \, dv$ over that volume of the experimental electron density (i.e., ρ_o) corresponding to the "molten zone." The protein molecule essentially is parameterized as a system of coupled rigid bodies. The Diamond procedure provides no convenient global or local statistical indicators to use in assessing the quality of the fit obtained and provides no feedback mechanism for improving the experimentally derived x-ray phases.

9. In conclusion, the main item that gives this reviewer a "warm happy feeling" that a protein-structure manuscript is supplying credible information is a critical residue-by-residue analysis of the electron-density map with adequate illustrations of the density map and the molecular model to support the author's conclusions (e.g., see Wyckoff *et al.*[16]). An interesting account of some protein crystallographers' views on the subject of "model-building results" versus "x-ray results" is found in the recorded discussions following a paper presented by Lipscomb.[17] An excellent cross section of the protein crystallography literature is contained in the Cold Spring Harbor Symposium on Quantitative Biology.[18] A computer data bank for protein structure coordinates now is operational at Brookhaven National Laboratory.*

Research sponsored by the U.S. Atomic Energy Commission under contract with the Union Carbide Corporation.

REFERENCES

1. B. W. Matthews, in *The Proteins,* 3rd ed., Vol. III, H. Neurath and R. L. Hill, ed., Academic Press, N.Y. (in press).
2. D. Harker, *Acta Cryst.,* **9**, 1 (1956).
3. G. M. Edelman and W. E. Gall, *Proc. Natl. Acad. Sci., U.S.A.,* **68**, 1444 (1971).
4. D. M. Blow and B. W. Matthews, *Acta Cryst.,* **A29**, 56 (1973).

*The address for the Protein Data Bank is: Dr. Thomas Koetzle, Department of Chemistry, Brookhaven National Laboratory, Upton, New York 11973.

5. B. W. Matthews, in *Crystallographic Computing,* F. R. Ahmed, ed., Munksgaard, Copenhagen (1970), p. 146.
6. R. E. Dickerson, J. C. Kendrew, and B. E. Standberg, *Acta Cryst.,* **14**, 1188 (1961).
7. A. N. Barrett and M. Zwick, *Acta Cryst.,* **A27**, 6 (1971). Some information is given in an article by G. Kartha, in *Crystallographic Computing,* F. R. Ahmed, ed., Munksgaard, Copenhagen (1970), p. 345.
8. A. L. Spek and H. Krabbendam, *The Application of the Generalized Tangent Formulae to Phase Refinement in Protein Crystallography,* Report of CECAM Workshop, Paris (Orsay) (1972).
9. D. Sayre, *Acta Cryst.,* **A28**, 210 (1972).
10. D. M. Blow and F. H. C. Crick, *Acta Cryst.,* **12**, 794 (1959).
11. F. M. Richards, *J. Mol. Biol.,* **37**, 225 (1968).
12. M. Levitt and S. Lifson, *J. Mol. Biol.,* **46**, 269 (1969).
13. R. Diamond, *Contemp. Phys.,* **13**, 23 (1972).
14. K. D. Watenpaugh, in *Computational Needs and Resources in Crystallography,* National Academy of Sciences, Washington, D.C. (1973), p. 37.
15. R. Diamond, *Acta Cryst.,* **A27**, 436 (1971).
16. W. A. Wyckoff, D. Tsernoglou, A. W. Hanson, J. R. Knox, B. Lee, and F. M. Richards, *J. Biol. Chem.,* **245**, 305 (1970).
17. W. N. Lipscomb, in *Bio-Organic Chemistry and Mechanisms,* Proceedings of Welch Foundation Conference on Chemical Research, W. O. Milligan, ed., Houston, Texas (1972), p. 131.
18. *Structure and Function of Proteins at the Three-Dimensional Level,* Cold Spring Harbor Symposium on Quantitative Biology, Vol. 36, Cold Spring Harbor, New York (1972).

F. A. Bovey

Studies of Protein Structure by NMR

There is no doubt that x-ray diffraction is the prime method for the determination of the three-dimensional structure of proteins. There is at present no other technique capable of providing the thousands of parameters necessary to accomplish this spectacular result. For the study of proteins in solution, however, other methods must be used. In recent years high resolution NMR has emerged as a method that can also provide a large number of structural parameters, some of which are peculiar to itself and give information not otherwise accessible. It has the important distinction from x ray that it necessarily must be practiced on protein solutions, since, owing to the phenomenon of *dipolar broadening,* it is necessary for the molecules to tumble in order to show a high-resolution spectrum. It follows further that the line width, being proportional to the tumbling time, τ_c, commonly called the *correlation time,* is proportional to the molecular weight of the protein, provided it is in its native, folded state and is at least approximately spherical in form.

NMR studies have been concerned with three principal aspects of protein structure:

1. The direct observation of the protein molecule itself, with emphasis on the interactions of the chains, principally the effects of folding

and unfolding them, and on the pH titration of certain residues, notably the imidazole rings of histidines. These groups are of particular significance since they commonly occur at the active sites of enzymes and participate in the reactions they catalyze.

2. The observation of the binding of small molecules, including substrates, substrate analogs, inhibitors, cofactors, and the solvent itself, to the protein.

3. The study of paramagnetic sites in proteins, through their effect on the chemical shift of neighboring protein nuclei and in relaxing the nuclei of the solvent or other associated molecules. The paramagnetic sites may be inherent to the structure, as in the heme proteins myoglobin, hemoglobin, and the cytochromes and in other non-heme types such as the ferredoxins, or they may be introduced by the binding of paramagnetic ions or spin labels at sites normally diamagnetic.

Much but not all of this work has dealt with proteins the three-dimensional structures of which have been determined by x ray: lysozyme, ribonuclease, myoglobin, hemoglobin, cytochrome C, carboxypeptidase, chymotrypsin, concanavalin, trypsin, elastase, and subtilisin. The principal nucleus has been the proton, but more recently ^{13}C has been studied by several groups. Other nuclei, such as ^{19}F, ^{31}P, and ^{35}Cl, have found limited application in special studies.

DIRECT OBSERVATIONS OF PROTEIN STRUCTURE

Available Data

High resolution NMR as traditionally practiced is not inherently a structure-determining method in the same sense that x-ray diffraction is—that is, a direct means of determining interatomic distances. It does share with x ray, however, the ability to detect single atoms in a protein structure. In principle, all the individual protons and all the individual carbon nuclei should give separate resonances. (This would also hold for the oxygen, nitrogen, and sulfur nuclei if it were practical to observe them.)

NMR is very powerful for the determination of the covalent structures of relatively small molecules and even of long-chain synthetic macromolecules, provided they are repetitive and can be dealt with statistically. It is particularly powerful in the determination of *conformations* (i.e., rotational states about covalent bonds) and *molecular symmetry*. The single most valuable measure of conformation is the magnitude of *scalar nuclear couplings,* but these are functions of bond dihedral angles rather than of internuclear distances, and they are not always free of interpretive ambiguities. Such measurements have been used, along with other data, for the determination of conformations of polypeptides of 1–12 amino acid residues.[1] For globular proteins in the native, folded state the correlation times are sufficiently long (10^{-9}–10^{-8} s) that the spectral splittings needed for the measurement of these couplings are ordinarily not observable owing to the dipolar broadening (*vide supra*). Recent work by R. J. P. Williams and his group at Oxford has shown, however, that under optimal observing conditions at least some coupling information can be retrieved from protein spectra by appropriate computer processing of the line shapes. Not all of this will necessarily have direct structural relevance, but may aid in identifications of resonances.

Although it would clearly not be appropriate to enter into a detailed exposition here, it may be helpful to show what protein NMR spectra look like. They all bear a strong family resemblance, as might be expected from their general similarity in composition. It is the differences in detail that are important. In Figure 1 are shown 300 MHz proton spectra of native cytochrome C in the reduced state (Fe^{2+}, $S = 0$) and oxidized state (Fe^{3+}, $S = 1/2$) (D. J. Patel, manuscript in preparation). The spectra were observed in D_2O and therefore those protons that are readily exchangeable (CO_2H, OH, NH_2) do not appear. Tryptophan indole NH (ca. 10 δ) are the least shielded protons in the diamagnetic state of the protein. Aromatic and histidine C–H protons appear at 8.0–6.6 δ. Between 6.5 and 5 δ there is a pronounced "window," observed in all protein spectra, while between 6 and 4 δ are the α-CH protons of the main chain, often partially obscured, as here, by residual solvent (HDO) protons. A variety of side-chain methylene and methyl proton resonances appear between 4 and 2 δ, followed by methyl resonances of aliphatic side chains (1.5–1.0 δ). A very highly shielded resonance of particular interest appears 3.2 upfield from the reference, i.e., more than 4 ppm beyond the most shielded "normal" protons. This will be discussed below.

When cytochrome C is in the paramagnetic state (Figure 1b), there are small alterations in the positions of almost all of the resonances and very large shifts, both upfield and downfield,[2,3] of the protons of the heme ring. In Figure 1b the chemical shift scale is expanded tenfold over that of Figure 1a, enabling these strongly shifted peaks to be seen. There is structural information implicit in these shifted peaks but it is not easy to disentangle. We shall also say a few further words about them below.

In Figure 2 are shown the natural abundance ^{13}C spectra of oxidized cytochrome C, observed at 25 MHz (D. J. Patel, manuscript in prepara-

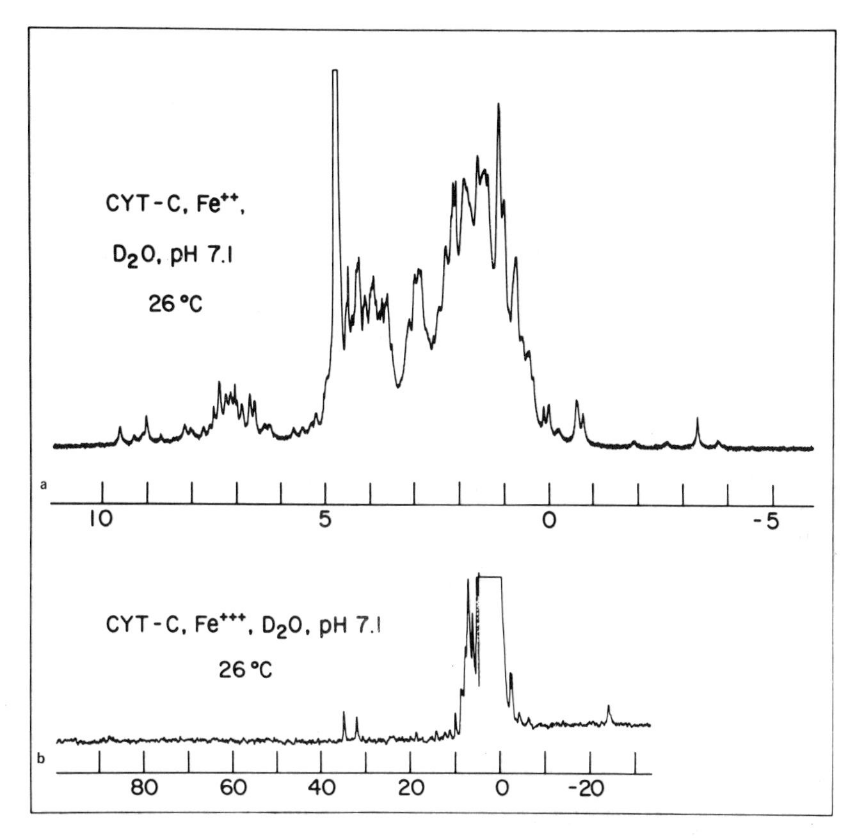

FIGURE 1 300-MHz proton spectra of (a) Fe^{2+} cytochrome C (diamagnetic) observed in D_2O at pH 7.1, 26 °C and (b) Fe^{3+} cyctochrome C (paramagnetic), same conditions with tenfold larger chemical shift scale. Chemical shifts expressed with respect to sodium 2,2-dimethyl-2-silapentane-5 sulfonate (DSS) reference as zero (D. J. Patel, manuscript in preparation).

tion). The carbon resonances are spread over a much greater chemical shift range than those of the protons and include carbonyl and quaternary carbons. Splittings caused by ^{13}C–^{1}H J couplings are removed by strong irradiation of the protons. By use of Fourier transform spectrum accumulation, sensitivity can be made sufficiently high so that at least some individual carbon nuclei can be observed, as here. A number of protein ^{13}C spectra have now been reported.[4] In Figure 2, as in all protein spectra, the region below 90 ppm (vs. $^{13}CS_2$ as zero) contains the resonances of all the unsaturated carbons–carbonyl (15–25 ppm) and aromatic (37–85 ppm)–and those of arginine $HN{=}C\langle^{NH_2}_{NH-}$ groups (ca. 36 ppm). At present, it is believed[4] that, because of dipolar broadening

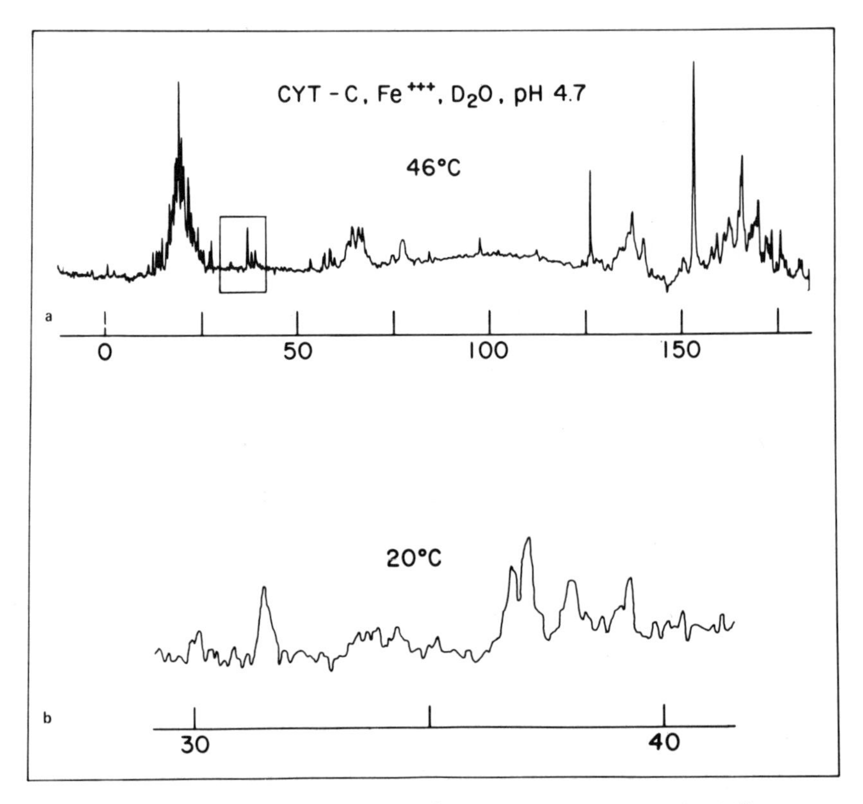

FIGURE 2 25-MHz carbon-13 spectra of Fe^{3+} cytochrome C in D_2O, pH 4.7, 46 °C: (a) complete spectrum; (b) expansion of 30–40 ppm region. The tyrosine-67 C-4 (hydroxyl-bearing) carbon appears at ca. 32 ppm. Chemical shifts expressed with reference to $^{13}CS_2$ as zero (D. J. Patel, manuscript in preparation).

by bonded protons, only quaternary carbons—i.e., the aromatic C-1 and C-4 of tyrosine, the C-1 of phenylalanine, the three quaternary carbons of the tryptophan indole side chain, histidine C-5, and the above arginine carbon—give discernible single-carbon resonances.[4]

Complete discrimination of all carbon resonances and their assignments can be accomplished at present only for relatively small peptides such as neurohypophyseal hormones containing 40–50 carbon nuclei.[5–7] For proteins, no assignments to particular carbons, as opposed to group assignments, have yet been made with certainty, although one tentative identification, which we shall not discuss further, is indicated in Figure 2. This is understandable, for such specific assignments clearly pose a very difficult task, although one that we do not feel to be insuperable. At present, however, protein ^{13}C NMR is still in the data-taking stage.

Some important structural clues can be obtained for larger, more complex proteins with subunit structure even though the resonances are seriously broadened at higher molecular weights. In the proton spectrum of hemoglobin, in addition to low- and high-field heme proton resonances,[8] Patel *et al.*[9] and Ogawa *et al.*[10] have observed significant low field peaks (at 12–14 δ), well removed from the main spectrum. Figure 3 shows the low field region of the proton spectrum of oxyhemoglobin, methemoglobin, and deoxyhemoglobin, all observed at 220 MHz. Such resonances are particularly important in cases such as this, where the main spectrum itself is almost completely intractable. We will discuss their interpretation below.

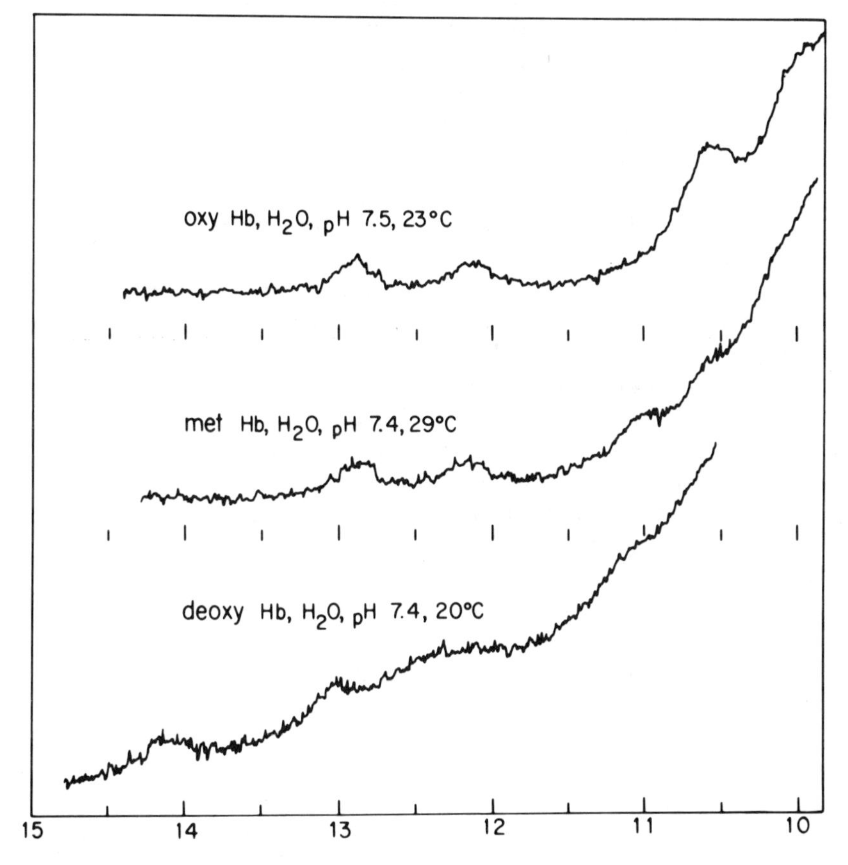

FIGURE 3 25-MHz proton spectra between 10 and 15 δ of human oxyhemoglobin, methemoglobin, and deoxyhemoglobin, observed in H_2O at pH 7.4.[9] Chemical shifts expressed with respect to sodium 2,2-dimethyl-2-silapentane-5 sulfonate (DSS) reference as zero.

Interpretation of the Data

To interpret protein chemical shifts in terms of structure, it is, of course, necessary to have some applicable theoretical framework or system of empirical correlations. One approximate distance-measuring device, long used by organic chemists, is the *ring-current shift.* This is a chemical shift effect arising from the presence of diamagnetically anisotropic groups, usually aromatic, the sign and magnitude of which have the approximate geometric dependence:

$$\Delta\nu = C \cdot r^{-3} (3 \cos^2 \theta - 1), \tag{1}$$

where C is related to the magnitude and degree of anisotropy of the group, r is the distance between the observed nucleus and the group, and θ is the angle between this vector and the symmetry axis of the group, which is taken here as a point dipole although more realistic models have been worked out. It is not a method of quantitative precision because of uncertainties in theory and because there are usually too many unknowns in the possible orientation of the anisotropic groups. Among the most interesting studies to make use of this concept are those of McDonald and Phillips[11] and Sternlicht and Wilson[12] on lysozyme, in which it was shown that a number of abnormally shielded high-field resonances can be assigned and explained in terms of the three-dimensional x-ray structure if proximity of aliphatic side chains to aromatic groups is taken into account. An even more striking instance is the above-mentioned three-proton peak at $-3.2\ \delta$ in the spectrum of reduced cytochrome C (Figure 1a). This is believed to be the methyl resonance[2,13] of methionine-80, which is known from x ray[14] to form the sixth ligand of the heme iron. The methyl group (and also the methylene) is known to be positioned in the shielding region ($\theta < 54.7°$) of the potent porphyrin ring current, and the upfield shift of over 5 ppm (compared with an unperturbed *S*-methyl group) is qualitatively reasonable although not yet tested quantitatively.

A class of potentially more precise and sensitive distance-measuring methods are those employing paramagnetic metal ions either as "shift reagents"[15] or as agents to enhance relaxation of the nuclei (^{1}H or ^{13}C) of bound ligands and the protein itself or of the protons of the aqueous solvent. For the latter, nitroxide spin labels may also be employed. If the paramagnetic metal ion has an axially symmetric, effectively anisotropic electron g tensor and can occupy a definite fixed site in the protein (preferably somewhere near the active site), it acts as a local magnetic dipole that perturbs the chemical shifts of the nuclei in its neighborhood. The magnitude of this so-called "pseudocontact" effect follows the geometrical relationship given above for diamagnetic groups, except that the constant C, which now contains the components of the electron g tensor and possibly crystal field terms as well,[16] is generally larger and may be either positive or negative depending on the nature of the ion and its binding site. The metal ion may replace a naturally occurring metal (Mn^{2+} for Mg^{2+}, for example) or may be bound to non-metalloenzymes, as Gd^{3+} is to lysozyme (*vide inferior*). The simple geometrical function implies some sort of effective symmetry axis passing through the ion. For the interaction of lanthanide ions (usually Eu^{3+} or Pr^{3+}) or their complexes with small substrate molecules, such a symmetry axis can usually be identified.[15] For the asymmetric environment that a protein binding site necessarily provides, the geometrical question clearly may be much more complicated. There is in addition a possibility of mixing in "real" contact interactions. These arise from electron spin delocalization and are responsible for at least some of the strongly shifted peaks in the spectra of cytochrome C and hemoglobin mentioned above. This possibility makes interpretation still more ambiguous, as such shifts do not obey simple distance relationships. In addition, one must be able to specifically assign the shifted peaks–no light task, since at present virtually the only assigned resonances of nuclei at or near active sites are those of histidine imidazole rings and of certain tryptophan resonances in lysozyme and ribonuclease.[17] In view of these difficulties, it is perhaps not surprising that there have as yet been few published results of such studies despite preliminary indications.[18]

If the paramagnetic ion is effectively isotropic rather than anisotropic and has a relatively long longitudinal (i.e., spin lattice) electron relaxation time T_{1e}, it acts as a line "broadener" rather than a "shifter." Gd^{3+} falls into this class. The broadening (in hertz) is given by $(\pi T_2)^{-1}$, where T_2^{-1}, the transverse or spin–spin relaxation rate of the observed nucleus i, is given by:

$$T_2^{-1} = \frac{1}{15}\frac{\mu^2\gamma^2}{r_i^6}\left[7\tau_c + \frac{13\tau_c}{1+\omega_s^2\tau_c^2}\right]. \tag{2}$$

Here, μ is the electron moment, γ the magnetogyric ratio of the species of nucleus being observed (^1H, ^{13}C, or possibly ^{19}F); ω_s is the angular precessional frequency of the electron in the field employed, and τ_c the characteristic average time of the modulation of the electron–nucleus interaction; r_i is the distance from the metal to the observed nucleus i. When the ion is tightly bound to the protein, τ_c becomes the rotational correlation time of the protein, usually of the order of 10^{-8} s.

Up to the present, this method has been principally applied to ternary complexes of metal, protein, and small-molecule ligand, which is commonly an inhibitor or substrate analog. The ligand resonances are usually observed rather than those of the protein. For a given complex all quantities, except r_i, are constant, and relative metal–nucleus distances should be determinable with great sensitivity owing to the strong distance dependence. Absolute distances, however, require a knowledge of τ_c. It must be assumed that the motion of the system is sufficiently isotropic that a single correlation time is appropriate. Even accepting this, the accuracy of estimates of τ_c, which can be made in several different ways in addition to the Stokes–Einstein approximation,[20] is not high and such estimates do not always agree well. One way of estimating τ_c is by measuring the ratio of T_1 to T_2, using the well-known theoretical development of Bloembergen, Purcell, and Pound[19] (somewhat modified by later authors). In such cases, it may not necessarily have a clear physical significance, but presumably may be "plugged in" anyway. Fortunately, such uncertaintics should not seriously affect the accuracy of distance measurements, because the calculated values of r_i should depend only on the one-sixth power of τ_c [Eq. (2)].

Examples of systems to which this approach has been applied include the ternary complex of lysozyme, Gd^{3+} (the binding site of which is known from x ray) and β-methyl-*N*-acetylglucosamine, a substrate analog[18,20]; the binding of pyruvate by pyruvate carboxylase and pyruvate kinase[21]; and the binding of α-methyl-*D*-glucopyranoside by concanavalin A.[22] In contrast to lysozyme, the last three are natural metalloproteins and the observations were of ^{13}C nuclei in the ligand rather than ^{1}H. It is of particular interest to note that the mean distance of the sugar nuclei from the metal (Mn^{2+}) in conconavalin A is found by the relaxation method (in this study T_1 was measured rather than T_2) to be 10 Å, whereas the value from x-ray crystallography is 20 Å. A real difference between the solution and crystal structures may be indicated here, possibly corresponding to different binding sites.

Wien, Morrisett, and McConnell[23] have employed nitroxide free-radical spin labels instead of paramagnetic metal ions. They found that lysozyme covalently spin-labeled at histidine-15, which is not at the active site, broadens the proton resonances of bound *N*-acetyl-α-D-glucosamine, enabling a distance (of the order of 20 Å) to be estimated, in agreement with the x-ray structure.

Another approach to the study of the active site structures of metalloenzymes, as indicated earlier, is the observation of the relaxation of the protons of the aqueous solvent when the natural divalent diamagnetic metal of the protein, usually Mg^{2+} or Ca^{2+}, is replaced by a paramagnetic ion that fits well and does not impair enzymic activity; Mn^{2+} commonly serves this purpose. Extensive studies of the large class of phosphoryl transfer enzymes have been carried out by this method.[24-26] Such studies depend on the observation that the spin lattice relaxation rates of water protons due to paramagnetic ions are very much enhanced when the ion is bound to a macromolecule,[27] owing to the greatly increased correlation time. The water molecules are in equilibration between the bound and free states; the paramagnetic ion may be in both bound and free (i.e., aquo) states, and the enzyme substrate may itself bind the metal. The situation can become so complicated and so many interactions may have to be measured that a computer must be used to sort things out. The appropriate τ_c may well not be the rotational time of the protein–ligand complex. Specifically bound nitroxide spin labels can be used here also, allowing use of electron spin resonance (ESR) as well as NMR.[23] In this way, active sites have been "mapped," and qualitative and quantitative features of the active sites of this class of enzymes have been delineated.

In these studies of phosphoryl transferases, new structural information is provided, which x ray for the most part has not provided or, at any rate, not yet. This is clearly a healthy sign for the employment of NMR. For most protein NMR work that has appeared so far, the final product of the experiment has been a comparison of a very limited number of distance parameters to the corresponding values obtained from the x-ray structure. This is acceptable in the testing stages of the NMR methods, but clearly the correctness and reasonableness of the NMR structural conclusions may be substantially aided by the fact that the investigator can, so to speak, look up the answer in the back of the book. One hopes (and in general believes) that the investigator's intellectual honesty will impel him to report discrepancies, for such discrepancies may well be real.

The interpretation of the low-field peaks observed by Patel *et al.*[9] in the spectrum of hemoglobin is of particular interest in that they so far remain unassigned to specific residues or even types of residues, yet have been shown to provide important information concerning the quaternary structure of this protein. It is, of course, well known that the steric relationship of the four subunits of hemoglobin changes slightly but significantly depending on whether the heme iron is liganded or nonliganded, for example, by oxygen. By use of spin labeling and paramagnetic ion binding, the approximate locations of these resonances within the hemoglobin structure have been determined.[10] The peaks at 12 and 13 δ are characteristic of the liganded structure, while that at 14 δ is characteristic of the unliganded structure.[9] By this means, it was shown that certain chemically modified and variant hemoglobins exhibit a single quaternary structure whether oxygen is bound or not. Further estimates of the rate of interconversion between the two quaternary structures can be made; rate constants are of the order of 10^3 s^{-1}.[10] This appears to be an instance in which NMR observations give no direct measure of any single structural parameter, but rather signal an entire structure.

CONCLUSION

NMR need not necessarily trudge along a trail already well blazed by x ray. Perhaps one of the strengths of NMR in protein studies, in addition to those we have discussed, may prove to be in mechanistic investigations, where x ray is at a serious disadvantage. Many years ago, Pauling proposed that enzymes function by stabilizing activated states of the molecules participating in the reactions they catalyze. X ray requires hours or days to accumulate structural information from the specimen. NMR is under a time disadvantage too, but not as severe. In time, as instrumental methods develop, direct observations of reacting substrates and their interactions with the active site may become possible; thus, NMR may truly complement x ray and even partially supplant it for obtaining such knowledge.

REFERENCES

1. F. A. Bovey, A. I. Brewster, D. J. Patel, A. E. Tonelli, and D. A. Torchia, *Acc. Chem. Res.*, **5**, 193 (1972).
2. K. Wüthrich, *Proc. Natl. Acad. Sci., U.S.A.*, **63**, 1071 (1969).
3. C. C. McDonald, W. D. Phillips, and S. N. Vinogradov, *Biochem. Biophys. Res. Commun.*, **36**, 442 (1969).
4. A. Allerhand, R. F. Childers, and E. Oldfield, *Biochemistry*, **12**, 1335 (1973); this paper gives references to earlier work, in most of which sensitivity was not high enough to observe single carbons.
5. A. I. Richard Brewster, V. J. Hruby, A. F. Spatola, and F. A. Bovey, *Biochemistry*, **12**, 1643 (1973).
6. R. Deslauriers, R. Walter, and I. C. P. Smith, *Biochem. Biophys. Res. Commun.*, **48**, 4 (1972).
7. I. C. P. Smith, R. Deslauriers, and R. Walter, in *Proceedings of the Third American Peptide Symposium* (Boston), J. Meienhofer, ed., Ann Arbor Science Publishers, Inc., Ann Arbor, Mich. (1972).
8. K. Wüthrich, R. G. Shulman, and T. Yamane, *Proc. Natl. Acad. Sci., U.S.A.*, **61**, 1199 (1968).
9. D. J. Patel, L. Kampa, R. G. Shulman, T. Yamane, and M. Fujiwara, *Biochem. Biophys. Res. Commun.*, **40**, 1224 (1970).
10. S. Ogawa, D. J. Patel, and S. Simon, in preparation.

11. C. C. McDonald and W. D. Phillips, *J. Am. Chem. Soc.*, **89**, 6332 (1967).
12. H. Sternlicht and D. Wilson, *Biochemistry*, **6**, 2881 (1967).
13. K. Wüthrich, *Struct. Bonding*, **8**, 53 (1970).
14. T. Takano, R. Swanson, O. B. Kallai, and R. E. Dickerson, *Cold Spring Harbor Symposia on Quantitative Biology*, Vol. 36, 397, Cold Spring Harbor Laboratory, Cold Spring Harbor, N.Y. (1971).
15. J. K. M. Sanders and D. H. Williams, *Nature*, **240**, 385 (1972); a review.
16. B. Bleaney, C. M. Dobson, B. A. Levine, R. B. Martin, R. J. P. Williams, and A. V. Xavier, *Chem. Commun.*, 791 (1972).
17. F. A. Bovey, *High Resolution NMR of Macromolecules*, Academic Press, New York (1972), Ch. XIV.
18. K. G. Morallee, E. Nieboer, F. J. C. Rossotti, R. J. P. Williams, A. V. Xavier, and R. A. Dwek, *Chem. Commun.*, 1132 (1970).
19. N. Bloembergen, E. M. Purcell, and R. V. Pound, *Phys. Rev.*, **73**, 679 (1948).
20. R. A. Dwek, R. J. P. Williams, and A. V. Xavier, in *Metal Ions in Biological Systems*, ed. by H. Sigel, Marcel Dekker, New York (1973).
21. C. H. Fung, A. S. Mildvan, A. Allerhand, R. Komoroski, and M. C. Scrutton, *Biochemistry*, **12**, 620 (1973).
22. C. F. Brewer, H. Sternlicht, D. M. Marcus, and A. P. Grollman, *Proc. Natl. Acad. Sci., U.S.A.*, **70**, 1007 (1973).
23. R. W. Wien, J. D. Morrisett, and H. M. McConnell, *Biochemistry*, **11**, 3707 (1972).
24. A. S. Mildvan and M. Cohn, *Adv. Enzymol.*, **33**, 1 (1970).
25. M. Cohn and J. Reuben, *Acc. Chem. Res.*, **4**, 214 (1971).
26. R. A. Dwek, *Adv. Mol. Relaxation Proc.*, **4**, 1 (1961).
27. J. Eisinger, R. G. Shulman, and W. E. Blumberg, *Nature*, **192**, 963 (1961).

Discussion Leader: HELEN BERMAN

Reporters: JAMES W. EDMONDS
CLARA B. SHOEMAKER

Discussion

BERMAN initiated the discussion with a brief summary of the three papers. She pointed out that a chief objective of crystallographic and NMR investigation of macromolecules is to explain biochemical phenomena. Thus, structural studies center on the active sites of enzymes, the prosthetic groups of proteins, the relationships among the catalytic and regulatory units of allosteric enzymes, and the elucidation of molecular genetics. X-ray studies provide static models from which aspects of function can be inferred. The quality of the models is critical and yet these are based on measurements and analyses beset by many difficulties. The phases obtained by multiple isomorphous replacement (MIR) techniques have rather large errors and yet the Fourier maps obtained show apparently convincing structural features. It is possible at 3 Å resolution to trace the polypeptide chains. Detailed fitting of the maps is based on information obtained from small-molecule structures. Refinement has been achieved by Diamond's method, by differential synthesis, and by unconstrained least squares refinement, which for the first time provided a good check of the MIR phases. The features one should look for in reports of protein structures are some objective test of the quality of phases such as the figure of merit, a statement of the resolution, and a careful residue-by-residue analysis of the protein map. NMR studies show properties of the proteins in solution and thus have the potential for observing kinetic properties of complexes in solution.

BERMAN commented that the two examples given by Bovey on NMR structures of proteins also had been studied by x-ray crystallography. She asked his opinion of studying systems that have *not* been studied by x rays. BOVEY mentioned the fine work by Mildred Cohn who, using the NMR nuclear relaxation techniques, has mapped the active site of several phosphoryl transferases. In general, however, he felt that most studies reported to date depended on knowing the x-ray results.

MULLER said about Bovey's paper that

"When NMR is used to determine the position of a small molecule bound to a protein carrying a paramagnetic atom, it should be borne in mind that the line broadening yields values of r^{-6} averaged over all values that this coordinate may take. It is not always clear *a priori* that one has rigid binding at a single binding site. If instead there is rapid exchange between two or more sites with quite different r values then r^{-6} will depend almost exclusively on the smallest possible r, even if the relative residence time at the corresponding site is appreciably smaller than 50 percent. Disturbing discrepancies between x-ray and NMR estimates of the location of the bound species could arise in this way."

BOVEY responded that this was true but that in the case of conconavalin A it was assumed that the sugar was bound at one site.

ZERBI asked whether there is any hope of detecting conformational changes in proteins by using NMR. BOVEY felt that the answer to this was yes and cited the example of oxy- and deoxyhemoglobin, where there were correlations between the changes in the NMR spectra and the quaternary structure of the protein.

ROBERT remarked that the proton NMR spectra of proteins are extremely difficult to interpret. Recently, however, a natural-abundance deuterium NMR spectrum has been published by Randall in which, of course, each deuteron is showing a single peak. Can one hope, in spite of the low gyromagnetic ratio of the deuteron, that the deuterium magnetic resonance will be a helpful tool in the near future? BOVEY agreed that this could be done but the difficulties are great and one would require spectral accumulation times of weeks or months.

VOS asked whether NMR techniques will rule out other techniques, such as circular dichroism, for finding out about structures in solution. BOVEY felt that optical rotatory dispersion and circular dichroism are techniques complementary to NMR; BUCKINGHAM pointed out that NMR does not provide information about enantiomorphs.

TEN EYCK made the following remarks about protein crystal structure:

"I agree with the comment that protein crystallographers are vast consumers of crystallographic information obtained from small-molecule studies; there are easier ways to determine the geometry of peptide bonds than by solving the structure of hemoglobin. The proper procedure is to use the available geometric information to solve the conformation of the biologically important portions of the structure, on which Johnson gave us a rating of 'speculative' (unfortunately correct). Considering the quality of the data, and the fact that the diffraction pattern is often dead by 2 to 2.5 Å resolution, most protein structures are not overdetermined sufficiently for proper refinement unless the bond lengths and some of the angles are taken as data. Jensen's unconstrained refinement of rubredoxin confirms this; in addition to demonstrating that convergent refinement is possible and desirable, it shows fairly large uncertainties in the parameters."

JOHNSON responded with his opinion that in the past most protein crystallographers believed they could not do refinements profitably, but that techniques are changing and that the field will soon reach a state where the precision of the geometrical parameters can be evaluated objectively. WILLIAMS asked just how much does the protein crystallographer rely on results of small-molecule crystallographic structures, and JOHNSON replied that heavy reliance is placed on these results in model building.

JENSEN pointed out that the structures of the bound molecules are also important in understanding enzyme mechanisms. He mentioned that with neutron diffraction it should be possible to locate the hydrogen atoms in the hydrogen bonds between the bound molecules and the protein.

DECAMP asked for a more quantitative idea of the accuracy of phasing and of the necessary precision in the measurements of the structure amplitudes. JENSEN said the figure of merit is the cosine of the error in phase angle so that if the figure of merit is, for example, 0.7, the angular error is about 45°. To find out the reliability of structure amplitudes, one can measure different crystals and symmetry-related reflections. JOHNSON mentioned Schoenborn's numerical results where, using either unit or correct amplitudes and the correct phases from a model, it is possible to see the molecule.

ABRAHAMS questioned whether there is information on the nature of the damage caused by x radiation to proteins (e.g., Does this cause conformation changes, bond splitting, or other types of degradation in long-range or short-range order?). JENSEN said it should be possible to use difference-map techniques to detect the type of damage being done and also pointed out that low-temperature experiments decrease the amount of decomposition by a factor of 10.

ABRAHAMS asked for comments on the Tukey–Schomaker proposal to measure overlapping data blocks, or measure each reflection at least twice, so that one can extrapolate back to zero exposure and thus correct for the effects of radiation damage on the x-ray intensity measurement. SCHOMAKER replied as follows:

"If the supply of crystal specimens is limited, if they decay seriously in the beam, and if, as seems to be typical, different reflections decay at significantly different rates, we are forced to conclude that each reflection measured on a particular specimen has to be measured twice (at least) on that specimen, so that the intensity can be extrapolated back to zero time. If counting time is the only issue, as in effect it is, and if radiation damage dominates the situation, the double counting costs little (divide the original time between the two counts) except for the extra counting time that will have to be spent to measure the slope of each I vs. t line. Because background always reduces the effective counting precision, this will be a problem only for weak reflections, whereas the precision at the strong counts will already be more than enough to sustain the extrapolation. For each reflection, however, it is easy (by least squares) to add a bit more information to the two net counts (and their weights); namely, the decay rate (and weight) that follows from the average decay rate (and spread of decay rates) of the representative set of standard reflections that will have to be followed carefully in any case. For a strong reflection the actual individual decay rate will still determine the extrapolation for a very weak reflection, for which the raw extrapolation from two measurements would be useless; the estimated slope from the standard will come in to rescue the situation. For intermediates, a reasonable mixture of information will be effected. If any significant correlation of decay rate with parameters such as sin θ/λ, odd or even index, etc., were to emerge, these results could be used for making a wiser choice of standards and a better extrapolation for each reflection."

ABRAHAMS asked for a comparison of the virtues of neutron vs. x-ray diffraction for protein crystallography. JOHNSON cited the neutron diffraction studies done by Schoenborn at Brookhaven. He pointed out that with neutrons one can differentiate hydrogen and deuterium atoms and that radiation damage is markedly decreased. Using deuterium exchange, one can locate the labile protons in the protein structure. It is also important that, using neutrons and an appropriate heavy atom such as Sm, one can use the differing anomalous dispersion at two wave lengths as a method of solving structures. Another effect that may be important is the distribution of hydrogen and deuterium between "short" and "other" hydrogen bonds, the lighter isotope tending to concentrate in the short hydrogen bonds.

BERMAN asked whether methods other than MIR have been successfully employed for x-ray crystal structures of proteins. JENSEN replied that, using the coordinates for tetragonal lysozyme, the Oxford group had solved triclinic lysozyme by the use of the rotation function. He felt there was no reason why the method could not be employed with molecular fragments. VOS wanted to know to what extent anomalous scattering has been used for phase determination in the x-ray case; JENSEN said it was possible to solve flavodoxin using a single Sm derivative by taking advantage of anomalous dispersion.

CAUGHLAN commented on the fact that a protein contains 20–40 percent water, a point earlier emphasized by TEN EYCK, and wondered how this would affect and limit the refinement. JENSEN pointed out that in rubredoxin about 130 water molecules have been put in, and in lysozyme there is a better chance of refining a greater fraction of the waters. EVANS pointed out that Jensen was able to locate many water molecules in the protein crystal and asked what can be learned about the structure of water from this information. JENSEN said that the water structure is indeed very interesting but that because of limitations of time most of the work has been concentrated on the structure of the protein itself. So far his group has not done a detailed analysis of the water structure. He also pointed out that *in vivo* many proteins may not be in solution but are bound, for example, to membranes within the cell.

SCHOMAKER was concerned about whether there are examples of wrong protein structure determinations and whether there is any relation to the low publication rate. JOHNSON replied that it is difficult to know when you are finished with a protein structure but that the confidence level is much higher with results at 2 Å than at, say, 6 Å. He felt that there was no great risk of completely wrong structures using MIR phases.

LORD pointed out that nothing had been said so far about the possibilities of laser–Raman spectroscopy as a technique for the study of molecular structure. He illustrated some of these possibilities by discussing the spectrum of the enzyme lysozyme:

"Raman spectroscopy provides the same kind of information as infrared spectroscopy, i.e., intensities, frequencies, and polarization of spectral lines due to molecular vibrations. Since these vibrational frequencies are usually characteristic of small groups of atoms and are furthermore dependent on the environment and specific conformation of these groups, they offer a source of qualitative and, sometimes, quantitative structural information.

"Figure 1 shows the Raman spectra of both native lysozyme and aqueous lysozyme denatured after 2 hours of heating at 100 °C. The band at 1250 cm^{-1} in native lysozyme (which is an unresolved collection of at least three frequencies) arises from a vibration in which the N–H bond of the peptide backbone bends parallel to the plane defined by the peptide group. The N–H group can be hydrogen-bonded in different ways; the frequency of this vibration is slightly different for α-helical and β-pleated sheet structures, and both of these differ from the frequency of the solvent-bonded NH. The shape and intensity of this band is invariant when the solution is heated from 32 to 76 °C. However, when lysozyme is denatured by chemical cleavage of the disulfide cross-links,[1,2] by high ionic strength (6M LiBr),[3] or by heating at 100 °C for 2 hours,[4] the effect of disrupting the helical and pleated-sheet structures shows up in an increase of the intensity of the 1250 cm^{-1} band together with a shift of its center of gravity to lower frequencies (~1245 cm^{-1}).

"The maximum at 509 cm^{-1} arises from the S–S bond-stretching vibrations of the four disulfide cross-links, and its intensity is proportional to the number of such cross-links. Since this peak is relatively sharp, we conclude that the conformations of the atoms bound to the disulfide groups are very similar for all four such groups in native lysozyme. Because the frequency, sharpness, and intensity of the peak change little upon heating from 32 to 76 °C, it appears that the number of disulfide bonds is unchanged and that the effect of the increase in temperature on the conformation of the protein in the immediate neighborhood of the disulfide groups is small. In the

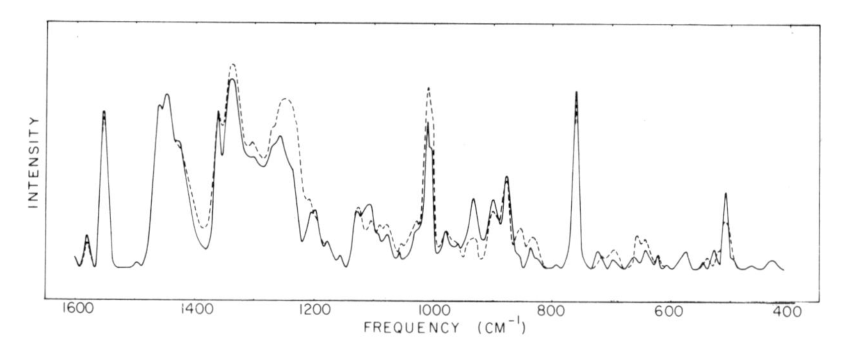

FIGURE 1 Solid curve: Raman spectrum of 7 percent aqueous lysozyme at 32 °C, pH5, 0.1 M NaCl, slit with 6 cm^{-1}. Dashed curve: Raman spectrum of 7 percent aqueous lysozyme at 32 °C after denaturation at 100 °C for 2 hours.

denatured compound, however, the peak intensity of this band drops substantially, perhaps enough to imply that one of the four S–S bridges has been permanently broken. A more likely explanation is that the conformations of the four disulfide groups have changed to different extents, leading to slight shifts in the S–S frequency and thus to a broader line with reduced peak intensity but constant area. This interpretation is reinforced by the increase in the intensity of the cystine C–S frequency at 661 cm^{-1}, which is known to be sensitive to the conformation of the cystine residue.[5]

"Finally when good crystals of a protein are available, it is generally possible to get good Raman spectra of the crystal; thus, one can compare spectra of crystalline and solute molecules; this has been done recently for lysozyme by Yu and Jo,[6] who concluded that the main-chain conformation of lysozyme is the same in the crystal as in solution. Also, the insensitivity of the lysozyme spectrum to temperature change in the range 68–76 °C is not necessarily in conflict with the changes in UV absorption, ORD, and NMR spectra over this range. These latter spectra depend on the environment and conformation of the aromatic chromophores and their associated protons in the amino acid side chains. If the conformations of the side chains alter without affecting the backbone structure significantly, there should be little effect on the disulfide and NH Raman bands."

ABRAHAMS commented that Lord's account of the detailed information on the main chain of protein molecules given by the Raman spectra, together with the complementary information provided by the NMR spectrum on the side-chain protons discussed by Bovey, leads to the following suggested experiment. We are concerned with the nature of the radiation damage suffered by proteins in the course of x-ray crystallographic measurement. The experiment is to follow by Raman and NMR the possible main-chain and/or side-chain changes that take place on irradiation. The protein sample most conveniently is in solution and contained within a cell fitted with a window transparent to x rays and would be measured in a Raman and then in an NMR spectrometer. The sample would be irradiated with CuK_α x rays for, say, 24 hours and the spectra remeasured. This would be repeated for several more stages of irradiation. No change in the spectra presumably would lead to the inference that radiation damage is primarily associated with loss of order in the solid. The more likely possibility of spectral change could be appropriately investigated.

REFERENCES

1. R. Mendelsohn, Ph.D. thesis, Massachusetts Institute of Technology, Cambridge (1972).
2. R. C. Lord, Proc. 23rd Congr. IUPAC, *Pure Appl. Chem.* (Suppl.), **7**, 179 (1971).
3. R. C. Lord and R. Mendelsohn, *J. Am. Chem. Soc.*, **94**, 2133 (1972).
4. M. C. Chen, R. C. Lord, and R. Mendelsohn, *Biochim. Biophys. Acta*, **328**, 252 (1973).
5. R. C. Lord and N. T. Yu, *J. Mol. Biol.*, **50**, 509 (1970).
6. N. T. Yu and B. H. Jo, *Arch. Biochem. Biophys.*, **156**, 469 (1973).

V

Vibrational Force Fields

I. M. Mills

Harmonic Force Field Calculations

MOTIVES FOR FORCE FIELD CALCULATIONS

The observed vibration frequencies of a molecule depend on two features of the molecular structure: the masses and equilibrium geometry of the molecule and the potential energy surface, or force field, governing displacements from equilibrium. These are described as kinetic and potential effects, respectively; for a polyatomic molecule the form and the frequency of each of the $3N-6$ normal vibrations depend on the two effects in a complicated way. The object of a force field calculation is to separate these effects. More specifically, if the kinetic parameters are known and the vibration frequencies* are observed spectroscopically, the object is to deduce the potential energy surface. A major difficulty in this calculation is that the observed frequencies are often insufficient to determine uniquely the form of the potential energy surface, and it is necessary to use data on the frequency shifts observed in isotopically substituted molecules or data on vibration/rotation interaction constants observed in high resolution spectra in order to obtain a unique solution.

The amplitudes of motion associated with one or two quantum excita-

* It would be better to use the phrase "vibration wave numbers," since data are generally observed and listed as wave numbers (usually in units of cm^{-1}, but occasionally in m^{-1}). However, it is common practice to use the word "frequency" in general discussions.

tion in the normal modes, as observed spectroscopically, are small compared with the bond lengths, typically around 0.1 Å, so we must recognize that information on the form of the potential surface can only be obtained for small displacements from equilibrium. An exception occurs for large-amplitude, low-frequency vibrations observed in nonrigid molecules, from which the potential energy surface can sometimes be obtained over large displacements, but such cases are not discussed here.

The potential energy surface $V(\mathbf{R})$, where $\mathbf{R}$ represents collectively some set of $3N-6$ internuclear coordinates, determines all the details of a vibration/rotation spectrum within the Born–Oppenheimer model; moreover, if $\mathbf{R}$ are geometrical coordinates, the function $V(\mathbf{R})$ is the same for different isotopic species. In our attempt to determine $V(\mathbf{R})$ from observed spectra we hope to achieve:

1. Data reduction: the consistent interpretation of all details of vibration/rotation spectra for many isotopic species in terms of a small number of force constants;
2. The provision of a model for use in analyzing other data, e.g., force constants to be transferred to more complex molecules containing similar groups, or normal coordinates for interpreting infrared intensity data or mechanisms of vibronic interaction; and
3. The provision of a meeting ground for comparing calculations of energy as a function of nuclear coordinates with the results of electronic wavefunction calculations.

It is sometimes necessary to make approximations or to use simplified models in making calculations on complex molecules, and it is then important to remember these motives in choosing the method of calculation.

REPRESENTATION OF THE POTENTIAL ENERGY

The potential energy is expanded as a power series in internal displacement coordinates $\mathbf{R}$; the coefficients are the force constants* in the chosen coordinate representation.

* If the force constants are to be regarded as derivatives of the potential energy V with respect to the coordinates $\mathbf{R}$, it is clearly important to include the coefficients $1/n!$ and to regard the summations as unrestricted as in Eq. (1). Although this is commonly done in harmonic calculations, anharmonic force constants are often defined in a different form in which the numerical factors are omitted and the summations are restricted.

$$V(\mathbf{R}) = \frac{1}{2}\sum_{i,j} f_{ij}R_iR_j + \frac{1}{6}\sum_{i,j,k} f_{ijk}R_iR_jR_k + \frac{1}{24}\sum_{i,j,k,l} f_{ijkl}R_iR_jR_kR_l + \ldots. \tag{1}$$

The leading terms involve the quadratic force constants f_{ij}, and for small displacements these are clearly dominant. In harmonic force field calculations all higher order terms are neglected, and this is the case in this paper.

For a given force field, the force constants depend on the choice of coordinates, and this choice deserves attention. First, it is usual to choose valency coordinates, bond stretches and angle bends, such that displacement of a single coordinate while holding the others undisplaced leads to the stretching of a single bond or the bending of a single angle in the molecule (see, for example, Wilson *et al.*,[1] p. 55). These are thought to provide the most meaningful representation of the force field. Second, there is a choice between "rectilinear" or "curvilinear" valency coordinates: The former are related to cartesian displacements by a linear transformation, such that the atoms move in straight lines through a distance proportional to the coordinate displacement, however large this may be, whereas the latter represent true angle bends or true bond stretches *holding all other coordinates undisplaced,* even for large displacements in the one coordinate (Figure 1). The difference between these representations, which is discussed in Hoy *et al.*,[2] does not affect the quadratic force constants, but appears in the cubic and quartic constants, so that the choice might seem to be unimportant in a harmonic force field calculation; however, in effect rectilinear coordinates are always used in setting up the vibrational Hamiltonian because it is only in this way that the effective masses in the kinetic energy can be treated as constants. (In anharmonic calculations, where the difference is important, curvilinear coordinates are always used to represent V, and this introduces the complication of a nonlinear coordinate transformation.)

FIGURE 1 Distinction between "rectilinear" (*left*) and "curvilinear" (*right*) representations of valency bond-angle displacements.

A third difficulty arises in molecules involving many bond angles around a central atom, or in ring structures, in that the valency coordinates number more than $3N-6$ and are not all independent. For example in CH_4 there are six interbond angles, and it is not possible to displace just one of these while holding the other five undisplaced. Such coordinates are said to include a redundancy, and it is usual to transform to a smaller number of independent coordinates (which may be symmetry coordinates) and to represent the force field in these independent coordinates before setting up the vibrational problem. Many force fields in the literature are quoted in a redundant coordinate representation, but it is important to note that when this is done the force constants are *never unique;* that is, the same potential function can be represented by many different sets of force constants.[3] For this reason I am opposed to presenting the results of force constant calculations in terms of redundant coordinates (with one exception related to model force fields discussed below), although it is common practice among many workers in the field. It is my view that force constants in a nonredundant representation are fewer in number, just as transferable between related molecules, and are altogether more meaningful.

A fourth difficulty arises from the dimensional difference between angle-bending and bond-stretching coordinates. Unless the units are chosen with great care, bond-stretching and angle-bending force constants have quite different magnitudes as well as different dimensions, and this causes confusion. Some authors scale their angles to give them the dimensions of length, others do not, and yet others do not say which they have done. I would like to propose a solution to this problem that would also simplify the units of all force constants: Make the bond-stretching coordinates dimensionless by writing them as $\delta r/r$, rather than scale the angle-bending coordinates. All force constants then have the dimension of energy, and a convenient unit is the attojoule, aJ = 10^{-18} J, equal to 1 millidyne angstrom (mdyn Å), so that for a bond length of 1 Å the familiar force constant in mdyn Å^{-1} is numerically equal to the proposed new constant in aJ. The effect on the force constants of the hydrogen halides is shown in Tables 1 and 2, and on the

TABLE 1 Force Constants for HX Molecules in Terms of Coordinate δr

Molecule	r_e/Å	f_{rr}/(mdyn Å^{-1})	f_{rrr}/(mdyn Å^{-2})	f_{rrrr}/(mdyn Å^{-3})
HF	0.917_1	+9.658	−71.1	+478
HCl	1.274_6	+5.162	−28.7	+140
HBr	1.414_3	+4.116	−21.2	+94
HI	1.609_0	+3.140	−15.0	+59.3

TABLE 2 Force Constants for HX Molecules in Terms of Coordinate $\delta r/r$

Molecule	r_e/Å	f_{rr}/aJ	f_{rrr}/aJ	f_{rrrr}/aJ
HF	0.917_1	$+8.11_6$	−54.8	+338
HCl	1.274_6	$+8.38_8$	−59.5	+369
HBr	1.414_3	$+8.23_4$	−60.0	+376
HI	1.609_0	$+8.13_0$	−62.4	+397

harmonic force constants of bent XY_2 molecules in Table 3. I suggest that this representation of the force field gives a more meaningful comparison between related molecules than that in common use, as well as overcoming many of the troublesome problems of dimensions and units.*

Calculation of Force Constants from Observed Spectra

The relationship between the potential function $V(\mathbf{R})$ and the observable spectroscopic parameters is summarized in Figure 2. The harmonic vibration frequencies are obtained as the eigenvalues of a secular determinant involving the quadratic force constants and the atomic masses and molecular geometry (the **F** and **G** matrices of Wilson's well-known formalism) by a calculational procedure discussed in detail by Wilson, Decius, and Cross.[1] The eigenvectors determine the normal coordinates **Q** in terms of which the kinetic and quadratic potential energy terms are both diagonal ($\mathbf{R} = \mathbf{LQ}$). The various anharmonicity constants and vibration/rotation interaction constants are obtained in terms of the

TABLE 3 Force Fields of Symmetric Bent Triatomic Molecules in Terms of Coordinates $\delta r_1/r$, $\delta r_2/r$, and $\delta\alpha$

Molecule	r_e/Å	α_e/°	f_{rr}/aJ	$f_{rr'}$/aJ	$f_{r\alpha}$/aJ	$f_{\alpha\alpha}$/aJ
H_2O	0.957	104.5	7.74_6	-0.09_2	$+0.21_0$	0.697
H_2S	1.336	92.1	7.64_2	−0.03	+0.07	0.75_8
H_2Se	1.460	91.0	7.4_8	−0.05	$+0.1_9$	0.71
O_3	1.272	116.8	9.9_7	$+2.4_2$	$+0.6_8$	2.1_2
SO_2	1.431	119.3	21.3	$+0.2_5$	$+0.7_5$	1.6_8
SeO_2	1.608	113.8	11.1_1	$+0.0_7$	$+0.0_2$	1.2_6
OF_2	1.405	103.1	8.3_6	$+1.7_5$	$+0.3_5$	1.4_6

* Since writing this I have learned that Professor Y. Morino is independently proposing this system to the IUPAC Commission on Molecular Structure and Spectroscopy.

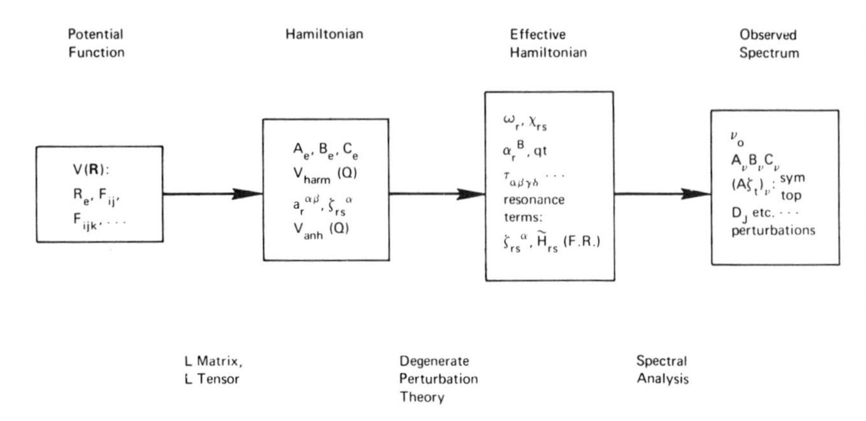

FIGURE 2 Relationship between the potential function and observable spectroscopic parameters.

eigenvectors (the **L** matrix) and the higher order force constants appearing in the Hamiltonian as formulated in terms of the normal coordinates. The calculation of spectroscopic observables from $V(\mathbf{R})$ is a complex but straightforward procedure, made painless by the use of electronic computers.

In practice, however, we wish to calculate the force constants from the spectroscopic observables, and this is more difficult:

1. In general, the data tend to be insufficient to determine uniquely all the force constants.
2. The data generally include anharmonic effects that are only approximately known, for which it is not possible to correct.
3. These difficulties both become complicated in the technicalities of the calculation, due to the very nonlinear relationship between the force constants and the data.

The first difficulty can be understood as follows: In a symmetry species of n vibrations there are $\frac{1}{2}n(n+1)$ independent quadratic force constants. Thus, given the n observed vibration frequencies, there are

$$\tfrac{1}{2}n(n+1) - n = \tfrac{1}{2}n(n-1) \qquad (2)$$

degrees of freedom in the force field. Thus to attempt to determine the force field from the observed frequencies of a single isotopic species is a process of data expansion rather than data reduction; the solution is clearly not unique.

The first and most important source of additional data to overcome this problem is the changes in vibration frequencies on making an isotopic substitution. The frequencies and normal coordinates change due to the change in mass, but they are functions of the *same force constants* (provided that the coordinates are geometrically defined). The two B_{2u} species vibrations of benzene may be used to illustrate the use of such data. The observed data are as follows:

	ν_{14}/cm^{-1}	ν_{15}/cm^{-1}
C_6H_6	1309	1146
C_6D_6	1282	824

If we use a symmetry coordinate representation defined by:

$$S_{14} = (-R_1 + R_2 - R_3 + R_4 - R_5 + R_6)/\sqrt{6}$$

$$S_{15} = R_0(-\beta_1 + \beta_2 - \beta_3 + \beta_4 - \beta_5 + \beta_6)/\sqrt{6},$$

where R_i and β_i denote stretching of one of the C–C bonds and bending of one of the C–H bonds, respectively (Figure 3), the force constants $F_{14,14}$ and $F_{15,15}$ may be calculated as a function of the interaction constant $F_{14,15}$ to give the graphs illustrated in Figure 4. Any vertical section of the graph defines a force field for C_6H_6, through the points of intersection with the solid lines, that *exactly* reproduces the frequency data; the graph thus displays possible force fields as a function of the interaction constant $F_{14,15}$, regarded as an independent variable, and similarly for C_6D_6. The points of intersection of the solid (C_6H_6) and dashed (C_6D_6) lines lie on two vertical sections, and define two different force fields *either of which* will exactly reproduce the fre-

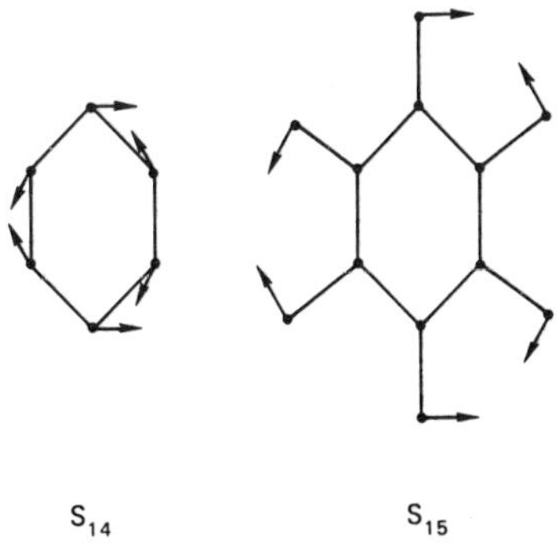

FIGURE 3 Symmetry coordinate representations S_{14} and S_{15} for benzene.

S14 S15

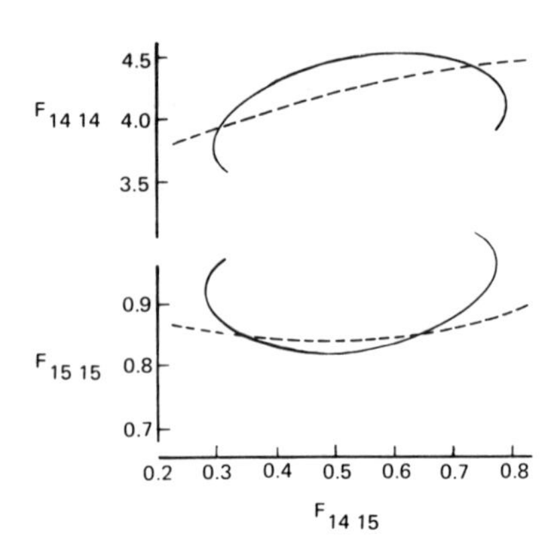

FIGURE 4 Force constants $F_{14,\,14}$ (*upper*) and $F_{15,\,15}$ (*lower*) plotted against $F_{14,\,15}$ for benzene corresponding to the symmetry coordinate representations in Figure 3. The solid curves are for C_6H_6 and the dashed curves for C_6D_6.

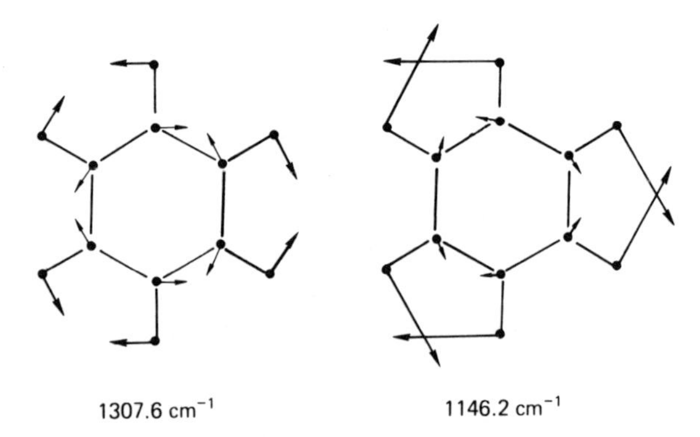

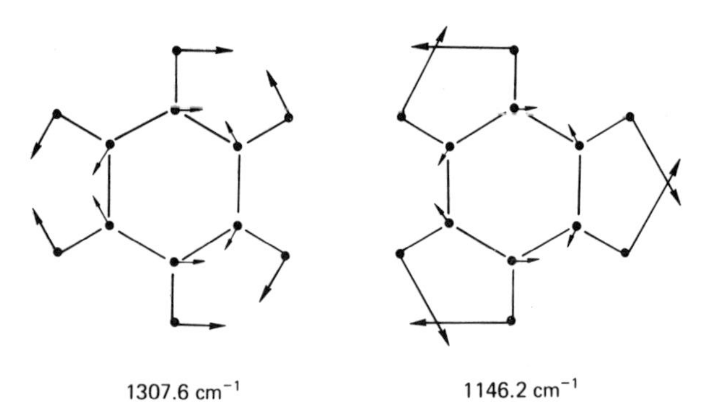

FIGURE 5 Forms of the normal coordinates for the two solutions for benzene corresponding to the higher and the lower value of the interaction constant $F_{14,\,15}$ given by the comparison of C_6H_6 with C_6D_6 shown in Figure 4.

quency data on both isotopic species. Thus in this case the isotopic data give *two* well-defined solutions to the force field. This situation arises from the nonlinear relation between frequencies and force constants and is not uncommon; usually, it is easy to choose between the two solutions by chemical intuition, but not always, as this example shows. The forms of the normal coordinates for the two solutions for C_6H_6 are illustrated in Figure 5, and the **L** matrices are as indicated below; it will be seen that the two force fields give quite different forms for the normal coordinates.

First Solution

$$F/(\text{mdyn Å}^{-1}) \quad \begin{matrix} & S_{14} & S_{15} \\ S_{14} & 3.93 & +0.24 \\ S_{15} & \text{sym} & 0.823 \end{matrix}$$

$$\nu_{\text{calc}}/\text{cm}^{-1} \qquad 1307.6 \qquad 1146.2$$

$$L/u^{-\frac{1}{2}} \quad \begin{matrix} S_{14} \\ S_{15} \end{matrix} \begin{bmatrix} +0.45 & +0.21 \\ -0.69 & +0.77 \end{bmatrix}$$

Second Solution

$$F/(\text{mdyn Å}^{-1}) \quad \begin{matrix} & S_{14} & S_{15} \\ S_{14} & 4.34 & +0.656 \\ S_{15} & \text{sym} & 0.823 \end{matrix}$$

$$\nu_{\text{calc}}/\text{cm}^{-1} \qquad 1307.6 \qquad 1146.2$$

$$L/u^{-\frac{1}{2}} \quad \begin{matrix} S_{14} \\ S_{15} \end{matrix} \begin{bmatrix} +0.45 & -0.21 \\ +0.16 & +1.02 \end{bmatrix}$$

A similar situation for pyramidal XY_3 molecules, in particular for NF_3, has recently been discussed by Hoy, Stone, and Watson.[4]

In general, both H/D isotopic substitution and relatively small isotopic substitution ($^{10}B/^{11}B$, $^{12}C/^{13}C$, $^{14}N/^{15}N$, $^{16}O/^{18}O$, etc.) provide useful information.[5] The sensitivity of a frequency to a small mass change may be thought of as providing information on the *magnitude of displacement* at the substituted atom for the normal mode concerned, without giving any information on the *direction* of displacement.

Further information on the force field comes from the observation of vibration/rotation interaction constants. These are obtained from infrared and Raman spectra at high resolution and from microwave spectra observed in vibrationally excited states. They are of the following types:

1. Coriolis zeta constants, $\zeta_{r,s}^{(\alpha)}$
2. Centrifugal distortion constants, $\tau_{\alpha\beta\gamma\delta}$
3. Inertia defects of planar molecules, Δ
4. l-type doubling constants of linear molecules, q_t.

Occasionally, useful information has also been obtained from one further source, namely,

5. Mean square amplitudes of vibration determined from gas-phase electron diffraction studies.

Coriolis Constants

These characterize the Coriolis interaction, due to rotation about an axis α, between two normal modes of vibration Q_r and Q_s; they are defined by the formula[6]

$$\zeta_{r,s}^{(\alpha)} = \sum_k |\ell_{k,r} \times \ell_{k,s}|_\alpha, \tag{3}$$

where the vectors $\ell_{k,r}$ define the form of the normal coordinate Q_r in terms of the mass-adjusted vector displacements of the atoms:

$$m_k^{\frac{1}{2}}\, \delta \mathbf{r}_k = \sum_r \ell_{k,r} Q_r. \tag{4}$$

Thus the zeta constants are sensitive to the *eigenvectors* of the vibrational secular equation, and it is clear that this is potentially useful information to add to the observed vibration frequencies or *eigenvalues.* Unfortunately, it is only when the two vibrations ν_r and ν_s are degenerate or nearly degenerate that the rotational structure is sufficiently perturbed to allow one to deduce the value of the zeta constant from rotational analysis. The zetas are dimensionless and of magnitude $1 > \zeta > -1$ [from Eq. (3)]. Their values for the interaction between two components of a degenerate vibration in a symmetric top or spherical top molecule may be relatively easily determined[7] with a precision of ±0.02, and for accidentally near-degenerate vibrations, for which $|\nu_r - \nu_s| < 20\, B^{(\alpha)} \zeta_{r,s}^{(\alpha)}$ (where $B^{(\alpha)}$ is the rotational constant about the α axis), with a precision of about ±0.05 to ±0.1. (See, for example, Smith and Mills[8] and di Lauro and Mills.[9]) In other cases the Coriolis perturbations cannot generally be separated from other interactions in the rotational analysis, in particular from anharmonic contributions to the vibrational dependence of the rotational constants, so that it is not possible to obtain useful information.

Figure 6 illustrates the use of an observed zeta constant for ethylene, $\zeta_{7,10}^{(z)}$, to determine the B_{2u} species force field.[8] Since the form of Q_7 is in this case determined by symmetry, the observed zeta may be regarded as a source of information on the form of Q_{10}, which is one of the two B_{2u} modes Q_9 and Q_{10}. The graph directly illustrates the accuracy with which the force field is determined by the observation. Other examples of molecules where zeta constants for degenerate vibra-

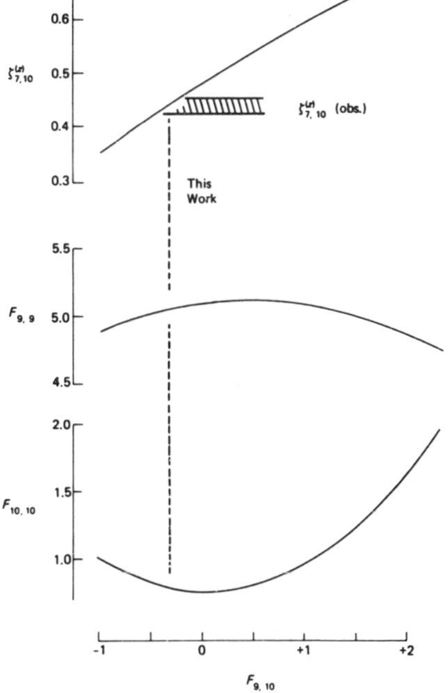

FIGURE 6 Use of observed zeta constant $\zeta_{7,10}^{(z)}$ for ethylene to determine the force field.

tions have been used in force field calculations include many C_{3v} methyl group molecules[10]—cyclopropane,[11] allene,[12] and similar molecules, the largest molecule to date probably being benzene.[13] For nondegenerate vibrations examples are O_3,[14] SO_2,[15] OF_2,[16] and many similar triatomics. Rather few zetas relating nondegenerate vibrations have been determined for polyatomic molecules and even fewer have been directly used in force constant calculations. The best example is probably a recent calculation[17] on the force field of ethylene.

Centrifugal Distortion Constants

These are determined from an analysis of rotational structure for high values of the rotational quantum numbers; a sophisticated analysis is generally required, as discussed by W. H. Kirchhoff in a following paper. For an asymmetric top, five independent linear combinations of the τ's may be determined[18] and for a symmetric top, only three. The general relation to the force field is given by the following formula[19]:

$$\tau_{\alpha\beta\gamma\delta} = -\frac{1}{2}\sum_i \sum_j \frac{J_i^{(\alpha\beta)} J_j^{(\gamma\delta)} (F^{-1})_{ij}}{I_{\alpha\alpha}^{(e)} I_{\beta\beta}^{(e)} I_{\gamma\gamma}^{(e)} I_{\delta\delta}^{(e)}}, \tag{5}$$

where

$$J_i^{(\alpha\beta)} = (\partial I_{\alpha\beta}/\partial R_i), \tag{6}$$

$I_{\alpha\beta}$ is an element of the inertia tensor ($\alpha, \beta = x, y,$ or z), and the R_i are the coordinates used to define the force constants **F**. Thus the τ's are sensitive to the elements of the inverse **F** matrix, $\mathbf{F}^{-1}$, sometimes called the compliance matrix and denoted **C**.

In practice, centrifugal distortion constants have proved most useful in determining the force fields of small asymmetric top molecules, particularly bent triatomics, as described by Kirchhoff (this volume). For larger molecules there is often a tendency for the τ's to be sensitive only to the most important diagonal force constants governing the lowest frequency "breathing" vibration, which may often be quite well determined from other data. However, if centrifugal distortion constants have been observed, they should certainly be used to test any calculated force field even if they are not used directly in deriving the force field. In general, the centrifugal constants in the ground vibrational state should be reproduced within 2 or 3 percent, the limitation arising from small zero-point contributions that generally cannot be determined.

Inertia Defects in Planar Molecules

As discussed by Laurie in an earlier paper, the observed rotational constants in any one vibrational state can be converted into effective moments of inertia through the relation

$$I_v = h/(8\pi^2 B_v). \tag{7}$$

For a planar molecule, the effective moments of inertia show a small positive inertia defect

$$\Delta_v = (I_c - I_a - I_b)_v. \tag{8}$$

The vibrational dependence of the inertia defect is given by formulas developed by Laurie and Herschbach[20] and also by Oka and Morino[21] involving the harmonic vibration frequencies and the Coriolis zeta constants:

$$\Delta_{\text{vibrational}} = \underset{\substack{\text{in plane}\\ \text{modes}}}{\sum_r \Delta_r (v_r + \tfrac{1}{2})} + \underset{\substack{\text{out-of-plane}\\ \text{modes}}}{\sum_s \Delta_s (v_s + \tfrac{1}{2})} \tag{9}$$

$$\Delta_r = (h/\pi^2 c) \sum_t \frac{\omega_t^2}{\omega_r(\omega_r^2 - \omega_t^2)} \left\{ \left(\zeta_{r,t}^{(a)}\right)^2 + \left(\zeta_{r,t}^{(b)}\right)^2 - \left(\zeta_{r,t}^{(c)}\right)^2 \right\} \tag{10}$$

$$\Delta_s = (h/\pi^2 c) \sum_t \left[\frac{\omega_t^2}{\omega_s(\omega_s^2 - \omega_t^2)} \left\{ \left(\zeta_{s,t}^{(a)}\right)^2 + \left(\zeta_{s,t}^{(b)}\right)^2 \right\} + \frac{3}{2\omega_s} \right] \tag{11}$$

Thus if the vibrational dependence of the inertia defect can be determined experimentally, either from the observation of microwave spectra in the ground and excited vibrational states or from high resolution vibration/rotation spectra, then this provides information on the Coriolis constants and hence on the forms of the normal coordinates and the harmonic force field.

In practice, inertia defects have been used successfully as a source of information on the harmonic force field only for bent triatomic molecules, particularly for O_3, SO_2, SeO_2, OF_2, and similar molecules.[14-16] For bent triatomics the vibrational inertia defect may be written in terms of the single zeta constant $(\zeta_{1,3}{}^{(c)})^2$, by using the zeta sum rule,[15] so that a measurement of the inertia defect may be regarded as a measurement of this one zeta constant, whose relation to the force field has already been discussed. For larger molecules the inertia defect tends to be insensitive to those degrees of freedom in the force field that remain after the vibration frequencies have been used as data; in other cases, the inertia defect has only been used as a check on the force field finally obtained.

ℓ-Type Doubling Constants

Excited vibrational states of linear molecules show a vibrational degeneracy that is split by rotation, the splitting being characterized by an ℓ-type doubling constant denoted q_t (where v_t is the doubly degenerate vibration involved). These ℓ-type doubling constants are readily determined, with high precision, from excited vibrational state microwave spectra, and vibration/rotation theory shows that they are given by a formula involving the harmonic force field only. Thus, they are a potential source of information on the harmonic force field. (Similar comments apply to symmetric top molecules, except that in this case the formula also involves the cubic anharmonic force field.)

In practice, however, the ℓ-doubling constants have always proved to be insensitive functions of the unknown parameters in the force field, to such an extent that they have not proved to be a useful source of information. (The converse of this statement is that ℓ-type doubling constants can be reliably calculated from poorly determined force fields.)

Mean Square Amplitudes of Vibration

Finally, there is one further source of information on the harmonic force field that has been used occasionally, namely *mean square amplitudes of vibration* in the various internuclear distances, as observed by gas-phase, electron-diffraction techniques. These can be measured experimentally from the widths of the peaks observed in the radial distribution function obtained from the Fourier transform of the observed diffraction pattern. They are related to the harmonic force field as follows.[22] If $\langle \ell_{mn}^2 \rangle$ denotes the mean square displacement in the distance between atoms m and n, then the mean amplitudes $\langle \ell_{mn}^2 \rangle$ are given as the diagonal elements of a matrix Σ, where

$$\mathbf{\Sigma} = \mathbf{V}\,\mathbf{L}\,\mathbf{\Delta}\,\mathbf{L}^\dagger\,\mathbf{V}^\dagger \qquad (12)$$

Here **V** is a matrix giving the change in the interatomic distances as a function of the internal coordinates **R**,

$$\ell_{mn} = \sum_i V_{mn,i} R_i, \tag{13}$$

L is the usual matrix relating **R** to **Q**, **R** = **LQ**, and **Δ** is a diagonal matrix of the thermally averaged mean square amplitudes of vibration in the normal modes:

$$\Delta_{rr} = \langle Q_r^2 \rangle = \coth(hc\omega_r/2kT). \tag{14}$$

The difficulty in using mean square amplitudes as a source of information on the force field is that the precision with which the quantities $\langle \ell_{mn}^2 \rangle$ can be determined experimentally is, in general, barely sufficient to make the information useful, even with the most precise data currently available. Exceptions are the molecules PCl_3 [ref. 23] and some C, Si, and Ge tetrahalides,[24] for which the electron diffraction patterns have been particularly well analyzed and useful information on the force field has been obtained. Mean square amplitudes may also sometimes be used to distinguish between two alternative discrete solutions to the force field of the type discussed earlier for the B_{2u} species of benzene; such examples have been discussed recently by Hoy, Stone, and Watson.[4]

ANHARMONIC EFFECTS AND METHODS OF CALCULATION

The previous section consisted mainly of a discussion of the possible sources of data on the harmonic force field. We must now emphasize an important limitation on all the data discussed there: namely, the effect of small higher order terms in both the analysis of the data and the theory arising from the effects of anharmonicity. Thus the observed vibration wave numbers of a molecule (ν_r) differ from the harmonic wave numbers (ω_r) with which it would vibrate in the *absence of anharmonic terms* in the force field by terms involving the anharmonicity constants x_{rs}[7]; the difference typically amounts to 2 or 3 percent of ω_r, rising to perhaps 5 percent for hydrogen-stretching vibrations. Similar comments apply to Coriolis zeta constants, centrifugal distortion constants, and all other sources of data: There are always anharmonic effects due to higher order terms, which generally remain unknown and uncorrected, but which typically amount to between 3 and 5 percent of the observed parameters. It is usually these model errors, rather than experimental uncertainties, that limit the uncertainty of the observed data used to obtain a force field.

The general method of using observed data to calculate the force constants is the method of nonlinear least squares. Thus, from a first trial set of force constants the data are calculated and compared with those observed, and the weighted sum of squares of errors is formed. The derivatives of the calculated data with respect to the force constants are then formed, and these are used to construct the normal equations from which corrections to the force constants are calculated in such a way as to minimize the sum of weighted squares of residuals. Because the relations between the data and the force constants are often very nonlinear, it is necessary to cycle this calculation until the changes in the force constants drop to zero when the calculation will have converged and the sum of weighted squares of errors will be minimized. The usual statistical formulas are then used to obtain the variance/covariance matrix in the derived best estimates of the force constants, and the estimated standard errors in the force constants are usually quoted along with their values. The whole procedure is referred to as a "force constant refinement calculation."

There are several difficulties in this calculation. The choice of weights is not easy, and the results obtained are sometimes sensitive to this choice. In general, the formula $w_i = 1/\sigma_i^2$ is used, where w_i is the weight and σ_i is the estimated uncertainty in the ith datum; σ_i is estimated to represent the unknown anharmonic corrections to the data rather than the experimental uncertainties. In work in our own laboratory we generally use $\sigma_i = 0.01\ \omega_i$ for vibration wave numbers, $\sigma_i = 0.02$–0.10 for zetas, and $\sigma_i = 0.03\ \tau_i$ for distortion constants, unless there is a genuine experimental uncertainty that exceeds this value. Another important point arises with small isotopic frequency shifts (resulting from ^{12}C to ^{13}C isotopic substitution, etc.); in such cases, the *shift* is generally known with greater precision than either of the individual vibration wave numbers, because the uncertainty arises from an unknown anharmonic correction, which can, however, be assumed to be essentially unchanged between the two isotopic species. Thus it is important *either* to use the wave number of the parent isotope ω_r and the wave number shift $\Delta\omega_r$ as data, using a much lower uncertainty for $\Delta\omega_r$ than for ω_r, *or* to use the two observed wave numbers with a nondiagonal weight matrix representing the correlation between the two data. The former procedure is more usual; see, for example, papers by McKean and coworkers,[5,17] who originally drew attention to this point, and Robiette *et al.*[25] on SiH_3F. A similar problem may arise in using τ's determined from analysis of microwave spectra. Since these are generally correlated, the input to the force constant refinement calculation should be made with a nondiagonal weight matrix obtained as the inverse of the variance/covariance matrix of the best estimates of the τ's. This procedure has not been followed in any calculation published to this date–that I am aware of–other than those described by Kirchhoff (this volume).

Another difficulty arises from the nonlinear nature of the calculation, which may often cause the "surface" of sum of weighted squares of residuals to have multiple minima as a function of the force constants, so that it may be possible to converge onto several different minima by starting from different trial force fields in the refinement. This can be particularly troublesome when the data are only just sufficient to determine the force field, so that the normal equations are somewhat ill-conditioned. The nature of the calculation is reminiscent of S. D. Colson's (private communication) description of a many-parameter nonlinear least squares calculation as: "comparable to the problem of finding the lowest point in the north American continent, given a horse and a barometer." The difficulty of multiminima in force-constant refinement calculations becomes more acute in ill-conditioned calculations, in which the data are only barely sufficient to determine the force field; in well-conditioned calculations it is less of a problem.

One must finally add a word of caution about quoted standard errors on calculated force constants. Experience comparing force fields obtained for the same molecule in different laboratories, using different methods of calculation and different data, suggests that the standard errors obtained from a statistical analysis are generally optimistic by a factor of 3 to 5 in assessing the reliability with which we truly know the force field. This is probably due to an optimistic assessment of the "standard error of an observation of unit weight" in the calculation, arising from the fact that the residuals obtained in fitting the data are often much smaller than the known uncertainties (arising from anharmonicity) in the data. Thus quoted standard errors in forces constants obtained in least squares calculations should not be taken too literally.

CONSTRAINTS AND MODEL FORCE FIELDS

When a force constant refinement calculation is ill-conditioned because of a lack of sufficient data, the only way to obtain a satisfactory refinement is to impose constraints on the force field. These may be summarized as follows:

1. Certain force constants may be constrained to zero, particularly when they represent interactions between remote coordinates in the

molecule. The choice of such constraints is a matter of intuition and judgment.

2. Certain force constants may be transferred from the known force fields of similar groups in chemically related molecules. Again, the success of this procedure depends on the judgment of the individual. Snyder and Schachtschneider's[26] force field for saturated hydrocarbons is an example of the success of the skilled use of constraints of the type 1 and 2 to obtain force fields on molecules for which there would otherwise be too few data.

3. One may use a "model force field," based on some simple assumptions about the nature of the forces between atoms in the molecule, to constrain the force field. The best known of these is the Urey–Bradley force field[27]; another that has been used with some success is the hybrid-orbital model.

The difficulty with constrained force field calculations is that from their very nature it is not possible to be certain of the extent to which the constraints imposed are justified. If one's motive is to obtain the true force field—for comparison with *a priori* calculations, for example—they should be treated with suspicion; if, on the other hand, one wishes to calculate the expected vibration frequencies of a related molecule as an aid to assignment, then there is no doubt that model force field calculations are often successful. Perhaps the most important point is that constraints—when imposed—should be clearly specified in reporting the force field.

SUMMARY

Table 4 summarizes the state of the art today in harmonic force field calculations, and Tables 5 and 6 summarize an example of a good recent

TABLE 4 Reliability of the Best Published Force Fields

1. Bent triatomics: very good (see, for example, Table 3); linear triatomics ($C_{\infty h}$): less satisfactory, unless corrected for anharmonic effects.

2. Molecules with no more than two vibrations in any one symmetry species: generally very good, assuming that some extra data have been used in addition to parent isotope frequencies.

3. Slightly larger molecules, 3 × 3 symmetry species: good results for selected molecules (e.g., CH_3X, CX_3H, C_2H_4, C_2H_6, . . .), but often one or two constraints are imposed to obtain a unique solution.

4. Larger molecules still, more than seven or eight atoms: satisfactory force fields are available only in selected cases and for saturated paraffins, benzene, and some simple derivatives.

5. General rule: multiply all standard errors by four.

TABLE 5 Data Used in Calculating the General Harmonic Force Field of Ethylene[a]

The following 57 data were used to obtain the best estimates of the 13 harmonic force constants:

Data	Range and fit
12 ω's for C_2H_4; 12 ω's for C_2D_4; 12 ω's for CH_2CD_2	3000 to 300 cm^{-1}: fit to ± 2 cm^{-1}
3 $\Delta\omega$'s for $H_2{}^{12}C^{13}CH_2$; 3 $\Delta\omega$'s for $D_2{}^{12}C^{13}CD_2$; 3 $\Delta\omega$'s for $H_2{}^{12}C^{13}CD_2$; 3 $\Delta\omega$'s for $H_2{}^{13}C^{12}CD_2$	22 to 4 cm^{-1}: fit to ± 0.2 cm^{-1}
3 more $\Delta\omega$'s for other species	
6 ζ's for C_2H_4	0.1 to 0.9: fit to ± 0.03

[a] Data from Duncan, McKean, and Mallinson.[17]
NOTE: The 5 τ's and the ground-state inertia defect were calculated and compared with those observed, but were not used directly in the force constant refinement.

unconstrained calculation of the harmonic force field of ethylene by Duncan, McKean, and Mallinson.[17] This is about as large a molecule as there is for which the harmonic force field has truly been determined without constraint, although—as Table 4 indicates—force fields have been determined for many larger molecules using only a few reasonably well-justified constraints.

TABLE 6 Ethylene Force Field in the Usual Symmetry Coordinate Representation[a]

$F(i,j)$		$F/(\text{mdyn } \text{Å}^{-1})$	$F(i,j)$		$F/(\text{mdyn } \text{Å}^{-1})$
1	1	9.395(38)[b]	7	7	0.950(5)
1	2	+0.365(56)			
1	3	−0.289(8)	8	8	0.702(4)
2	2	5.638(27)			
2	3	−0.052(24)	9	9	5.493(24)
3	3	0.414(2)	9	10	−0.161(42)
			10	10	0.413(4)
4	4	0.646(5)			
			11	11	5.603(25)
5	5	5.657(34)	11	12	+0.085(11)
5	6	+0.365(33)	12	12	0.384(2)
6	6	0.560(3)			

[a] Data from Duncan, McKean, and Mallinson.[17]
[b] Standard errors from the force constant refinement given in parentheses (digits correspond to final listed digits of the force constants).

REFERENCES

1. E. B. Wilson, J. C. Decius, and P. C. Cross, *Molecular Vibrations,* McGraw-Hill, New York (1955).
2. A. R. Hoy, I. M. Mills, and G. Strey, *Mol. Phys.,* **24**, 1265 (1972).
3. J. Overend and B. L. Crawford, *J. Mol. Spectr.,* **12**, 307 (1964).
4. A. R. Hoy, J. M. R. Stone, and J. K. G. Watson, *J. Mol. Spectr.,* **42**, 393 (1972).
5. D. C. McKean, *Spectrochim. Acta,* **22**, 251, 269 (1966).
6. J. H. Meal and S. R. Polo, *J. Chem. Phys.,* **24**, 1119 (1956).
7. G. Herzberg, *Infrared and Raman Spectra,* Van Nostrand, New York (1945).
8. W. L. Smith and I. M. Mills, *J. Chem. Phys.,* **40**, 2095 (1964).
9. C. di Lauro and I. M. Mills, *J. Mol. Spectr.,* **21**, 386 (1966).
10. J. L. Duncan, *Spectrochim. Acta.,* **20**, 1197 (1964).
11. J. L. Duncan and G. R. Burns, *J. Mol. Spectr.,* **30**, 253 (1969).
12. L. Nemes, J. L. Duncan, and I. M. Mills, *Spectrochim. Acta,* **23A**, 1803 (1967).
13. J. C. Duinker and I. M. Mills, *Spectrochim. Acta,* **24A**, 417 (1968).
14. T. Tanaka and Y. Morino, *J. Mol. Spectr.,* **33**, 538 (1970).
15. Y. Morino, Y. Kikuchi, S. Saito, and Y. Morino, *J. Mol. Spectr.,* **13**, 95 (1964).
16. Y. Morino and S. Saito, *J. Mol. Spectr.,* **19**, 435 (1966).
17. J. L. Duncan, D. C. McKean, and P. D. Mallinson, *J. Mol. Spectr.,* **45**, 221 (1973).
18. J. K. G. Watson, *J. Chem. Phys.,* **45**, 1360 (1966).
19. D. Kivelson and E. B. Wilson, *J. Chem. Phys.,* **21**, 1229 (1953).
20. V. W. Laurie and D. R. Herschbach, *J. Chem. Phys.,* **40**, 3142 (1964).
21. T. Oka and Y. Morino, *J. Mol. Spectr.,* **6**, 472 (1961).
22. S. J. Cyvin, *Molecular Vibrations and Mean Square Amplitudes,* Elsevier, Amsterdam (1968).
23. K. Hedberg and M. Iwasaki, *J. Chem. Phys.,* **36**, 594 (1962).
24. Y. Morino, Y. Nakamura, and T. Iijima, *J. Chem. Phys.,* **32**, 643 (1960).
25. A. G. Robiette, G. J. Cartwright, A. R. Hoy, and I. M. Mills, *Mol. Phys.,* **20**, 541 (1971).
26. R. G. Snyder and J. H. Schachtschneider, *Spectrochim. Acta,* **21**, 169 (1965).
27. T. Shimanouchi, *J. Chem. Phys.,* **17**, 245 (1949).
28. J. Overend and J. R. Scherer, *J. Chem. Phys.,* **32**, 1289 (1960).

Josef Plíva

Anharmonic Force Fields

Anharmonicity of molecular vibrations presents one of the most vexing problems in studies of molecular structure. Anharmonic corrections (of first order, involving the cubic force constants) are required in accurate determinations of the equilibrium structures of molecules from rotational spectra, as well as from electron diffraction measurements. To obtain accurate harmonic force fields, it is necessary first to correct the vibrational data for anharmonicity (using second-order corrections, involving cubic and quartic force constants). Information on anharmonic force fields obtained from experimental data is also important as a basis for comparison in quantum chemical investigations of molecular forces as well as in studies of high-temperature thermodynamic properties and of rate and dissociation processes. Yet detailed studies of anharmonic force fields have hitherto been limited to small molecules with $N = 2$–4 atoms (in isolated cases to $N = 6$).

In this report the attention is focused on the anharmonic vibrations of polyatomic molecules ($N \geqslant 3$) rather than on diatomic molecules for which studies of potential functions involve a rather different range of problems. Interest in the anharmonicity of polyatomic molecules has been steadily increasing in the past 15 years. Several factors, however, have so far impeded more rapid development and limited such investigations to small molecules.

Calculations of the anharmonic force constants become increasingly complex with increasing numbers of atoms. Technically, this is not an insurmountable problem since improved computational techniques have become available[1-10] that make calculations to the second order of accuracy (involving quadratic, cubic, and quartic force constants) practicable for molecules with $N \geqslant 4$; a fourth-order calculation (including up to sextic force constants) has recently been reported for CO_2.[11] A more fundamental limitation is posed by the omnipresent anharmonic and rovibrational interactions ("resonances") whose number increases with the size of the molecule and makes it necessary to handle increasingly large volumes of data and to diagonalize very large energy matrices in the calculations.

The principal obstacle in anharmonic force field studies today is the lack of sufficiently complete, accurate, and internally consistent experimental data. Our main sources of data are high-resolution measurements of infrared vibrational/rotational spectra (involving not only the fundamentals but as many overtone and combination bands as possible) and microwave rotational spectra of molecules in excited vibrational states. Very high resolution combined with high absolute accuracy is required in the infrared measurements, and apparatus capable of such performance necessarily represents a large investment. The handling and analysis of the thousands of lines observed is an exceedingly tedious process, even though automation has recently been introduced in several laboratories to alleviate this problem (personal communications from W. E. Blass, T. H. Edwards, P. A. Jansson, J. W. C. Johns and A. G. Maki; see also ref. 12). Infrared absorption measurements are generally limited to fairly low vibrational states, and since information on the high vibrational states is essential for anharmonic studies, measurements of infrared emission spectra at high resolution have recently been initiated to provide such data.[13] The necessity of measurements of the spectra of two or more isotropic species of a given molecule to determine its force field adds to the expense of the experimental work.

Other sources of data have generally been of much less use for anharmonic force field studies than high-resolution infrared and microwave measurements. Solid-state infrared measurements using the matrix-isolation technique have recently provided useful vibrational data for some species inaccessible to gas-phase measurements, e.g., the cyanate (OCN^-)[14,15] and nitrate (NO_3^-)[16] anions. Gas-phase Raman measurements of vibration/rotation bands are very difficult to perform with adequate resolution, and Raman data have mainly been utilized for locating some of the infrared-inactive fundamentals in symmetrical molecules such as CO_2, CS_2,[17] C_2H_2,[18] CH_4,[19,20] C_2H_6.[21-23] High-resolution data on electronic spectra,[24] too, have only been used in rather special cases, mostly to provide information on vibrational levels inaccessible in the infrared, e.g., for C_2H_2 (J. K. G. Watson and A. E. Douglas, unpublished data and ref. 25), C_2N_2,[26,27] or for molecules in excited electronic states, e.g., CH_2O,[28,29] ClO_2,[30–33] or for unstable species such as the free radicals.[34] Thus the status of vibration/rotation data remains less than satisfactory at present.

Among the simple polyatomic molecules for which sufficiently complete and reliable infrared data (including data on isotopic molecules) exist to warrant calculation of the cubic and quartic force constants are the linear triatomics CO_2 [13,17,35–39] and N_2O.[40–52] Slightly less extensive are the data for CS_2,[53–59] HCN,[60–73] COS,[74–79] H_2O,[80–86] and C_2H_2.[18,87–97] Less complete or less accurate, yet still usable for calculations of simplified anharmonic force fields, are the data for H_2S,[98–104] H_2Se,[105–109] SO_2,[110–116] NO_2,[117–123] ClO_2,[30–33] ClCN,[124,125] NH_3.[126–136] Various pieces of good data do exist for a number of additional simple molecules, but the information is not yet sufficiently complete to allow more than the simplest of the anharmonic model potentials to be obtained. This is the case, for example, for the cyanogen halides FCN,[137,138] BrCN,[139,140] for HCP,[141,142] C_2N_2,[26,27,143,144] HCNO,[145–148] BF_3,[149–153] NF_3,[154–160] PF_3 and AsF_3,[161,162] CH_2O,[28,29,163–166] CH_4,[19,20,167–176] the methyl halides CH_3F,[177–179] CH_3Cl,[180–184] CH_3Br,[185–190] CH_3I,[191–205] for SiH_4,[206–208] ethylene (C_2H_4),[209,210] glyoxal ($C_2H_2O_2$),[211,212] ethane (C_2H_6).[21–23,213–219]

In the microwave studies of rotational transitions in excited vibrational states the situation is slightly more favorable even though investigators in this field have problems of their own in performing measurements at high temperatures to attain sufficient populations of the higher lying vibrational states. This again tends to limit the investigations to small stable polar molecules. The microwave studies provide very accurate data on the vibrational dependence of the rotational constants needed to obtain equilibrium values of the rotational constants and equilibrium structures. The data can also yield values of the cubic (but not quartic) force constants and anharmonic corrections for structures derived from electron diffraction experiments.

Microwave measurements of the rotational transitions in vibrationally excited states from which the cubic force constants could be determined[220] are available, e.g., for SO_2,[115,116] SeO_2,[221,222] O_3,[223] OF_2,[224] SiF_2,[225] GeF_2,[226] HCP,[141,142] NF_3,[159,160] $HSiF_3$,[227] as well as for many of the molecules mentioned above.

Another important limiting factor in anharmonicity studies has been the inavailability of a reliable and generally applicable model for the potential function. The most general quartic function,*

$$V = \sum_{i \leqslant j} f_{ij} R_i R_j + \sum_{i \leqslant j \leqslant k} f_{ijk} R_i R_j R_k + \sum_{i \leqslant h \leqslant k \leqslant l} f_{ijkl} R_i R_j R_k R_l, \tag{1}$$

where R_i are internal valence coordinates, contains too many force constants f to be determined from the existing experimental data—except for some of the triatomics—and some simplification of the potential energy expression is necessary for most molecules.

A physically acceptable potential must be isotopically invariant in a high order of approximation, and it was shown that curvilinear valence coordinates, i.e., bond stretchings and angle bendings that always point exactly along the lines joining the nuclei (not their projections) should be used to set up the potential function to satisfy this requirement.[3,4] This is borne out by equipotential contour maps such as the one for CO_2 shown in Figure 1. The "valley" of the potential surface is seen to project very nearly into a semicircle that is the path defined by the curvilinear bending coordinate γ; a rectangular bending coordinate, which would correspond to a tangent to the circle, would obviously cross many more equipotential lines and yield much larger stretch–bend interaction terms.

From calculations made for a number of simple molecules, it has become clear that in the cubic and quartic part of the potential written in curvilinear valence–force coordinates, the diagonal bond-stretching force constants (f_{rrr} and f_{rrrr}) are much larger than the bending and interaction constants. On this observation is based the simplest model potential, the anharmonic simple valence–force (SVF) model that consists of a complete harmonic potential† with only the diagonal cubic and quartic stretching constants f_{rrr} and f_{rrrr} added,

$$V = \sum_{i \leqslant j} f_{ij} R_i R_j + \sum_{r} \left[f_{rrr} r^3 + f_{rrrr} r^4 \right] . \tag{2}$$

* Some authors[7] prefer to write the potential energy expansion with unrestricted sums and with factors $1/n!$,

$$V = \frac{1}{2!} \sum_i \sum_j f_{ij} R_i R_j + \frac{1}{3!} \sum_i \sum_j \sum_k f_{ijk} R_i R_j R_k + \dots . \tag{1a}$$

Although there are some advantages in this formulation, most of the force constants available today are based on Eq. (1).

† A harmonic potential with all symmetry-permitted terms is meant here. Unless the harmonic part of the potential is represented quite accurately, the anharmonic part can hardly be adjusted to fit the data on the higher vibrational states.

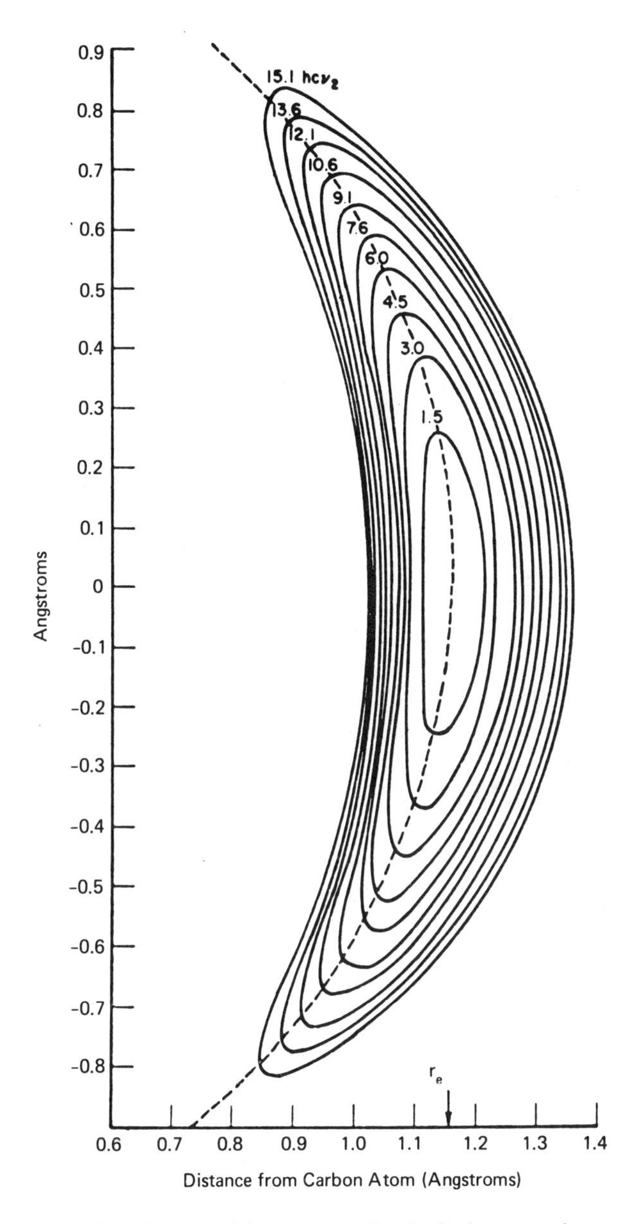

FIGURE 1 Equipotential contour map for the displacement of one of the O atoms in the CO_2 molecule.[5]

Further simplification of this model can be attained by assuming that the stretching of a bond in a polyatomic molecule can be represented by a diatomic-type potential function[1–4,228]: Predictions of the values of the constants f_{rrr} and f_{rrrr} from the quadratic constant f_{rr} have been made with the aid of the Morse function,[229] the Lippincott function,[230,231] or other diatomic potentials[232–238] with parameters only slightly adjusted from their diatomic values to reflect the change in geometry in the polyatomic bond. The SVF model has been explored on a number of triatomic molecules[239] where the results can be checked against more accurate treatments, and it has been successfully used to predict anharmonic corrections in determining equilibrium structures from rotational spectra and/or from electron diffraction measurements for various molecules, e.g., NO_2, ClO_2, BO_2,[239] NH_3, CH_4,[1,2,240] SO_3.[241,242] It has also been shown that this model with simple modifications can predict anharmonic corrections for fundamental vibrational frequencies ν_s to provide zero-order harmonic frequencies ω_s necessary in harmonic force field calculations.[243]

More general potentials are needed to correctly represent all the anharmonic constants x_{ss} and vibration/rotation coupling constants α_s. Several different approaches to the construction of such potentials have been explored. The most straightforward method is to simply modify the SVF by adding some cubic and quartic constants in an estimated order of their importance. The diagonal bending force constants, $f_{\gamma\gamma\gamma}$ (if permitted by symmetry) and $f_{\gamma\gamma\gamma\gamma}$, and some of the stretch–bend interaction constants, such as $f_{rr'\gamma}$, $f_{rr\gamma\gamma}$, $f_{rr\gamma}$, $f_{r\gamma\gamma}$, $f_{rr'\gamma\gamma}$, appear to be most important (in that order) as their inclusion frequently improves the results quite considerably. The stretch–stretch interactions, $f_{rrr'}$, $f_{rr'r'}$, $f_{rrrr'}$, etc., probably came next, and so on. Much is to be said in favor of this flexible approach, in particular for molecules for which at least some data on the overtone and combination bands are available; however, it tends to introduce too many parameters that must be adjusted to fit the experimental data. Unless very extensive sets of data are available for a given molecule, it is undesirable to use a function with too many parameters since addition of each constant increases the uncertainty in the determination of all of the constants: Only such constants should be included that significantly improve the fit to the experimental data. For this reason attempts have been made to model the cubic and quartic constants from the quadratic constants f_{ij} with as few additional parameters as possible. The bending potential $V(\gamma)$ has been modeled with the aid of double-minimum functions[3,243,244] of the form shown in Figure 2. Using a diatomic-type stretching potential $V(r_i)$ for each bond and a bending potential $V(\gamma_m)$ for each valence

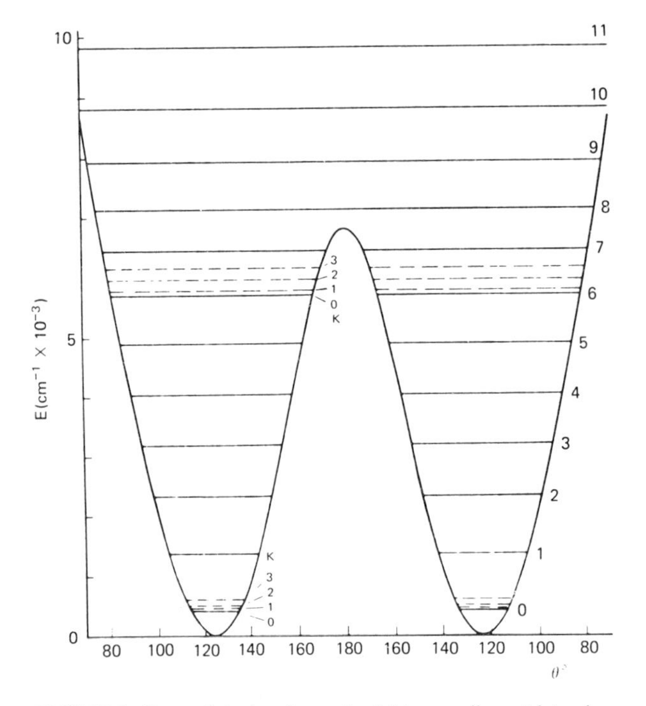

FIGURE 2 Form of the bending potential in a nonlinear triatomic molecule (drawn to scale for the $\tilde{A}\,^1A''$ electronic state of HCN).[244]

angle, it is possible to model the cubic and quartic interactions in various ways[3,243,245–248] from contributions originating in terms such as $V(r_i)[1 + \alpha_i R_j + \ldots]$, $V(\gamma_m)[1 + \alpha_m R_j + \ldots]$, or $V(r_i)\exp(\alpha_i R_j)$, etc., where α is adjustable. These models, of course, are quite empirical and, though they seem to work for the triatomics on which they have been tested, little can be said about their performance for other molecules.

In all the studies reviewed here the assumption has been made that all the vibrational amplitudes are small. For the bending motions in some molecules, e.g., molecules having a wide valence angle (so-called "quasi-linear" molecules), this is not a good approximation, and in the conventional treatment of these cases the stretch–bend interaction terms become very large and hard to deal with by perturbation techniques.[249,250] An alternative treatment has been developed[251,252] in which the large-amplitude bending motion is combined with the rotational rather than vibrational part of the Hamiltonian. The results thus far obtained in the lowest order of approximation for several molecules indicate that with this formalism, using only a double-minimum potential for the bending motion with no explicit stretch–bend interaction terms, the existing data can be accounted for rather satisfactorily.[251,252]

Another approach to the problem of introducing the bending and interaction terms[1,2,253] has been based on anharmonic extensions of the Urey–Bradley[254,255] function (UBF), which has been widely used for harmonic calculations. The appeal of this approach is that the introduction of a simple Lennard-Jones type 6-12 (or exp-6) term involving just one or two parameters for the interaction between a pair of nonbonded atoms automatically yields a variety of interaction terms after transforming the nonbonded coordinate into the valence–force coordinate space. There are, however, difficulties with such functions, similar to those experienced in harmonic force field studies, in that some of the interaction terms thus obtained turn out to be disproportionately large while others are too small.[256] Additional *ad hoc* terms are then necessary to correct the behavior of the UBF, and their introduction defeats the simplicity of the model. Some of the modified valence–force functions mentioned above appear to do at least as good a job in fitting the data as a modified UBF with a comparable number of parameters.

In a recent series of papers, Parr and his co-workers[233–238,257–260] have used the quantum mechanical virial theorem or the Poisson equation with a semiempirical charge distribution to obtain a theoretically justifiable form of a potential $V(r)$ for bonds in diatomic and polyatomic molecules and proposed models of polyatomic potentials the latest form of which[259,260] will be referred to as the Parr–Anderson function (PAF). This model in its basic form introduces the bending and stretch–bend interaction terms through functions of the type $A/(r_{jk})^n$ where r_{jk} is the distance between nonbonded atoms j and k and n is an integer between 1 and 6 depending on the type of molecule. The appealing feature of this function, besides its quantum-chemical background, is that it introduces very few parameters that have to be adjusted to fit the data. For example, the basic form of the function for a linear XY_2 molecule contains only one adjustable parameter, A, of the term $A/(r_{13})^n$; the values of other parameters involved in the stretching function $V(r)$ can be predicted from differences between the bond lengths in the diatomic and polyatomic situations. This function has been tested with fairly encouraging results on a number of triatomic molecules.[257–261] However, a critical evaluation of these results does reveal some shortcomings reminiscent of those encountered with the UBF to which the PAF is formally analogous. It appears that the PAF, like the UBF, needs some modification terms to correct its detailed behavior. Most essential, it should be made to generate a complete harmonic force field (see footnote p. 292), but it was found that other terms such as $f_{r\gamma\gamma}$ have to be added in order to correctly reproduce the data for some triatomic molecules.[261]

Two kinds of basic data can be used for calculating the anharmonic force constants. The zero-order frequencies ω_s and anharmonic constants x_{ss}, in the second-order formula for the vibrational energy E_V,

$$E_\mathrm{V}/hc = \sum_s \omega_s (v_s + d_s/2) + \sum_{s \leqslant s'} \left\{ x_{ss'}(v_s + d_s/2)(v_{s'} + d_{s'}/2) + x_{\ell_s \ell_{s'}} \ell_s \ell_{s'} \right\}, \quad (3)$$

and the vibration/rotation coupling constants α_s representing the first-order vibrational dependence of the rotational constants, B_V (and A_V, C_V for nonlinear cases),

$$B_\mathrm{V} = B_\mathrm{e} - \sum_s \alpha_s (v_s + d_s/2), \quad (4)$$

can be extracted directly from sufficiently extensive experimental data. From these quantities the harmonic and anharmonic force constants can be calculated using Nielsen's formulas[262–264] and the nonlinear transformation relations between valence and normal coordinates.[1–10] The advantage of this method (referred to here as method A) is that some of the molecular constants ω_s, $x_{ss'}$, α_s can frequently be determined from the experimental data with great accuracy and, in turn, may yield accurate values for some of the force constants.[220,265] Also, if sufficient data are available, it is possible to use a fourth-order energy formula that can usually represent the data to within the experimental uncertainty—which the second-order expression is incapable of—and the inclusion of at least some of the higher order constants $y_{ss's''}$, γ_{ss}, etc., tends to absorb possible residual higher order effects and inaccuracies so that the lower order constants ω_s, x_{ss}, and α_s needed for the calculation of a quartic force field become more reliable. The disadvantage of this method is that the quantities ω_s, $x_{ss'}$, α_s are not readily available for many molecules, in particular for isotopic molecules that furnish essential information for calculating the force field. Another objection to this approach has been that, for two or more isotopic molecules, the set of molecular constants ω_s, $x_{ss'}$, α_s and ω_s^*, $x_{ss'}^*$, α_s^* (asterisks are for the isotopically substituted species) is, in fact, redundant since all of these constants can be reduced to one smaller set of force constants. However, there is nothing wrong with using a redundant set of constants as intermediate data provided these constants can be extracted from the experiment in an unambiguous way and provided the theoretically known dependences among the constants (in the normal coordinate space) are taken into account.

A second approach is to adjust the force constants to fit directly the observed band centers and rotational constants[5] or, better yet, the vibration/rotation energy levels. The main advantage of this computationally more demanding method (referred to as method B) is that all the available pieces of data on all isotopic species of a molecule can be utilized at the same time, without first having to extract the intermediate molecular constants ω_s, $x_{ss'}$, α_s from the data. A major disadvantage is that the data cannot be fitted to within experimental uncertainties unless a fourth-order calculation is undertaken—a feat that has hitherto been accomplished only for CO_2.[11] Therefore, some of the quantities that can be directly determined from the data with great accuracy are not actually utilized in this method. It would appear that a compromise is feasible in which those constants known with accuracy could be included in the data with appropriately high statistical weights.

Ideally, the two methods should yield the same results, but, in practice, they rarely do. In fact, even calculations made by two different investigators using the same method with the same set of basic data are not always exactly identical because, e.g., differences in weighting the data may lead to somewhat different results. In assigning the statistical weights to the individual pieces of data two factors should be considered. First, the experimental factor—the experimental error limits of the data—should be clearly stated by the experimentalist: For quantities such as band centers or rotational constants a statement of, say, a 90 percent confidence interval σ_{90} is preferred to an "estimated error." The experimental statistical weight can be made proportional to $1/(\alpha^2 + \sigma_{90}^2)$ with suitably chosen α.

Second, a consideration of the vibrational and rotational dependence of the data calls for a corresponding adjustment of the weight: For the higher excited vibrational/rotational states it is found that the experimental accuracy is best matched if the experimental statistical weight is scaled down by an appropriate factor like $1/v_s[J(J+1)]^{1/2}$. The actual computations of the force constants are generally made by the method of least squares, and in order to yield correct confidence intervals for the resulting constants, the statistical weights w_i of the data should be further scaled so that $\sum w_i = N$, the total number of data points.

In a critical evaluation of the results of anharmonic force constant calculations it is necessary to consider all the factors discussed above concerning the completeness, internal consistency, and accuracy of the data, the possible idiosyncrasies of the model potential used, the method of reducing the data, as well as the goodness of fit the force constants yield for the data points. It may be in order to mention that careful high-resolution infrared measurements of line positions routinely attain absolute accuracy of about ±0.01 cm^{-1} (with resolution of the order of 0.03 cm^{-1}) while a second-order energy expression [Eq. (3) and (4)] provides a fit with a standard deviation typically 0.10–0.25 cm^{-1}. Therefore, a set of quadratic, cubic, and quartic force constants that fits an extensive set of data points (using method B) with a standard deviation of 0.20 cm^{-1} or so can be considered good, provided it also fits well those intermediate molecular constants ω_s, $x_{ss'}$, α_s that are known with accuracy.

Fairly reliable model-independent values of the force constants in the most general quartic force field [Eq. (1)] exist at present only for the linear triatomics CO_2,[5,266] CS_2,[267,268] and N_2O,[269,270] for which the calculations were based on rather extensive sets of experimental data for several isotopic species; in the first two cases it is encouraging to see that the values of the quadratic and cubic constants obtained in a hand-calculation made some 15 years ago[3,271] agree quite well with recent least squares computer calculations based on considerably enlarged sets of data.[266–268] Tables 1 and 2 summarize what are considered the best available values of quadratic, cubic, and quartic constants in the valence-force representation for a number of triatomic molecules. The dispersions quoted in parentheses (unless otherwise stated) are standard errors as reported by the authors of the constants. It should be remembered that these were generally obtained in least squares calculations involving many adjustable constants, e.g., $n = 17$ for CO_2 and CS_2 and $n = 23$ for N_2O. This means that in order to obtain, say, 90 percent confidence limits for the constants, the standard errors should be multiplied by a factor $[nF_{90}(n,N)]^{1/2}$ involving the value of the F distribution for a given number of constants (n) and data points (N): The minimum value of this factor for $n = 17$ and large N is ~5.0 (for tables of the F distribution see Dixon and Massey[272]). The values of the cubic force constants are in most cases considerably more reliable than those of the quartic constants as evidenced by the standard errors listed. In particular, microwave data on the excited vibrational states have provided most valuable information for determining the cubic constants. The results for the molecules COS,[273] HCN (G. Strey and I. M. Mills, unpublished data, and ref. 274, 275), H_2O,[1–3,7,239,276] H_2S and SO_2[239,253] are also largely model-independent although, as the dispersions listed in Tables 1 and 2 indicate, these must be considered less reliable and, in particular, the values of some of the quartic constants are rather uncertain. This uncertainty can be traced back, on the one hand, to the lack of experimental data for these molecules or, on the other hand, to the insensitivity of the observables to variations in some

TABLE 1 Anharmonic Force Constants of Some Linear Triatomics[a]

	CO_2[b,c]	CS_2[b,c]	N_2O[b,c]	COS[b]	HCN[d]	ClCN[c,e]	FCN[f]	BrCN[f]	ICN[f]	HCP[g]
f_{rr}	8.0112(24)[h]	3.9337(17)	6.015(15)	3.722(20)	3.126(1)	2.642(7)	4.666	2.195	1.696	3.125
$f_{rr'}$	1.2613(43)	0.6360(34)	1.024(16)	1.040(65)	−0.200(1)	0.395(45)	−0.248	−0.221	−0.188	−0.060
$f_{r'r'}$	–	–	9.095(23)	8.070(55)	9.352(2)	8.991(40)	8.480	8.597	8.597	4.550
$f_{\gamma\gamma}$	0.3925(2)	0.2849(1)	0.3330(2)	0.3256(2)	0.1298(1)	0.1751(5)	01.95	0.152	0.130	0.128
f_{rrr}	−18.989(28)	−7.441(14)	−17.15(7)	−6.67(56)	−5.89(8)	−4.84(5)	−10.94	−3.68	−2.51	−6.67
$f_{rrr'}$	−1.954(104)	−0.793(43)	−0.96(12)	−2.93(1.9)	0.02(10)	−0.61(1)	0.17	0.12	0.10	0.0[i]
$f_{rr'r'}$	–	–	−0.24(23)	5.24(3.6)	0.20(34)	−1.16(1)	0.46	0.38	0.32	0.0[i]
$f_{r'r'r'}$	–	–	−22.44(10)	−26.12(4.5)	−20.93(22)	−19.48(34)	−16.53	−16.87	−16.58	−8.03
$f_{r\gamma\gamma}$	−0.609(3)	−0.367(4)	−0.79(6)	−0.61(6)	−0.10(6)	−0.30(1)	−0.47	−0.24	−0.22	−0.05
$f_{r'\gamma\gamma}$	–	–	−0.84(10)	−0.09(17)	−0.32(4)	−0.38(3)	−0.55	−0.39	−0.33	−0.30
f_{rrrr}	26.25(17)	8.83(15)	25.62(54)	8.2(5.2)	7.56(41)	5.17(6)	16.73	3.98	2.41	7.94
$f_{rrrr'}$	3.68(21)	1.38(52)	−3.22(72)	4.2(4.7)	0.0[i]	0.71(1)	−0.09	−0.05	−0.04	0.0[i]
$f_{rrr'r'}$	3.02(40)	1.67(79)	−2.17(69)	−14.0(26.0)	0.35(87)	−3.31(4)	−0.36	−0.25	−0.19	0.0[i]
$f_{rr'r'r'}$	–	–	1.74(44)	−53.4(33.0)	0.0[i]	1.19(2)	−0.66	−0.51	−0.43	0.0[i]
$f_{r'r'r'r'}$	–	–	31.05(41)	74.2(40.0)	24.18(1.25)	24.61(75)	19.23	19.88	19.37	9.44
$f_{rr\gamma\gamma}$	0.50(6)	0.43(4)	0.80(21)	−0.2(0.6)	0.03(0.18)	0.25(1)	0.001	0.016	0.013	0.0[i]
$f_{rr'\gamma\gamma}$	1.87(12)	1.14(10)	2.51(32)	3.2(2.0)	0.0[i]	1.30(10)	0.48	0.32	0.25	0.0[i]
$f_{r'r'\gamma\gamma}$	–	–	0.75(30)	0.0(1.7)	0.07(0.18)	0.42(6)	0.93	0.60	0.50	0.0[i]
$f_{\gamma\gamma\gamma\gamma}$	0.046(2)	0.032(3)	0.076(2)	0.05(0.01)	0.0013(12)	0.0232(3)	0.043	0.038	0.033	0.002
r ; r'	–	–	NO ; NN	CS ; CO	CH ; CN	CCl ; CN	CF ; CN	CBr ; CN	CI ; CN	CH ; CP
Method	B	B	B	B	A	B	B	B	B	A
Reference	266	268	269, 270	273	j	277	261	261	261	j

[a] Units: f_{rr}, mdyn/Å; $f_{\gamma\gamma}$, mdyn Å; f_{rrr}, mdyn/Å^2; $f_{r\gamma\gamma}$, mdyn; f_{rrrr}, mdyn/Å^3; $f_{rr\gamma\gamma}$, mdyn/Å; $f_{\gamma\gamma\gamma\gamma}$, mdyn Å.

[b] General quartic force field.

[c] Including four third-order constants for vibrational and rotational dependence of Fermi resonance.

[d] General quartic force field with three quartic constraints.

[e] Tentative values based on a modified valence-force model with 11 parameters.

[f] Tentative values based on a modified Parr–Anderson model.

[g] Tentative values based on a simple valence-force model with cubic stretch–bend interactions.

[h] Standard errors; digit(s) within parentheses generally correspond to last digit(s) in numerical values tabulated.

[i] Constrained value.

[j] G. Strey and I. M. Mills, unpublished data.

of the force constants. The lack of data can, in fact, give rise to apparent correlations among the force constants. The values of the constants quoted in Tables 1 and 2 for the molecules ClCN,[227] FCN, BrCN, ICN,[261] HCP (G. Strey and I. M. Mills, unpublished data), NO_2,[239,247,248,256] and ClO_2[256] should be regarded as tentative since they are based on various model potentials (as indicated in the footnotes) and on much less extensive sets of data. Table 3 lists the values of the cubic constants in the valence–force coordinate space (f_{ijk}) and in the dimensionless normal coordinate space ($k_{ss's''}$) as obtained for

TABLE 2 Anharmonic Force Constants of Some Nonlinear Triatomics[a]

	H_2O[b]	H_2S[b]	SO_2[b]	NO_2[c]	ClO_2[d]
f_{rr}	4.227	2.142	5.166	5.524	3.528
$f_{rr'}$	−0.101	−0.012	0.081	2.134	−0.102
$f_{r\gamma}$	0.219	0.134	0.325	0.650	0.071
$f_{\gamma\gamma}$	0.3485	0.377	0.835	0.810	0.712
f_{rrr}	−9.89(50)[e]	−3.89(5)[f]	−11.85(6)[f]	−15.81	−7.65
$f_{rrr'}$	0.13(75)	0.11(2)	−0.57(16)	−3.21	−0.22
$f_{rr\gamma}$	0.20(10)	−0.40(3)	−0.68(5)	−0.59	−0.23
$f_{rr'\gamma}$	−0.40(29)	−0.15(2)	−0.76(3)	−1.17	−0.50
$f_{r\gamma\gamma}$	−0.11(5)	−0.05(10)	−0.90(7)	−1.08	−0.61
$f_{\gamma\gamma\gamma}$	−0.146(17)	−0.021(20)	−0.41(2)	−0.40	−0.25
f_{rrrr}	16.0(2.6)	4.8(2)	16.5(6)	26.4	7.1
$f_{rrrr'}$	−0.8(2.9)	−0.1(3)	−1.1(1.3)	4.1	0.4
$f_{rrr'r'}$	0.1(1.8)	0.2(6)	−0.1(2.1)	−7.1	0.6
$f_{rrr\gamma}$	0.0[g]	0.0[g]	0.0[g]	0.3	0.3
$f_{rrr'\gamma}$	0.0[g]	0.0[g]	0.0[g]	1.0	1.2
$f_{rr\gamma\gamma}$	0.35(2)	−0.7(4)	0.7(3.2)	1.8	0.8
$f_{rr'\gamma\gamma}$	0.30(25)	−0.3(8)	2.8(6.2)	3.5	1.8
$f_{r\gamma\gamma\gamma}$	0.0[g]	0.0[g]	0.0[g]	0.6	1.1
$f_{\gamma\gamma\gamma\gamma}$	−0.003(8)	0.0(3)	−0.6(2)	0.17	−0.04
Method	A	A	A	B	A
r_e	0.9572	1.328	1.4308	1.1934	1.49
φ_e	104°31′	92°12′	119°19′	134°15′	118°31′
Reference	7	239	239	247,248	256

[a] For units see Table 1.
[b] General quartic force field with three quartic constraints.
[c] Tentative values based on a modified valence–force model with nine parameters.
[d] Tentative values based on a modified Urey–Bradley model with nine parameters.
[e] Statistical standard errors; digit(s) within parentheses correspond generally to last digit(s) in values tabulated.
[f] Estimated uncertainties; digit(s) within parentheses correspond generally to last digit(s) in values tabulated.
[g] Constrained value.

TABLE 3 General Cubic Force Fields of Bent XY_2 Molecules from Microwave Data[a]

	SO_2	SeO_2	O_3	OF_2	SiF_2	GeF_2
f_{rr}	5.166	3.51	3.08	1.99	2.51_5	2.04
$f_{rr'}$	0.081	0.16	1.50	0.83	0.31	0.26
$f_{r\gamma}$	0.325	0.15	0.42	0.21	0.22	−0.02
$f_{\gamma\gamma}$	0.835	0.67	0.65_5	0.71	0.55_5	0.47_5
f_{rrr}	−11.85	−7.47	−9.03	−4.53	−4.97	−3.78
$f_{rrr'}$	−0.57	−0.29	−1.19	−0.82	−0.20	−0.13
$f_{rr\gamma}$	−0.68	−0.30	−1.60	−0.67	−0.45	−0.03
$f_{rr'\gamma}$	−0.76	−0.20	−1.16	−0.41	−0.72	−0.36
$f_{r\gamma\gamma}$	−0.90	−0.62	−1.92	−1.31	−0.60	−0.40
$f_{\gamma\gamma\gamma}$	−0.41	−0.32	−0.65	−0.55	−0.48	−0.37
k_{111}	−43.9	−35.2	−47.6	−40.0	−29.6	−23.9
k_{112}	−19.1	−6.2	−29.5	−22.0	−19.3	−9.2
k_{122}	+8.0	+9.6	−24.3	−9.8	+9.9	+8.4
k_{222}	−6.2	−5.3	−19.2	−16.0	−9.4	−7.4
k_{133}	−157.0	−111.5	−221.8	−139.3	−110.5	−77.5
k_{233}	+4.7	+2.5	−59.3	−31.8	+5.1	+1.1
r_e	1.4308	1.6076	1.2717	1.4053	1.5901	1.7321
φ_e	119°19′	113°50′	116°47′	103°04′	100°46′	97°10′
Reference	239	221,222	223	224	225	226

[a] f_{ij} and f_{ijk} are the quadratic and cubic constants of the valence-force representation; $k_{ss's''}$ are the cubic constants in the normal coordinate representation (in cm^{-1}).

the molecules of SO_2,[239,256] SeO_2,[221,222] O_3,[223] OF_2,[224] SiF_2,[225] and GeF_2[226] from microwave data.[220] In Table 4 are quoted tentative values of the valence–force constants for C_2H_2 (I. M. Mills and G. Strey, private communication). Anharmonic calculations carried out thus far for other tetratomic and more complex molecules, e.g., BF_3,[278] NF_3,[159] SO_3,[241,242] NH_3 and CH_4,[1,2] $HSiF_3$,[227] were based on the SVF or UBF models (with simple modifications) and their aim was to obtain anharmonic corrections for equilibrium structure determinations and/or harmonic (zero-order) frequencies. The values of the anharmonic force constants obtained in these calculations are not claimed by the authors to be more than informed guesses and are, therefore, not quoted here.

From the review of the existing information on anharmonic force constants one must conclude that our knowledge of the anharmonicity of force fields in polyatomic molecules is still very incomplete. Only the gross features of the molecular potentials have been established with some confidence. We know now that the potential opposing the stretch-

TABLE 4 Tentative Valence–Force Field for Acetylene (C_2H_2)[a,b]

Quadratic constants							
f_{rr}	3.185[c]	$f_{rr'}$	−0.019	f_{rR}	−0.095	f_{RR}	8.170
$f_{\gamma\gamma}$	0.126	$f_{\gamma\gamma'}$	0.093				
Cubic constants							
f_{rrr}	−5.80 (0.33)[d]	$f_{rrr'}$	0.30 (0.10)	f_{rrR}	−0.18 (0.25)	$f_{rr'R}$	−0.59 (0.90)
f_{rRR}	0.35 (0.65)	f_{RRR}	−16.28 (1.1)	$f_{r\gamma\gamma}$	−0.34 (0.03)	$f_{r\gamma\gamma'}$	−0.02 (0.01)
$f_{r\gamma'\gamma'}$	−0.26 (0.02)	$f_{R\gamma\gamma}$	−0.17 (0.02)	$f_{R\gamma\gamma'}$	0.27 (0.01)		
Quartic constants[e]							
f_{rrrr}	6.73 (1.5)	f_{rrRR}	0.86 (1.8)	f_{RRRR}	18.18 (5.7)	$f_{rr\gamma\gamma}$	1.30 (0.18)
$f_{rR\gamma\gamma}$	0.54 (0.25)	$f_{RR\gamma\gamma}$	−0.84 (0.20)	$f_{\gamma\gamma\gamma\gamma}$	0.017 (0.002)	$f_{\gamma\gamma\gamma'\gamma'}$	0.040 (0.005)

[a] I. M. Mills and G. Strey, personal communication.
[b] $r \equiv r(CH)$; $R \equiv R(CC)$.
[c] For units see Table 1.
[d] Statistical standard errors given in parentheses.
[e] Values of 14 quartic interaction constants were set equal to zero.

ing of bonds is responsible for the dominant part of the anharmonic terms and that this stretching potential behaves essentially as in a diatomic molecule so that it can be represented by a diatomic-type potential function. The potential opposing the deformation of valence angles can be represented by double-minimum functions, at least in unbranched situations. Little is known about the anharmonic interaction terms. The cubic valence–force interaction constants all appear to be either negative or small in magnitude. The available values of the quartic interaction constants are too uncertain to enable any conclusions to be made.

In view of the role of the vibrational anharmonicity in molecular structure studies it seems important to extend investigations of the anharmonic force fields to many more molecules. To do this one needs primarily more good high-resolution data. Only then can reliable comparative studies of the various model potentials and of the transferability of the parameters in these potentials be made. It is the hope that such studies could eventually help construct model potentials that could be used with some confidence to obtain anharmonicity corrections for more complex molecules of interest.

The author would like to express his thanks to Drs. K. Kuchitsu, I. M. Mills, Y. Morino, and J. Overend for providing preprints and other information on some of their work prior to publication.

Support from the National Science Foundation and from the Office of Naval Research is gratefully acknowledged.

REFERENCES

1. K. Kuchitsu and L. S. Bartell, *J. Chem. Phys.*, **36**, 2460, 2470 (1962).
2. Y. Morino, K. Kuchitsu, and S. Yamamoto, *Spectrochim. Acta*, **24A**, 335 (1968).
3. J. Plíva, *Coll. Czech. Chem. Commun.*, **23**, 777, 1839, 1846 (1958).
4. Z. Cihla and J. Plíva, *Coll. Czech. Chem. Commun.*, **28**, 1232 (1963).
5. M. A. Pariseau, I. Suzuki, and J. Overend, *J. Chem. Phys.*, **42**, 2335 (1965).
6. K. Machida, *J. Chem. Phys.*, **44**, 4186 (1966).
7. A. R. Hoy, I. M. Mills, and G. Strey, *Mol. Phys.*, **24**, 1265 (1972).
8. A. Ya. Tsaune, N. T. Storchai, L. V. Belyavskaya, and V. P. Morozov, *Opt. Spectrosc.*, **26**, 502 (1969).
9. Z. Cihla, *Stud. Czech. Acad. Sci.*, No. 9 (1972).
10. A. Chedin and Z. Cihla, *J. Mol. Spectrosc.*, **45**, 475 (1973).
11. Z. Cihla and A. Chedin, *J. Mol. Spectrosc.*, **40**, 337 (1971).
12. R. N. Jones, *Appl. Optics*, **8**, 597 (1969).
13. T. K. McCubbin, J. Plíva, R. E. Pulfrey, W. B. Telfair, and T. Todd, *J. Mol. Spectrosc.*, **49**, 136 (1974).
14. V. Schettino and I. C. Hisatsune, *J. Chem. Phys.*, **52**, 9 (1970).

15. D. F. Smith, J. Overend, J. C. Decius, and D. J. Gordon, *J. Chem. Phys.*, **58**, 1636 (1973).
16. M. Tsuboi and I. C. Hisatsune, *J. Chem. Phys.*, **57**, 2087 (1972).
17. B. P. Stoicheff, *Adv. Spectrosc.*, **1**, 91 (1959).
18. H. Fast and H. L. Welsh, *J. Mol. Spectrosc.*, **41**, 203 (1972).
19. J. Herranz and B. P. Stoicheff, *J. Mol. Spectrosc.*, **10**, 448 (1963).
20. H. Berger, M. Faivre, J. P. Champion, and J. Moret-Bailly, *J. Mol. Spectrosc.*, **45**, 298 (1973).
21. D. W. Lepard, D. M. C. Sweeney, and H. L. Welsh, *Can. J. Phys.*, **40**, 1567 (1962).
22. D. W. Lepard, D. E. Shaw, and H. L. Welsh, *Can. J. Phys.*, **44**, 2353 (1966).
23. D. E. Shaw and H. L. Welsh, *Can. J. Phys.*, **45**, 3823 (1967).
24. G. Herzberg, *Molecular Spectra and Molecular Structure. 3. Electronic Spectra and Electronic Structure of Polyatomic Molecules,* Van Nostrand, N.Y., 1966.
25. K. K. Innes, *J. Chem. Phys.*, **22**, 863 (1954).
26. G. J. Cartwright, D. O'Hare, A. D. Walsh, and P. A. Warsop, *J. Mol. Spectrosc.*, **39**, 393 (1971).
27. G. B. Fish, G. J. Cartwright, A. D. Walsh, and P. A. Warsop, *J. Mol. Spectrosc.*, **41**, 20 (1972).
28. V. A. Job, V. Sethuraman, and K. K. Innes, *J. Mol. Spectrosc.*, **30**, 365 (1969).
29. V. Sethuraman, V. A. Job, and K. K. Innes, *J. Mol. Spectrosc.*, **33**, 189 (1970).
30. J. B. Coon and E. Ortiz, *J. Mol. Spectrosc.*, **1**, 81 (1957).
31. A. W. Richardson, R. W. Redding, and J. C. D. Brand, *J. Mol. Spectrosc.*, **29**, 93 (1969).
32. J. C. D. Brand, R. W. Redding, and A. W. Richardson, *J. Mol. Spectrosc.*, **34**, 399 (1970).
33. A. W. Richardson, *J. Mol. Spectrosc.*, **35**, 43 (1970).
34. G. Herzberg, *The Spectra and Structures of Simple Free Radicals*, Cornell, Ithaca, N.Y., 1971.
35. C. P. Courtoy, *Can. J. Phys.*, **35**, 608 (1957).
36. C. P. Courtoy, *Ann. Soc. Sci. Brux.*, **73**, 5 (1959).
37. H. R. Gordon and T. K. McCubbin, *J. Mol. Spectrosc.*, **18**, 73 (1965).
38. H. R. Gordon and T. K. McCubbin, *J. Mol. Spectrosc.*, **19**, 137 (1966).
39. R. Oberly, K. N. Rao, L. H. Jones, and M. Goldblatt, *J. Mol. Spectrosc.*, **40**, 356 (1971).
40. E. D. Tidwell, E. K. Plyler, and W. S. Benedict, *J. Opt. Soc. Am.*, **50**, 1243 (1960).
41. D. H. Rank, D. P. Eastman, B. S. Rao, and T. A. Wiggins, *J. Opt. Soc. Am.*, **51**, 929 (1961).
42. H. R. Gordon and T. K. McCubbin, *J. Opt. Soc. Am.*, **54**, 956 (1964).
43. R. P. Grosso and T. K. McCubbin, *J. Mol. Spectrosc.*, **13**, 240 (1964).
44. J. Plíva, *J. Mol. Spectrosc.*, **12**, 360 (1964).
45. J. Plíva, *J. Mol. Spectrosc.*, **25**, 62 (1968).
46. J. Plíva, *J. Mol. Spectrosc.*, **27**, 461 (1968).
47. J. Plíva, *J. Mol. Spectrosc.*, **33**, 500 (1970).
48. J. L. Griggs, K. N. Rao, L. H. Jones, and R. M. Potter, *J. Mol. Spectrosc.*, **18**, 212 (1965).
49. J. L. Griggs, K. N. Rao, L. H. Jones, and R. M. Potter, *J. Mol. Spectrosc.*, **25**, 34 (1968).

50. A. W. Mantz, K. N. Rao, L. H. Jones, and R. M. Potter, *J. Mol. Spectrosc.*, **30**, 513 (1969).
51. W. J. Lafferty and D. R. Lide, *J. Mol. Spectrosc.*, **14**, 407 (1964).
52. R. Pearson, T. Sullivan, and L. Frenkel, *J. Mol. Spectrosc.*, **34**, 440 (1970).
53. A. H. Guenther, T. A. Wiggins, and D. H. Rank, *J. Chem. Phys.*, **28**, 682 (1958).
54. D. Agar, E. K. Plyler, and E. D. Tidwell, *J. Res. Natl. Bur. Stand.*, **66A**, 259, (1962).
55. D. F. Smith and J. Overend, *Spectrochim. Acta,* **26A**, 2269 (1970).
56. D. F. Smith, J. Overend, *et al.*, *Spectrochim. Acta,* **27A**, 1979 (1971).
57. D. F. Smith and J. Overend, *J. Chem. Phys.*, **54**, 3632 (1971).
58. G. Blanquet and C. P. Courtoy, *Ann. Soc. Sci. Brux.*, **84**, 293 (1970).
59. G. Blanquet and C. P. Courtoy, *Ann. Soc. Sci. Brux.*, **86**, 259 (1972).
60. R. E. Kagarise, H. D. Rix, and D. H. Rank, *J. Chem. Phys.*, **20**, 1437 (1952).
61. D. H. Rank, G. Skorinko, D. P. Eastman, and T. A. Wiggins, *J. Mol. Spectrosc.*, **4**, 518 (1960).
62. D. H. Rank, G. Skorinko, D. P. Eastman, and T. A. Wiggins, *J. Opt. Soc. Am.*, **50**, 421 (1960).
63. A. E. Douglas and D. Sharma, *J. Chem. Phys.*, **21**, 448 (1953).
64. A. G. Maki and L. R. Blaine, *J. Mol. Spectrosc.*, **12**, 45 (1964).
65. A. G. Maki, E. K. Plyler, and R. Thibault, *J. Opt. Soc. Am.*, **54**, 869 (1964).
66. A. G. Maki, B. Olson, and R. L. Sams, *J. Mol. Spectrosc.*, **36**, 433 (1970).
67. W. W. Brim, J. M. Hoffman, H. H. Nielsen, and K. N. Rao, *J. Opt. Soc. Am.*, **50**, 1208 (1960).
68. B. D. Alpert, A. W. Mantz, and K. N. Rao, *J. Mol. Spectrosc.*, **39**, 159 (1971).
69. P. L. Yin and K. N. Rao, *J. Mol. Spectrosc.*, **42**, 385 (1972).
70. V. K. Wang and J. Overend, *Spectrochim. Acta,* **29A**, 687 (1973).
71. C. A. Burrus and W. Gordy, *Phys. Rev.*, **101**, 599 (1956).
72. A. G. Maki and D. R. Lide, *J. Chem. Phys.*, **47**, 3206 (1967).
73. G. Winnewisser, A. G. Maki, and D. R. Johnson, *J. Mol. Spectrosc.*, **39**, 149 (1971).
74. G. D. Saksena, T. A. Wiggins, and D. H. Rank, *J. Chem. Phys.*, **31**, 839 (1959).
75. A. G. Maki, E. K. Plyler, and E. D. Tidwell, *J. Res. Natl. Bur. Stand.*, **66A**, 163 (1962).
76. A. G. Maki, *J. Mol. Spectrosc.*, **23**, 110 (1967).
77. A. E. Triaille and C. P. Courtoy, *J. Mol. Spectrosc.*, **18**, 118 (1965).
78. A. Fayt, *Ann. Soc. Sci. Brux.*, **84**, 69 (1970).
79. Y. Morino and C. Matsumura, *Bull. Chem. Soc. Japan,* **40**, 1095, 1101 (1967).
80. W. S. Benedict, N. Gailar, and E. K. Plyler, *J. Chem. Phys.*, **24**, 1139 (1956).
81. N. M. Gailar and F. P. Dickey, *J. Mol. Spectrosc.*, **4**, 1 (1960).
82. W. S. Benedict, S. A. Clough, L. Frenkel, and T. E. Sullivan, *J. Chem. Phys.*, **53**, 2565 (1970).
83. P. E. Fraley, K. N. Rao, and L. H. Jones, *J. Mol. Spectrosc.*, **29**, 312, 348 (1969).
84. J. G. Williamson, K. N. Rao, and L. H. Jones, *J. Mol. Spectrosc.*, **40**, 372 (1971).
85. J. M. Flaud, C. Camy-Peyret, and A. Valentin, *J. Phys.*, **33**, 741 (1972).
86. R. A. Carpenter, N. M. Gailar, H. W. Morgan, and P. A. Staats, *J. Mol. Spectrosc.*, **44**, 197 (1972).

87. T. A. Wiggins, E. K. Plyler, and E. D. Tidwell, *J. Opt. Soc. Am.*, **51**, 1219 (1961).
88. E. K. Plyler, E. D. Tidwell, and T. A. Wiggins, *J. Opt. Soc. Am.*, **53**, 589 (1963).
89. W. J. Lafferty and R. J. Thibault, *J. Mol. Spectrosc.*, **14**, 79 (1964).
90. J. F. Scott and K. N. Rao, *J. Mol. Spectrosc.*, **16**, 15 (1965).
91. S. Ghersetti and K. N. Rao, *J. Mol. Spectrosc.*, **28**, 27, 373 (1968).
92. K. F. Palmer, S. Ghersetti, and K. N. Rao, *J. Mol. Spectrosc.*, **30**, 146 (1969).
93. S. Ghersetti, J. Plíva, and K. N. Rao, *J. Mol. Spectrosc.*, **38**, 53 (1971).
94. A. Baldacci, S. Ghersetti, and K. N. Rao, *J. Mol. Spectrosc.*, **41**, 222 (1972).
95. A. Baldacci, S. Ghersetti, S. C. Hurlock, and K. N. Rao, *J. Mol. Spectrosc.*, **42**, 327 (1972).
96. K. F. Plamer, M. E. Mickelson, and K. N. Rao, *J. Mol. Spectrosc.*, **44**, 131 (1972).
97. J. Plíva, *J. Mol. Spectrosc.*, **44**, 145, 165 (1972).
98. H. C. Allen and E. K. Plyler, *J. Chem. Phys.*, **25**, 1132 (1956).
99. R. E. Miller and D. F. Eggers, *J. Chem. Phys.*, **45**, 3028 (1966).
100. R. E. Miller, G. E. Leroi, and D. F. Eggers, *J. Chem. Phys.*, **46**, 2292 (1967).
101. T. H. Edwards, N. K. Moncur, and L. E. Snyder, *J. Chem. Phys.*, **46**, 2139 (1967).
102. L. E. Snyder and T. H. Edwards, *J. Mol. Spectrosc.*, **31**, 347 (1969).
103. P. Helminger, R. L. Cook, and F. C. DeLucia, *J. Chem. Phys.*, **56**, 4581 (1972).
104. R. L. Cook, F. C. DeLucia, and P. Helminger, *J. Mol. Spectrosc.*, **41**, 123 (1972).
105. E. D. Palik, *J. Mol. Spectrosc.*, **3**, 259 (1959).
106. R. A. Hill and T. H. Edwards, *J. Mol. Spectrosc.*, **14**, 203 (1964).
107. R. A. Hill, *J. Mol. Spectrosc.*, **24**, 100 (1967).
108. R. A. Hill and T. H. Edwards, *J. Chem. Phys.*, **42**, 1391 (1965).
109. T. Oka and Y. Morino, *J. Mol. Spectrosc.*, **8**, 300 (1962).
110. R. D. Shelton, A. H. Nielsen, and W. H. Fletcher, *J. Chem. Phys.*, **21**, 2178 (1953).
111. A. H. Nielsen, R. D. Shelton, and W. H. Fletcher, *J. Phys. Radium*, **15**, 604 (1954).
112. S. R. Polo and M. K. Wilson, *J. Chem. Phys.*, **22**, 900 (1954).
113. A. Barbe and P. Jouve, *J. Mol. Spectrosc.*, **38**, 273 (1971).
114. A. Barbe, C. Secroun, and P. Jouve, *J. Phys.*, **33**, 209 (1972).
115. Y. Morino, Y. Kikuchi, S. Saito, and E. Hirota, *J. Mol. Spectrosc.*, **13**, 95 (1964).
116. S. Saito, *J. Mol. Spectrosc.*, **30**, 1 (1969).
117. G. E. Moore, *J. Opt. Soc. Am.*, **43**, 1045 (1953).
118. E. T. Arakawa and A. H. Nielsen, *J. Mol. Spectrosc.*, **2**, 413 (1958).
119. M. D. Olman and C. D. Hause, *J. Mol. Spectrosc.*, **26**, 241 (1968).
120. R. E. Blank, M. D. Olman, and C. D. Hause, *J. Mol. Spectrosc.*, **33**, 109 (1970).
121. R. E. Blank and C. D. Hause, *J. Mol. Spectrosc.*, **34**, 478 (1970).
112. G. R. Bird, J. C. Baird *et al.*, *J. Chem. Phys.*, **40**, 3378 (1964).
123. G. R. Bird, G. R. Hunt, H. A. Gebbie, and N. W. B. Stone, *J. Mol. Spectrosc.*, **33**, 244 (1970).
124. W. J. Lafferty, D. R. Lide, and R. A. Toth, *J. Chem. Phys.*, **43**, 2063 (1965).
125. C. B. Murchison and J. Overend, *Spectrochim. Acta*, **26A**, 599 (1970); **27A**, 2407 (1971).

126. W. S. Benedict, E. K. Plyler, and E. D. Tidwell, *J. Res. Natl. Bur. Stand.*, **61**, 123 (1958).
127. W. S. Benedict and E. K. Plyler, *Can. J. Phys.*, **35**, 1235 (1957).
128. W. S. Benedict, E. K. Plyler, and E. D. Tidwell, *J. Chem. Phys.*, **29**, 829 (1958).
129. W. S. Benedict, E. K. Plyler, and E. D. Tidwell, *J. Chem. Phys.*, **32**, 32 (1960).
130. J. S. Garing, H. H. Nielsen, and K. N. Rao, *J. Mol. Spectrosc.*, **3**, 496 (1959).
131. K. N. Rao, W. W. Brim *et al.*, *J. Mol. Spectrosc.*, **7**, 362 (1961); **11**, 389 (1963).
132. F. O. Shimizu and T. Shimizu, *J. Mol. Spectrosc.*, **36**, 94 (1970).
133. T. Urisu, T. Tanaka, E. Hirota, and Y. Morino, *J. Mol. Spectrosc.*, **35**, 345 (1970).
134. P. Helminger, F. C. DeLucia, and W. Gordy, *J. Mol. Spectrosc.*, **39**, 94 (1971).
135. J. M. Dowling, *J. Mol. Spectrosc.*, **27**, 527 (1968).
136. R. E. Walker and B. F. Hochheimer, *J. Mol. Spectrosc.*, **34**, 500 (1970).
137. W. J. Lafferty and D. R. Lide, *J. Mol. Spectrosc.*, **23**, 94 (1967).
138. A. R. H. Cole, L. Isaacson, and R. C. Lord, *J. Mol. Spectrosc.*, **23**, 86 (1967).
139. A. G. Maki and C. T. Gott, *J. Chem. Phys.*, **36**, 2282 (1962).
140. A. G. Maki, *J. Chem. Phys.*, **38**, 1261 (1963).
141. J. W. C. Johns, H. F. Shurvell, and J. K. Tyler, *Can. J. Phys.*, **47**, 893 (1969).
142. J. W. C. Johns, J. M. R. Stone, and G. Winnewisser, *J. Mol. Spectrosc.*, **38**, 437 (1971).
143. A. G. Maki, *J. Chem. Phys.*, **43**, 3193 (1965).
144. A. Picard, *Spectrochim. Acta,* **29A**, 423 (1973).
145. B. P. Winnewisser and M. Winnewisser, *J. Mol. Spectrosc.*, **29**, 505 (1969).
146. B. P. Winnewisser, *J. Mol. Spectrosc.*, **40**, 164 (1971).
147. M. Winnewisser and B. P. Winnewisser, *J. Mol. Spectrosc.*, **41**, 143 (1972).
148. W. D. Sheasley, C. W. Matthews, E. L. Ferretti, and K. N. Rao, *J. Mol. Spectrosc.*, **37**, 377 (1971).
149. S. G. W. Ginn, C. W. Brown, J. K. Kenney, and J. Overend, *J. Mol. Spectrosc.*, **28**, 509 (1968).
150. S. G. W. Ginn, D. Johansen, and J. Overend, *J. Mol. Spectrosc.*, **36**, 448 (1970).
151. S. G. W. Ginn, J. K. Kenney, and J. Overend, *J. Chem. Phys.*, **48**, 1571 (1968).
152. C. W. Brown and J. Overend, *Can. J. Phys.*, **46**, 977 (1968).
153. C. W. Brown and J. Overend, *Spectrochim. Acta,* **25A**, 1535 (1969).
154. R. J. L. Popplewell, F. N. Masri, and H. W. Thompson, *Spectrochim. Acta,* **23A**, 2797 (1967).
155. F. N. Masri and W. E. Blass, *J. Mol. Spectrosc.*, **39**, 98 (1971).
156. A. Allan, J. L. Duncan, J. H. Holloway, and D. C. McKean, *J. Mol. Spectrosc.*, **31**, 368 (1969).
157. J. M. R. Stone and I. M. Mills, *J. Mol. Spectrosc.*, **35**, 354 (1970).
158. S. Reichman and S. G. W. Ginn, *J. Mol. Spectrosc.*, **40**, 27 (1971).
159. M. Otake, C. Matsumura, and Y. Morino, *J. Mol. Spectrosc.*, **28**, 316 (1968).
160. M. Otake, E. Hirota, and Y. Morino, *J. Mol. Spectrosc.*, **28**, 325 (1968).
161. S. Reichman and J. Overend, *Spectrochim. Acta,* **26A**, 379 (1970).
162. S. Reichman, *J. Mol. Spectrosc.*, **35**, 329 (1970).
163. H. H. Blau and H. H. Nielsen, *J. Mol. Spectrosc.*, **1**, 124 (1957).
164. T. Oka, K. Takagi, and Y. Morino, *J. Mol. Spectrosc.*, **14**, 27 (1964).
165. T. Nakagawa, Y. Morino *et al.*, *J. Mol. Spectrosc.*, **31**, 436 (1969).

166. T. Nakagawa, Y. Morino *et al.*, *J. Mol. Spectrosc.*, **38**, 70, 84 (1971).
167. E. K. Plyler, E. D. Tidwell, and L. R. Blaine, *J. Res. Natl. Bur. Stand.*, **64A**, 201 (1960).
168. W. L. Barnes, J. Susskind, R. H. Hunt, and E. K. Plyler, *J. Chem. Phys.*, **56**, 5160 (1972).
169. D. H. Rank, D. P. Eastman, G. Skorinko, and T. A. Wiggins, *J. Mol. Spectrosc.*, **5**, 78 (1960).
170. L. H. Jones and M. Goldblatt, *J. Mol. Spectrosc.*, **2**, 103 (1958).
171. R. S. McDowell, *J. Mol. Spectrosc.*, **21**, 280 (1966).
172. W. E. Blass and T. H. Edwards, *J. Mol. Spectrosc.*, **24**, 116 (1967).
173. L. Henry, N. Husson, R. Andia, and A. Valentin, *J. Mol. Spectrosc.*, **36**, 511 (1970).
174. J. Botineau, *J. Mol. Spectrosc.*, **41**, 182 (1972).
175. J. C. Deroche, G. Graner, and C. Alamichel, *J. Mol. Spectrosc.*, **43**, 175 (1972).
176. W. B. Olson, *J. Mol. Spectrosc.*, **43**, 190 (1972).
177. W. L. Smith and I. M. Mills, *J. Mol. Spectrosc.*, **11**, 11 (1963).
178. J. L. Duncan, D. C. McKean, and G. K. Speirs, *Mol. Phys.*, **24**, 553 (1972).
179. M. Betrencourt and G. Graner, *Spectrochim. Acta*, **28A**, 1019 (1972).
180. T. M. Holladay and A. H. Nielsen, *J. Mol. Spectrosc.*, **14**, 371 (1964).
181. E. W. Jones, R. J. L. Popplewell, and H. W. Thompson, *Spectrochim. Acta*, **22** 669 (1966).
182. J. L. Duncan and A. Allan, *J. Mol. Spectrosc.*, **25**, 224 (1968).
183. M. Morillon-Chapey and G. Graner, *J. Mol. Spectrosc.*, **31**, 155 (1969).
184. R. W. Peterson and T. H. Edwards, *J. Mol. Spectrosc.*, **38**, 524 (1971).
185. D. H. Rank, H. D. Rix, and T. A. Wiggins, *J. Opt. Soc. Am.*, **43**, 157 (1953).
186. R. G. Brown and T. H. Edwards, *J. Chem. Phys.*, **37**, 1029 (1962).
187. E. W. Jones, R. J. L. Popplewell, and H. W. Thompson, *Spectrochim. Acta*, **22**, 639, 647 (1966).
188. D. R. Anderson and J. Overend, *Spectrochim. Acta*, **27A**, 2013 (1971).
189. R. W. Peterson and T. H. Edwards, *J. Mol. Spectrosc.*, **41**, 137 (1972).
190. Y. Morino and C. Hirose, *J. Mol. Spectrosc.*, **24**, 204 (1967).
191. T. A. Wiggins, E. R. Shull, and D. H. Rank, *J. Chem. Phys.*, **21**, 1368 (1953).
192. R. G. Brown and T. H. Edwards, *J. Chem. Phys.*, **37**, 1035 (1962).
193. R. G. Brown and T. H. Edwards, *J. Chem. Phys.*, **40**, 2740 (1964).
194. T. L. Barnett and T. L. Edwards, *J. Mol. Spectrosc.*, **23**, 302 (1967).
195. R. W. Peterson and T. H. Edwards, *J. Mol. Spectrosc.*, **38**, 1 (1971).
196. R. J. L. Popplewell and H. W. Thompson, *Spectrochim. Acta*, **25A**, 287 (1969).
197. H. Kurlat, M. Kurlat, and W. E. Blass, *J. Mol. Spectrosc.*, **38**, 197 (1971).
198. H. Matsuura and J. Overend, *Spectrochim. Acta*, **27A**, 2165 (1971).
199. H. Matsuura and J. Overend, *Spectrochim. Acta*, **28A**, 1203 (1972).
200. H. Matsuura and J. Overend, *J. Chem. Phys.*, **55**, 1787 (1971).
201. H. Matsuura and J. Overend, *J. Chem. Phys.*, **56**, 5725 (1972).
202. D. R. Anderson and J. Overend, *Spectrochim. Acta*, **28A**, 1225, 1231, 1637 (1972).
203. Y. Morino and C. Hirose, *J. Mol. Spectrosc.*, **22**, 99 (1967).
204. Y. Morino, J. Nakamura, and S. Yamamoto, *J. Mol. Spectrosc.*, **22**, 34 (1967).
205. T. E. Sullivan and L. Frenkel, *J. Mol. Spectrosc.*, **39**, 185 (1971).
206. G. R. Wilkinson and M. K. Wilson, *J. Chem. Phys.*, **44**, 3867 (1966).
207. R. W. Lovejoy and W. B. Olson, *J. Chem. Phys.*, **57**, 2224 (1972).

208. A. Cabana, L. Lambert, and C. Pépin, *J. Mol. Spectrosc.*, **43**, 429 (1972).
209. D. Van Lerberghe, I. J. Wright, and J. L. Duncan, *J. Mol. Spectrosc.*, **42**, 251, 463 (1972).
210. C. DiLauro, *J. Mol. Spectrosc.*, **30**, 266 (1969).
211. J. Paldus and D. A. Ramsay, *Can. J. Phys.*, **45**, 1389 (1967).
212. F. W. Birss, D. A. Ramsay *et al.*, *Can. J. Phys.*, **48**, 1230 (1970).
213. L. G. Smith, *J. Chem. Phys.*, **17**, 139 (1949).
214. G. E. Hansen and D. M. Dennison, *J. Chem. Phys.*, **20**, 313 (1952).
215. H. C. Allen and E. K. Plyler, *J. Chem. Phys.*, **31**, 1062 (1959).
216. W. J. Lafferty and E. K. Plyler, *J. Chem. Phys.*, **37**, 2688 (1962).
217. H. C. Allen and E. K. Plyler, *J. Res. Natl. Bur. Stand.*, **67A**, 225 (1963).
218. A. R. H. Cole, W. J. Lafferty, and R. J. Thibault, *J. Mol. Spectrosc.*, **29**, 365 (1969).
219. I. Nakagawa and T. Shimanouchi, *J. Mol. Spectrosc.*, **39**, 255 (1971).
220. Y. Morino, *Pure Appl. Chem.*, **18**, 323 (1969).
221. H. Takeo, E. Hirota, and Y. Morino, *J. Mol. Spectrosc.*, **34**, 370 (1970).
222. H. Takeo, E. Hirota, and Y. Morino, *J. Mol. Spectrosc.*, **41**, 420 (1972).
223. T. Tanaka and Y. Morino, *J. Mol. Spectrosc.*, **33**, 538, 552 (1970).
224. Y. Morino and S. Saito, *J. Mol. Spectrosc.*, **19**, 435 (1966).
225. H. Shoji, T. Tanaka, and E. Hirota, *J. Mol. Spectrosc.*, **47**, 268 (1973).
226. H. Takeo, R. F. Curl, and P. W. Wilson, *J. Mol. Spectrosc.*, **38**, 464 (1971).
227. A. R. Hoy, M. Bertram, and I. M. Mills, *J. Mol. Spectrosc.*, **46**, 429 (1973).
228. D. R. Herschbach and V. W. Laurie, *J. Chem. Phys.*, **35**, 458 (1961).
229. P. M. Morse, *Phys. Rev.*, **34**, 57 (1929).
230. E. R. Lippincott, *J. Chem. Phys.*, **21**, 2070 (1953).
231. E. R. Lippincott and R. Schroeder, *J. Chem. Phys.*, **23**, 1131 (1955).
232. D. Steele, E. R. Lippincott, and J. T. Vanderslice, *Rev. Mod. Phys.*, **34**, 239 (1962).
233. R. G. Parr and R. F. Borkman, *J. Chem. Phys.*, **46**, 3683 (1967).
234. R. F. Borkman and R. G. Parr, *J. Chem. Phys.*, **48**, 1116 (1968).
235. R. G. Parr and R. F. Borkman, *J. Chem. Phys.*, **49**, 1055 (1968).
236. A. B. Anderson and R. G. Parr, *J. Chem. Phys.*, **53**, 3375 (1970).
237. A. B. Anderson and R. G. Parr, *J. Chem. Phys.*, **55**, 5490 (1971).
238. A. B. Anderson and R. G. Parr, *J. Chem. Phys.*, **56**, 5204 (1972).
239. K. Kuchitsu and Y. Morino, *Bull. Chem. Soc. Japan,* **38**, 805, 814 (1965).
240. K. Kuchitsu and L. S. Bartell, *J. Phys. Soc. Japan,* **17**, B II, 23 (1962).
241. A. Kaldor, A. G. Maki, A. J. Dorney, and I. M. Mills, *J. Mol. Spectrosc.*, **45**, 247 (1973).
242. A. J. Dorney, A. R. Hoy, and I. M. Mills, *J. Mol. Spectrosc.*, **45**, 253 (1973).
243. J. Plíva, V. Špirko, and D. Papoušek, *J. Mol. Spectrosc.*, **23**, 331 (1967).
244. J. W. C. Johns, *Can. J. Phys.*, **45**, 2639 (1967).
245. K. Machida and J. Overend, *J. Chem. Phys.*, **50**, 4429, 4437 (1969).
246. K. Machida and J. Overend, *J. Chem. Phys.*, **51**, 2537 (1969).
247. D. F. Smith and J. Overend, *J. Chem. Phys.*, **55**, 1157 (1971).
248. D. F. Smith and J. Overend, *Spectrochim. Acta,* **28A**, 2387 (1972).
249. D. R. Lide and R. L. Kuczkowski, *J. Chem. Phys.*, **46**, 4768 (1967).
250. D. R. Lide and C. Matsumura, *J. Chem. Phys.*, **50**, 71, 3080 (1969).
251. J. T. Hougen, P. R. Bunker, and J. W. C. Johns, *J. Mol. Spectrosc.*, **34**, 136 (1970).
252. P. R. Bunker and J. M. R. Stone, *J. Mol. Spectrosc.*, **41**, 310 (1972).

253. K. Kuchitsu and Y. Morino, *Spectrochim. Acta,* **22**, 33 (1966).
254. H. C. Urey and C. A. Bradley, *Phys. Rev.*, **38**, 1969 (1931).
255. T. Shimanouchi, *J. Chem. Phys.*, **17**, 245 (1949).
256. D. Papoušek and J. Plíva, *Coll. Czech. Chem. Commun.*, **29**, 1973 (1964).
257. R. G. Parr and J. E. Brown, *J. Chem. Phys.*, **49**, 4849 (1968).
258. J. E. Brown and R. G. Parr, *J. Chem. Phys.*, **54**, 3429 (1971).
259. A. B. Anderson, *J. Chem. Phys.*, **56**, 4228 (1972).
260. A. B. Anderson, *J. Chem. Phys.*, **57**, 4143 (1972).
261. V. K. Wang and J. Overend, *Spectrochim. Acta,* **29A**, 1623 (1973).
262. H. H. Nielsen, *Rev. Mod. Phys.*, **23**, 90 (1951).
263. H. H. Nielsen, *Handbuch der Physik,* **XXXVII/1**, 173 (1959).
264. G. Amat, H. H. Nielsen, and G. Tarrago, *Rotation-Vibration of Polyatomic Molecules*, M. Dekker, New York, 1971.
265. D. R. Lide, *J. Mol. Spectrosc.*, **33**, 448 (1970).
266. I. Suzuki, *J. Mol. Spectrosc.*, **25**, 479 (1968).
267. D. F. Smith and J. Overend, *J. Chem. Phys.*, **54**, 3632 (1971).
268. J. Giguère, V. K. Wang, J. Overend, and A. Cabana, *Spectrochim. Acta,* **29A**, 1197 (1973).
269. I. Suzuki, *J. Mol. Spectrosc.*, **32**, 54 (1969).
270. D. F. Smith, J. Overend, R. C. Spiker, and L. Andrews, *Spectrochim. Acta,* **28A**, 87 (1972).
271. J. Plíva, in *Symposium on Molecular Structure and Spectroscopy*, Ohio State University, Columbus, 1960.
272. W. J. Dixon and F. J. Massey, *Introduction to Statistical Analysis,* McGraw-Hill, N.Y., 1969.
273. Y. Morino and T. Nakagawa, *J. Mol. Spectrosc.*, **26**, 496 (1968).
274. I. Suzuki, M. A. Pariseau and J. Overend, *J. Chem. Phys.*, **44**, 3561 (1966).
275. T. Nakagawa and Y. Morino, *Bull. Chem. Soc. Jap.*, **42**, 2212 (1969).
276. D. F. Smith and J. Overend, *Spectrochim. Acta,* **28A**, 471 (1972).
277. C. B. Murchison and J. Overend, *Spectrochim. Acta,* **27A**, 1509, 1801, 2407 (1971).
278. S. G. W. Ginn, S. Reichman, and J. Overend, *Spectrochim. Acta,* **26A**, 291 (1970).

William H. Kirchhoff

Determination of Force Fields by Analysis of Centrifugal Distortion in Microwave Spectra

It is the purpose of this paper to discuss the accuracy of molecular force fields derived from centrifugal distortion constants. This discussion will not involve a general review of all studies of force fields derived partially or wholly from centrifugal distortion data but rather will focus on work done in recent years at the National Bureau of Standards. The reason for this narrow approach is twofold. First, it is the work with which the author is most familiar and, therefore, best able to describe (although, perhaps, least able to criticize). More importantly, in this work alone (with one notable exception) attempts have been made to discriminate between the effects of measurement and model errors. The one exception has been the case of diatomic molecules, where a great deal is known about the theoretical interpretation of the spectral constants. The case of the diatomic molecule will be mentioned briefly but only as a basis for comparison with more complicated molecules.

As a further restriction on the scope of this discussion, only bent, symmetric, triatomic molecules will be covered. This is because the force field of XYX-type molecules is overdetermined by the microwave data; thus, the reliability of the model can be tested by studying the internal consistency of the force constants calculated from different combinations of centrifugal distortion constants.

The accuracy of the force field can best be judged by calculating from the force constants the harmonic, fundamental vibrational frequencies and comparing these with the harmonic frequencies obtained from the infrared spectrum (or, if these are not available, with the observed infrared band origins). Thus, greater emphasis will be placed on the calculation of the vibrational fundamentals than on the force constants themselves.

Much but not all of the discussion in this paper represents a summary and extension of previously published work,[1] and the reader who wishes greater mathematical detail should turn to that work. In addition to SO_2, OF_2, and SiF_2, which were covered there, the molecules O_3, OCl_2, CF_2, and SF_2 are included here.

It will be demonstrated that problems encountered in the calculation of force fields from centrifugal distortion constants are analogous to those encountered in the calculation of molecular structures from rotational constants. Bond distances can be calculated from rotational constants with a reproducibility of 0.01 percent or about 0.0001 Å. That is, remeasurement of the microwave spectrum (but not necessarily the same specific transitions) and reanalysis of the data will yield bond distances that agree with the original values within the above limits. The interpretability of these bond distances as distances between two constantly moving nuclei is less accurate. Unless special care is taken to account for the effects of molecular vibrations, the accuracy of a microwave-determined bond distance as a measure of either the equilibrium distance or the average distance between two nuclei is on the order of 1 percent. Similar statements can be made of the force field as determined from centrifugal distortion constants. The force constants can be determined with a reproducibility or precision between about 0.1 and 1 percent. However, comparison of these force constants to those determined from either infrared or electronic spectroscopy shows differences on the order of 1–10 percent. It appears that, if greater accuracy is desired from force constants determined from centrifugal distortion constants, the effects of molecular vibrations on the values of the distortion constants must be included in the analysis.

MODEL ERRORS AND MEASUREMENT ERRORS

To demonstrate the conclusions of the preceding paragraph, it is necessary to show that the effects of model errors and measurement errors can be distinguished. The observed transition frequencies in the microwave region represent differences between molecular rotational energy levels, which are eigenvalues of the following Hamiltonian:

$$\mathcal{H} = AP_a^2 + BP_b^2 + CP_c^2 + \frac{1}{4} \sum_{\substack{\alpha,\beta,\gamma,\delta \\ = a,b,c}} \tau_{\alpha\beta\gamma\delta} P_\alpha P_\beta P_\gamma P_\delta + \sum_{i=1}^{7} H_i P_i^6 + \ldots, \tag{1}$$

where P_α is a component of the angular momentum vector along the αth principal axis. A, B, and C are the rigid rotor rotational constants, which are functions of the molecular geometry and masses. The τ's are related to the molecular force field through equations of the form

$$\tau_{\alpha\beta\gamma\delta} = -\frac{\hbar^3}{4\pi} \frac{1}{I_\alpha I_\beta I_\gamma I_\delta} \sum_{i=1}^{3N-6} \sum_{j=1}^{3N-6} \frac{\partial I_{\alpha\beta}}{\partial R_i} \frac{\partial I_{\gamma\delta}}{\partial R_j} (f^{-1})_{ij}, \tag{2}$$

where I_α ($\alpha = a, b, c$) is a principal moment of inertia of the molecule, $I_{\alpha\beta}$ is a product of inertia, R_i and R_j are internal displacement coordinates (not necessarily but usually normal coordinates) and $(f^{-1})_{ij}$ is the corresponding element of the inverse force constant matrix. The third term of Eq. (1) represents contributions from terms of sixth power in angular momentum and, as indicated, the entire series is an infinite one in even powers of angular momentum.

From the analysis of a microwave spectrum it is not possible to obtain A, B, C and all 81 $\tau_{\alpha\beta\gamma\delta}$'s independently but rather eight linear combinations of these quantities.[2–5] Thus, the algebra implied by Eq. (2) for extracting force field information is complicated by the fact that only a few linear combinations of these constants can be obtained from the microwave spectrum.

The first point to be made concerns the accuracy with which the model of Eq. (1) can be used to fit the microwave spectrum. The typical measurement accuracy of most microwave spectroscopists ranges from 0.01 to 0.2 MHz. It has been found that the model can reproduce the observed microwave spectrum with an accuracy on the order of 10 kHz, except in situations where molecular vibrations are large such as in the bending vibration of H_2O and H_2S[6,7] or the nearly free internal rotation in CH_3OH.[8] Effects such as spin–rotation interactions and spin–spin interactions contribute splittings on the order of 10 kHz and are observed only under exceptionally high resolution. The effects of nuclear electric quadrupole interactions present difficulties but these can be accounted for by inclusion of a few more terms in Eq. (1). The microwave spectra of over 30 molecules have been fit to this model within experimental error.[1,9] It has been possible to use the model to locate misprints or mismeasurements in published spectra. The statistics associated with a least squares fit of the measured microwave spectra have been used to calculate standard deviations of the calculated transition frequencies. Newly measured transitions almost always fall within two standard deviations of their predicted frequencies. This is a strong indication that the errors encountered are those arising from random measurement errors.

It should be mentioned that inclusion of the sextic terms of Eq. (1) is often quite important. For molecules containing more than five heavy atoms, the quadratic and quartic terms of Eq. (1) are sufficient to fit the microwave spectrum within experimental uncertainty. The small effect of the sextic terms can be absorbed by the quartic terms. However, the quartic terms thus obtained can be in error by more than three standard deviations. "In error" means that predictions of new transitions can be in error by three standard deviations or, alternatively, if better measurements are made and the P^6 terms included, the new values of the τ's will differ from the old by as many as three standard deviations. If the τ's are to be interpreted from the fit of the spectrum, it is important that the effects of inclusion of sextic terms be investigated.

It has been possible to use the statistics from the least squares fit of the microwave spectrum through propagation of errors formulas (including the correlation of errors between the determined rotational and distortion constants) to calculate standard deviations for the force constants calculated from the distortion constants. These standard deviations, which represent measurement errors, can be compared with the lack of internal consistency between force constants calculated from different distortion constants. This comparison gives the relative magnitude of measurement vs. model errors.

As mentioned above, the microwave spectrum yields only five independent distortion constants: τ_{aaaa}, τ_{bbbb}, and τ_{cccc}, which can be applied directly to Eq. (2), and two combinations τ_1 and τ_2. For the simple case of XYX-type molecules, τ_1 and τ_2 have relatively simple forms:

$$\tau_1 = (\tau_{aabb} + 2\tau_{abab}) + \tau_{bbcc} + \tau_{ccaa} \tag{3}$$

and

$$\tau_2 = \frac{A}{S}\,\tau_{bbcc} + \frac{B}{S}\,\tau_{ccaa} + \frac{C}{S}(\tau_{aabb} + 2\tau_{abab}), \tag{4}$$

where $S = A + B + C$. For XYX molecules, there are only four harmonic force constants. Three of these, F_{11}, F_{12}, and F_{22}, relate to the symmetric stretching and bending motions of the molecule. The fourth, F_{33}, relates to the antisymmetric stretch. Because five τ's are obtained from the microwave spectrum, the force constants are overdetermined. Thus, F_{11}, F_{12}, and F_{22} can be calculated from τ_{aaaa}, τ_{bbbb}, and either τ_{cccc} or τ_1 and τ_2, giving two values for each of these constants. On the other hand, F_{33} can be calculated from τ_{aaaa}, τ_{bbbb}, and any pair of τ_1, τ_2, or τ_{cccc}, giving three values for F_{33}.

It should now be stated that these calculations are not unique. For example, F_{33} could be calculated from any two independent combinations of τ_1, τ_2, and τ_{cccc} although the calculation would be more complex. Thus, the range in values of F_{33} may be larger than that indicated by the range obtained from pairs of τ_1, τ_2, and τ_{cccc} alone. Alternatively, there may be one combination that is a better predictor of F_{33} than any other, but such a combination is not yet known.

ACCURACY OF THE FORCE FIELD CALCULATION

The force constants are calculated from the distortion constants by means of Eq. (2). This expression was derived assuming an equilibrium configuration for the molecules. The effects of molecular vibrations cause Eq. (2) not to be obeyed exactly so that force constants calculated using the spectroscopic constants of the ground state will be somewhat in error. One way to assess this error is to compare the harmonic vibrational fundamentals calculated from the centrifugal distortion constants with those calculated from the infrared spectrum (or, when the harmonic frequencies are unknown, with the observed infrared band origins). This comparison is presented in Table 1 for a variety of bent, symmetric triatomic molecules.

It can be seen from Table 1 that the value of ω_2 obtained from the centrifugal distortion (cd) analysis is in rather good agreement (better than 4 percent) with the harmonic frequencies obtained from the infrared (ir) spectrum. In all cases, ω_2 (cd) is less than ω_2 (ir) and closer in frequency to ν_2 (ir) than to ω_2 (ir). The results for ω_1 are similar with the differences between ω_1 (cd) and ω_1 (ir) or ν_1 (ir) being less than 4 percent. In every case ω_1 (cd) calculated from τ_1 and τ_2 was greater than ω_1 (cd) calculated from τ_{cccc}. With one exception, ω_1 (cd) from τ_1 and τ_2 agreed with ω_1 (ir) or ν_1 (ir) within 2 percent. The one exception was for SiF_2, where the uncertainty (one standard deviation) in ω_1 was 4 percent. The results for ω_3 showed the greatest variation between the microwave and infrared results, with differences reaching

TABLE I Comparison of Harmonic Vibrational Fundamentals Calculated from Centrifugal Distortion Constants with Those Calculated from the Infrared Spectrum[a]

	ω_1 (cm^{-1})		ω_2 (cm^{-1})		ω_3 (cm^{-1})		References for Spectra
	cd	ir	cd	ir	cd	ir	
SO_2	1165.0 ± 1.0 1139.7 ± 1.0	1167.60 (1151.4)	507.75 ± 0.03 509.94 ± 0.03	526.27 (517.8)	1287.3 ± 1.2 1386.4 ± 1.7 1402.6 ± 2.0	1380.91 (1360.5)	mw: 10, 11 ir: 12
OF_2	916.5 ± 3.2 897.9 ± 3.0	(928)	455.9 ± 0.3 458.3 ± 0.5	(461)	807.5 ± 2.2 851.0 ± 6.5 858.3 ± 7.3	(831)	mw: 13, 14 ir: 15
O_3	1106.7 ± 1.2 1090.1 ± 1.2	1124 (1103.157)	690.5 ± 0.2 694.7 ± 0.3	719 (701.42)	1018.6 ± 1.9 1062.5 ± 3.5 1067.8 ± 4.1	1102 (1042.096)	mw: 16, 17 ir: 18–20
SiF_2	920 ± 40 832 ± 32	(855)	340.3 ± 1.0 345.3 ± 2.0	(343)	864 ± 36 1110 ± 135 1180 ± 189	(872)	mw: 21, 22 ir: 23–25
OCl_2	631 ± 16 612 ± 11	(640)	295.6 ± 2.5 297.7 ± 3.7	(300)	752 ± 48 964 ± 232 990 ± 267	(685.9)	mw: 26 ir: 27
CF_2	1234 ± 5 1222 ± 4	(1224)	661.1 ± 1.0 663.0 ± 1.4	(668)	1091 ± 5 1124 ± 13 1128 ± 14	(1112)	mw: 28 ir: 29
SF_2	852 ± 4 827 ± 4	*b*	350.0 ± 0.2 351.9 ± 0.3	*b*	777 ± 4 820 ± 9 831 ± 13	*b*	mw: 30

[a] For ω_1 and ω_2, two values for the fundamental frequencies are obtained from the microwave spectrum, and for ω_3, three values are obtained. If harmonic frequencies have not been determined from the infrared spectrum, the band origins are given in parentheses. The uncertainties quoted for the microwave-determined harmonic frequencies represent one standard deviation based on the least squares fit of the microwave spectrum.
[b] The infrared spectrum of SF_2 has not yet been observed.

values as large as 7 percent (except for SiF_2 and OCl_2, where ω_3 was ill determined by the microwave spectrum). In all cases ω_3 (cd) from τ_1 and τ_2 was less than ω_3 (cd) from τ_1 and τ_{cccc}, which, in turn, was less than but nearly equal to ω_3 (cd) from τ_2 and τ_{cccc}. Although it may be coincidental, it can be seen that the average of the three centrifugal distortion values for ω_3 agrees with ν_3 (ir) [not ω_3 (ir)] within 1 percent for all cases where the ω_3's (cd) were well determined.

Another point that should be noted on inspecting Table 1 is that the ω's are rather sensitively determined by the microwave data. With two exceptions, the ω's have been determined with precisions of better than 1 percent and sometimes better than 0.1 percent. The two exceptions are SiF_2 and OCl_2. For SiF_2, the assignment of the microwave spectrum is incomplete in that there exist microwave transitions involving $J \leqslant 40$, which cannot be predicted with standard deviations less than 50 MHz. Measurement of an additional (but carefully selected) dozen or so transitions could greatly improve the precision of ω_1 and ω_3. For OCl_2, the assignment is more complete, with all transitions with $J \leqslant 40$ being predicted with standard deviations less than 5 MHz. Although measurement of a few more transitions could be expected to improve the precision of ω_1 and ω_3, this precision would still not be comparable with the other species. The main reason for this is that the overall measurement accuracy for OCl_2 is lower than for the other molecules because of the effects of quadrupole splitting and broadening from the two chlorine nuclei. The standard deviation for the fit of the microwave spectrum of OCl_2 was 0.24 MHz, whereas for the other molecules the standard deviation was on the order of 0.1 MHz.

The above discussion summarizes the main conclusions about accuracy and precision of vibrational fundamentals calculated from centrifugal distortion data. The same discussion could have been applied to the force field itself, but comparison of vibrational fundamentals seems to be more illustrative. However, a comparison of this discussion with a similar one for diatomic molecules can serve to make one more point.

For a diatomic molecule, the energy levels of the nonrigid rotor are given, to the order of perturbation considered here, by

$$E = B_v J(J+1) - D_v J^2 (J+1)^2, \tag{5}$$

where D_v is the centrifugal distortion constant. The equilibrium values, B_e and D_e, are related to the harmonic vibrational frequency ω_e by

$$\omega_e^2 = \frac{4B_e^3}{D_e}. \tag{6}$$

For bent, symmetric triatomic molecules, there is an analogous expression relating $\omega_3^{(e)}$ to $\tau_{abab}^{(e)}$. The contribution of $\tau_{abab}^{(v)}$ to the rotational energy in the vth vibrational state is $+(1/4)\,\tau_{abab}^{(v)}(P_a{}^2P_b{}^2 + P_b{}^2P_a{}^2)$ as contrasted to $-D_vJ^2(J+1)^2$ of Eq. (5). The relationship corresponding to Eq. (6) is

$$[\omega_3^{(e)}]^2 = 4\,\frac{A_eB_eC_e}{-\tau_{abab}^{(e)}} \tag{7}$$

We now consider how Eq. (6) behaves for CO when we use B_o and D_o or B_e and D_o instead of B_e and D_e. The appropriate spectroscopic constants for CO are[31] as follows:

$$\begin{aligned}
\omega_e &= 2169.8233\ \text{cm}^{-1}\\
B_e &= 1.931271\ \text{cm}^{-1}\\
D_e &= 6.1198 \times 10^{-6}\ \text{cm}^{-1}\\
B_o &= 1.922521\ \text{cm}^{-1}\\
D_o &= 6.1193 \times 10^{-6}\ \text{cm}^{-1}.
\end{aligned}$$

Using B_o and D_o in Eq. (6) one obtains

$$\omega_e = 2155.1910\ \text{cm}^{-1},$$

which differs from the correct ω_e by 0.6 percent. Using B_e and D_o in Eq. (6) one obtains

$$\omega_e = 2169.9211,$$

which differs by 0.004 percent. Thus, the largest error in using B_o and D_o in Eq. (6) comes from substituting B_o for B_e. The error arising from D_o is two orders of magnitude smaller. For SO_2, OF_2, and O_3, equilibrium values have been obtained for A_e, B_e, and C_e. Using these in Eq. (7) for calculating ω_3 caused an increase in ω_3 over using A_o, B_o, and C_o by about 10 cm^{-1}—roughly the same as in the case of CO. However, the spread in the values of ω_3 corresponding to the spread in values of τ_{abab} is still as great. Moreover when A_e, B_e, and C_e are used for calculating τ_{abab} from τ_1, τ_2, and τ_{cccc}, this spread diminishes, the ordering $\omega_3(\tau_1 \text{ and } \tau_2) < \omega_3(\tau_1 \text{ and } \tau_{cccc}) < \omega_3(\tau_2 \text{ and } \tau_{cccc})$ no longer holds, and the agreement with $\omega_3^{(e)}$ is somewhat improved, but *only* somewhat.

Clearly, the triatomic case is not entirely analogous to the diatomic case. The major difficulty may involve the complexity of the calculation of τ_{abab} from τ_1, τ_2, and τ_{cccc}. To see this, consider the following equations for calculating τ_{abab} from τ_1 and τ_2:

$$\tau_{abab} = \tfrac{1}{2}\tau_1 - \tfrac{1}{2}\frac{C^2}{A^2}\tau_{aaaa} - \tfrac{1}{2}\frac{C^2}{B^2}\tau_{bbbb} - \tfrac{1}{2}\left(1 + \frac{C^2}{A^2} + \frac{C^2}{B^2}\right)\tau_{aabb}, \tag{8}$$

where τ_{aabb} is given by

$$\left[(C-A)\frac{C^2}{A^2} + (C-B)\frac{C^2}{B^2}\right]\tau_{aabb} = C\tau_1 - (A+B+C)\tau_2 + (B-C)\frac{C^2}{A^2}\tau_{aaaa} + (A-C)\frac{C^2}{B^2}\tau_{bbbb}. \tag{9}$$

The effects of vibrational averaging in expressions as complex as these are likely to be much more complicated than in the case of a diatomic molecule. Moreover, combining large terms to give smaller terms may magnify the inaccuracies due to vibrational averaging.

CONCLUDING REMARKS

It has been shown that for bent, symmetric triatomic molecules centrifugal distortion constants can provide force constants with precisions ranging from 0.1 to 1 percent and accuracies ranging from 1 to 10 percent. A final question about combining infrared spectra with centrifugal distortion data to give the force field is worth considering. There are four approaches to this, and the merits of each will now be discussed.

In the first method, vibrational frequencies are fit exactly and the microwave spectrum is fit as well as possible in order to determine a "best" force field (for example, see Pierce *et al.*[13]). This has several drawbacks. It is assumed that somehow the effects of vibrational motions on the values of the distortion constants will be "averaged out." This will not necessarily be true. What will be lost will be information concerning the magnitude of this vibrational averaging. Moreover, since the microwave data are being fit to a constrained minimum, the values obtained for the force field may be sensitive to which microwave transitions are being fit. Before results from such calculations are accepted, attempts should be made to determine the magnitude of this sensitivity by varying the microwave data set used in the fit. Finally, because anharmonic contributions to the infrared frequencies can be as high as 5 percent, it is not clear that the force constants should be constrained to fit the infrared data exactly. This procedure should therefore not be used unless the harmonic frequencies are known and the sensitivity of the force constants to the choice of microwave transitions is determined.

A second method would be to fit the force field by least squares to the infrared data and to the τ's. This technique has the advantage that the force field will be independent of the microwave data base. Again, it is not obvious that the values obtained for the force constants will be close to the true values, but the force field obtained will come closest to reproducing the observed data. The method would be enhanced if each parameter (centrifugal distortion constant or vibrational fundamental) were to be weighted by the inverse square of its estimated model uncertainty (which will usually be considerably larger than its measurement uncertainty). The force field thus obtained would probably provide a close approximation to the true force field and could be used for intermolecular comparisons, but it would have the drawback that model error information would be lost.

A third method would be to choose one or more force constants that are insensitive to the choice of τ's from which they are calculated and calculate the remainder of the force field from the infrared data (see, for example, Kirchhoff *et al.*[30]). This method would also be insensitive to the microwave data set. However, it also suffers from ignoring the anharmonic contributions to the infrared frequencies by forcing the force constants to fit the vibrational fundamentals exactly. The effects of vibrational averaging of the τ's is not lost by virtue of exclusion of those τ's with large vibrational inconsistencies; these effects are simply ignored. The method is probably best used when harmonic vibrational frequencies are available.

The final method involves theoretically investigating the manner in which molecular vibrations affect the distortion constants, much in the same way that the effects of molecular vibrations on molecular structural information have been studied. Once these effects can be accounted for, even in only an approximate way, a truly representative force field could be calculated. It can be concluded that the best method is, as usual, the most difficult one but perhaps also the most interesting.

REFERENCES

1. W. H. Kirchhoff, *J. Mol. Spectrosc.*, **41**, 333 (1972).
2. J. K. G. Watson, *J. Chem. Phys.*, **45**, 1360 (1966).

3. J. K. G. Watson, *J. Chem. Phys.*, **46**, 1935 (1967).
4. J. K. G. Watson, *J. Chem. Phys.*, **48**, 181 (1968).
5. J. K. G. Watson, *J. Chem. Phys.*, **48**, 4517 (1968).
6. F. C. DeLucia, P. Helminger, R. L. Cook, and W. Gordy, *Phys. Rev. A*, **5**, 487 (1972).
7. P. Helminger, R. L. Cook, and F. C. DeLucia, *J. Chem. Phys.*, **56**, 4581 (1972).
8. R. M. Lees and J. G. Baker, *J. Chem. Phys.*, **48**, 5299 (1968).
9. K.-M. Marstokk and H. Møllendal, *J. Mol. Struct.*, **8**, 234 (1971).
10. G. Steenbeckeliers, *Ann. Soc. Sci. Brux.*, **82**, 331 (1968).
11. Y. Morino, Y. Kikuchi, S. Saito, and E. Hirota, *J. Mol. Spectrosc.*, **13**, 95 (1964).
12. R. D. Shelton, A. H. Nielsen, and W. H. Fletcher, *J. Chem. Phys.*, **21**, 2178 (1953).
13. L. Pierce, R. H. Jackson, and N. DiCianni, *J. Chem. Phys.*, **35**, 2240 (1961).
14. Y. Morino and S. Saito, *J. Mol. Spectrosc.*, **19**, 435 (1966).
15. E. A. Jones, J. S. Kirby-Smith, P. J. H. Woltz, and A. H. Nielsen, *J. Chem. Phys.*, **19**, 337 (1951).
16. M. Lichtenstein, J. J. Gallagher, and S. A. Clough, *J. Mol. Spectrosc.*, **40**, 10 (1971).
17. T. Tanaka and Y. Morino, *J. Mol. Spectrosc.*, **33**, 538 (1970).
18. T. Tanaka and Y. Morino, *J. Mol. Spectrosc.*, **33**, 552 (1970).
19. S. A. Clough and F. X. Kneizys, *J. Chem. Phys.*, **44**, 1855 (1966).
20. D. J. McCaa and J. H. Shaw, *J. Mol. Spectrosc.*, **25**, 374 (1968).
21. V. M. Rao and R. F. Curl, Jr., *J. Chem. Phys.*, **45**, 2032 (1966).
22. H. Shoji, T. Tanaka, and E. Hirota, *J. Mol. Spectrosc.*, **47**, 268 (1973).
23. V. M. Khanna, R. Hauge, R. F. Curl, Jr., and J. L. Margrave, *J. Chem. Phys.*, **47**, 5031 (1967).
24. J. W. Hastie, R. H. Hauge, and J. L. Margrave, *J. Am. Chem. Soc.*, **91**, 2536 (1969).
25. D. E. Milligan and M. E. Jacox, *J. Chem. Phys.*, **49**, 4269 (1968).
26. G. E. Herberich, R. H. Jackson, and D. J. Millen, *J. Chem. Soc. (A)*, 336 (1966).
27. M. M. Rochkind and G. C. Pimentel, *J. Chem. Phys.*, **42**, 1361 (1965).
28. W. H. Kirchhoff, D. R. Lide, Jr., and F. X. Powell, *J. Mol. Spectrosc.*, **47**, 491 (1973).
29. D. E. Milligan, D. E. Mann, M. E. Jacox, and R. A. Mitsch, *J. Chem. Phys.*, **41**, 1199 (1964).
30. W. H. Kirchhoff, D. R. Johnson, and F. X. Powell, *J. Mol. Spectrosc.*, **48**, 157 (1973).
31. P. H. Krupenie, *The Band Spectrum of Carbon Monoxide*, National Standard Reference Data System–National Bureau of Standards Publ. No. 5, Washington, D.C. (1966).

Discussion Leader: W. V. F. BROOKS

Reporters: FOIL A. MILLER
ROBIN S. McDOWELL

Discussion

CRAWFORD made the following remarks on the general status of force field calculations:

"I should like to enter a vigorous dissent with the general impression given by Professor Mills. In our desire for a 'warm happy feeling' we long for the state of bliss when the cup runneth over, and we have ideal precision and ideal information. It seems to me that Professor Mills in his talk notes that the cup is half empty and leaves the impression of a rather unsatisfactory state of affairs. I should prefer to note that it is half full and that we do have a significant amount of information.

"Indeed, I believe that too often we have tended to stress and emphasize our ignorance and our quite legitimate dissatisfaction, and I think that we are perhaps using the words 'critical evaluation' in the wrong sense. If Professor Mills would permit me to quote Cyrano, I would assert that long-nosed scholar's belief, '*et que le fin du fin ne soit la fin des fins*.' We do, as men of integrity seeking pure truth, long for better information and a more ideal knowledge; but as men of expediency seeking a general understanding that will permit us to make shrewd and productive guesses with regard to tomorrow's experiment, we should note the quite useful knowledge that we have.

"I think those of us who work with force constants should convey to our fellow scientists, not only those working in structures but all those chemists who seek some sort of guidance and help in solving their problems, that the understanding of force fields has, in fact, progressed to a point where, if we cannot split the wave number every time, we can by and large supply useful guidance with regard to vibrational spectra and vibrational motions. I would stress the point that, using the Urey–Bradley model and transferring force constants, Snyder and Schachtschneider were able to make coordinated good sense of all of the spectra of the hydrocarbons, both aliphatic and aromatic, not to mention chlorinated aromatics; and I note also that quite recently Peticolas has been able to make valuable use of Raman spectra of DNA and RNA, using as his pilot and guide normal-coordinate calculations of the spectra to be expected.

"If I were asked to sum up the state of force constant knowledge in this broad-brush, useful-guideline sense, I would like to convey the impression that, using the Urey–Bradley field with the few special additional constants that are known, we can do quite a decent job on molecules made up of the 'covalent atoms'—in which I include H, C, N, O, F, Cl, Br, I, S, and perhaps P and Se—held together by normal electron pair bonds, or involving aromatic rings. By and large, in this class of molecules, it is possible for us to calculate in advance the vibrational frequencies and the expected spectra, not with great precision, but with enough confidence to supply useful guidance in using vibrational information to understand chemical problems.

"So while I would agree with Professor Mills that we have a long way to go, and that for really accurate knowledge of force constants we have only a pitiful handful of molecules available, I would like to enter this dissent stressing the other side of the balance, which I suspect he has included in his written paper, but which in my opinion needs stressing here also."

In his reply, MILLS stated:

"I think we are not disagreeing to quite the extent our remarks might suggest. It is partly that we have a slightly different philosophy in our approach to the normal-coordinate problem: His is more chemical; mine is more physical.

"If one's motive is to obtain a sufficiently reliable force field to enable one to predict the vibrational spectra of large molecules—such as hydrocarbons, aliphatic and aromatic and substituted, and even DNA and RNA as mentioned by Professor Crawford—then there is no doubt that such calculations can be made with considerable success. I would emphasize, however, that the actual form of the normal coordinates is sensitive to small changes in the force field, and there is evidence that in such large-molecule calculations we really do not know the forms of the normal coordinates with precision. Our optimism or pessimism in approaching this problem depends on which of the motives summarized at the beginning of my talk one regards as most important. Perhaps my motive has always been to obtain a force field that is sufficiently reliable to compare with *a priori* calculations of the electronic energy, whereas Professor Crawford's motive has been more to achieve an assignment and understanding of the vibrational spectra of larger molecules."

RANDIČ observed that the general tenor of the discussion is that only small molecules can be successfully treated. However, large molecules are not hopeless. Two factors help: The number of force constants does not increase as rapidly as the square of the number of atoms since many interactions are zero or negligibly small; and more isotopic species are available for large molecules. However, POLO remarked that data such as centrifugal distortion constants and mean amplitudes may depend on only a few of the force constants. Thus these data may have little effect on the determination of other force constants, and the accumulation of such data (e.g., centrifugal distortion constants for a number of isotopic species) may not improve the overall accuracy of the force field.

In the more detailed discussion LIDE questioned whether the use of dimensionless bond-stretching constants would lose the direct intuitive correlation between the magnitude of the force constant and the frequency. MILLS replied that his proposal is part of the attempt to separate the kinetic and potential effects on the frequency. If one wishes to correlate force constants to frequencies in different molecules, one must be careful to consider the mass factor, which may change dramatically. The simple intuitive correlation can be misleading. PLÍVA pointed out that the dimensionless force constants are analogous to the reduced potential functions for diatomic molecules.

RAMSAY asked about the effect of the neglect of anharmonic corrections in the calculation quoted by Mills on ethylene (Duncan, McKean, and Mallinson). MILLS was not certain, but noted that the authors "harmonized" their data and found remarkably small differences with a calculation in which anharmonicity was neglected. Anharmonic corrections are made in his laboratory, but the effect on the force field is rather small, even on the interaction constants. MILLER expressed more caution about mixing data on different isotopic species if anharmonicity is neglected.

In regard to the use of centrifugal-distortion data for determining the harmonic force field, CRAWFORD suggested that the five τ's and the harmonic vibrational frequencies should be combined into a single least squares calculation. However, KIRCHHOFF pointed out practical problems because of the high correlation between the τ's and the ω's and the difficulty in deciding on the proper weighting. It appears that the optimum strategy for this type of calculation is not yet clear. In response to other questions, KIRCHHOFF stated that sixth-order effects appear to be small in the molecules he discussed (which exclude water and other hydrides); he he has not investigated the τ's in excited vibrational states, but this has been discussed in a thesis by a student of Bellet. KIRCHHOFF agreed with RAMSAY that uniform notation and units should be adopted for centrifugal distortion, but until that is done each author should define his terms clearly. The two systems in common use now are the τ notation of E. B. Wilson and the system recently introduced by Watson.

In the discussion on anharmonic constants, BUCKINGHAM asked for clarification of the statement that anharmonic constants in Cartesian or rectilinear coordinate systems are mass dependent, whereas in curvilinear coordinates they are mass independent. MILLS replied that the mass dependence of force constants in rectilinear coordinates arises from the Eckart conditions. Normal CO_2 and $C^{16}O^{18}O$ provide a simple example. Consider the bending coordinates in a rectilinear system. In normal CO_2, the bending introduces (in terms higher than second order) a Δr for the CO distances, but both Δr's are the same (Figure 1). The bending of $C^{16}O^{18}O$ must satisfy the Eckart conditions; the rectilinear bending must have less displacement of the ^{18}O atom than the ^{16}O atom; therefore, the two Δr's are un-

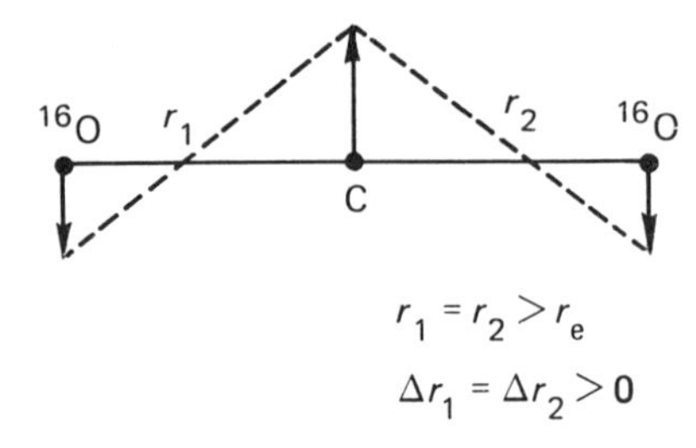

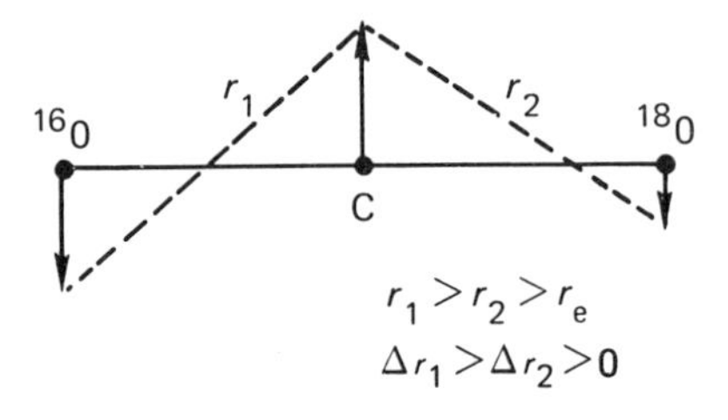

FIGURE 1 Valency bond-bending displacements for CO_2 in rectilinear and in curvilinear coordinate systems. In rectilinear coordinates, the C–O distances are increased by bending, equally in the case of $C^{16}O_2$ (upper diagram) but unequally in the case of $C^{16}O^{18}O$ (middle diagram). They remain unchanged in curvilinear coordinates (lower diagram).

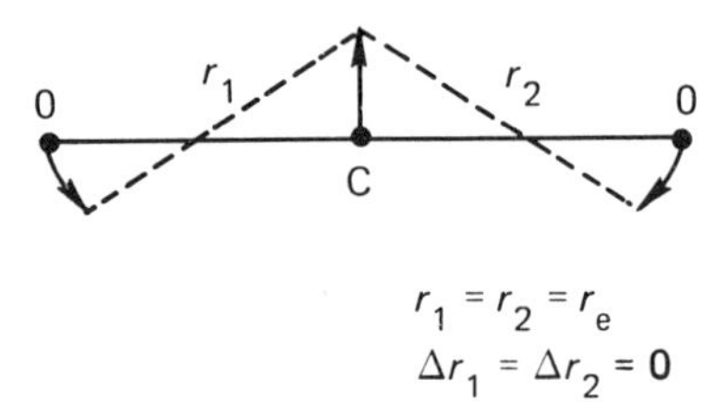

equal. Thus the cubic and higher potential constants of the two CO_2 molecules will not be identical. A curvilinear coordinate system can have the Δr's both zero in the bending mode so that the potential constants will be the same for both species. PLÍVA added that there are two main reasons for using curvilinear coordinates: They minimize stretch bend interaction, as illustrated by Mills, and the chemist visualizes them readily.

PLÍVA explained that his results on OCS are poor compared with those on CO_2, despite the availability of microwave data for the former molecule, because of problems with resonances. The OCS calculations were done some years ago and do not include the most recent data. In response to a question from WEBER, PLIVA replied that he always combines infrared and microwave data in his calculations, with the result that he is unable to compare results obtained from the two types of data.

BUCKINGHAM asked whether *ab initio* calculations of anharmonic force constants are of any help at present. PLÍVA felt that they are not sufficiently accurate, but SCHAEFER pointed out that a large configuration interaction calculation on H_2O is being done by Kern and Shavitt at Ohio State. This should determine cubic and quartic terms to a precision comparable with experimental error. He has calculated the potential for the bending mode of C_3; this appears to be nearly a square well, in agreement with some results of Strauss and Robiette.

KUCHITSU asked about procedures for estimating anharmonic stretch–bend interactions. PLÍVA replied that various models have been tried, but he has little confidence in any of them. MILLS pointed out a helpful way of visualizing some anharmonic interactions by writing the cubic constant as:

$$f_{r\alpha\alpha} = \frac{\partial^3 V}{\partial r \partial \alpha^2} = \frac{\partial}{\partial r}\frac{\partial^2 V}{\partial \alpha^2} = \frac{\partial}{\partial r} f_{\alpha\alpha}.$$

We may consider the dependence of the harmonic bending constant, $f_{\alpha\alpha}$, on r in a bent triatomic molecule. Plotting $f_{\alpha\alpha}$ vs. r, $f_{\alpha\alpha}$ should go to zero for very large r, so the plot might have the form shown in Figure 2. The $f_{r\alpha\alpha}$ term is the slope of the curve at r_e. One may thus be able to guess the sign and possibly even the order of magnitude of $f_{r\alpha\alpha}$.

In a discussion of mean amplitudes of vibration, HEDBERG questioned their sensitivity to the specific force field. MILLS felt they were not very sensitive, although he had not made many calculations. CRAWFORD concurred that, although the normal coordinates themselves are uncertain, the mean square amplitudes (which involve summation) are not sensitive to the particular force field. LORD pointed out that an exception occurs for molecules having highly anharmonic, large

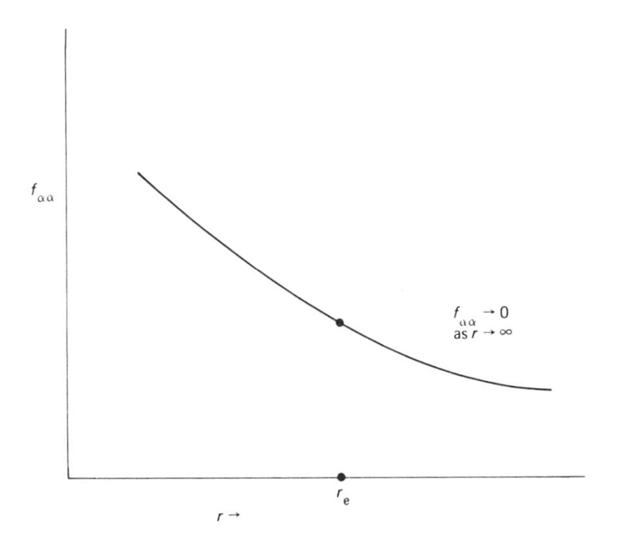

FIGURE 2 Schematic plot of harmonic bending constant $f_{\alpha\alpha}$ against interatomic distance r for a bent triatomic molecule.

amplitude vibrations, such as ring puckering modes where a quartic term is dominant. A harmonic force field is not adequate for mean amplitude calculations in such cases. KUCHITSU agreed and mentioned the lowest frequency bending mode of carbon suboxide (C_3O_2) as another example. The consensus was that, fortunately, harmonic force fields are adequate for the needs of electron diffraction workers except for molecules with highly anharmonic motions. On a semantic point, CRAWFORD noted that electron diffraction workers sometimes use "anharmonic" to mean a non-Gaussian peak in a radial distribution function, but this is quite distinct from the anharmonic character of the potential function.

VI

Large-Amplitude Motions

Aksel A. Bothner-By

Internal Motions Studied by High Resolution NMR

The explosive growth of structural determination by nuclear magnetic resonance (NMR) can have escaped no chemist's notice. The May 30, 1973, issue of the *Journal of the American Chemical Society* contained 53 articles and 24 "Communications to the Editor"; of these, 26 articles and 14 communications reported explicit uses of high resolution NMR for structural determination. The method is clearly in use in every corner of chemistry, and its value and power are universally recognized.

Yet, high-resolution NMR cannot be described as a structural tool in the same way as, for example, x-ray diffraction or microwave spectroscopy, because the theory linking interatomic distances or geometrical arrangements and the fundamental spectrometric parameters–chemical shifts and coupling constants–is almost unmanageable. Thus the great power of NMR is a qualitative one. It is extremely useful in selecting between qualitatively different structural possibilities, e.g., structures with symmetry versus those without, or *cis* versus *trans* isomers.

Much effort has been expended on the formulation of more quantitative approaches, including the deduction of precise distances, bond angles, conformations, and potential energy surfaces for internal motions. These approaches may rely on *a priori* calculation of expected spectral parameters from hypothesized structures, or empirical correlation with compounds of "known" structure. In either case there is hazard and uncertainty involved that may be considerable.

In this paper, we consider in sequence typical procedures for determining geometries of single species and equilibria between interconverting species, and for calculating barriers to interconversion. In each case we shall examine the steps and formulate critical questions that may be asked to prove the reliability of the conclusions.

DETERMINATION OF STRUCTURES OF SINGLE SPECIES

The typical determination proceeds in these steps:

1. Observation and calibration of high-resolution NMR spectrum of the single species.
2. Analysis of the spectrum to yield chemical shifts for each equivalent set of nuclei and spin–spin coupling constants between pairs of nuclei.
3. Comparison of parameters obtained with those calculated theoretically for hypothesized structures or with those observed in analogous compounds of known structure.

NMR instrumentation available commercially today is capable of such resolution, stability, and sensitivity that the determination of transition frequencies can be carried out with accuracy much better than is ordinarily required for the interpretation. Line widths for transitions in the proton magnetic resonance spectra of low molecular weight compounds are ordinarily determined by instrumental limitations. Reasonable care provides line widths of 0.2–0.5 Hz, while lines are spread over a region of 1–2 kHz. Other commonly observed nuclei, such as ^{19}F, ^{13}C, and ^{31}P yield transitions with similar line widths, but the spectral range of frequencies is 5 to 20 times greater. The measurement of the frequencies of transitions may be carried out directly[1] or by using an audio-sideband calibration[2]: each of these methods can give transition frequencies with probable errors of a few hundredths of a hertz. Special techniques have been developed to measure frequencies to 0.001 Hz or less.[3]

In the last years the technique of Fourier transform NMR spectroscopy has become widely appreciated for its speed and increased sensitivity; it has especially become the method of choice for studies of the less sensitive nuclei, such as ^{13}C or ^{15}N. In principle, the same information is contained in the slow passage and Fourier-transformed, pulse-excited spectra, and each may be optimized for resolution or for studies of particular features of the spectrum. Fourier transformation is normally done numerically on digitized accumulated responses, so that the final spectrum is obtained in digitized form, sorted into some finite number of channels. The spectral range covered divided by the number of channels available places a mechanical lower limitation on the resolution, which is defined by the pulse repetition rate, and this precision may be considerably lower than that attainable by slow-passage methods. On the other hand, if exchange broadened lines cover a substantial part of the observed range, the availability of the complete line shape in digitized form can provide a much better data base for analysis, as eloquently outlined by Harris.[4]

Traps in making accurate frequency determinations are almost absent. Aside from the obvious question of instrument calibration, two may be worth mentioning. First, the presence of additional radio-frequency fields used for locking, decoupling, or calibrating purposes may give rise to measurable Bloch–Siegert shifts.[5] In general, the frequency at which a particular transition occurs, ν_A, will be shifted from its normal position ν_A^0 when a second field of strength H_2 is applied at ν_2. The shift is given by

$$\nu_A = \nu_A^0 + \tfrac{1}{2}(\gamma H_2/2\pi)^2/(\nu_A - \nu_2).$$

Second, the audio-side-band method applied in a field sweep spectrometer can give different frequency separations depending on whether the high-field or low-field line is observed. This is obvious from the fact that the *ratio* of the frequencies of the two lines is constant, i.e., $w_o - \rho w_o \neq (w_o/\rho) - w_o$. Both these effects are likely to be small. The only critical question on this subject is "Have the experimenters observed reasonable care in the calibration of their frequency sources and instrumentation?"

The analysis of the spectrum in terms of the high-resolution Hamiltonian [Eq. (1)] has been tested thoroughly[6,7]:

$$\mathcal{H} = \gamma H_o \sum_i (1 - \sigma_i) I_{zi} + \sum_i \sum_{j>i} J_{ij} \vec{I}_i \cdot \vec{I}_m \tag{1}$$

In this Hamiltonian, the σ_i are the shieldings, or chemical shifts, of the nuclei, J_{ij} the indirect spin–spin coupling constants between pairs of nuclei, I_{zi} the z component of the spin operator $\vec{I}_i$, γ the gyromagnetic ratio, and H_o the strength of the static applied field. Within the experimental accuracy of measurements achieved thus far, no additional terms are required in the Hamiltonian to attain an exact fit of theoretical and calculated spectra.[3] For many simple spin systems explicit solutions of the quantum mechanical problems have been found and expressions for all transition frequencies and intensities in terms of the chemical shifts and coupling constants obtained. In cases for which closed solutions are not available, numerical methods may be applied. Diverse programs are available for digital computers for iterative adjustment of chemical shift and spin–spin coupling constants to provide best fit between calculated and observed spectra based on a particular assignment.[8] In principle, such programs may be expanded to handle systems of any complexity, but practical limitations of time and memory size limit one to systems of no more than about eight coupled nuclei.

An example of the degree of complexity that may be unraveled and the precision of the analysis is given in Figure 1 and Table 1, where experimental and calculated spectra for the compound thiepin-1,1-dioxide are reproduced, as obtained by Williamson, Mock, and Castellano.[9]

Incorrect assignments are possible: In tightly coupled spectra they may sometimes be detected by departures from the expected intensity patterns. In first-order spectra the effect will be to reverse the sign of one or more coupling constants. Assignments may be checked in a variety

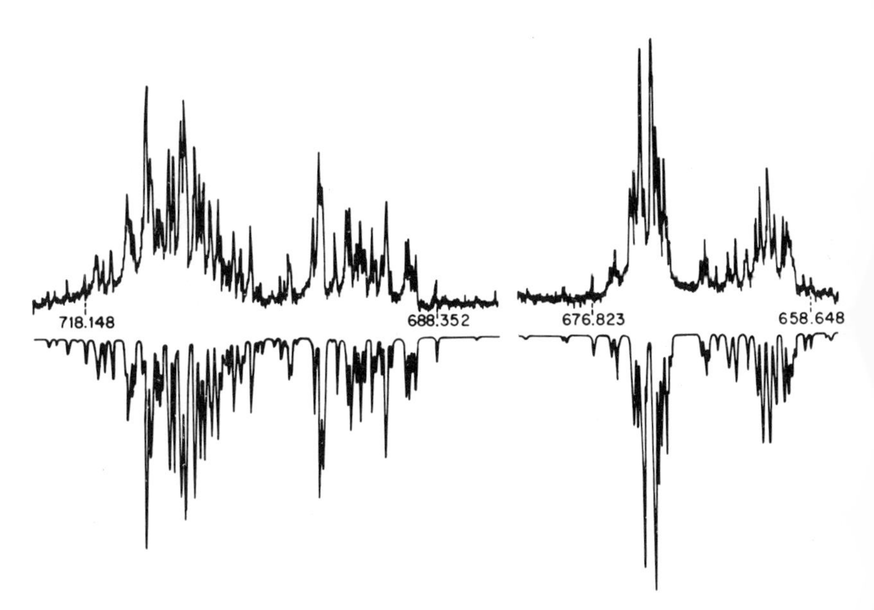

FIGURE 1 Experimental (upper section) and calculated (lower section) 100-MHz proton NMR spectrum of thiepin 1,1-dioxide. Frequencies (hertz) are referred to tetramethyl silane (TMS) used as an internal standard.

TABLE 1 NMR Spectral Parameters of Thiepin 1,1-Dioxide

$W(2) = W(7)$ (α)	667.708(0.004)
$W(3) = W(6)$ (β)	699.394(0.005)
$W(4) = W(5)$ (γ)	712.502(0.005)
$J(2,3) = J(6,7)$	10.700(0.008)
$J(2,4) = J(5,7)$	0.571(0.011)
$J(2,5) = J(4,7)$	0.942(0.011)
$J(2,6) = J(3,7)$	−0.535(0.006)
$J(2,7)$	2.290(0.007)
$J(3,4) = J(5,6)$	6.957(0.007)
$J(3,5) = J(4,6)$	0.350(0.006)
$J(3,6)$	0.444(0.007)
$J(4,5)$	11.983(0.009)

of ways: spin-tickling; selective decoupling; observations at a second spectrometer frequency; changing external conditions, such as solvent or temperature to affect some of the σ_i's and/or J_{ij}'s; and isotopic substitution. Assignments now are very often based on analogy with the large number of known values in the literature. The question "Is the assignment correct?" is always an appropriate one.

There is a special trap that lurks in analysis. Many spectra appear to be clearly first order, with insignificant mixing of states. These spectra invite analysis by first-order theory. It is, however, well known that this may be deceptive.[6,7] Figure 2 illustrates this point. Displayed is the calculated spectrum for the XCH_2CH_2Y molecule with the parameters $J_{AX} = 5.0$, $J_{AX'} = 7.0$, $J_{AA} = -10.0$, $J_{XX} = -16.0$, an entirely reasonable set of parameters for such a molecule. The spectrum, however, clearly invites the interpretation $J_{AX} = J_{AX'} = 6.0$ Hz. In those cases where first-order analysis is applied, it is appropriate to ask, "Would exact analysis yield a different answer, or at least much larger error limits?"

Still, the technique and instrumentation of measurement, the mathematical techniques of analysis, and the available tests for these are such that the competent NMR spectroscopist is very unlikely to make a sizable error in the measurements of chemical shifts and coupling constants.

Once the shifts and coupling constants have been obtained, the riskiest part of the venture begins: interpretation of observed shifts and coupling constants in terms of structure.

Normal spectroscopic practice is to compare the observed parameters with those predicted from hypothesized structures. The *a priori* calculation of shifts and coupling constants for moderately complicated molecules is extremely difficult. Great advances have been made in the past

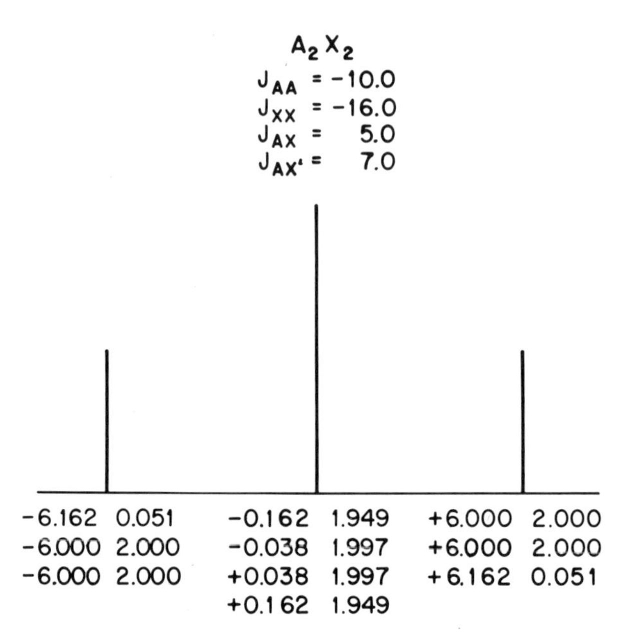

FIGURE 2 Calculated NMR spectrum of molecule A_2X_2 for specified values of parameters.

years, and the state of the art is reviewed by R. Ditchfield elsewhere in this volume. As of now, however, the best calculations barely place the chemical shifts of various groups within their observed experimental ranges, and the detailed deduction of bond angles, distances, etc., from NMR data is not practical by this method. Theory is very useful in establishing trends and as a guide to the search for useful empirical correlations.

Thus, the chemical shift of a proton attached to an organic compound is known to be affected by primary structure, i.e., the identity of the atom to which it is directly attached, the identities of those one atom removed, etc., by hybridization changes on these atoms, by van der Waals interactions with nearby nonbonded atoms, by electric fields associated with permanent moments in the molecule, and by "neighboring magnetic anisotropy." Chemical shifts of substances in the liquid state may also be affected sizably by the medium. The magnitudes of these effects are comparable, and each is not readily calculated. Thus only a bold NMR spectroscopist will be confident in predicting the proton chemical shift to within 0.5 ppm.

At the same time a tremendous amount of data on shifts has been accumulated, systematized, and cataloged, and this accumulation continues vigorously. Turning again to the most recent *Journal of the American Chemical Society*, two papers may be cited: Eggert and Djerassi[10] give ^{13}C chemical shifts for 103 aliphatic amines and provide empirical rules for predicting these shifts, while Dalling, Grant, and Paul[11] provide ^{13}C chemical shift data on *cis*- and *trans*-decalin and eleven methyl or dimethyl decalins, subjecting the results to multiple regression analysis to obtain substituent and conformational parameters applicable to hydrocarbons in general.

In the study of coupling constants a large lore has also been accumulated. The qualitative behavior of the H–H coupling constant in saturated methylene groups is predicted by molecular orbital theory,[12] and the effects of electronegative element substitution, angle deformation, and adjacent π-bonded groups observed experimentally confirm the theory. The prediction of Karplus[13] many years ago that the HCCH coupling constant would depend on dihedral angle in a way now familiar as the "Karplus curve" has been especially fruitful; Karplus found it necessary to point out, however, that other factors than the dihedral angle were involved and that dangers lay in accepting the functional dependence uncritically.[14] In recent years these effects have been more specifically recognized. It has been shown moreover that many four-atom systems, such as HCCF, FCCF, HCCP, $HCO^{31}P$, $HCC^{15}N$, $HCC^{13}C$, show functional dependence of ^3J on ϕ similar to that in the HCCH case.[15] The extent of lore now available is such that it is often possible to choose with assurrance between alternative hypothesized structures on the basis of analogy. Confidence may be attached to the choice, provided a statistically significant number of cases analogous to each hypothetical structure have been studied, that the predicted parameters for the alternatives differ sufficiently and that it is felt that no unsuspected structural effects in the known compounds alter the observed parameters. The last is a matter of faith and numerical estimates are inappropriate.

The least ambiguous decision that can be made is the one dealing with the symmetry of the compound. If the alternative structures are such that one of them contains nuclei equivalent by virtue of symmetry, a firm choice for or against that structure is usually possible.

INTERCONVERTING SPECIES: EQUILIBRIA

The behavior of the NMR spectrum of a mixture of two interconverting species depends on the rate of interconversion. Simple arguments based on the uncertainty principle provide a crude guide. If the frequency of interconversion of the species is much less than the frequency separations of corresponding NMR transitions of the two species, the resulting spectrum will closely approximate a superposition of the spectra of the two species. At intermediate rates, complex behavior with broadened, coalescing lines is expected. If the interconversion is fast, then a single spectrum will be observed. The virtual molecule giving rise to this spectrum has shifts and coupling constants that are population-weighted averages of the shifts and coupling constants of the individual species.

Thus for the case shown in the following diagram:

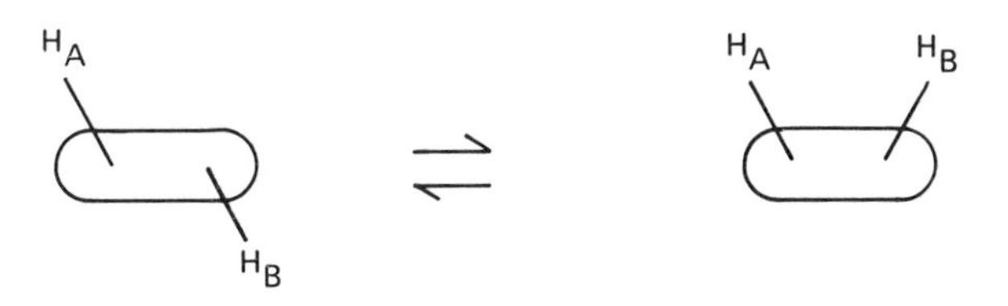

II

the parameters are given by:

$$\sigma_A = \chi_I \sigma_{AI} + \chi_{II} \sigma_{AII}$$

$$\sigma_B = \chi_I \sigma_{BI} + \chi_{II} \sigma_{BII}$$

$$J_{AB} = \chi_I J_{ABI} + \chi_{II} J_{ABII}$$

where χ_I, χ_{II} are the mole factions of species I and II and, in general,

$$\vec{P} = \sum_i \chi_i \vec{P}_i \qquad (2)$$

where $\vec{P}$ is the vector of parameters and summation is over all rapidly interconverting species.

For slowly interconverting species the problem of analysis is just that of identifying transitions belonging to the separate species and performing an analysis on each subspectrum. The parameters obtained are used to assign structures, and the proportion of each determined from the integrated intensities of the two subspectra.

Rates of interconversion of many species may be slowed by lowering the temperature, and this approach has been extremely powerful in revealing details of conformational equilibria in hundreds of diverse systems, including axial equatorial interconversion in substituted cyclohexanes and cyclohexenes, internal rotation in heavily halogen-substituted ethanes, ring inversions in numerous heterocyclic six-membered rings, large carbocyclic rings, paracyclophanes, 1,2-disubstituted octatetraenes, nitrogen inversions in aziridines, oxaziridines, and diaziridines, *cis–trans* interconversion in amides, and many, many more.[4]

For many cases, for example, for lightly substituted ethanes, propenes, and aldehydes, sufficiently low temperatures are not attainable to slow the interconversion rate such that the NMR spectrum is appreciably broadened. In this case it is necessary to study the spectrum while changing external variables (as temperature, pressure, solvent) and attempt to interpret the variation in parameters in terms of a model. This kind of study is beset with great difficulties. The method of analysis usually adopted is as follows: Assume a trial set of values of ΔH, ΔS for the equilibria involved. Then, calculate χ_{ij}, the mole fractions of species i at temperature j. The overdetermined set of linear equations

$$\vec{P}_j = \sum_i \chi_{ij} \vec{P}_i \tag{3}$$

is then solved for $\vec{P}_i$, and the residuals for Eq. (3) are calculated. The process is repeated with different values of ΔH and ΔS until a minimum in the residuals is found. While this method has been used by many investigators (including the author[16]), we will here leave them anonymous because it is so dubious. Among the problems that can arise are these: ΔH and ΔS need not be constant over the temperature range studied, and in most cases probably are not, because of the changing solvent properties, especially specific solvating power, association, and dielectric constant.[17] Both chemical shifts and coupling constants are often subject to variation with temperature and solvent properties,[18] so that $\vec{P}_i$ is not likely to be constant. The minimum in residuals of Eq. (3) as a function of trial values of ΔH, ΔS is often very flat making large ranges of ΔH, ΔS allowable; relatively small changes in the input parameter vectors $\vec{P}_j$ can cause large changes in the deduced ΔH, ΔS values. Frequently, statistics are poor, and the parameter vector $\vec{P}_i$ may be selected to be a single coupling constant, or shift.

Because of these difficulties, a second method that is much less complicated and probably gives equally good or better results has been used. This is simply to estimate the $\vec{P}_i$ from analogous compounds, and to solve Eq. (3) directly for the χ_{ij}, calculating ΔH and ΔS for them.[19]

Abraham and co-workers[17] have pointed out an interesting test that may be used in the case of binary interconverting systems. For such systems

$$\vec{P}_j = \vec{P}_m + \chi_{nj} (\vec{P}_n - \vec{P}_m), \tag{4}$$

so that if $\vec{P}_j$ is observed for several temperatures (pressures, solvents, etc.) the parameter vectors should be linearly related, provided $\vec{P}_m$ and $\vec{P}_n$ are independent of the temperature changes. Stated another way, if the observations are plotted as points in the space of parameters, they should lie on a straight line that passes through $\vec{P}_m$ and $\vec{P}_n$. This provides a test that can easily be applied to all such data.

INTERCONVERSION RATES

In the measurements of interconversion rates, hence, barriers, the steps followed include:

1. Observation of spectra with changes of external variable, usually temperature, such that spectra corresponding to slow, intermediate, and fast exchange are recorded.
2. Deduction of species and amounts present from slow and fast exchange limits.
3. Calculation of rates from line shapes in intermediate cases.
4. Thermodynamic deductions.

Observation of spectra over a temperature range from -100 to $+200$ °C can be achieved in many laboratories with no impairment in accuracy of frequency determination. Assignment of subspectra in the low-temperature limit to structures suffers from the difficulties already discussed.

Two basic methods of analysis of line shape as a function of interconversion rate are in use. The first, based on the phenomenonological equations of Bloch[20] was developed by McConnell,[21] and is derived by inserting in the equations terms representing essentially instantaneous transfers of nuclear magnetization from one species to the other. It is successful in dealing with interconversion in those cases where the interconverting nuclei are not coupled with each other. It can be generalized to many-site exchange and provides a fairly rapid numerical method for these simple cases. Fast computer programs to generate predicted line shapes are available.[22]

The second method of analysis is that based on the density matrix formalism of Wangsness and Bloch,[23] Bloch,[24] and Redfield.[25] The theory and application has been set forth simply by Slichter,[26] and the application to line shape analysis has been beautifully reviewed by Hoffman.[27] Intra- and intermolecular exchange may be treated, and coupled systems are also treated exactly. Specific relaxation processes may be included. Binsch[8,22] has constructed elegant, economical computer programs DNMR2 and DNMR3 for the calculation of exchange broadened spectra using this method. The latter program contains provision for calculation of spin systems with magnetic equivalence or symmetry. Calculations of line shapes for varying rates of exchange in two simpler systems are shown in Figures 3 and 4. Thus the experimental capabilities for accurate observation and the mathematical techniques and apparatus for exact analysis are available.

We will now consider two cases of applications that exemplify good

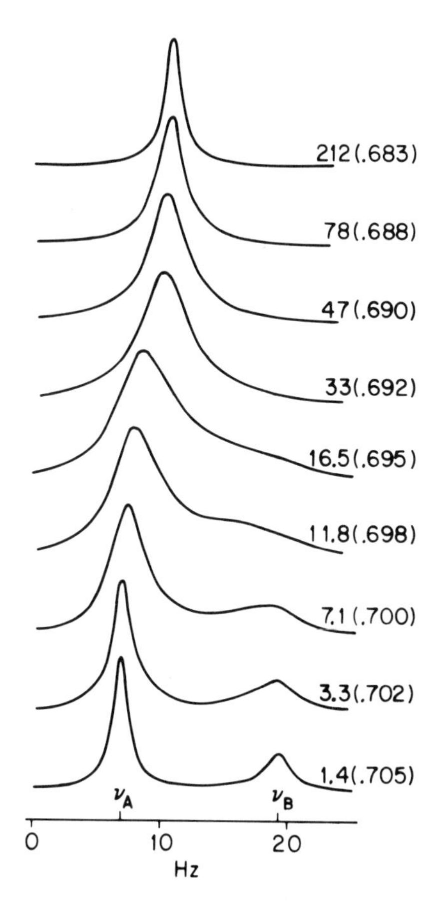

FIGURE 3 Calculated line shapes for exchange of nuclei between two unequally populated sites (A and B). The rate constants for site exchange (k_{AB} in s^{-1}) are shown by each spectrum; the figures in brackets refer to the value of p_A; ν_A = 6.9 Hz, ν_B = 19.4 Hz, w_{AO} = 1.1 Hz, w_{BO} = 1.3 Hz.

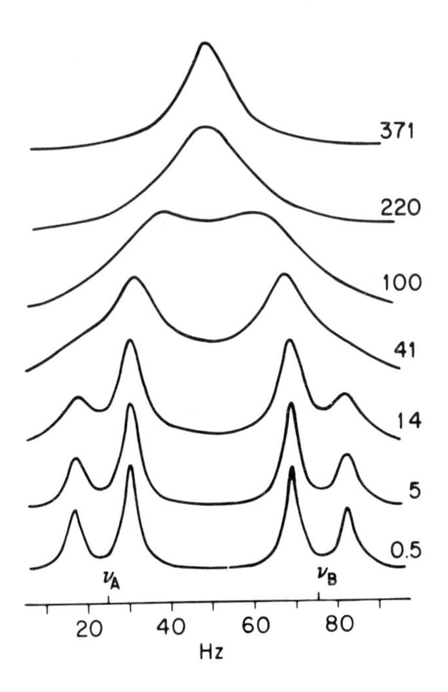

FIGURE 4 Calculated line shapes for the exchange of mutually coupled nuclei between two sites (A and B). The rate constants for site exchange (s^{-1}) are shown by each spectrum; $\nu_A = 24.6$ Hz, $\nu_B = 75.4$ Hz, $J_{AB} = 13.5$ Hz, $w_{AO} = w_{BO}$ is varied from 4.0 to 3.15 Hz as the exchange rate is increased.

practice in determination of barriers in simple systems and that demonstrate the points at which difficulties are encountered.

The first example is a study of internal rotation in polyhaloethanes by Weigert *et al.*[28] Figures 5, 6, and 7 depict the experimentally observed spectra and the computer-predicted spectra for specific sets of conformer energies and barrier heights. Measurements at several temperatures below the slow-exchange limit were in each case used to estimate the intrinsic dependence of chemical shift on temperature for each conformer, and those extrapolated shifts were used as input parameters in the line-shape calculations. Changes of 100 calories per mole in input values for relative conformer energies or barrier heights significantly reduced the agreement between experiment and calculation.

Two possible sources of error remain. First, the assumption was made that the shape of the potential energy curve itself remained constant over the temperature range studied. Since the conformers are probably characterized by significantly different dipole moments and the dielectric constant of the medium is changing with temperature, this is probably not strictly true.[17] Second, the assignment of the individual rotamers to the appropriate subspectra cannot be unambiguously determined, as is carefully pointed out by the authors themselves. A plausible and appealing case is made for the assignments resulting in the data of Table 2; however, the question cannot be regarded as settled.

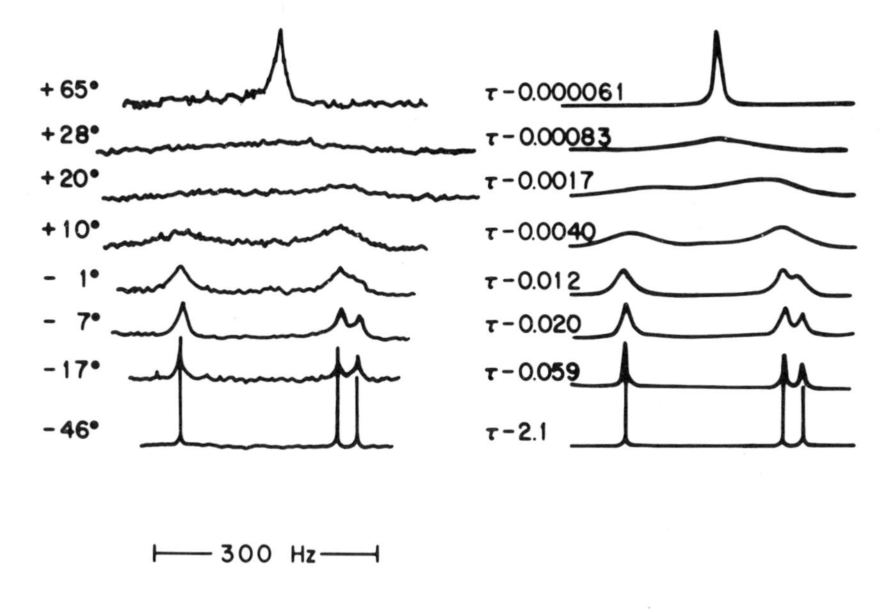

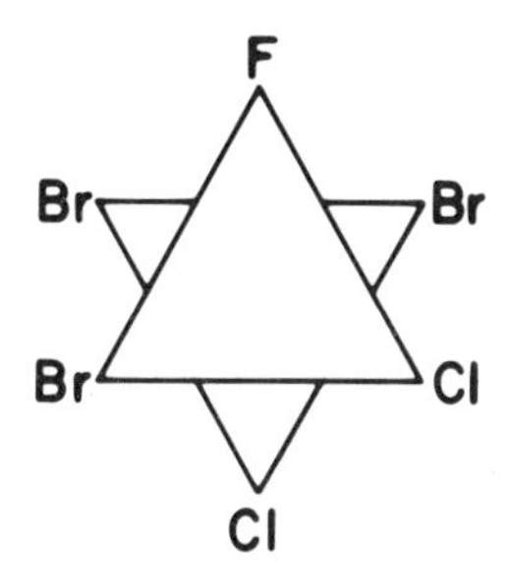

FIGURE 5 Fluorine NMR spectra of 1,1,2-tribromo-1,2-dichloro-2-fluoroethane at 56.4 MHz. *Left,* experimental spectra as a function of temperature. *Right,* Calculated spectra as a function of mean lifetime before rotation. The τ values beside each calculated curve represent the mean lifetime (in seconds) of the predominant rotational isomer before it is converted to rotation about its carbon–carbon bond to either of the lesser isomers. These values do not by themselves define the calculated spectra but are appended to show the general order of magnitude of lifetimes of the isomers before rotation occurs.

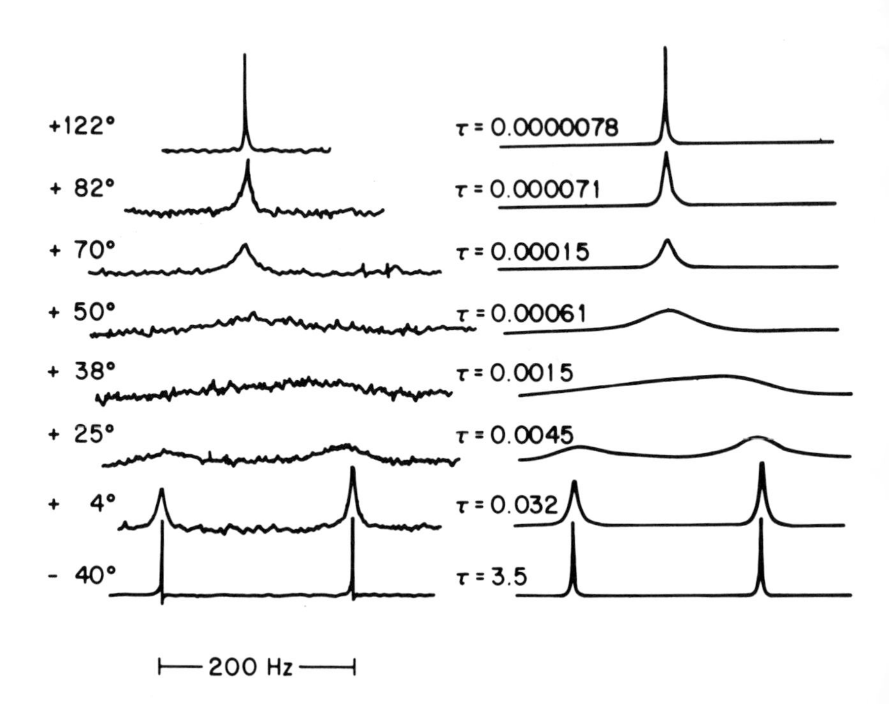

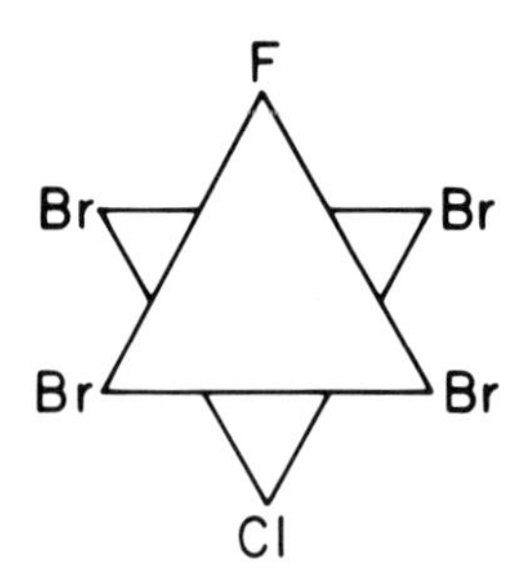

FIGURE 6 Fluorine spectra at 56.4 MHz of 1,1,2,2-tetrabromo-1-chloro-2-fluoroethane. *Left,* experimental spectra as a function of temperature. *Right,* calculated spectra as a function of mean lifetime before rotation (seconds). (See caption to Figure 5.)

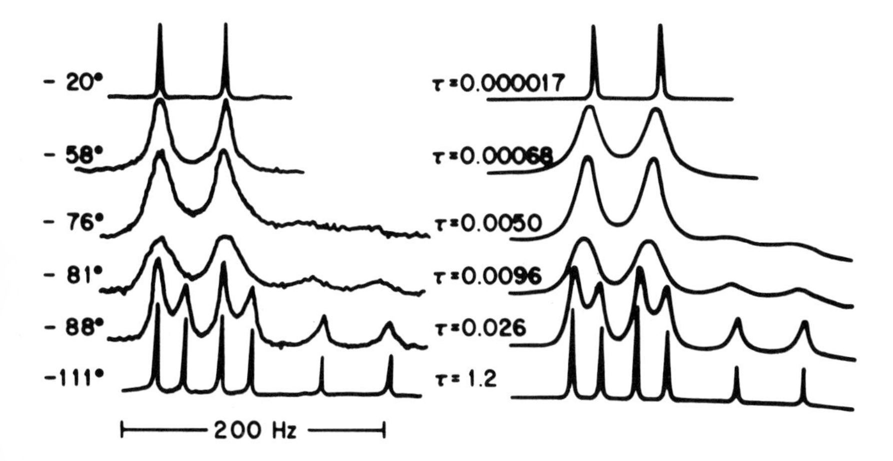

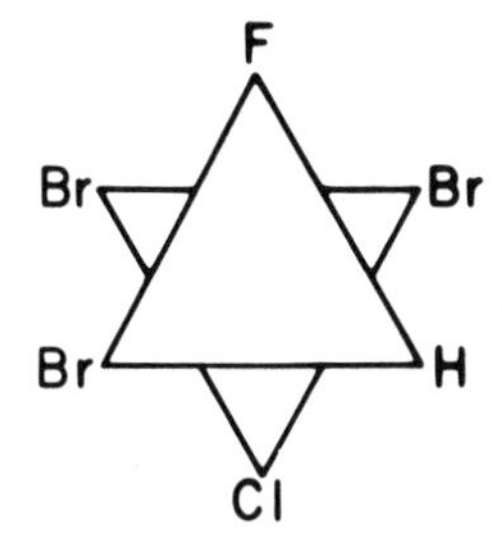

FIGURE 7 Fluorine spectra at 56.4 MHz of 1,1,2-tribromo-1-chloro-2-fluoroethane. *Left*, experimental spectra as a function of temperature. *Right*, calculated spectra as a function of mean lifetime before rotation (seconds). (See caption to Figure 5.)

TABLE 2 Chemical Shifts and Equilibria and Free Energies of Activation for Interconversion of Rotational Isomers of Halogenated Ethanes*

Compound in Most Stable Conformation	$\theta = 0°$[a]	$\Delta G^{\ddagger}(0 \rightarrow 120°)$[b,c]	$\theta = 120°$[a]		$\Delta G^{\ddagger}(120 \rightarrow 240°)$[b,c]	$\theta = 240°$[a]		$\Delta G^{\ddagger}(0 \rightarrow 240°)$[b,c]
	δ_F[e]		ΔG[b,d]	δ_F[e]		ΔG[b,d]	δ_F[e]	
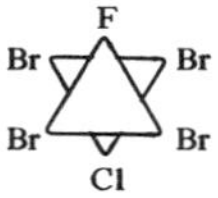 F; Br, Br; Br, Br; Cl	2652	14.8	0.26	2863	*f*	0.26	2863	14.8
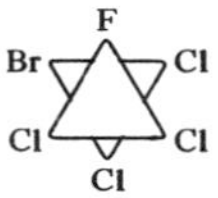 F; Br, Cl; Cl, Cl; Cl	3031	*f*	0.0	3031	13.1	0.20	3299	13.1
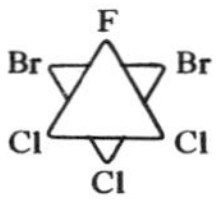 F; Br, Br; Cl, Cl; Cl	2868	13.2	0.14	3111	*f*	0.14	3111	13.2
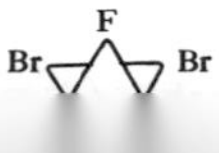 F; Br, Br	[illegible]	13.9	0.16	2938	14.7	0.28	2965	14.0

Cl, Br; Cl	2899	13.8	0.17	2923	14.5	0.30	3153	14.2
CF_3; Br, Cl; Cl, Br; Cl	3311	13.7	0.13	3360	14.1	0.25	3363	13.9
F; Br, Br; Br, H; Cl	6573 396[g] 46.0[h]	9.6	0.07	6595 408[g] 47.8[h]	10.2	0.30	6699 392[g] 48.7[h]	10.0
F; Br, Cl; H, Cl; Cl	6713 389[g] 48.8[h]	9.0	0.12	6766 368[g] 50.2[h]	9.4	0.22	6876 388[g] 48.0[h]	9.1

[a] Angle of clockwise rotation of the *rear* carbon of the structure with respect to the front carbon.
[b] In kilocalories per mole.
[c] Free energy of activation for indicated rotational isomer interconversion.
[d] Free energy relative to the most stable conformation as measured below the coalescence point.
[e] Fluorine chemical shift in hertz at 56.4 MHz upfield from dichlorodifluoromethane as measured below the coalescence point, normally rather temperature dependent.
[f] This barrier is between rotational isomers which are mirror-image isomers and cannot be measured by the nmr method.
[g] Proton chemical shift in hertz at 60.0 MHz downfield from TMS.
[h] J_{HF} in hertz as determined from the low-temperature fluorine spectra.

The second example is that of the interconversion shown in the system:

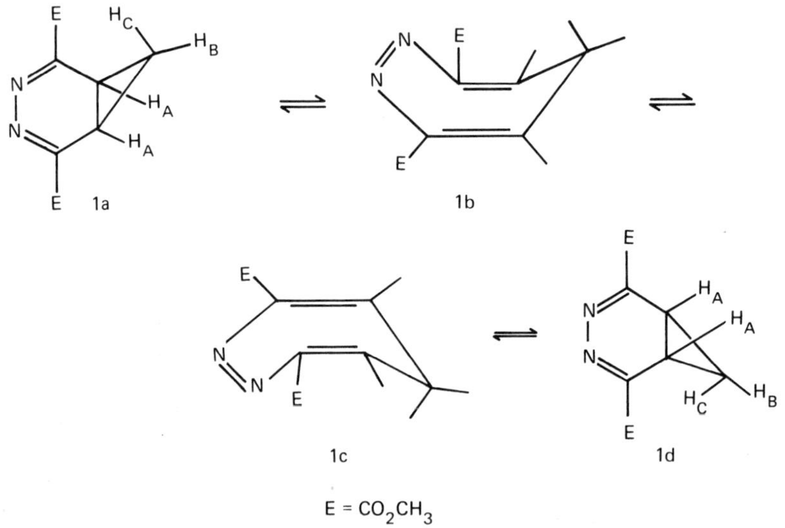

$E = CO_2CH_3$

studied by Kleier *et al.*[29] In this case several of the experimental difficulties encountered in other studies are missing. Since the interconversion is between species of the same energy, ΔE is identically zero, and the mole fractions of the two species exactly 0.5 at all temperatures, simplifying the analysis. The assignments of proton resonances of the bridgehead protons are dictated by symmetry considerations. The assignment of the CH_2 protons is not clear, but does not affect the result, since in the high temperature limit they become equivalent. Finally, the complex nature of the spectrum, involving spin–spin coupling, provides a larger temperature range over which the rate may be precisely studied. The experimental and simulated spectra (using the density-matrix approach) are displayed in Figure 8. In this case also a study of the spectra at temperatures below the slow-exchange limit provided data on temperature dependence of shifts and coupling constants, yielding input data for line-shape analysis at higher temperatures.

Pulse methods may be used to get at some of the relaxation processes involved in species interconversion directly. Recently, Carver and Richards[30] have performed calculations showing that the dependence of observed T_2 on spectrometer frequency and on pulse–repetition rate in

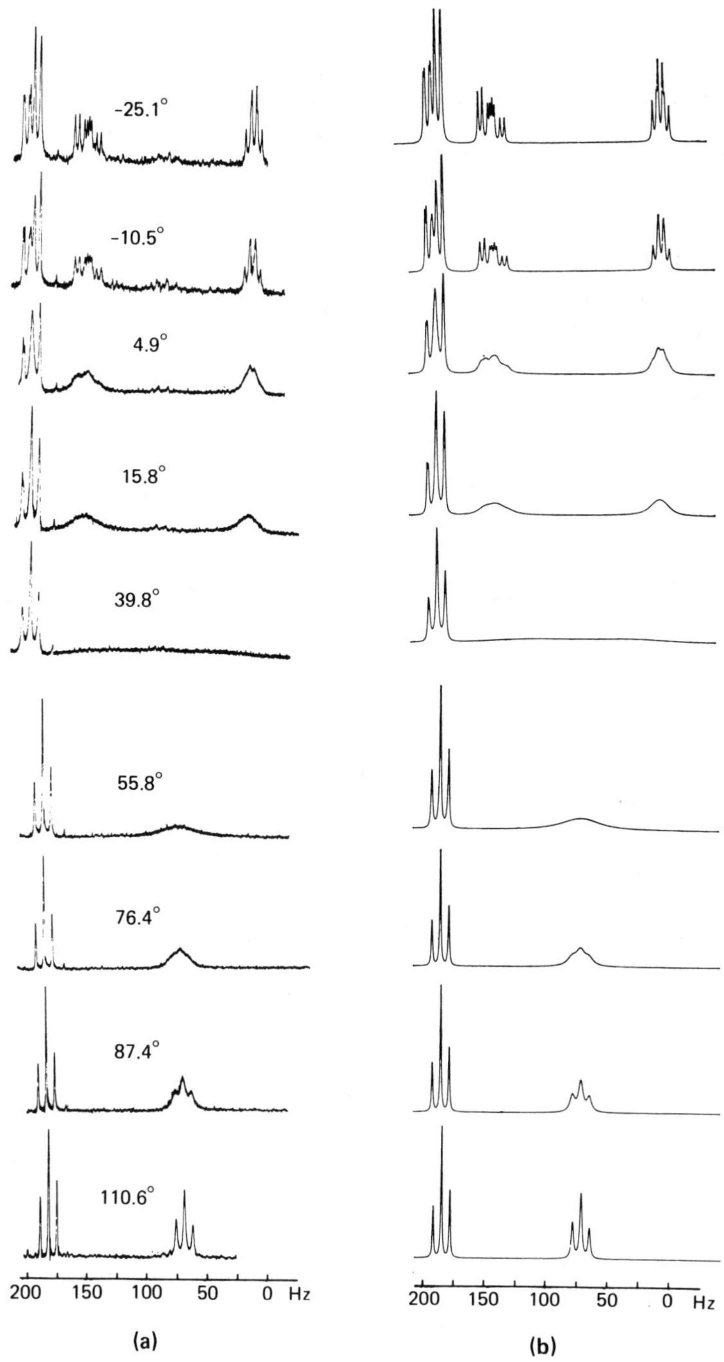

FIGURE 8 Experimental (A) and computed (B) NMR spectra of interconversion system illustrated on p. 348.

the Carr–Purcell experiment can be interpreted directly to give information about populations and exchange rate in a two-species system. Extension of this technique to more complicated systems could provide a powerful general method for the investigation of internal motions.

SUMMARY

The questions that are usually appropriate to ask in studies of internal motion by NMR include the following:

1. Has care been taken in calibration of instruments and frequency sources?
2. Has appropriate quantum mechanical analysis been performed? Is there deceptive simplicity in spectra? Are line assignments correct?
3. Does adequate analogy and experience guide assignment of structures from shifts and coupling constants?
4. Has adequate temperature range been covered?
5. Do NMR parameters of single species vary with temperature, and is this accounted for?
6. Do conformational energies, barriers, or complete potential energy curve vary with temperature?

REFERENCES

1. P. Hampson and A. Mathias, *Mol. Phys.*, **11**, 541 (1966).
2. J. T. Arnold and M. G. Packard, *J. Chem. Phys.*, **19**, 1608 (1951).
3. R. Freeman and B. Gestblom, *Chem. Phys.*, **48**, 5008 (1968).
4. R. K. Harris, *Nuclear Magnetic Resonance–Specialist Periodical Reports*, Vol. 1. The Chemical Society, London (1972), pp. 211–256.
5. F. Bloch and A. Siegert, *Phys. Rev.*, **57**, 522 (1940).
6. J. W. Emsley, J. Feeney, and L. H. Sutcliffe, *High Resolution Nuclear Magnetic Resonance Spectroscopy*, Pergamon Press, Oxford (1965), pp. 280–480.
7. R. J. Abraham, *The Analysis of High Resolution NMR Spectra*, Elsevier, Amsterdam (1971).
8. R. K. Harris and J. Stokes, *A Library of Computer Programs for NMR Spectroscopy*, Science Research Council, Didcot, Berks, U.K. (1971); Quantum Chemistry Program Exchange, Chemistry Department, Indiana University, Bloomington.
9. M. P. Williamson, W. L. Mock, and S. M. Castellano, *J. Magn. Res.*, **2**, 50 (1970).
10. H. Eggert and C. Djerassi, *J. Am. Chem. Soc.*, **95**, 3710 (1973).
11. D. K. Dalling, D. M. Grant, and E. G. Paul, *J. Am. Chem. Soc.*, **95**, 3718 (1973).
12. J. A. Pople and A. A. Bothner-By, *J. Chem. Phys.*, **42**, 1339 (1965).
13. M. Karplus, *J. Am. Chem. Phys.*, **30**, 11 (1959).
14. M. Karplus, *J. Am. Chem. Soc.*, **85**, 2870 (1963).

15. R. Grinter, *Nuclear Magnetic Resonance–Specialist Periodical Reports,* Vol. I. The Chemical Society, London (1972), pp. 88–99.
16. A. A. Bothner-By and D. F. Koster, *J. Am. Chem. Soc.*, **90**, 2351 (1968).
17. R. J. Abraham, L. Cavalli, and K. G. R. Pachler, *Mol. Phys.*, **11**, 471 (1966).
18. M. Barfield and M. D. Johnston, *Chem. Rev.*, **73**, 53 (1973).
19. A. A. Bothner-By, S. Castellano, S. J. Ebersole, and H. Günther, *J. Am. Chem. Soc.*, **88**, 2466 (1966).
20. F. Bloch, *Phys. Rev.*, **70**, 460 (1946).
21. H. M. McConnell, *J. Chem. Phys.*, **28**, 430 (1958).
22. G. Binsch, *J. Am. Chem. Soc.*, **91**, 1304 (1969).
23. R. K. Wangsness and F. Bloch, *Phys. Rev.*, **89**, 728 (1953).
24. F. Bloch, *Phys. Rev.*, **102**, 104 (1956); *ibid.*, **105**, 1206 (1957).
25. A. G. Redfield, *Adv. Magn. Res.*, **1**, 1 (1965).
26. C. P. Slichter, *Principles of Magnetic Resonance,* Harper & Row, New York (1963), pp. 127–159.
27. R. A. Hoffman, *Adv. Magn. Res.*, **4**, 87 (1970).
28. F. J. Weigert, M. B. Winstead, J. I. Garrels, and J. D. Roberts, *J. Am. Chem. Soc.*, **92**, 7359 (1970).
29. D. A. Kleier, G. Binsch, A. Steigel, and J. Sauer, *J. Am. Chem. Soc.*, **92**, 3787 (1970).
30. J. P. Carver and R. E. Richards, *J. Magn. Res.*, **6**, 89 (1972).
31. F. J. Weigert, M. B. Winstead, J. I. Garrels, and J. D. Roberts, *J. Am. Chem. Soc.*, **92**, 7362 (1970).

H. Dreizler

Determination of Barriers to Internal Rotation by Microwave Techniques

Since detailed investigation of properties of molecules began, the motions of the atomic nuclei within a molecule have been of great interest because they reflect the dynamics of the molecule. Usually the internal motions are adequately described as vibrations, but if there is not a one-to-one correspondence of potential well and nucleus, we speak of inversion or hindered internal rotation. In the latter case, which is the subject of this paper, the change of sites of the nuclei is performed by a rotation of a part of the molecule (top) against the rest of the molecule (frame). Internal rotation and torsion are used synonymously in this paper.

Infrared spectroscopy[1] gave only a limited amount of information on the problem of internal rotation, as the transitions are mostly in the far infrared region. The experimental techniques of this spectral region have been developed rather recently. In addition, the internal rotation is nearly inactive if one of the parts of the molecule is symmetric, such as a methyl group.

Microwave spectroscopy[2–6] was able to contribute more information on the problem of hindered internal rotation, as it is possible by the high resolving power to observe an interaction effect in the fine structure of the rotational spectrum. This fine structure is related to the dynamics of the hindered internal rotation.

This paper is limited to the application of microwave spectroscopy to internal-rotation problems in molecules with one or two symmetric internal rotors.[7,8]

MOLECULAR MODEL

The analysis of the rotational spectrum is based on a model in which the many degrees of freedom of a real molecule are reduced. The first step is the use of the Born–Oppenheimer[9] approximation, which introduces a potential for the N nuclei dependent on their $3N - 6$ internal coordinates. By this procedure the electrons are included collectively, so that the problem is reduced to $3N - 6$ internal degrees of freedom and three degrees of freedom of overall rotation. Translation is not of interest here.

In a subsequent step, all internal degrees of freedom are usually neglected except those of internal rotation. Thus the remaining internal degrees of freedom are one or two if molecules with one or two methyl groups are considered. A consequence of this step is that the parts of the molecule that rotate against each other are represented by a rigid arrangement of mass points, which are given the mass of atoms. These parts are called frame and top (or tops). The geometry of the molecular model is contained collectively in several parameters. As all other vibrations are neglected, no vibration/internal rotation interaction can exist in the model.

The potential surface in a $(3N - 6)$-dimensional space is reduced by the above assumption to a "surface" in one or two dimensions, respectively. As the molecule reproduces its configuration by a rotation of the top through $2\pi/3$, we know that the potential energy is periodic in $2\pi/3$. A mathematical description of such a periodic potential-energy "surface" is a one- or two-dimensional Fourier series. The most general form of a one-dimensional Fourier series with the internal rotation angle α as the variable may be reduced to

$$V(\alpha) = \sum_{n=1}^{\infty} \frac{V_{3n}}{2}(1 - \cos 3n\alpha) + \sum_{n=2}^{\infty} \frac{V'_{3n}}{2} \sin 3n\alpha. \qquad (1)$$

If we identify the origin of the α scale with one of the minima of Eq. (1), the first term is the most important. An adjustment of a constant potential term and recognition of the fact that the method cannot determine a "phase" angle δ of the potential have been introduced. The latter consideration reflects that it is impossible to determine the angular position of the potential minima unless another method, e.g., an r_s

structure determination, is applied. For molecules with a plane of symmetry, which necessarily contains the symmetry axis of the one methyl top, Eq. (1) reduces further to

$$V(\alpha) = \sum_{n=1}^{\infty} \frac{V_{3n}}{2} (1 - \cos 3n\alpha). \tag{2}$$

If the rest of the molecule has C_2 symmetry about the internal-rotation axis, the first coefficient is V_6.

The two-top case is a little more complicated by the two-dimensional Fourier series, but the same steps of reduction are possible. The reduction by symmetry has to allow for additional cases. Taking for example a two-top molecule with C_{2v} configurational symmetry, such as $(CH_3)_2S$, the first potential terms are

$$V(\alpha_1, \alpha_2) = \frac{V_3}{2} (1 - \cos 3\alpha_1) + \frac{V_3}{2} (1 - \cos 3\alpha_2) + V_{12} \cos 3\alpha_1 \cos 3\alpha_2 + V'_{12} \sin 3\alpha_1 \sin 3\alpha_2. \tag{3}$$

The coefficients of the truncated Fourier series of Eq. (2) or Eq. (3) are the molecular parameters for the hindering potential. The use of a limited number of those potential parameters implicitly means an assumption of the barrier shape. Usually, the first term (V_3 or V_6) is the only term that is determined from the experiment. Using only one term of Eq. (2) means the assumption of a pure sinusoidal shape of the potential with the height of V_3. If also higher terms V_{3n} are included, the shape and height change.

HAMILTONIAN OF THE MOLECULAR MODEL

In the usual procedure a classical Hamiltonian function for the model is formulated. Attention should be given to the choice of the molecular coordinate system. As the frame and top are rigid, a convenient choice is the principal axis system of the whole molecule.[8] As a consequence of the symmetry of the top, the orientation of the coordinate system within the molecule is independent of the torsion angle. Another choice, which is called the internal axis system, is defined in such a way that the angular momenta produced by internal rotation of the top and frame compensate each other.[10] The Hamiltonian functions in both coordinate systems are related by a contact transformation, which guarantees the invariance of Poisson brackets[11] and, subsequently, of the commutation relations.

Translated to quantum mechanics, the Hamiltonian operator provides the tool to analyze the spectra. The form of the operator in the principal axis system, called the PAM form, is

$$H = H_T + H_R + H_{RT} \tag{4}$$

with

$$H_T = F p_\alpha^2 + V(\alpha) \tag{5}$$

$$H_R = AP_a^2 + BP_b^2 + CP_c^2 + \frac{1}{2}\sum_{gg'} D_{gg'}(P_g P_{g'} + P_{g'} P_g) \tag{6}$$

$$H_{RT} = -2F \sum_g \rho_g p_\alpha P_g \tag{7}$$

$$A = \frac{\hbar^2}{2I_a} + F\rho_a^2, \text{ etc.} \tag{8}$$

$$D_{gg'} = F\rho_g \rho_{g'} \tag{9}$$

$$F = \frac{\hbar^2}{2rI_\alpha} \tag{10}$$

$$\rho_g = \lambda_g I_\alpha / I_g \tag{11}$$

$$r = 1 - \sum_g \lambda_g^2 I_\alpha / I_g \tag{12}$$

where P_g is the component of total angular momentum with respect to principal axis $g = a, b, c$; p_α, angular momentum of internal rotation; I_g, moments of inertia of the whole molecule; I_α, moment of inertia of the top about the internal rotation axis; λ_g, direction cosine between internal rotation and principal axis g, and $\sum_g \lambda_g^2 = 1$.

In the operator of Eq. (4) the potential function is taken from Eq. (2). Thus $V_3, V_6, \ldots, I_a, I_b, I_c, I_\alpha, \lambda_a, \lambda_b$ are the parameters of the molecular model. As mentioned above, the structural data are collectively contained in the inertial parameters. Also, the λ_g are not purely dependent on the geometry of the molecule; they are defined with respect to inertia axes, which in turn depend on the molecular geometry and the atomic masses.

DEPENDENCE OF ROTATIONAL SPECTRA ON THE HINDERING POTENTIAL

In the next step it will be shown how the information on the hindering potential may be inferred from the fine structure of rotational spectra. We start with the pure internal-rotation Hamiltonian H_T [Eq. (5)], which contains one mass–geometry-dependent constant, F, and the potential parameters. First, assume V_3 alone is important. Both parameters may be incorporated in a reduced potential barrier s, defined as

$$s = \frac{4}{9}\frac{V_3}{F}. \tag{13}$$

The energy states of H_T are functions of s scaled by F:

$$E = \frac{9}{4} F b(s). \tag{14}$$

The energy states of H_T are illustrated in Figure 1. The splitting of the levels into sublevels of symmetry species A and E is very sensitive, via the reduced barrier height s, to V_3 and F. If this level scheme could be measured directly, it would give the best information for the hindering potential, but with an α-independent dipole moment, which is a consequence of a strict interpretation of the model, direct transitions are forbidden.

Next we consider H_R, which is a pure rotational Hamiltonian of an asymmetric top. It is independent of internal rotation. This means that

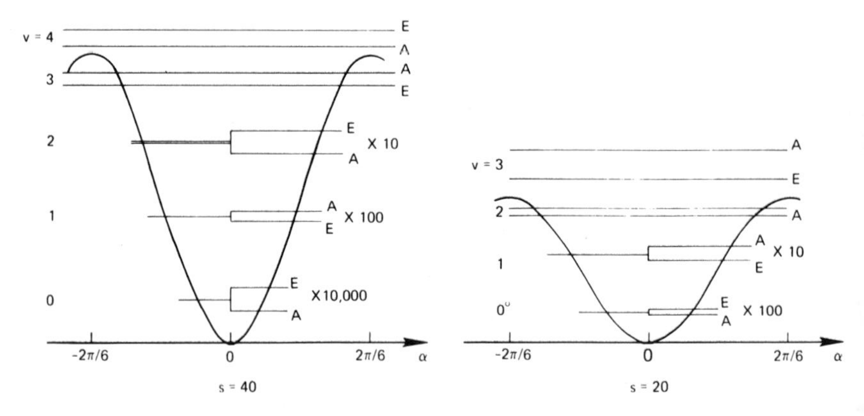

FIGURE 1 Energy levels and potential $V(\alpha)$ of the torsional Hamiltonian of Eq. (5) for a medium (s = 40) and a low (s = 20) barrier case. Only one of the three potential wells is drawn. The splitting of the lower states is enlarged by the factor indicated.

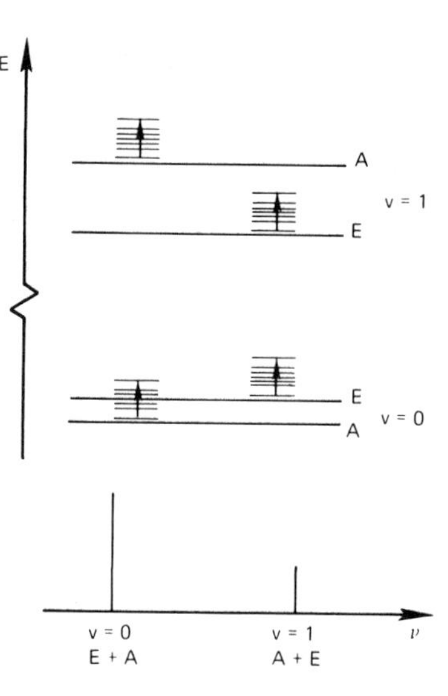

FIGURE 2 Energy level $v = 0$ and $v = 1$ with sublevels A and E for the torsional and rotational Hamiltonian $H_T + H_R$ without interaction. The spectrum consists of single lines, as torsional and rotational energies are purely additive. $v = 0$ and $v = 1$ states differ in rotational constants.

$H_T + H_R$ provides energy levels that result from addition of internal rotation and overall rotation energies. Rotational transitions would provide in this case no information on internal rotation as is demonstrated in Figure 2.

H_{RT}, the interaction between internal rotation and overall rotation, is the essential term for this method, since through it the rotational energy levels are modified in a manner specific to the internal rotation state. Thus, the splitting of internal rotation levels is in a sense transferred to the rotational levels. The rotational-transition splitting is illustrated in Figure 3. These splittings are the key source of information.

The addition of the next potential coefficient V_6 changes the splitting of the torsional level specific to the torsional state. The splitting of one torsional state alone or the splitting of rotational lines in one torsional state is not sufficient to determine more than one potential coefficient.

It should be mentioned that the difference between the energy levels of H_T for different torsional states represented by quantum number ν can also be determined by another method.[12–15] The intensity ratio of equivalent rotational lines depends, via Boltzmann's law, on the energy levels E_v of H_T. This point is of interest especially in the case where no splitting is observable. In detail we have for the ratio of the peak intensities[16]:

$$\frac{I_0}{I_v} = \frac{L_0\, g_0\, |\mu_{ij}|_0^2\, \nu_0^2\, \Delta\nu_v}{L_v\, g_v\, |\mu_{ij}|_v^2\, \nu_v^2\, \Delta\nu_0}\, e^{(E_v - E_0)/kT} = f e^{(E_v - E_0)/kT} \qquad (15)$$

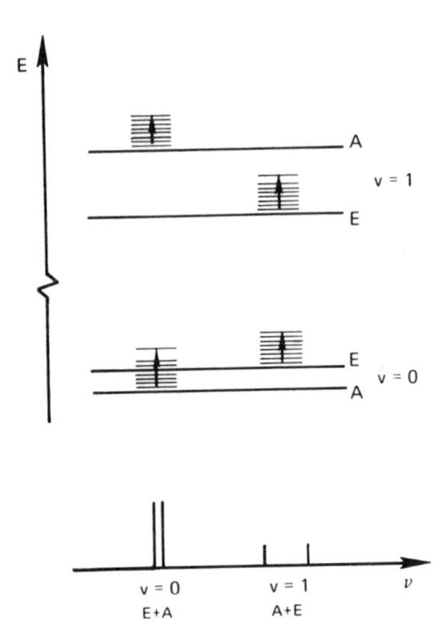

FIGURE 3 Energy levels $v = 0$ and $v = 1$ with sublevels A and E for the complete Hamiltonian of Eq. (4). The spectrum consists of doublets in consequence of the interaction H_{RT}.

Here L_0 and L_v are the effective absorption path lengths at line frequencies ν_0 and ν_v, respectively; g is a statistical weight factor; $|\mu_{ij}|$ is the dipole matrix element of the rotational transition $i \to j$, and $\Delta\nu$ is the half-width of the line at half intensity.

However, the accuracy of a barrier determined by intensity measurement has to compete with that from a frequency measurement if the line splittings can be resolved. With well-resolved torsional multiplets of rotational lines, the higher accuracy results from the measurement of the splittings.

REMARKS ON THE EVALUATION OF MOLECULAR PARAMETERS

Let us assume that the rotational spectra of the torsional states $v = 0$ and $v = 1$ have been measured and assigned. The information available is sets of line frequencies ν, splittings $\Delta\nu$, and an intensity ratio I_0/I_1. From the frequencies ν of the component lines of the multiplets, frequencies of fictitious unsplit lines are calculated by a weighted mean using formulas resulting from the Hamiltonian of Eq. (4). These "lines" are used to fit by a least squares method the rotational constants A, B, C or, implicitly, the effective principal moments of inertia $I_a^0 = \hbar^2/2A$, etc., of H_R, which are different from I_g [compare Eq. (8)]. Experience shows that for the two rotational spectra, $\nu = 0$ and $\nu = 1$, two different sets of rotational constants must be used. Here a limitation of the model shows up since one set of rotational constants should reproduce the spectra of all torsional states. If centrifugal distortion may be neglected, a fit of the lines within some tens of kilohertz may usually be achieved. The gross features of the rotational spectra follow the asymmetric rotor pattern, unless the barrier is very low.

The accuracy with which the rotational constants can be determined from microwave transitions depends on the type of spectrum. Thus with a μ_a spectrum alone, the accuracy of A is poor.

In a second step, which aims mainly at the determination of the potential parameters, the line splittings are used. First we assume that only V_3 is essential. The moments of inertia I_g are taken to be approximated by effective I_g^0. There remain besides s the parameters F and λ_g ($g = a, b, c$), or the equivalent set I_α and λ_g, to be determined. In many cases difficulties arise at this point. One possibility of overcoming these is to derive I_α and λ_g from structural data. Strictly speaking, the λ_g are the direction cosines of the internal rotation axis referred to the principal axes, but from structural data only the λ_g of the adjacent bond are available, which are not necessarily identical as a consequence of the possible tilt of the methyl group. Thus the λ_g should be included in the fitting procedure as unknowns. I_α, the moment of inertia of the methyl group, depends on the determination of structural data on the CH bond length and HCH bond angle. These are in most cases the least accurately determined structural parameters. Since an inclusion of I_α in the fitting procedure is not always successful, a "reasonable" I_α is usually assumed. But this means that the error of I_α transfers to V_3 via Eq. (10) and Eq. (13).

To determine the additional potential parameter V_6, it is necessary to analyze the splittings of rotational spectra in at least two internal rotation states. For one state V_3 and V_6 are correlated. This reflects the situation with an energy level system that results from H_T alone. In only a few cases has V_6 been successfully determined.[17-19] Sometimes an ambiguity of two sets V_3, V_6 has arisen.[20,21]

In principle the two steps of analysis may be incorporated into one fitting procedure. This or the stepwise procedure may be repeated iteratively. In the last case the error of taking I_g^0 for I_g may be eliminated.

In the investigated cases the spectra are reproduced well with moments of inertia specific to the torsional state and I_α, λ_g, V_3 (V_6). The fitting of the frequencies depends more on the choice of the moments of inertia than the line splittings. This limitation is a consequence of the approximation included in the Hamiltonian itself and not a consequence of the approximate numerical treatment. The usual van Vleck perturbation treatment has been checked against direct diagonalization of the energy matrix in the case of a two top molecule, and both numerical methods resulted in the same calculated line frequencies (V. Laurie, personal communication, and ref. 22, 23).

Some low-barrier, one-top molecules have been analyzed by fitting the parameters to the frequencies of the lines only.[20,21] In this case the doublet components are widely separated from each other. The picture of a split line is no longer valid.

The intensity ratio may be used to determine V_3 independently of the analysis of line frequencies ν and line splittings $\Delta\nu$. The measurement of intensity ratios is a delicate experimental problem. After the determination of I_0/I_v there still remain difficulties with some factors contained in the coefficient in Eq. (15). By measurement of the temperature dependence of I_0/I_v the coefficient f may be eliminated and a value of $E_v - E_0$ obtained. As this energy difference is a function of V_3 and F, an error in F influences V_3 also.

The general method for analyzing the spectra is the same for molecules with two methyl groups or tops, although now a multiplet pattern appears instead of a doublet. The Hamiltonian contains, in addition, terms for torsion–torsion interaction [V_{12}, V'_{12}, of Eq. (3)], which are additional constants to be determined.

ESTIMATE OF ERRORS

The evaluation of the data in a general case is quite complicated. Consequently, computer programs are used, which are not very transparent for the study of the susceptibility of the molecular parameters to experimental uncertainties. In the special case of a high barrier, an approximate analytical formulation is possible, which may give an impression of general trends. Let us assume that a molecule is investigated where second-order van Vleck perturbation treatment is sufficient. Further we assume a molecule with an a,b-symmetry plane ($\lambda_c=0$), such as CH_3SCN. In this case the splittings $\Delta\nu$ may be expressed in terms of increments of the rotational constants of the two torsional substates or inversely

$$\Delta A_v = f(\{\Delta\nu\}) \qquad \Delta B_v = g(\{\Delta\nu\}), \tag{16}$$

where ν indicates the torsional state and $\{\Delta\nu\}$ that a set of line splittings has been measured. ΔA_v and ΔB_v are connected with the other parameters by

$$\Delta A_v = \frac{\hbar^2}{2I_a^2} \cdot \frac{\lambda_a^2 I_\alpha}{r} \Delta W_v^{(2)}(s) \tag{17}$$

$$\Delta B_v = \frac{\hbar^2}{2I_b^2} \cdot \frac{\lambda_b^2 I_\alpha}{r} \Delta W_v^{(2)}(s). \tag{18}$$

The quotient of the direction cosines is simply

$$\left(\frac{\lambda_a}{\lambda_b}\right)^2 = \frac{\Delta A_v B_v^2}{\Delta B_v A_v^2} = K = \frac{1}{\overline{K}} \qquad (K, \overline{K} > 0). \tag{19}$$

We first consider the errors of the λ_g, which are

$$\frac{\delta(\lambda_a^2)}{\lambda_a^2} = \frac{1}{1+K} \frac{\delta K}{K} \tag{20}$$

$$\frac{\delta(\lambda_b^2)}{\lambda_b^2} = \frac{1}{1+\overline{K}} \frac{\delta \overline{K}}{\overline{K}} \tag{21}$$

$$\frac{\delta K}{K} = \frac{\delta \overline{K}}{\overline{K}} = \left\{ \left(\frac{\delta(\Delta A_v)}{A_v}\right)^2 + \left(\frac{\delta(\Delta B_v)}{B_v}\right)^2 + 4\left(\frac{\delta A_v}{A_v}\right)^2 + 4\left(\frac{\delta B_v}{B_v}\right)^2 \right\}^{\frac{1}{2}} \tag{22}$$

The first two terms in Eq. (22) depend on the relative accuracy of the line splittings, the last two on the accuracy of the rotational constants. Usually the first two contribute most.

Next we calculate the error of the reduced potential s. Here Eq. (17) and Eq. (18) should be used in the inverse form:

$$\Delta W_v^{(2)}(s) = \Delta A_v \frac{2I_a^2 r}{\hbar^2 \lambda_a^2 I_\alpha} \tag{23}$$

$$\Delta W_v^{(2)}(s) = \Delta B_v \frac{2I_b^2 r}{\hbar^2 \lambda_b^2 I_\alpha} \tag{24}$$

The error of s is given by:

$$\frac{\delta s}{s} = \Delta W_v^{(2)}(s) \left\{ \frac{\partial[\Delta W_v^{(2)}(s)]}{\partial s} s \right\}^{-1} \cdot \frac{\delta[\Delta W_v^{(2)}(s)]}{\Delta W_v^{(2)}(s)} \tag{25}$$

For $W_v^{(2)}(s)$ an empirical interpolation formula was given in the form[24,25]:

$$W_{vA}^{(2)}(s) = a\, s^{-b}\, e^{-c\sqrt{s}} \tag{26}$$

with appropriate constants a, b, and c. An approximation sufficient for our purpose is

$$W_{vA}^{(2)} \approx -2\, W_{vE}^{(2)}$$

and, consequently,

$$\Delta W_v^{(2)}(s) = W_{vA}^{(2)}(s) - W_{vE}^{(2)}(s) = \tfrac{3}{2}\, a\, s^{-b}\, e^{-c\sqrt{s}}\,. \tag{27}$$

We get

$$\Delta W_v^{(2)}(s) \left\{ \frac{\partial [\Delta W_v^{(2)}(s)]}{\partial s}\, s \right\}^{-1} = -\left(b + \frac{c}{2}\sqrt{s} \right)^{-1}. \tag{28}$$

Replacing $\delta I_a/I_a$ by $\delta A_v/A_v$ and $\delta I_b/I_b$ by $\delta B_v/B_v$ we get

$$\frac{\delta s}{s} = \frac{1}{b + \frac{c}{2}\sqrt{s}} \left\{ \left[1 + \frac{1}{(1+K)^2}\right] \left[\frac{\delta(\Delta A_v)}{\Delta A_v}\right]^2 + \frac{1}{(1+K)^2} \left[\frac{\delta(\Delta B_v)}{\Delta B_v}\right]^2 \right.$$
$$\left. + \left(\frac{\delta I_\alpha}{I_\alpha}\right)^2 + 4\left[1 + \frac{1}{(1+K)^2}\right] \left(\frac{\delta A_v}{A_v}\right)^2 + \frac{4}{(1+K)^2} \left(\frac{\delta B_v}{B_v}\right)^2 + \left(\frac{\delta r}{r}\right)^2 \right\}^{\frac{1}{2}} \tag{29}$$

and a second one with K, A_v, B_v replaced by $\overline{K}$, B_v, A_v, respectively. The resulting errors may be different, depending on the weight factors.

In Eq. (29) the first two expressions reflect the accuracy of the line splittings, the next one the accuracy of I_α, which may give a large contribution, whereas the errors of A_v and B_v and especially r may be neglected. With typical values[5] $b \approx 0.87$, $c \approx 0.88$ for the $v = 0$ state the error $\delta s/s$ decreases with increasing s. But with increasing s $\delta(\Delta A_v)/A_v$ and $\delta(\Delta B_v)/B_v$ generally increase. Finally we use Eq. (13) and Eq. (10) and get

$$\frac{\delta V_3}{V_3} = \left[\left(\frac{\delta s}{s}\right)^2 + \left(\frac{\delta I_\alpha}{I_\alpha}\right)^2 + \left(\frac{\delta r}{r}\right)^2 \right]^{\frac{1}{2}}. \tag{30}$$

Summarizing this sample calculation for the error of the potential coefficient V_3, we may say that the uncertainty of I_α contributes heavily if the measurements of splittings result in only a minor error. If we assume a measuring accuracy of 40 kHz for splittings and $\delta I_\alpha / I_\alpha \approx 5$ percent, we conclude that both errors should be carefully investigated if the mean splittings are less than 800 kHz. This last estimate is conservative, as usually many lines are measured, which reduces the error of ΔA_ν and ΔB_ν.

Further in the last step, Eq. (30), $\delta I_\alpha / I_\alpha$ contributes again. Studying the structure of the formula, one observes that I_α contributes at two points in such a way that an error compensates only partially, as shown in Figure 4. The larger contribution comes from Eq. (30).

The error calculation for the intensity method also involves several steps: First is the determination of the energy difference ΔE of the torsional levels by Eq. (15), preferably using measurements over a range of temperatures according to Ruitenberg,[16] which eliminates practically the factor f of Eq. (15); second, the determination of s by Eq. (14); and, finally, the determination of V_3 by Eq. (13). For the calculation of the error we use an approximation for $b(s)$ in Eq. (14), which is valid for high s:

$$b(s) \rightarrow 2\sqrt{s}\,(v + \tfrac{1}{2}). \tag{31}$$

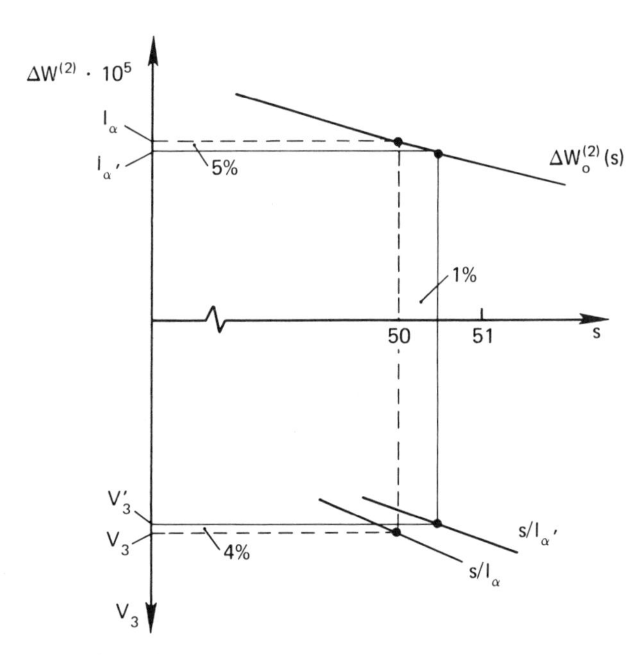

FIGURE 4 Influence of the value of I_α on the final value of V_3. The erroneous I'_α is taken 5 percent larger than I_α. With $s = 50$ there is an error $\delta s/s \approx$ 1 percent but $\delta v_3/v_3 \approx 4$ percent.

For $s > 64$ and $v = 0$ the error is less than 4 percent; for $v = 1$, the error is less than 6 percent. The error $\delta V_3/V_3$ is

$$\frac{\delta V_3}{V_3} = \left\{4\left(\frac{kT}{\Delta E}\right)^2 \left[\frac{\delta(I_0/I_v)}{I_0/I_v}\right]^2 + 5\left(\frac{\delta I_\alpha}{I_\alpha}\right)^2 + 4\left(\frac{\delta T}{T}\right)^2 + 5\left(\frac{\delta r}{r}\right)^2\right\}^{\frac{1}{2}}. \tag{32}$$

Here the errors of the intensity measurement and the error of I_α are generally of the same order. They usually exceed those of T and r. Thus I_α is again a critical parameter.

ERROR OF THE MODEL

The analysis of the rotational spectra in the case of the frequency method is essentially based on the fitting of constants contained in the effective rotational Hamiltonian produced by a perturbation method from the Hamiltonian of Eq. (4), which is mathematically a polynomial in the components of the angular momentum operator

$$H_R\,(\text{eff}) = \sum_{gg'\ldots} (a_g P_g + b_{gg'} P_g P_{g'} + c_{gg'g''} P_g P_{g'} P_{g''} + \ldots). \tag{33}$$

Different physical models always lead to an effective Hamiltonian of the general form given in Eq. (33). The coefficients are different for the different torsional states. They are all related to the basic parameters of a model in a way specific to that model. But as these coefficients are fitting parameters, a good fit to a limited number of states is not a sufficient indication of the validity of the model. Conversely, a bad fit indicates limitations of the model.

As mentioned above a limitation of the present model is obvious when one tries to find a single set of parameters for fitting rotational spectra in several torsional states. The moments of inertia present part of the difficulty. Another limitation has been observed in trying to fit rotational spectra in excited states of a low-energy (nontorsional) vibration. If internal rotation is *totally independent* of vibrations, the given model should apply, since in this case the different energy level schemes should be additive. The experimental results indicate some breakdown of the basic assumption of the hitherto used model by which vibrations are neglected. A special consequence is that determination of V_3 from an excited torsional state may be erroneous if possible interaction with another low energy vibration is neglected. The quality of the fit will give no indication of the principal error.

In this connection an extended Hamiltonian has been developed that includes an additional vibrational degree of freedom. This extended model was tested with three molecules. It was possible to fit the splittings in the rotational spectra of the lowest three internal rotation/vibration states with one set of coefficients. There is an indication that for higher states an even more extended Hamiltonian is necessary.

Another approach, which aims at the interpretation of rotational spectra in different internal rotation states only, is the r_e-relaxation method proposed by Günthard[26] and Bauder.[27] Here a deformation of the molecule during internal rotation is related to the internal rotation angle α. Thereby the introduction of additional degrees of freedom is avoided, but additional parameters describing the special type of deformation with α are introduced, which seem to be sensitive to the difference between torsional doublets of different states. The tilt angle then results without any further assumptions.

It is difficult to give a general estimate of the influence of the model on the results. However, Table 1 gives a comparison of the results on CH_3CH_2CN obtained by the two models. The difference of V_3 is about 1 percent, which is within the error range usually introduced by the error of I_α.

TABLE 1 Microwave and Infrared Results on Ethyl Cyanide, CH_3CH_2CN

Parameter	Microwave, Torsion Only[a]	Microwave, Torsion Plus ν_{13}[b]	Infrared Measurements
V_3 (kcal/mole)	3.05[c]	3.075 ± 0.002	
V_6 (kcal/mole)		−0.126 ± 0.001	
s	81.5	81.94	
I_α (u Å^2)	3.195	3.194	
λ_z	0.6676	[d]	
ν_{13} (cm^{-1})			206.5[e]
Derived values (cm^{-1})			
ν_T (0 → 1)[f]	222	210.5	211[g]
ν_T (1 → 2)		201.2	202[g]
ν_T (2 → 3,A)		189.4	188[g]
ν_T (2 → 3,E)		188.6	?

[a] Data from Laurie.[28]
[b] Data from Mäder[29] and Heise.[30]
[c] V_3 = 2.85 from μ_b lines only.[29]
[d] $\lambda_z = 0.671 + 0.329q + 0.064q^2$, where q is the coordinate for the ν_{13} mode.
[e] Far-infrared measurement of ν_{13}, the in-plane CCN bending mode.
[f] Torsional frequency.
[g] From combination bands with ν_1 (CH_3 stretching mode) at 2999.3 cm^{-1} and ν_5 (CH_3 deformation) at 1470.9 cm^{-1}.

COMPARISON WITH OTHER METHODS

It is interesting to compare the results of microwave investigations of internal rotation with the results of other methods. For some molecules far-infrared spectra have been measured. Besides experimental difficulties in this spectral region, the transition moment for torsional transitions is very small, at least when methyl groups are involved. Further, the assignment of the spectra is difficult, although some assignments have been made with the help of parameters obtained by microwave methods. These parameters were then modified to fit the infrared data. There are some features of the spectra that are not well understood. Further, it is not finally proved that an A–E splitting has even been observed in a torsional band. Comparison for some examples is made in Table 2.

Another possibility is the investigation of combination bands of CH stretch and deformation, both with torsion. Confirmation of the assignment is also a severe problem in this case. By using the spectroscopic combination principle, the distance between torsional levels may be determined. Both of these infrared methods finally have to use the Hamiltonian H_T, with the consequence that I_α influences the results for V_3 in a way discussed under the microwave intensity method. The infrared results on CH_3CH_2CN are compared in Table 1.

More recently some results obtained by neutron inelastic scattering have been presented.[36] Since this is mainly a solid state method, it is not considered here.

SUMMARY

The best data describing the hindering potential of methyl torsions in the range from 0 to 3500 cal/mol may be obtained by a sophisticated analysis of microwave measurements. The data of other fields are useful as a supplement. Although a combined use of microwave splitting and infrared data (or microwave intensity data) is promising in principle, it has not yet been widely used. The most severe problem concerning accuracy of the barrier height is in most cases the determination of the moment of inertia of the methyl group.

I wish to thank all members of the Kiel microwave group, especially D. Sutter and A. Guarnieri for criticizing the manuscript and L. Charpentier and Ch. Martens for the help in preparing the manuscript.

TABLE 2 Comparison of Microwave and Infrared Results

Molecule	Microwave I_α (u Å^2)	F (cm^{-1})	V_3 (cm^{-1})	V_6 (cm^{-1})	s	Infrared[a] I_α (u Å^2)	F (cm^{-1})	V_3 (cm^{-1})	V_6 (cm^{-1})	s
CH_3CH_2Cl	(3.1613)[b]	6.06	1289 ± 4	(0)	94.43[c]	3.14	6.085	1291 ± 10	0 ± 1	94.29
$CH_3CF{=}CH_2$	(3.11)		854 ± 6	(0)	[d]		5.620	818 ± 4	4 ± 1	65.12
$CH_3CH{=}CH_2$										
$v_T = 0$	(3.160)	7.101	686.7	(0)	42.9[e]		6.972	711 ± 10	−16 ± 1	44.55
$v_T = 1$	(3.160)	7.101	697.4	(0)	43.65[e]					
$v_T = 2$	(3.160)	7.106	703.6	(0)	44.0[e]					
$v_T = 0,1,2$	(3.160)		698.4 ± 0.5	−13 ± 2	[e]					
$CH_3CH{-}CH_2$ (epoxide, O bridging)	(3.194)	5.856	895 ± 25	< 10	[f]		5.768	900 ± 8	−9 ± 1	68.73
$(CH_3)_2SiH_2$	(3.16)	5.836	575.3 ± 1	(0)	43.83 ± 0.08[g]		(5.83)	513.8		[h]

[a] Infrared results from Fately and Miller,[32] except h. The microwave value of s was used in the determination of F.
[b] Assumed values in parentheses.
[c] From Schwendeman and Jacobs.[33]
[d] From Pierce and O'Reilly.[34]
[e] From Hirota.[17]
[f] From Herschbach and Swalen.[35]
[g] From Trinkaus *et al.*[22]; $F' = -0.373$ cm^{-1}, $V'_{12} = -13 \pm 1$ cm^{-1}, $V_{12} = (0)$ cm^{-1}; calculated from microwave: $\nu_{0\to1} = 151.3$ cm^{-1}, $\nu_{0\to1} = 167.6$ cm^{-1}.
[h] From Grant *et al.*[36]; $\nu_{0\to1}$ (obs) = 150 cm^{-1}.

REFERENCES

1. K. D. Möller and W. G. Rothschild, *Far-Infrared Spectroscopy*, Wiley-Interscience, New York (1971), Ch. 1–4; Ch. 8 and App. VII, Far-Infrared-Bibliography by E. D. Palik.
2. B. Starck, *Landolt-Börnstein*, Neue Serie Vol. II/4, Springer-Verlag, Berlin-Heidelberg-New York (1967).
3. J. Demaison, W. Hüttner, B. Starck, I. Buck, R. Tischer, and M. Winnewisser, *Landolt-Börnstein*, Neue Serie Vol. II/6, Springer-Verlag, Berlin-Heidelberg-New York (1973).
4. E. B. Wilson, *Proc. Natl. Acad. Sci., U.S.A.*, **43**, 816 (1957).
5. C. C. Lin and J. D. Swalen, *Rev. Mod. Phys.*, **31**, 841 (1959).
6. H. Dreizler, *Fortschr. Chem. Forsch.*, **10**, 59 (1968).
7. R. W. Kilb, C. C. Lin, and E. B. Wilson, *J. Chem. Phys.*, **26**, 1695 (1957).
8. D. R. Herschbach, *J. Chem. Phys.*, **31**, 91 (1959).
9. M. Born and F. R. Oppenheimer, *Ann. Phys.*, **84**, 457 (1927).
10. H. H. Nielsen, *Phys. Rev.*, **40**, 445 (1932).
11. H. Goldstein, *Classical Mechanics*, Addison-Wesley Publishing Co., Reading, Mass. (1959).
12. P. H. Verdier and E. B. Wilson, *J. Chem. Phys.*, **29**, 340 (1958).
13. D. H. Baird and G. R. Bird, *Rev. Sci. Instr.*, **25**, 319 (1954).
14. A. S. Esbitt and E. B. Wilson, Jr., *Rev. Sci. Instr.*, **34**, 901 (1963).
15. A. Dymanus, H. A. Dijkerman, and G. R. D. Zijderveld, *J. Chem. Phys.*, **32**, 717 (1960).
16. G. Ruitenberg, *J. Mol. Spectrosc.*, **42**, 161 (1972).
17. E. Hirota, *J. Chem. Phys.*, **27**, 283 (1962); *J. Chem. Phys.*, **45**, 1984 (1966).
18. M. K. Kemp and W. H. Flygare, *J. Am. Chem. Soc.*, **91**, 3163 (1969).
19. T. S. Huang and R. A. Beaudet, *J. Chem. Phys.*, **52**, 935 (1970).
20. H. D. Rudolph and A. Trińkaus, *Z. Naturforsch.*, **23a**, 68 (1968).
21. H. Dreizler, H. D. Rudolph, and H. Mäder, *Z. Naturforsch.*, **25a**, 25 (1970).
22. A. Trińkaus, H. Dreizler, and H. D. Rudolph, *Z. Naturforsch.*, **28a**, 750 (1973).
23. H. Dreizler and H. Legell, *Z. Naturforsch.*, **28a**, 1414 (1973).
24. J. D. Swalen, *J. Chem. Phys.*, **24**, 1072 (1956).
25. K. T. Hecht and D. M. Dennison, *J. Chem. Phys.*, **26**, 48 (1957).
26. M. Ribeaud, A. Bauder, and Hs. H. Günthard, *Mol. Phys.*, **23**, 235 (1972).
27. A. Bauder, *Microwave Spectroscopy Conference*, Paper 1 A3, Bangor (Sept. 1972).
28. V. Laurie, *J. Chem. Phys.*, **31**, 1500 (1959).
29. H. Mäder, Dissertation, Kiel (1972), to be published in *Z. Naturforsch.*
30. H. Heise, Diplomarbeit, Kiel (1972).
31. R. C. Livingston, D. M. Grant, R. J. Pugmire, K. A. Strong, and R. M. Brugger, *J. Chem. Phys.*, **58**, 1438 (1973).
32. W. G. Fately and F. A. Miller, *Spectrochim. Acta*, **19**, 611 (1963).
33. R. H. Schwendeman and G. D. Jacobs, *J. Chem. Phys.*, **36**, 1245 (1962).
34. L. Pierce and J. M. O'Reilly, *J. Mol. Spectrosc.*, **3**, 536 (1959).
35. D. R. Herschbach and J. D. Swalen, *J. Chem. Phys.*, **29**, 761 (1958).
36. D. M. Grant, R. J. Pugmire, and R. C. Livingstone, *J. Chem. Phys.*, **52**, 4424 (1970).
37. J. R. Durig and C. W. Hawley, *J. Chem. Phys.*, **58**, 237 (1973).

J. J. Rush

Study of Large-Amplitude Vibrations in Molecules by Inelastic Neutron Scattering

Neutron inelastic scattering techniques have been widely applied to the study of vibrational and rotational dynamics in hydrogenous molecular systems.[1] The bulk of this research has been concerned with the study of intermolecular and interionic motions in solids, but a limited yet significant amount of effort has been directed toward the study of large-amplitude intramolecular vibrations, most notably torsional vibrations and hydrogen-bond modes.[2] The present paper is restricted primarily to a discussion of the application of neutron scattering to the study of torsional vibrations and barriers to rotation of methyl groups in molecules. We will present several examples in which neutron spectra have provided information complementary to that obtained by the more widely available and applicable infrared and Raman techniques. We will also discuss in simple terms some limitations and pitfalls of the neutron technique and the interpretation of neutron spectral results.

NEUTRON TECHNIQUE

A schematic view of a generalized neutron experiment is shown in Figure 1, along with the essential quantities defining the incident and scattered neutrons. A collimated "white" thermal neutron beam from a reactor is impinged on a single crystal monochromator or a mechanical

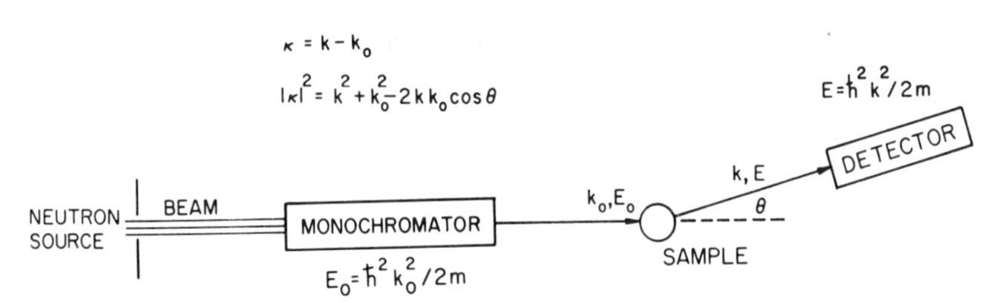

FIGURE 1 Generalized representation of a neutron-scattering experiment. Incident and scattered energies are designated by E_0, E and incident and scattered momenta (wave vector) by k_0, k. Expressions giving the relationship between moments and energy are also shown.

neutron chopper to obtain a monochromatic beam of energy E_0, generally between 20 and 1500 cm^{-1} (8.07 cm^{-1} ≡ 1 meV) and momentum k_0. The neutrons are then scattered from the sample of interest and the distribution of scattered energies and momenta (wave vector) E and k are measured at a neutron detector either by time-of-flight (velocity) analysis or by using a crystal analyzer in front of the detector for neutron energy analysis. The neutron technique is analogous to Raman scattering, with the important difference that due to the finite mass of the neutron, it can cover a much wider range of momentum transfer κ. Conventional neutron experiments generally utilize incident neutron wavelengths, λ, of ~0.7–7 Å and thus incident k_0's ($2\pi/\lambda$) of ~1–10 Å^{-1}. Momentum transfers are measurable between ~0.1 and 10 Å^{-1}, compared with a κ of $\lesssim$ 0.01 Å^{-1} in light scattering experiments. This widely variable κ, which provides a unique capability for neutrons in such areas as the measurement of phonon dispersion curves in crystals, can create problems in the measurement of molecular spectra as shown below.

We will now briefly discuss the inelastic scattering of neutrons by molecules. Our discussion will be limited to hydrogenous molecules (which cover the bulk of the cases thus far investigated by neutron scattering) and to the assumption of one-quantum scattering in the harmonic approximation.[3] Moreover, since the incoherent scattering cross section of hydrogen is larger by at least a factor of 10 than that of other nuclei,[2,3] only incoherent scattering by the hydrogen atoms will be considered. Under these conditions the differential cross section for inelastic scattering into solid angle Ω and with energy transfer $\hbar\omega$ is given by:

$$\left(\frac{\partial^2\sigma}{\partial\Omega\partial\omega}\right)_{inc,\mathrm{H}} = \frac{k}{k_0}\frac{1}{N}\sum_{j,n}(a_{\mathrm{H}}^2)\exp(-2W_n)\exp\left[\frac{-\hbar\omega}{2kT}\right]\frac{\hbar(\kappa\cdot c_j^n)^2}{4N\omega_j M_{\mathrm{H}}}\operatorname{csch}\left[\frac{\hbar\omega_j}{2kT}\right]\delta(\omega_j-\omega), \tag{1}$$

where a_H is the known hydrogen scattering amplitude, N is the number of hydrogen atoms in the molecule, M_H is the hydrogen mass, ω_j and c_j^n are the frequency and hydrogen displacement vectors for the nth hydrogen atom in the jth normal mode, $\exp(-2W_n)$ is the Debye-Waller factor of the nth hydrogen, and δ is a delta function.

An important feature to observe in Eq. (1) is that for a given temperature and normal mode frequency the scattering cross section, and thus the observed spectral peak intensity, is roughly proportional to the square of the hydrogen amplitude $(c_j^n)^2$. Thus in neutron spectra measured for hydrogenous molecules, vibrational transitions involving large-amplitude hydrogen motions (such as methyl torsions) will appear with considerably enhanced intensity compared with modes involving smaller hydrogen displacements. In principle this intensity "selection rule" can be quite useful for the observation and assignment of modes such as torsional vibrations that are often weak or "forbidden" in infrared and Raman spectroscopy. Unfortunately, however, the detailed study of molecular vibrations in gas or liquid phases by neutron scattering is almost always difficult due to the momentum-dependent "recoil resolution" inherent in the neutron scattering process.[4] Due to this effect, all vibrational peaks in a measured neutron spectrum will have an inherent width at half maximum $\Delta E_{1/2}$ (unrelated to instrumental resolution) proportional to $\kappa(T/M_{\text{eff}})^{1/2}$, where M_{eff} is the effective molecular mass.[4] Thus neutron measurements in fluid phases must generally be performed using fairly high incident energies and very small scattering angles (see Figure 1) to attain the lowest possible momentum transfer. This, in turn, lowers the cross section for observing vibrational transitions, which is roughly proportional to κ^2 [see Eq. (1)]. These experimental restrictions can be relaxed somewhat in the study of molecular vibrations in solid phases (very large M_{eff}), and most of the neutron studies thus far have been on molecular solids.

The effects of recoil broadening are illustrated in Figure 2, which shows the shapes of the neutron spectra theoretically predicted for gaseous neopentane under various conditions.[5] These spectra were generated assuming two methyl torsional modes (A_2 and F_2) at 210 and 265 cm^{-1} and carbon skeletal modes at 330 and 410 cm^{-1} (E and F_1). This figure clearly indicates that such vibrational transitions are difficult or impossible to resolve in fluid phases using "cold" ($E_0 \lesssim 40$ cm^{-1}) incident neutrons (a very common neutron experimental condition), even for a relatively heavy molecule such as neopentane. The figure dramatically demonstrates the advantages of small κ (high E_0, low θ) scattering, but shows that complete peak separation may require scattering conditions ($E_0 \geqslant 100$ meV, $\theta \sim 1°$, $\Delta E \sim 1$ meV) which are almost

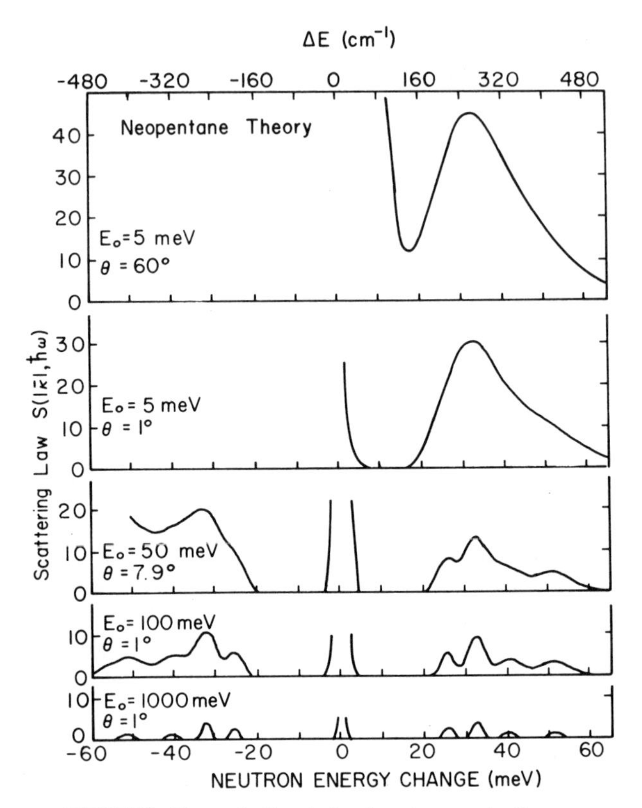

FIGURE 2 Theoretically calculated neutron spectra for gaseous neopentane, $C(CH_3)_4$.[5] Various conditions of incident neutron energy (E_0) and scattering angle (θ) are shown; instrumental resolution is assumed negligible.

prohibitive with existing neutron sources and instrumentation. At this point it may be useful for comparative purposes to point out that the best instrumental resolution now available in neutron scattering experiments varies roughly between 2 and 10 cm^{-1} in the range of vibrational transitions between 10 and 600 cm^{-1} (compared with resolutions of ~2 cm^{-1} or better attainable by infrared and Raman spectroscopy).

Thus the use of neutron spectroscopy to investigate intramolecular vibrations must be done under appropriate scattering conditions for appropriately chosen molecules. Very often this will require solid-phase experiments, and even then precise assignments of overlapping peaks due to vibrations of comparable amplitude can be difficult or misleading. In spite of this, however, neutron scattering (if only because the neutron interacts with the vibrating nuclei of a molecule and is not influenced by molecular dipoles or polarizabilities) can often be a powerful complement to optical techniques, as will be illustrated in the next section. In fact there have been recent and reasonably successful attempts to calculate neutron spectra of a number of molecules directly using Eq. (1) and appropriate molecular force fields[6,7] and thus to provide a more rigorous basis for vibrational assignments and possibly a meaningful test for various force models.

DISCUSSION OF SELECTED RESULTS

In this section we will discuss several examples of the study of large amplitude (primarily torsional) vibrations by neutron scattering. We will try to present an instructive evaluation of these results and their comparison to relevent optical data. We will also briefly discuss the derivation of barriers to internal rotation from the observed torsional peaks.

Simple Molecules

First, in order to illustrate the way in which neutron spectra can sometimes complement optical results in the study of large amplitude modes, we will compare the neutron and infrared spectral studies of hydrogen bond modes in a very interesting bihalide salt, $CsHCl_2$. The assignment of the H-bond bending (ν_2) and stretching (ν_1 and ν_3) modes is important in the study of hydrogen bonding in X–H–X^- ions and other systems. In Figure 3 we have sketched infrared[8] and neutron[9] spectra for this salt in the 400–1500-cm^{-1} region. It was assumed from the infrared spectrum that the strong, broad 1170-cm^{-1} band was due to the Cl–H–Cl^- ν_2 mode, while the weak feature at 600 cm^{-1} was unassigned. Subsequently, an extensive study of very low-temperature infrared spectra of $CsHCl_2$ and a number of other X–H–X^- species provided convincing evidence that this and other bands in the 1000–1200-cm^{-1} region for a number of bihalide salts should be reassigned to a $2\nu_2$ transition.[10] The neutron spectrum in Figure 3 provides dramatic confirmation of the reversed assignments. The band at 650 cm^{-1} is by far the most intense peak observed in both neutron energy gain and energy loss spectra[9] and must therefore be assigned to the large-amplitude ν_2 vibration; the weaker maximum around 1200 cm^{-1} has the expected lower intensity for a two-quantum transition. This clearly demonstrates a case where the more predictable neutron-nuclear scattering cross section [Eq. (1)] can provide valuable information in molecular systems, even under poor resolution conditions.

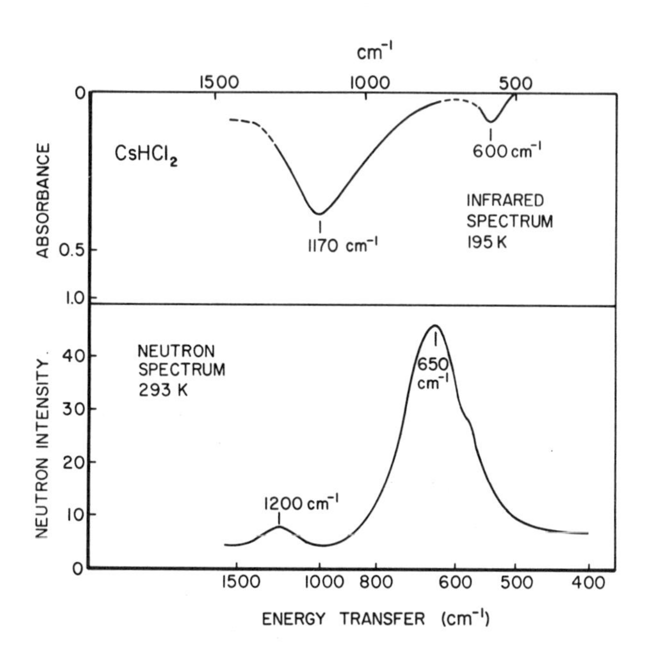

FIGURE 3 Infrared and neutron spectra for $CsHCl_2$.[9,10] The neutron spectral resolution in this measurement is ~50 cm^{-1}.

Next we will consider an example of a neutron study of the optically inactive torsional mode in a prototype single-methyl-top molecule, CH_3CCl_3. The barrier to methyl group rotation in this molecule was determined by thermodynamic measurements[11] and infrared combination bands[12] to be about 3.0 kcal/mol (12.5 kJ/mol), which is quite close to the value for ethane. The neutron time-of-flight energy gain spectra for CH_3CCl_3 [13] in the liquid and two solid phases are shown in Figure 4. It can be seen that the inelastic spectra above 150 cm^{-1} are dominated by a very strong peak at 300 ± 12 cm^{-1} in all phases, which can only be assigned to the large-amplitude methyl torsional vibration (neutron intensity calculations[14] provide additional confirmation of this assignment). The broad maxima below 100 cm^{-1} are due to whole-molecule hindered rotations (liquid and plastic phases) or librations (low-temperature crystal phase). If a simple cosine potential is assumed and sixfold and higher terms are neglected, a barrier to internal rotation can be calculated directly from the torsional frequency and appropriate molecular parameters, using Herschbach's[15,16] formulation for a symmetric top attached to a rigid frame. This calculation gave a barrier of 24 ± 2 kJ/mol (5.8 ± 0.5 kcal/mol), which is almost double the previous

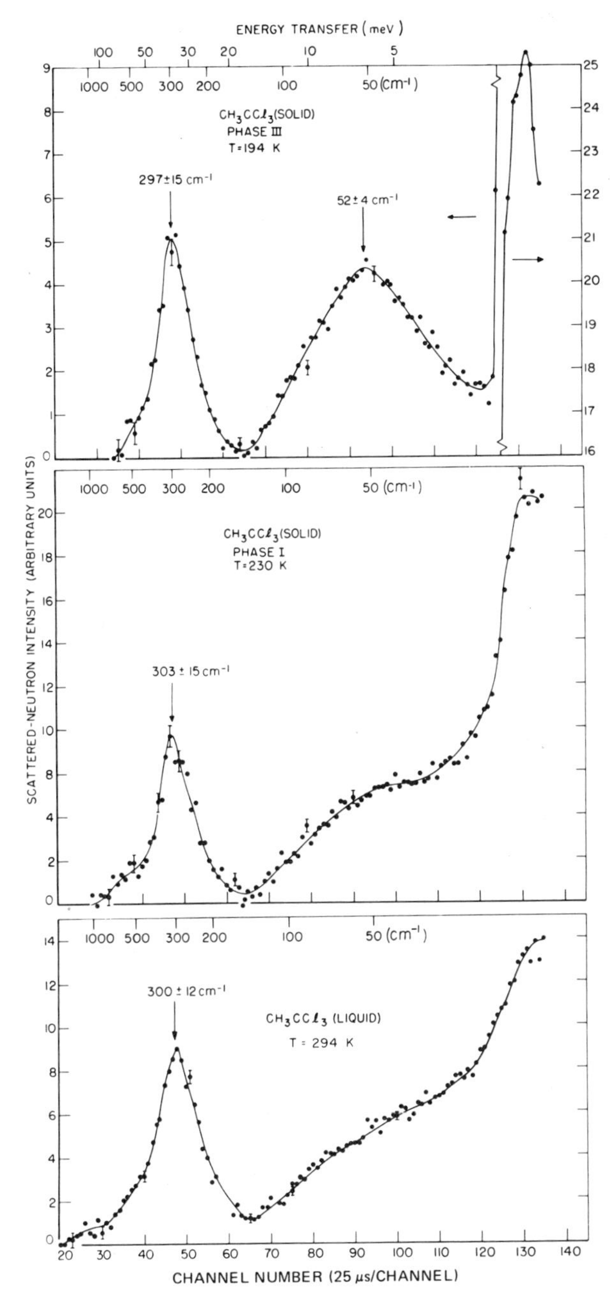

FIGURE 4 Time-of-flight neutron spectra in liquid and crystal phases of CH_3CCl_3. Energy-gain scales in millielectron volts and cm^{-1} are also shown.

value, and provides a key experimental result for the theoretical study of the origin of barriers in substituted ethanes.[17]

It should be pointed out here that these results were obtained using an incident cold neutron beam with $\Delta E_0 \approx 20$ cm^{-1} and that the instrumental resolution at the peak (aside from recoil broadening) was only about 40 cm^{-1}. This explains the uncertainty of ± 12 cm^{-1} on the assigned frequency. In interpreting other low-resolution data of this type (mostly before 1968), particularly for more complex molecules (e.g., two or more CH_3 groups), one should recognize that the torsional frequencies are generally average values for several overlapping modes[18] and can be complicated further by unresolved lower amplitude skeletal modes.[14] In addition the neutron time-of-flight spectra *must* be transferred to an energy scale ($dE \propto t^{-3} dt$) to assure meaningful peak assignments. (This was done in the neutron work cited here.)

It should also be noted here that the torsional mode assignment for CH_3CCl_3 has recently been confirmed in the low-temperature solid phase by Durig and co-workers,[19] who have demonstrated in an extensive series of infrared and Raman studies of molecular torsional vibrations that "forbidden" or weakly active modes can very often be definitively observed in crystal phases due to the lower symmetry of the crystal field.

Multitop Molecules

We shall now consider a few examples of the application of neutron spectroscopy (primarily the small κ method) to the study of methyl torsional modes in two or more top molecules. The measurement of differences in the frequency of in-phase and out-of-phase torsional modes can in principle yield independent evidence concerning the effects of top–top interaction on the potential governing internal rotation, thus providing an excellent complement to microwave results. The discussion in this paper will omit consideration of the splitting of the various torsional levels through the tunneling effect, since this can rarely be resolved in vibrational spectroscopy. As previously stated, the observation and definite assignment of these modes by infrared spectroscopy is often difficult due to weakly active or inactive modes.[16,20] This is particularly true of attempts to study "unperturbed" molecules in fluid phases.

Propane [$(CH_3)_2CH_2$] has been the object of considerable study by infrared[16,21] and microwave[22,23] spectroscopy. No direct observation of the symmetry allowed (symmetry species B_2) transition has been made, but frequencies have been extracted from infrared combination bands[24] and the microwave splitting.[22] In Figure 5 are shown "small κ" neutron spectra for gaseous, liquid, and solid propane.[4] It should be pointed out that these results (and those in Figures 6 and 7) are presented in "scattering law" form,[4] which adjusts the observed cross section for detailed balance and the k/k_0 dependence [Eq. (1)] to obtain

$$S(\kappa,\omega) = (k_0/k) \exp(h\omega/2kT) \frac{\partial^2 \sigma}{\partial\Omega\partial\omega} . \tag{2}$$

The energy loss ("Stokes") results in Figure 3 show an intense broad band in the gas phase, a doublet in the liquid phase, and two analogous sharp peaks in the solid phase that can be assigned with certainty to the in-phase (A_2) and higher frequency out-of-phase (B_2) torsional transitions. In fact, the intensity of these modes is very well predicted from cross-section calculations [Eq. (1)] by Hudson and Gordon[7] who used the consistent force field of Lifson and Warshel[25] to obtain the normal mode eigenvalues and (hydrogen) eigenvectors. The additional peaks at low frequency in the solid phase are due to lattice vibrations. The assigned torsional frequencies are listed in Table 1 and are compared with other results. The stated errors may be slightly optimistic in view of the instrumental resolution of the experiment (≈ 10 cm^{-1}) and the additional recoil broadening in the liquid result, but there is no doubt that the frequencies are well established. The gas-phase frequencies that were extracted from the broad peaks in Figure 5 by cyclical curve fitting are not of comparable accuracy. The frequencies derived from microwave data are in good agreement with the neutron results, while the infrared combination band values are questionable.

The derivation of quantitative information concerning the potential barrier to internal rotation from such measured frequencies in multitop molecules is a somewhat questionable procedure, which will not be treated in great depth here. For a two-top molecule the potential energy function can be expanded in a Fourier series (omitting sixfold and higher terms) to obtain

$$2V = V_0 - V_3 (\cos 3\phi_1 + \cos 3\phi_2) - V_{33} \cos 3\phi_1 \cos 3\phi_2 + V'_{33} \sin 3\phi_1 \sin 3\phi_2 , \tag{3}$$

where ϕ_1 and ϕ_2 are the angles of internal rotation of the two CH_3 groups, V is the potential energy, the V_3 terms represent independent internal rotation of the two tops, and the V_{33} and V'_{33} terms represent top–top interaction.

The form of the potential will obviously be even more complex for a molecule containing more than two tops.[21] The difference of the A_2

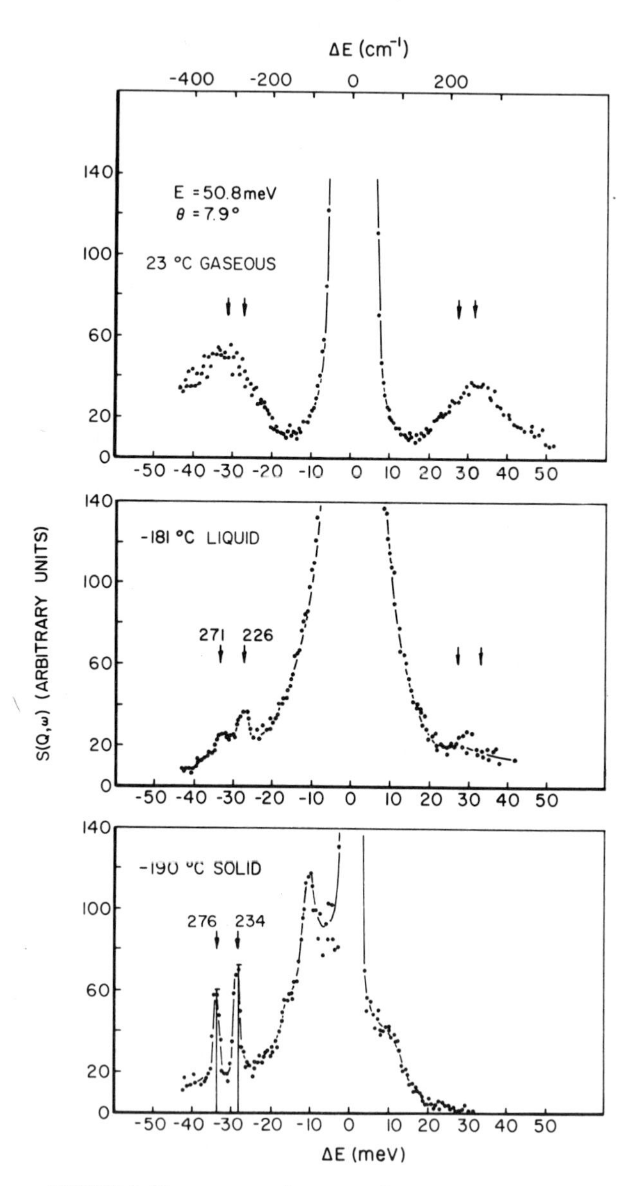

FIGURE 5 Neutron-scattering spectra for propane in gas, liquid, and solid phases. Torsional peak positions are indicated as are the theoretical peak intensities calculated on the basis of a molecular force field.[7]

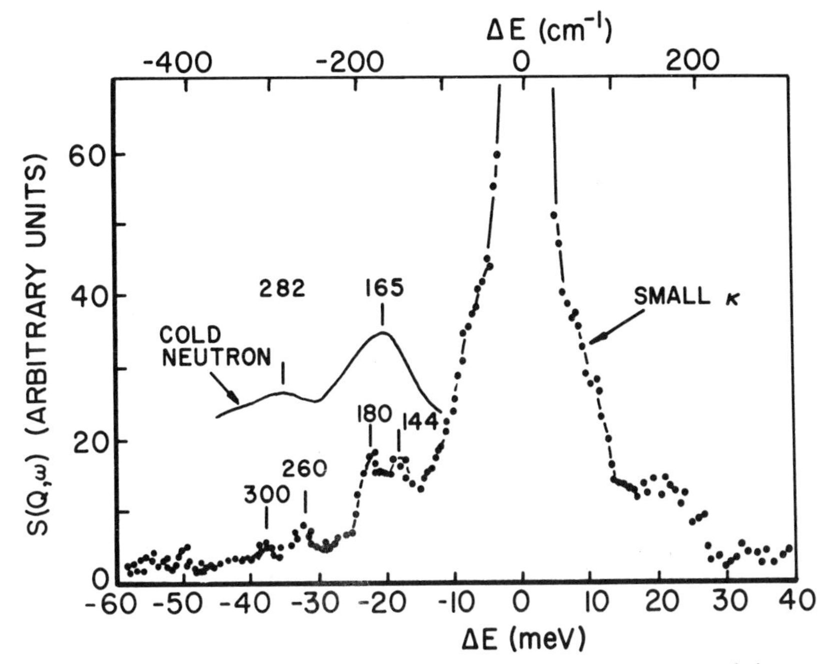

FIGURE 6 Neutron spectra for *o*-xylene measured by small κ and cold neutron techniques.

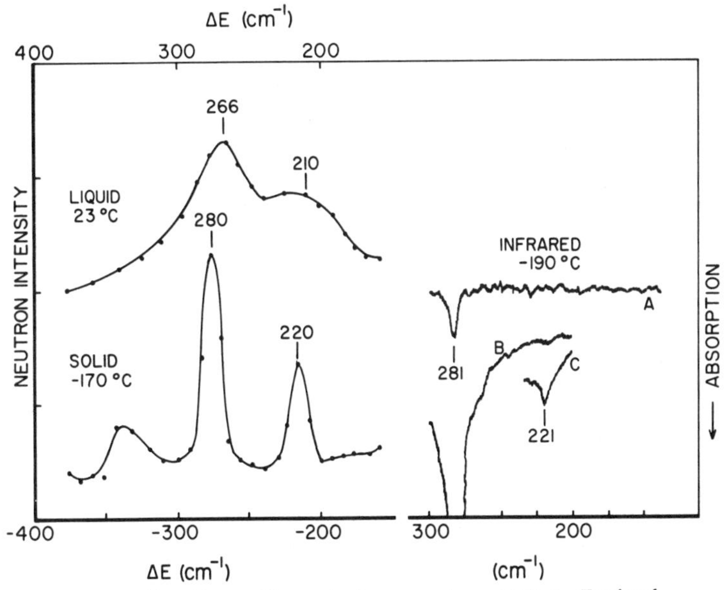

FIGURE 7 Neutron[30] and infrared[26] spectra for neopentane, $C(CH_3)_4$. Torsional peak frequencies in cm^{-1} are shown. The high-energy peak in the solid phase neutron spectrum is due to a skeletal molecular mode. A, B, and C in the infrared spectrum refer to increasing sample thickness.

TABLE 1 Torsional Frequencies and Potential Barrier Parameters Derived by Different Methods for Propane

Frequencies (A_2, B_2) (cm^{-1})		V_3 (kJ/mol)[a]	V_{33} (kJ/mol)[a]
Neutron	217 ± 8, 265 ± 8 gas		
	226 ± 3, 271 ± 3 liquid	15.90[b]	−0.84[b]
	234 ± 2, 276 ± 2 solid		
Micro-wave	216, 271[c] gas	13.90[d]	−0.71[d]
		14.96[e]	−1.55[e]
		12.32[f]	−1.17[f]
		13.97[g]	
Infrared	208, 216[h] gas		
	282[i] solid		

[a] 1 kJ/mol = 0.239 kcal/mol.
[b] From Grant *et al.*[4] These values are calculated from solutions of a two-dimensional Hill's equation, assuming $V_{33} = V'_{33}$.
[c] From Hirota *et al.*[22]
[d] From analysis of microwave splittings.[22]
[e] From Grant *et al.*,[4] calculated as in *b* using microwave torsional frequency values.
[f] From Weiss and Leroi[21] as derived from the harmonic potential method of Lide and Mann.[27]
[g] From Weiss and Leroi[21]; average of the barrier heights calculated from each frequency assuming a sinusoidal potential and no top-top interaction.
[h] From Gayles and King.[24]
[i] J. R. Durig (personal communication).

and B_2 frequencies in such molecules as propane contains contributions from both kinetic energy coupling terms and coupling terms in the potential energy of Eq. (3).[4,16,21] The first effect can be calculated rigorously, provided accurate structural parameters are available to obtain well-defined reduced moments of inertia associated with the A_2 and B_2 methyl torsions. The splitting of the torsional modes by potential energy coupling is more difficult to assess in detail in cases where at most two torsional transitions are observed (which is true thus far in all the neutron work and most of the optical work). Clearly, one has too few data (two torsional transitions) to obtain separate values for V_3, V_{33}, and V'_{33} in Eq. (3), and the situation becomes progressively worse for cases with three or more tops.

Several different approaches have been used to derive the potential energy parameters of Eq. (3) from experimental data. Grant *et al.*[4] have derived barriers from their neutron results by assuming $V_{33} = V'_{33}$ and then using their assigned frequencies in solving a two-dimensional Hill's equation describing the torsional motion to obtain values for V_3 and V'_{33}. Other workers[21,26] have applied the treatment of Lide and Mann[27] in which the potential energy is expanded in a Taylor series about the equilibrium position and only the quadratic terms retained (harmonic approximation). In this approximation the frequencies are expanded in terms of a principal potential constant K and an interaction constant L, each of which is a linear combination of the parameters in the Fourier expansion of the potential energy. In the propane-type molecule the relations are

$$K = (9/2)\,(V_3 + V_{33})$$

$$L = (9/2)\,V'_{33}.$$

In addition, if the coupling terms are much smaller than V_3, then $V_3 \simeq 2/9K$. A third approach[16,18,21,26] for estimating V_3 is to ignore coupling effects entirely and to derive an "average" V_3 by assuming a simple threefold potential for each top[15] and calculating the average of the barriers obtained from the two torsional frequencies.

The rather wide variation in barrier parameters derived from different methods is illustrated in Table 1. The neutron frequencies for liquid propane are used for these barrier comparisons since they are more reliable and, indeed, identical to the gas-phase results within the estimated uncertainty. The same molecular parameters used in the microwave study[22,23] are used in each calculation, so that uncertainties in kinetic coupling effects do not affect the relative barrier values. It can be seen (not unexpectedly) that the barrier terms obtained by different methods using identical frequencies differ by as much as 30 percent. It should be noted that, as pointed out by Weiss and Leroi,[21] the assumption of a harmonic potential in the Lide and Mann method causes an inherent decrease in the derived V_3 terms (~10 percent) compared with methods retaining the sinusoidal potential. These errors are quite similar for different molecules, however, so that comparisons between barrier results for different molecules by this method can be informative.[21] It must obviously be concluded that barrier parameters obtained for multitop molecules on the basis of one or two observed frequencies have limited quantitative significance, except in relative terms, and that one must carefully scrutinize the method of calculation in evaluating such results.

With these observations in mind, we shall now discuss two other examples of neutron spectra of multitop molecules: *o*-xylene (1,2-dimethylbenzene) and neopentane (tetramethylmethane). *o*-Xylene is an interesting prototype for internal rotation and top–top coupling in substituted aromatics.[29] The results for *o*-xylene by small κ[28] and cold neutron[18] experiments are shown in Figure 6. The strongest peaks in both these spectra are again assigned to the torsional modes. Neutron spectra for durene (1,2,4,5-tetramethylbenzene) show more intense peaks at almost identical frequencies[18,28] and thus provide further confirmation of this assignment. Comparison of the spectra again demonstrates the advantage of the small κ measurement; whereas the torsional modes at 146 ± 5 and 182 ± 5 cm^{-1} are clearly resolved in this spectrum (as are two higher frequency molecular modes), the cold neutron result, due to poor resolution and probably some κ-dependent (possibly multiphonon) broadening, provides only unresolved bands for both the torsional and higher frequency modes.

The large splitting of the torsional modes in solid *o*-xylene clearly indicates a strong top–top coupling term in the potential, since kinetic coupling effects are negligible. Under the assumption that the methyl–methyl interactions are repulsive (V_{33} negative), the torsional modes were assigned[28] as 146 cm^{-1} (A_2) and 182 cm^{-1} (B_2). These, in turn, were used to calculate V_3 and V_{33} barrier parameters of 10.0 ± 0.5 and -2.25 ± 0.30 kJ/mol. These results are consistent with the gas-phase thermodynamic barrier determination,[29] but this, or course, does not mean that these values can be attributed to the free molecule with certainty.

As a final case, we present a comparison of neutron and infrared spectra for a globular molecule, neopentane $C(CH_3)_4$. The A_2 and F_2 torsional transitions in this T_d molecule are inactive in infrared and Raman spectroscopy. They have been observed directly in liquid and solid phases by small κ neutron scattering.[30] Durig and co-workers have also observed weak but fairly sharp peaks in $C(CH_3)_4$ and $C(CD_3)_4$ in the low-temperature solid phase, which they assign to these torsional modes.[26] The neutron and infrared spectra are compared in Figure 7. The neutron spectra again have an instrumental resolution of about 10 cm^{-1}, and the liquid result is clearly broadened by "recoil" broadening. Again the torsional mode (0–1) transitions are clearly assigned by their absolute and relative intensities to give the following frequencies for the A_2 and F_2 (triply degenerate) modes: liquid, 210 and 266 cm^{-1} (± 4 cm^{-1}); solid, 220 and 280 cm^{-1}. These values show the often observed increase in frequency from the fluid to solid phases. The liquid phase neutron results must be considered the best estimates available for the free molecule since the available frequencies from infrared combination bands[21] for the gas phase (230 and 282 cm^{-1}) are *higher* than the neutron and infrared solid-phase frequencies in Figure 7 and, thus, must be suspect. The neutron frequencies were used as described above[4] to derive a V_3 parameter of 21.1 ± 0.7 kJ/mol (kinetic coupling effects are again small).

The agreement of the infrared and neutron solid-phase frequencies is excellent. Note that the neutron peak widths are significantly greater, which is due to both poorer resolution and possibly some dispersion of the torsional modes in the crystal lattice that would not show up in the infrared spectrum (κ=0). It should also be noted that Durig *et al.*,[26] using essentially identical frequencies to those from the neutron measurements on the solid phase, derived a V_3 = 18.0 kJ/mol in the harmonic approximation, as compared with V_3 = 23.6 kJ/mol obtained by Grant *et al.* (unpublished data).[30] This again illustrates the care with which the general reader must interpret and utilize such barrier values.

CONCLUDING REMARKS

The above illustrations and discussion lead us to several general conclusions concerning the use of neutron spectroscopy in the study of torsional vibrations (and other large-amplitude modes) in molecular systems. First, the neutron technique–since it involves the interaction of neutrons with vibrating nuclei and is especially sensitive to large amplitude motions–can for appropriate molecules be an ideal complement for optical spectroscopy. Neutron spectroscopy, however, is hampered somewhat by the available instrumental resolution (~10 cm^{-1}) and by the inherent "recoil" resolution broadening in fluid-phase spectra. In addition, present accessibility of instrumentation for the neutron method (for low κ molecular spectroscopy) is limited. For example, there are only a few reactors in the United States where appropriate instruments and intensity exist for such measurements (neutron sources and instrumentation amenable to the study of crystal and liquid structure and interatomic and intermolecular dynamics are more accessible). These factors make it imperative that studies of molecular systems be chosen with some care.

It should be noted again that for studies in solid phases, the neutron method, though often providing intense, readily interpretable peaks for large-amplitude modes, may be rivaled by careful, higher resolution infrared and Raman measurements. Moreover, assignment of intramolecular modes in crystals below ~150 cm^{-1} is often difficult for both techniques due to the presence of intermolecular lattice modes. There is a great deal of important work to be done, however, by a careful application of neutron spectroscopy. As one example, the torsional vibrations in CH_3CCl_3 can and should be redetermined much more accurately in both gas and liquid phases by small κ scattering. Attempts should also be made to observe 1 → 2 torsional transitions so that more definitive information on top–top coupling can be obtained. The possibility of testing assignments and force models by direct calculation[2,7] also provides an enticing, if unproven, prospect. Finally, in pursuing and interpreting neutron molecular spectroscopy, it would be well if the approach lay somewhere between the occasional euphoria of the neutron experimentalists and the sometimes ill-informed skepticism of the optical spectroscopists.

The author is particularly indebted to R. C. Livingston and D. R. Lide, Jr., for helpful discussions during the preparation of this chapter, and to D. M. Grant, R. M. Brugger, J. R. Durig, B. Hudson, and their co-workers for providing information on unpublished results.

REFERENCES

1. For a recent survey of neutron scattering research in molecular and complex ionic systems, see *Neutron Inelastic Scattering,* Proceedings, Fifth Symposium (Grenoble), International Atomic Energy Agency, Vienna (1972).
2. J. W. White, in *Neutron Inelastic Scattering,* Proceedings, Fifth Symposium (Grenoble), International Atomic Energy Agency, Vienna (1972), p. 315. This paper reviews some recent neutron studies on "molecular" vibrations.
3. For a detailed review of the theory of neutron scattering, see *Thermal Neutron Scattering,* P. A. Egelstaff, ed., Academic Press, New York (1965).
4. D. M. Grant *et al., J. Chem. Phys.,* **52**, 4424 (1970).
5. R. M. Brugger, K. A. Strong, and D. M. Grant, in *Neutron Inelastic Scattering,* Proceedings, Fourth Symposium (Copenhagen), International Atomic Energy Agency, Vienna (1968), Vol. II, p. 323.
6. P. A. Reynolds and J. W. White, *Discuss. Faraday Soc.,* **48**, 131 (1969).
7. B. Hudson, Ph.D. thesis, Harvard University, Cambridge, Mass. (1972).
8. J. C. Evans and G. Lo, *J. Phys. Chem.,* **70**, 11 (1966).
9. G. C. Stirling, C. J. Ludman, and T. C. Waddington, *J. Chem. Phys.,* **52**, 2730 (1970).
10. J. W. Nibler and G. C. Pimentel, *J. Chem. Phys.,* **47**, 710 (1967).
11. T. R. Rubin, B. H. Levedahl, and D. M. Yost, *J. Am. Chem. Soc.,* **66**, 279 (1944).
12. K. S. Pitzer and J. L. Hollenberg, *J. Am. Chem. Soc.,* **75**, 2219 (1953).
13. J. J. Rush, *J. Chem. Phys.,* **46**, 2285 (1967).
14. P. N. Brier, J. S. Higgins, and R. H. Bradley, *Mol. Phys.,* **21**, 721 (1971).
15. D. R. Herschbach, *J. Chem. Phys.,* **31**, 91 (1959).
16. W. G. Fateley and F. A. Miller, *Spectrochim. Acta,* **17**, 857 (1961); **18**, 977 (1962); **19**, 611 (1963).
17. R. A. Scott and H. A. Sheraga, *J. Chem. Phys.,* **42**, 2209 (1965).
18. J. J. Rush, *J. Chem. Phys.,* **47**, 3936 (1967).
19. J. R. Durig *et al., J. Chem. Phys.,* **54**, 479 (1971).
20. K. D. Moller *et al., J. Chem. Phys.,* **47**, 2609 (1967).
21. S. Weiss and G. E. Leroi, *Spectrochim. Acta,* **25A**, 1759 (1969).
22. E. Hirota, C. Matsamura, and Y. Morino, *Bull. Chem. Soc. Japan,* **40**, 1124 (1967).

23. D. R. Lide, *J. Chem. Phys.*, **33**, 1514 (1960).
24. J. N. Gayles and W. T. King, *Spectrochim. Acta*, **21**, 543 (1965).
25. S. Lifson and A. Warshel, *J. Chem. Phys.*, **49**, 5116 (1968); **53**, 582 (1970).
26. J. R. Durig, S. M. Craven, and J. Bragin, *J. Chem. Phys.*, **52**, 2046 (1970).
27. D. R. Lide and D. E. Mann, *J. Chem. Phys.*, **28**, 572 (1958).
28. R. C. Livingston *et al.*, *J. Chem. Phys.*, **58**, 1438 (1973).
29. K. S. Pitzer and D. W. Scott, *J. Am. Chem. Soc.*, **65**, 803 (1943).
30. D. M. Grant, K. A. Strong, and R. M. Brugger, *Phys. Rev. Lett.*, **20**, 983 (1968), and unpublished solid-phase results.

Walter J. Lafferty

Determination of Potential Functions and Barriers to Planarity for the Ring-Puckering Vibrations of Four-Membered Ring Molecules

The vibrational potential function of the ring-puckering of four-membered ring molecules is the result of a delicate balance between ring strain forces, which are minimized when the ring is planar, and repulsive torsional forces between hydrogen atoms on adjacent carbon atoms in the ring, which are at a minimum when the ring is puckered. As early as 1945 Bell[1] pointed out that the ring-puckering vibration of four-membered ring molecules should be nearly quartic in nature. The first determination of the ring-puckering potential for a four-membered-ring molecule was reported by Rathjens, Freeman, Gwinn, and Pitzer[2] for cyclobutane using thermodynamic methods.

This paper will be concerned with the more direct methods for the determination of the ring-puckering potential and barriers to planarity of four-membered ring molecules and "pseudo-four-membered" ring molecules, i.e., five-membered ring molecules such as cyclopentene containing a double bond which stiffens the ring. Space does not permit discussion of the interesting but more complex phenomenon of pseudorotation* that appears in flexible five-membered ring molecules. Many

*The field of pseudorotation has been recently reviewed by Laane,[3] and theoretical methods have been outlined by Harris and co-workers.[23]

of the comments to follow concerning experimental techniques and errors, however, can be extended to pseudorotation.

This is not meant to be a historical review. It should be mentioned, however, that ring-puckering transitions were first directly observed in the far-infrared spectrum of trimethylene oxide (TMO) by Lord and his co-workers at MIT.[4] Microwave studies on TMO were concurrently carried out by Gwinn and co-workers at Berkeley.[5-8] Theoretical methods were principally developed there, particularly by Chan.[6-8] An excellent review of the field has been written by Blackwell and Lord.[9]

THEORY AND MODEL

Figure 1 shows the potential curve for the ring-puckering vibration of silacyclobutane deduced by Laane and Lord[10] from the far-infrared spectrum. The reduced coordinate Z is proportional to half the perpendicular distance between the two ring diagonals. At $Z = 0$ the ring is planar. There is a sizable barrier to planarity in silacyclobutane, and it is obvious that the ring is nonplanar in the ground state, i.e., that the expectation

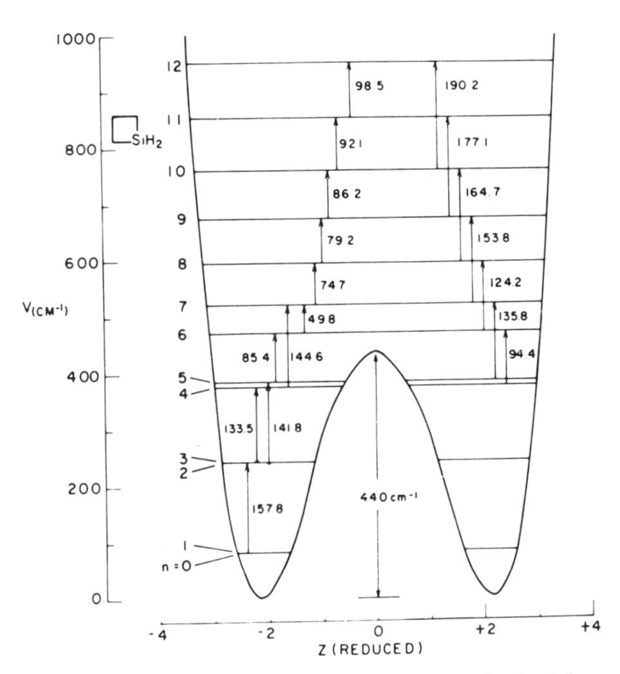

FIGURE 1 Potential function and vibrational energy levels of the ring-puckering mode of silacyclobutane. (Reproduced by permission from Laurie and Lord.[10])

value of Z in the ground state is near the potential minimum at $Z = 2.2$. The goal in studies of such molecules is to obtain a value for the barrier, which is defined as the energy difference between the bottom of the well and the maximum at $Z = 0$. In addition, if the dimensionless coordinate Z can be transformed to a real coordinate, the dihedral angle of the equilibrium ring conformation can be obtained.

The numbers labeled n along the left hand of the curve in Figure 1 are the quantum numbers of the vibrational energy levels indicated on the figure. As is typical for all inversion problems, the lower levels below the barrier occur as nearly degenerate pairs. The energy separations between the $n = 0$ and $n = 1$ levels and the $n = 2$ and $n = 3$ levels have been determined to be 75.75 and 7793 MHz, respectively.[11] As n increases, this splitting increases rapidly until above the barrier the energy separation is roughly that expected for a quartic oscillator. The observed infrared frequencies are also included in the figure.

In their study of TMO, Chan and co-workers[6–8] investigated a number of potential functions for the ring mode. The one that best fit both microwave and infrared data also had the advantage of extreme simplicity:

$$V = ax^4 + bx^2 . \tag{1}$$

Here V is the potential energy, a and b are constants, and x is the coordinate for the puckering vibration, which is usually defined as half the perpendicular distance between the ring bisectors. If b is negative, the potential function will have a central barrier.

The vibrational energy of the puckering mode may be calculated from the one-dimensional wave equation

$$-(\hbar^2/2\mu)(d^2\Psi/dx^2) + V\Psi = E\Psi, \tag{2}$$

where μ is the reduced mass of the vibration. The validity of this model depends on two assumptions: (1) Since ring-puckering modes fall at quite low frequencies in four-membered ring molecules, it is assumed that this vibration can be treated as an isolated mode. Mixing between the puckering mode and other vibrations, in particular, CH_2-rocking modes, will result in a breakdown of the model and introduce additional terms in both the kinetic energy and potential energy parts of the Hamiltonian. "Pseudo-four-membered" ring molecules have two out-of-plane ring modes. The ring-twisting vibration has generally a much higher frequency than the ring-puckering mode, but the possibility of some mixing exists. (2) The assumption is generally made that the reduced mass μ is constant, i.e., not a function of x. This assumption is not strictly valid. For four-membered ring molecules the variation of the reduced mass with puckering is small; for "pseudo-four-membered" ring molecules, the variation is much larger (Figure 2).

The following expressions can be used to transform to a dimensionless coordinate[12]:

$$Z = (2\mu/\hbar^2)^{1/6}\, a^{1/6}\, x$$

$$B = (2\mu/\hbar^2)^{2/3}\, a^{-1/3}\, b$$

$$A = (\hbar^2/2\mu)^{2/3}\, a^{1/3}, \tag{3}$$

where a and b are the potential constants of Eq. (1). This transformation gives

$$V = A(Z^4 + BZ^2) \tag{4}$$

and

$$H = A\,(p^2 + Z^4 + BZ^2). \tag{5}$$

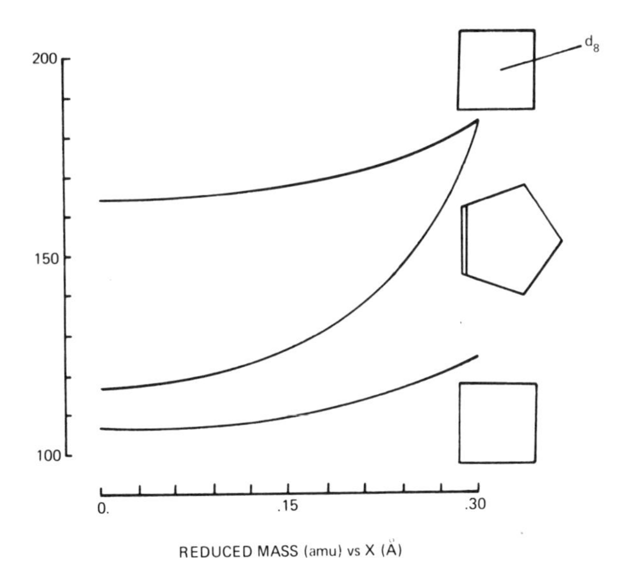

FIGURE 2 Plot of reduced mass of cyclobutane, cyclobutane-d_8, and cyclopentene as a function of the puckering coordinate. (From unpublished calculations of T. B. Malloy and W. J. Lafferty.)

The part of Eq. (5) within the parentheses gives the energy levels in reduced form; the constant A, conventionally expressed in units of cm^{-1}, is a scaling factor. Matrix elements for these calculations are readily available.[13]

This transformation has the advantage in that no knowledge is required of μ, although the constants A and B are mass-dependent. Note that the barrier to planarity is mass-independent since

$$V_{max} = \frac{AB^2}{4} = \frac{1}{4}\left(\frac{b^2}{a}\right). \tag{6}$$

In the case of molecules with no symmetry about the ring plane, such as monosubstituted cyclobutane derivatives, odd terms also appear in the potential function. Thus it is necessary to add at least one more term to Eq. (1), e.g.,

$$V = ax^4 + bx^2 + cx^3. \tag{7}$$

A linear term is excluded with the assumption that (dV/dx) is 0 or nearly 0 when $x = 0$.[14]

Although the potential function as defined in Eq. (4) is far from harmonic, it is possible to obtain quite accurate eigenvalues and eigenvectors, provided a sufficient number of harmonic oscillator basis sets are taken. Using 100×100 harmonic oscillator basis sets, Chan and Stelman[15] have calculated the first 20 energy levels and wave functions for a pure quartic oscillator and tabulated the matrix elements of x, x^2, x^3, x^4, and x^6 in the quartic oscillator representation. Reid[16] has noted slight computational errors in their values and calculated corrected values for some of the matrix elements. Since ring-puckering modes are more nearly quartic than harmonic in nature, Chan and Stelman[15] suggested that the use of a quartic oscillator representation leads to more rapid convergence. Computers are now both larger and much swifter, and it would seem more feasible now to choose the harmonic oscillator basis set approach, since these matrix elements can be easily generated exactly.

Computations for molecules for which the potential function is symmetric are facilitated by the fact that the Hamiltonian can be factored into two matrices—one containing even values of the vibrational quantum number n, the other odd values. Experience has shown that in this case basis sets of 30×30 to 35×35 for each matrix will be sufficient to obtain accurate energy levels up to at least $n = 15$. The accuracy of the eigenvalues, of course, should be checked by comparison of results obtained from basis sets of different sizes. Reid[16] outlines other methods that can be made to check the validity of the results. No such factoring is possible for unsymmetric potential functions. Double precision arithmetic should be used in the diagonalization process, especially if small splittings are to be calculated.

DETERMINING POTENTIAL FUNCTION FROM ENERGY LEVELS

Methods of obtaining energy levels via various techniques are given below in the section "Experimental Techniques." This section deals with methods and model limitations in the determination of potential constants and geometric parameters from the observed energy levels and other spectroscopic data.

Initially, far-infrared data were fit by trial and error methods. Since ratios of successive energy level spacings are independent of the potential constant A in Eq. (4), values of B can first be obtained that best fit the observed ratios. A value of the scaling factor A is then calculated for best fit of the observed frequencies. Laane[12] has tabulated the eigenvalues for the potential $V = Z^4 + BZ^2$ for 50 values of B in the range $-50 \leqslant B \leqslant 100$. These tables are very convenient for the determination of approximate potential constants.

A far better least squares approach has been developed in more recent years as computer storage capacity and speeds have improved. Ueda and Shimanouchi[13] have written a very sophisticated program for the fitting of observed energy levels using Jacoban techniques. This program also calculates transition probabilities. Most workers in the field use the Ueda–Shimanouchi program or variations of it.

There is always the possibility in a method of this nature of wandering into a false solution, and this possibility should be tested by using a number of differing initial estimates of A and B. In the case of symmetric molecules with only even terms in the potential function, bad estimates of the A and B constants merely lead to slow convergence. Unsymmetric molecules with odd terms in the potential are another matter, however. Often several physically feasible solutions are obtained. Cyclopentene oxide, a "pseudo-four-membered" ring molecule, is a good illustration of this point. Figure 3 gives two possible potential curves derived by Carreira and Lord[14] that fit almost equally well the observed far-infrared transitions. Obviously, in cases of this type some other method must be used to select the actual potential. In the case of cyclopentene oxide, it was possible to exclude the double minimum potential on the basis of a microwave study.[17]

Considerations of weighting are important in the application of these

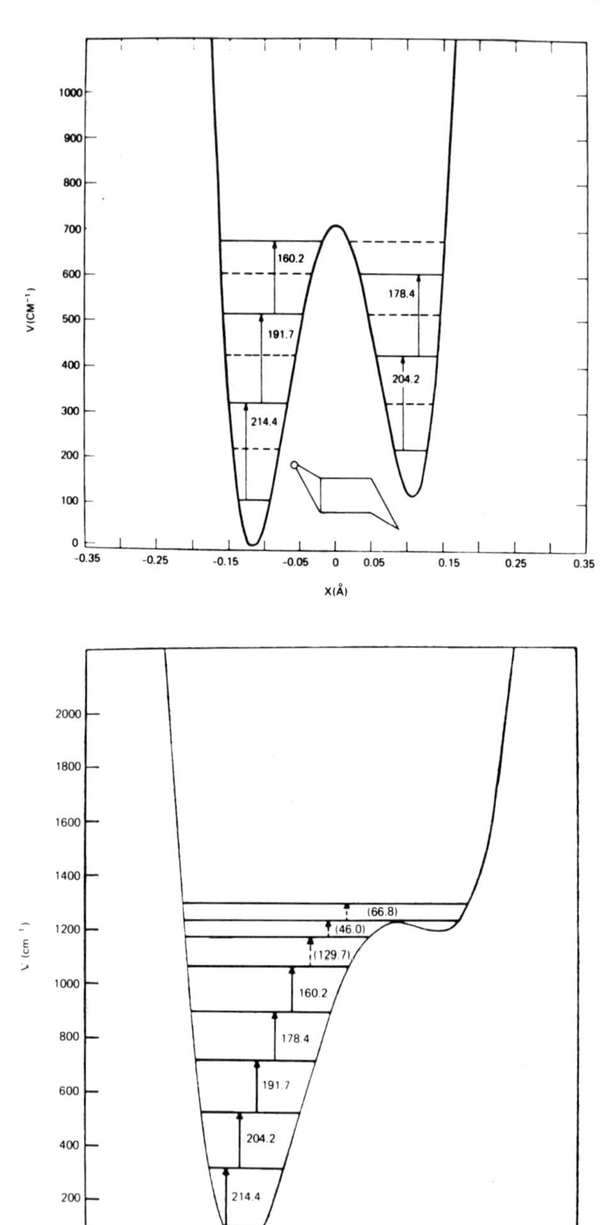

FIGURE 3 Two possible potential functions for the ring-puckering mode of cyclopentene oxide. (Reproduced by permission from Carreira and Lord.[14])

fitting techniques. Normally, frequencies are weighted proportionally to the square of the reciprocal of their estimated uncertainty. A number of molecules were refit during the preparation of this paper. In many cases it was not possible to reproduce the original author's results exactly because the weights were not specified in his paper. If weighting is used, the weights should be included in the tables in which the original data are reported.

Table 1 gives a list of barriers determined for a number of typical four-membered ring molecules and one "pseudo-four-membered" ring molecule, cyclopentene. The barrier for cyclobutane was obtained by fitting combination differences calculated from puckering vibrational transitions observed in the mid-infrared region by Stone and Mills[18] and Miller and Capwell,[19] as well as the Raman data of Miller and Capwell. Barriers from the other molecules were obtained from far-infrared data. All data were refit from the original data using a standard nonlinear least squares program in order to obtain a consistent set of values and uncertainties. In the case of cyclobutane, Raman transitions were given a weight of 1, while the combination differences were weighted $1/\sqrt{2}$. For the other molecules, all data were weighted equally.

One notes that for the four-membered ring molecules, the barrier heights are quite well determined, in most cases to better than 1 per-

TABLE 1 Barriers to Planarity for Some Four-Membered and "Pseudo-Four-Membered" Ring Molecules

Molecule	$V(\text{cm}^{-1})$
Cyclobutane	514.8 ± 4.4[a, b]
Silacyclobutane	438.2 ± 3.3[c]
Trimethylene sulfide	270.9 ± 2.2[d]
Cyclopentene	231.2 ± 9.0[e]
Trimethylene oxide	15.27 ± 0.58[f]

[a] Errors cited are three times standard deviations.
[b] Obtained from refitting the combined infrared data of Stone and Mills[18] and Miller and Capwell[19] and the Raman data of Miller and Capwell.
[c] Refit from far-infrared data of Laane and Lord[10]; levels noted as perturbed were omitted from fitting.
[d] Obtained by refitting far-infrared data of Borgers and Strauss.[33]
[e] Obtained by refitting far-infrared data of Laane and Lord.[36]
[f] Obtained by refitting far-infrared data of Wieser *et al.*[24]

cent. Residuals in the fitting are random and roughly within experimental error. The uncertainty in the value of the barrier height obtained from cyclopentene is much larger, however, and the fitting suggests systematic deviations. Transitions with low quantum numbers are higher than calculated; those with high quantum numbers are all low. This behavior is typical of "pseudo-four-membered" ring molecules and indicates that the model is not entirely adequate.

Malloy[20] has investigated a number of these molecules and has determined that the behavior stems largely from the assumption of constant reduced mass. Figure 2 shows a plot of reduced mass vs. the puckering coordinate, x, for cyclobutane, cyclobutane-d_8, and cyclopentene (see below for the details of calculations of the reduced mass). The reduced mass varies very slightly in the four-membered ring compounds, compared with the variation in the "pseudo-four-membered" ring molecule. By expanding the reciprocal of the reduced mass in a power series

$$g = g_0 + g_2 x^2 + g_4 x^4 + g_6 x^6 \cdots, \tag{8}$$

where g is $1/\mu$, and adding the terms $g_2 p x^2 p$, etc., to the Hamiltonian, much better agreement with the observed spectra was obtained. By empirically mixing CH_2 rocking motion with the puckering motion, a further improvement in the fitting was obtained. Malloy notes, however, that barrier heights obtained by this method are virtually the same as those obtained using the simpler model. Malloy also points out that the addition of a term, cpx^2p, where c is an empirical constant, results in a considerable improvement of the fitting.

Carreira, Mills, and Person[21] have considered the case of interaction between the ring-puckering and ring-twisting modes of 2,5-dihydrofuran. They calculate a sizable potential energy interaction between the two modes. Unfortunately, this molecule is planar; the effect of such mixing on barrier heights remains unknown.

Once a potential curve is obtained, the equilibrium value of the coordinate x can be determined from the first of Eq. (3), if the reduced mass μ is known. Ueda and Shimanouchi[13] have outlined a method for the calculation of μ. The reduced mass is expressed as

$$\mu = \sum_i m_i (\partial \vec{r}_i / \partial x) \cdot (\partial \vec{r}_i / \partial x), \tag{9}$$

where m_i is the mass of the ith atom and $\vec{r}_i$ is its coordinate. Values of $\partial \vec{r}_i / \partial x$ used must satisfy the following conditions

$$\sum_i m_i (\partial \vec{r}_i / \partial x) = 0 \tag{10}$$

and

$$\sum_i m_i \vec{r}_i \times (\partial \vec{r}_i / \partial x) = 0. \tag{11}$$

The calculation of μ is a laborious and error-prone task, especially for molecules lacking C_{2v} symmetry in the planar form. Most papers omit details of this calculation, and indeed, very few list the structural parameters used. In order for others to be able to reproduce the calculations, it is highly recommended that these parameters be tabulated.

There are two principal sources of error in the calculation of the reduced mass for the puckering vibration. Usually the structural parameters of these molecules are not known and reasonable estimates of the bond lengths and angles must be made. Obviously, errors in the assumed structural parameters will be reflected in the value of μ obtained. The reduced mass is particularly sensitive to the positions of the hydrogen atoms since they swing on a long arm and $\partial \vec{r}_i / \partial x$ is large. For example, an error in the estimate of ∠HCH in cyclobutane of ±2° will produce an error in μ of 0.5 u, about 0.5 percent.

A more important error arises from the fact that the form of the motion of the molecule during the vibration must be assumed since the exact form is usually unknown. For example, in a four-membered-ring molecule, does the molecule bend about the 1–3 ring diagonal with no change in the 143 and 123 ring angles, does it bend about the other diagonal, or is the actual motion a combination of both? Information can be derived on the puckering motion from the vibrational dependence of the rotational constants[6,22]; in most cases, however, this information is lacking and, normally, the form of the vibration must be assumed. Once the form of the vibration is chosen, best results are obtained by using a coordinate system in which all atoms move along curvilinear paths such that all bond lengths are kept constant. In his investigation of mass effects in "pseudo-four-membered" ring molecules, Malloy[20] has outlined methods for the evaluation of μ for these molecules. He has noted that, if even small amounts of CH_2- or SiH_2-rocking motion are mixed with the puckering motion, the vibrational reduced mass will be altered significantly.

Once μ–and thus a value for the puckering coordinate at the minimum of the well–is obtained, it is possible to extract some structural information about the molecule in the ground state, such as the ring dihedral angle. A number of papers have appeared in which dihedral angles are reported where neither geometric information nor the details of the reduced mass calculation are given. In some cases the reduced mass has been assumed by comparison with μ's calculated for other molecules. In many cases ring dihedral angles are reported with estimated errors of ±1°. In view of the numerous assumptions made in the calculation of μ, especially the unknown amount of mixing with CH_2- or SiH_2-rocking motion, it is this writer's feeling that the quoted errors should be regarded with great suspicion and multiplied at least fivefold.

One means of checking the reliability of a model is by studying isotopic species. If a potential function of the form given in Eq. (4) is used, ratios of the potential constants can be calculated to check the isotopic dependencies, which do not depend on the values taken for the reduced mass. Laane and Lord[10] have shown that

$$\left(\frac{\mu_D}{\mu_H}\right)^{2/3} = \frac{A_H}{A_D} = \left(\frac{B_D}{B_H}\right)^2, \tag{12}$$

where the subscript H refers to the normal and D to the isotopic species. Unfortunately, few isotopic molecules have been studied.

In the case of cyclobutane and cyclobutane-d_8, the ratio of the A potential constants is calculated to be 1.4063 ± 0.004, while the square of the ratio of the B potential constants is 1.367_5 ± 0.0080 (T. B. Malloy and W. J. Lafferty, unpublished calculations). There is a small but significant discrepancy. The far-infrared and Raman spectra of five deuterated species of trimethylene oxide (TMO) have been reported in an important series of papers by Wieser and his co-workers.[24-26] In the case of this molecule, which has a low barrier of about 15 cm^{-1}, the isotopic relations are found to fail totally.

Although the reasons for the breakdown of the isotopic relations are not entirely clear at this time, several can be suggested. The model is based on the assumption that the barrier is isotopically invariant. This assumption appears to be only approximately correct. The barrier in cyclobutane goes from 515 to 500 cm^{-1} upon deuteration. For TMO the percentage change is much larger; the barrier height shifts from 15.3 cm^{-1} for the normal compound to 11.2 cm^{-1} in the perdeuterated molecule. The relation given in Eq. (12) should not be affected by mixing with other modes, provided the degree of mixing is the same for both isotopic species. Since the degree of mixing is likely to be affected to some extent by isotopic substitution, this may also explain part of the discrepancies.

EXPERIMENTAL TECHNIQUES

Microwave, far-infrared, Raman, and conventional infrared techniques have been used in barrier height determinations. Each technique has its own advantages and limitations and will be discussed separately.

Microwave Spectroscopy

The conventional microwave spectrum extends from 8 to 40 GHz. Normally, only pure rotational spectra are observed in this region. A number of ring molecules with higher barriers, and therefore nearly degenerate lower energy levels, have possible vibrational transitions in this region; however, the transition moment for such vibrational transitions is quite small, and they have only been observed in a few instances where perturbation mixing between the two levels increases the line intensity.[22] The ring vibration modes, however, fall at quite low frequencies so that many vibrational states are populated at room temperatures. Because of this, each ground-state rotational transition will be accompanied by a number of vibrational satellite lines.

The following methods can be used to gain information about barrier heights:

1. *Relative Intensity Measurements* Once the ground state and vibrational satellite lines have been assigned (usually on the basis of observed Stark effect patterns), it is possible to estimate the energy difference between a given vibrational state and the ground state by measuring the intensity of an excited state line relative to a ground-state line. If lines involving the same rotational transition in both states are compared, the ratio of the line intensities will lead directly to the vibrational energy difference since the population of molecules in the excited states is governed by Boltzmann statistics. The dipole moment does change with vibrational state slightly, and for very accurate determinations, this vibrational dependence should be taken into account.

Intensity measurements are not a *forte* of microwave spectroscopy. Typically, cited errors range from 3 to 8 percent, provided the intensity ratios of several rotational transitions have been measured. Esbitt and Wilson[27] have discussed the problems of relative intensity measurements in some detail. Harrington[28] has described a system using a microwave bridge in which errors of about 1 percent are claimed, and this system has been incorporated into a commercial spectrometer. Errors of ±3 percent in intensity ratio measurement lead to an uncertainty of ±6 cm^{-1} for a given vibrational level at room temperature.

Other methods, such as the direct observation of far-infrared transitions, are preferable, but this method permits a quick check on the assignment of infrared or Raman spectra; if such spectra have not been obtained or are too weak to observe, it permits a moderately accurate estimate of the energy of the lower vibrational levels.

2. *Vibrational Dependence of Rotational Constants* If a barrier is pres-

ent in a molecule, a plot of the rotational constants against the puckering vibrational quantum number will not give a monotonically varying curve as in the case of a pure quartic oscillator, but rather an irregular "zigzag" plot will be produced. Such a plot taken from a paper by Scharpen and Laurie[29] for the A rotational constant of methylenecyclobutane is seen in Figure 4. In this case there is a Coriolis-type resonance between the states with $n = 0$ and $n = 1$ and the squares in the plot are the "effective" observed constants, while the dots for these levels represent the values of A_v corrected for the resonance.

The rotational constant dependence can be written[6,22]

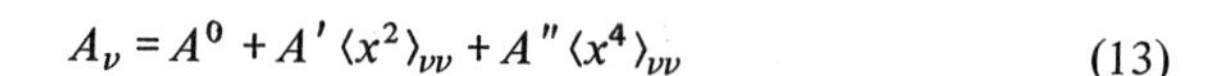

$$A_v = A^0 + A' \langle x^2 \rangle_{vv} + A'' \langle x^4 \rangle_{vv} \tag{13}$$

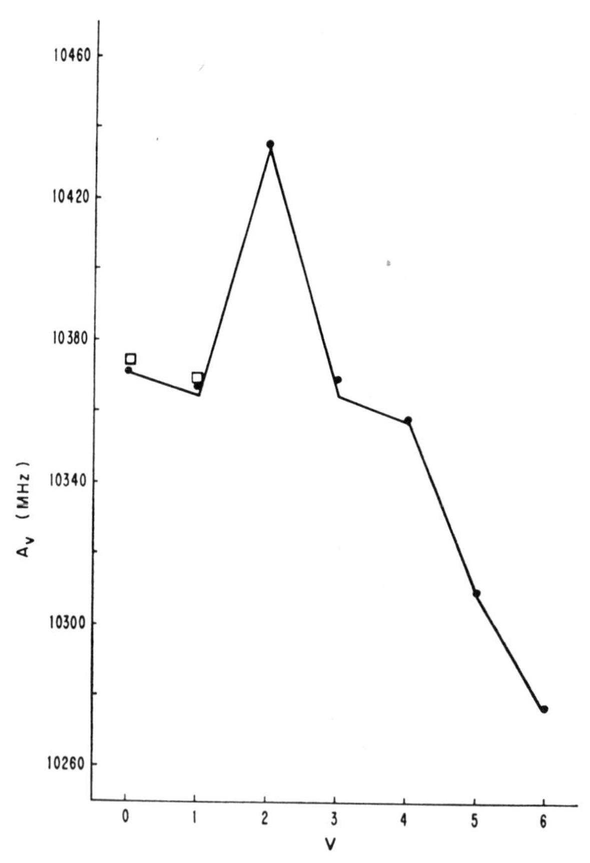

FIGURE 4 Plot of the A rotational constant of methylene cyclobutane against puckering vibration quantum number. (Reproduced by permission from Scharpen and Laurie.[29])

if the assumption is made that interaction with other vibrations can be neglected. Here A_v is a rotational constant, A^0, A', and A'' are usually treated as empirical constants, and x is the ring-puckering coordinate. Similar expressions apply to the remaining two rotational constants. The expectation values $\langle x^2 \rangle_{vv}$ and $\langle x^4 \rangle_{vv}$ depend only on the value of B in the potential, i.e., only on the reduced energy levels. A value of the potential constant B, which minimizes the rms deviation between observed and calculated rotational constants, is found by iteration.

It is difficult to assess the errors in the potential constant B determined by this technique. Microwave frequency measurement accuracy is quite high, and rotational constants, which typically range from 10 to 3 GHz for these molecules, can be obtained from the rotational spectrum with uncertainties less than ±0.3 MHz except in special cases. The rotational constants can be reproduced by Eq. (13) typically to within ±3 MHz or better. Errors in this case are almost completely determined by model failure, in particular the assumption that the puckering mode is a completely isolated vibration. Table 2 gives a comparison of B potential constants obtained by optical data only with those obtained from microwave studies using this method. The agreement between the two methods is quite good, the largest discrepancy being about 2 percent.

Once the potential constant B is determined, some other technique is necessary to determine the scaling factor A and thus the barrier height. In high-barrier cases quite often the splitting of a pair of nearly degenerate levels can be determined by treatment of the Coriolis resonances between the levels (see below). These splittings may be used to scale the energy levels; however, even though they are very well determined, this

TABLE 2 Comparison of B Potential Constants Obtained from Optical and Microwave Data[a]

	B_{Op}	B_{MW}
Cyclopentene	-6.16 ± 0.14[b]	-6.18 ± 0.20[c]
Trimethylene selenide	-8.28[d]	-8.36[e]
Methylene cyclobutane	-5.90[f]	-5.76 ± 0.20[g]

[a] Uncertainties where given are the authors' estimates.
[b] Data from Laane and Lord.[36]
[c] Data from Scharpen.[30]
[d] Data from Harvey *et al.*[34]
[e] Data from Petit *et al.*[31]
[f] Data from Durig *et al.*[37]
[g] Data from Scharpen and Laurie.[29]

procedure may lead to large errors in the barrier determination.[30] A more accurate barrier estimate can be made in such cases by scaling to the larger frequency separation between the $n = 1$ and $n = 2$ levels by using intensity measurements or by using directly observed infrared or Raman spectra. Table 3 gives a comparison of barriers determined in this manner with those obtained from optical data. In general, the agreement is quite good. Note, however, the large uncertainty resulting from scaling to the $0 \rightarrow 1$ vibrational spacing in the cases where this quantity was used.

3. *Determining Small Vibrational Splittings* For molecules with C_{2v} or lower symmetry when the molecule is planar the possibility exists of Coriolis resonance between states with even and odd values of the puckering quantum number. When the barrier is low, the energy separation is large, and the effects of this resonance are very small. For intermediate barriers, however, the lower levels merge into nearly degenerate pairs with even and odd quantum numbers as shown in Figure 1. When this spacing is in the order of 0.2–2 cm^{-1}, the possibility exists of accidental coincidences between rotational energy levels of appropriate symmetry. This produces large resonance shifts and leads to what is known as "nonrigid" behavior, where the rotational lines of the two interacting states are shifted in a nonuniform manner so that the resulting spectrum cannot be fit to the usual asymmetric rotor expressions. For small splittings, the matrix element connecting the even and odd

TABLE 3 Comparison of Barriers Derived from Microwave and Optical Data

	$V_{Op}(cm^{-1})$	$V_{MW}(cm^{-1})$
Trimethylene sulfide	270.9 ± 2.2[a]	274.2 ± 2[b]
Cyclopentene	231.2 ± 9.0[a]	236 ± 55[c] 232 ± 5[d]
Trimethylene selenide	378.1 ± 4[e]	383.1 ± 4[f]
Methylene cyclobutane	141 ± 5[g]	160 ± 40[h]

[a] For references see Table 1.
[b] From Harris *et al.*[22]; scaled to 2-1 separation (intensity measurements).
[c] From Scharpen[30]; scaled to 1-0 separation.
[d] From Scharpen[30]; scaled to 2-1 separation obtained from far IR data.
[e] From Harvey *et al.*[34]
[f] From Petit *et al.*[31]; scaled to far-infrared data.
[g] From Durig *et al.*[37]
[h] From Scharpen and Laurie[29]; scaled to 1-0 separation.

states vanishes and rigid rotor spectra are again observed. This is the case observed for the trimethylene selenide molecule,[31] where the separation between the $n = 0$ and $n = 1$ vibrational levels is quite small (0.0172 cm^{-1}), and normal rotational spectra are observed for the lines of these states. The $n = 2$ and $n = 3$ states, however, are separated by 1.14 cm^{-1}, and rotational lines of these states are perturbed.

Analysis of "nonrigid" rotors is by no means routine. The amount of perturbation depends on the size of the matrix element connecting the states and the frequency separation between the states, which is a function of the rotational constants and vibrational potential function. The Stark effect is very useful for verification of assignments. The resonance may be treated by using the following Hamiltonian

$$H = A_v P_a^2 + B_v P_b^2 + C_v P_c^2 + H_v + 2F p_x P_b \tag{14}$$

where A_v, B_v, and C_v are the rotational constants; P_a, P_b, and P_c are the angular momenta about the various molecular axes; H_v is the vibrational Hamiltonian; F is a constant; and p_x is the vibrational angular momentum operator. The last term in Eq. (14) is the perturbation coupling term. The assumption is usually made that the zero point effects of the remaining vibrational modes are absorbed in the coefficients of the Hamiltonian, so that there is no coupling of these modes with the puckering mode. Eq. (14) is appropriate for molecules where the coupling occurs through the angular momentum about the b principal axis.

An iterative procedure is usually used to solve for the vibrational energy separation. It is sometimes difficult to evaluate the uncertainty in the energy difference obtained because of the assumptions required. In the case of trimethylene sulfide, Harris *et al.*[22] were able to locate direct perturbation-induced transitions between the resonating $n = 0$ and $n = 1$ states. The value of the energy separation obtained agreed to within a few megahertz of that obtained using Eq. (14). A number of small vibrational splittings have been determined in this manner.[11, 22, 29, 31, 32]

In one case for silacylobutane, Pringle[11] was able to determine vibrational splitting between two pairs of levels. A very small splitting between the $n = 0$ and $n = 1$ levels was rather precisely determined from the analysis of Stark effect perturbations, while a much larger splitting between the $n = 2$ and $n = 3$ levels was determined by the methods given above. The splittings obtained, 75.75 ± 0.03 MHz for the $0 \rightarrow 1$ spacing and 7793 ± 7 MHz for the $2 \rightarrow 3$ spacing, agree quite well with the values 73.1 ± 4.5 and 8140 ± 447 predicted from the far-infrared spectrum.[10] One might think that two very accurately determined vibrational splittings should lead to a rather good barrier from the two parameter potential function given in Eq. (14). Unfortunately, these vibrational splittings appear to be somewhat sensitive to the barrier shape. Using the microwave data alone, Pringle obtained a barrier height of 227 cm^{-1} but was unable to account for the observed infrared spectrum. Only by adding an additional term $C \exp(-DZ)$ to the potential function was it possible to reconcile the two sets of data.

Far-Infrared and Raman Spectroscopy

While microwave spectroscopy must rely on rather indirect means to obtain barrier height information, transitions between energy levels of different n can be directly observed by infrared and Raman studies. Resolution is much poorer in these techniques; thus while microwave studies are limited primarily by model errors, the limit to optical studies is to a large extent due to experimental uncertainties.

Far-infrared spectroscopy has a large number of limitations. Intensity of far-infrared sources is low, and there is great difficulty in isolating the wanted radiation from very much more intense higher order radiation. The fact that the entire far-infrared region is covered by the intense water vapor pure rotational spectrum requires that instruments must be flushed with dry air or more preferably evacuated. The puckering vibration in most molecules is quite weak and very long absorption path lengths are required. Initial studies in this region were greatly hampered by these experimental difficulties. In fact, a controversy concerning discrepancies between the potential function of the trimethylene oxide molecule determined from infrared and microwave studies was not resolved until it was discovered that the rather weak $n = 0 \rightarrow 1$ transition had been missed in the original far-infrared work[4] because of an overlapping water vapor line.[8]

Instrumentation has greatly improved over the years, however, with the advent of larger instruments, improved detectors, scanning interferometers, and either evacuated or flushed double-beamed instruments. Spectra in this region can now be reliably and almost routinely obtained. Figure 5 shows a scan of the far-infrared spectrum of TMO-d_6 taken by Wieser *et al.*[24] This molecule is almost a pure quartic oscillator with a very small barrier. Fortunately, the $n \rightarrow n + 1$ transitions of the puckering vibration are C-type bands, which have very sharp, strong unresolved Q branches and much weaker P and R branches. The series of "lines" observed are, in fact, the Q branches of such C-type bands and what is observed is a "hot band" progression; the weak fundamental vibration ($n = 0 \rightarrow 1$) falls at about 41.2 cm^{-1}, and the stronger upper stage transi-

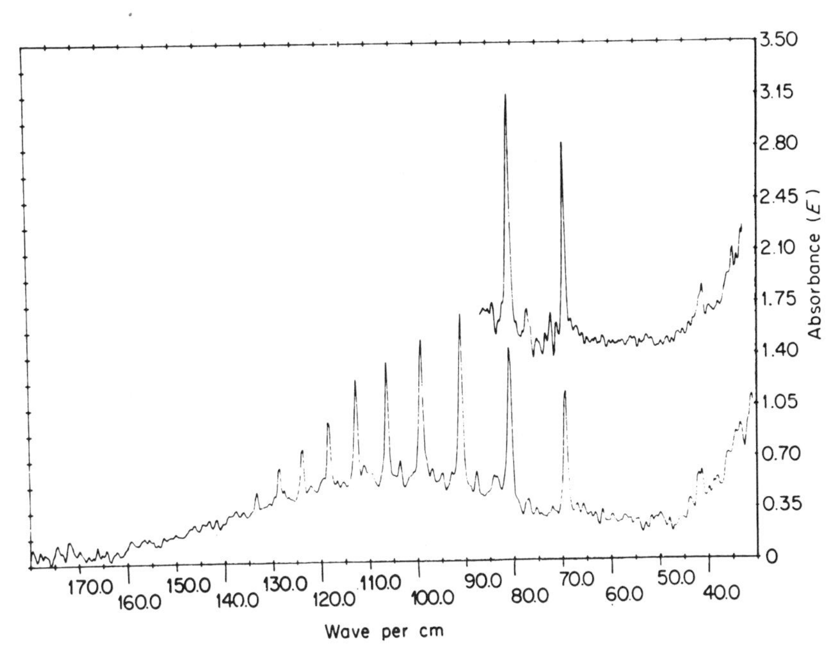

FIGURE 5 Far-infrared spectrum of trimethylene oxide-d_6. (Reproduced by permission from Wieser *et al.*[24])

tions, i.e., $n = 1 \rightarrow 2$, $n = 2 \rightarrow 3$, are observed at higher frequencies. While the signal-to-noise is moderately low here, the Q-branch peaks can be readily identified and measured to within ± 0.1 cm^{-1}. This series can be fit using the simple potential function of Eq. (4) with a standard deviation of ± 0.14 cm^{-1}. In this case the model is quite adequate.

Figure 6 shows the more complicated spectrum of the ring-puckering vibration of silacyclobutane as obtained by Laane and Lord.[10] This molecule has a higher barrier (438 cm^{-1}), and the spectrum shows no apparent regularities. The strongest peak at 157.8 cm^{-1} consists of the overlapped Q branches of the $n = 1 \rightarrow 2$ and $n = 0 \rightarrow 3$ transitions and the peaks at 133.5 cm^{-1} and 141.8 cm^{-1} are the Q branches of the $n = 3 \rightarrow 4$ and $n = 2 \rightarrow 5$ transitions. The myriad of remaining Q-branch peaks are due to higher quantum number transitions.

Since the development of the ionized argon laser as a Raman excitation source, the study of gas-phase Raman spectra has advanced rapidly, and many ring molecules have been found to have surprisingly intense Raman spectra. Figure 7 shows the Raman spectrum of TMO obtained by Kiefer *et al.*[26] Again, an unresolved Q-branch series is obtained; here, however, the selection rules are $\Delta n = +2$. The 130.7 cm^{-1} peak is the

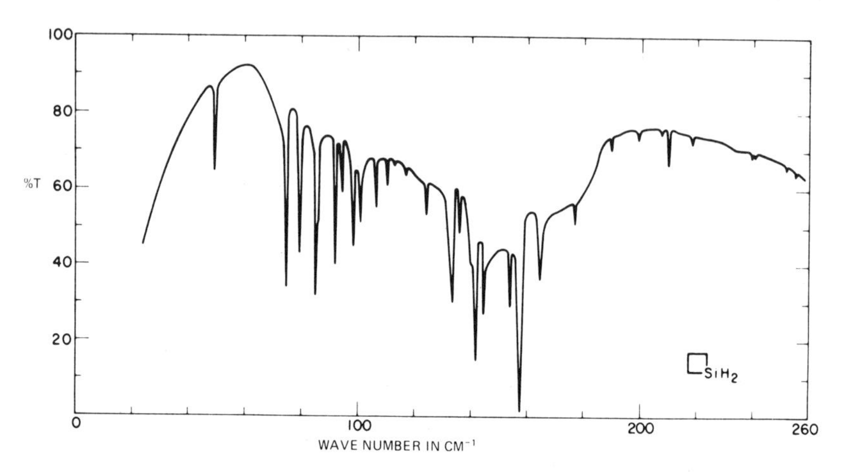

FIGURE 6 Far-infrared spectrum of the ring-puckering vibration of silacyclobutane. (Reproduced by permission from Laurie and Lord.[10])

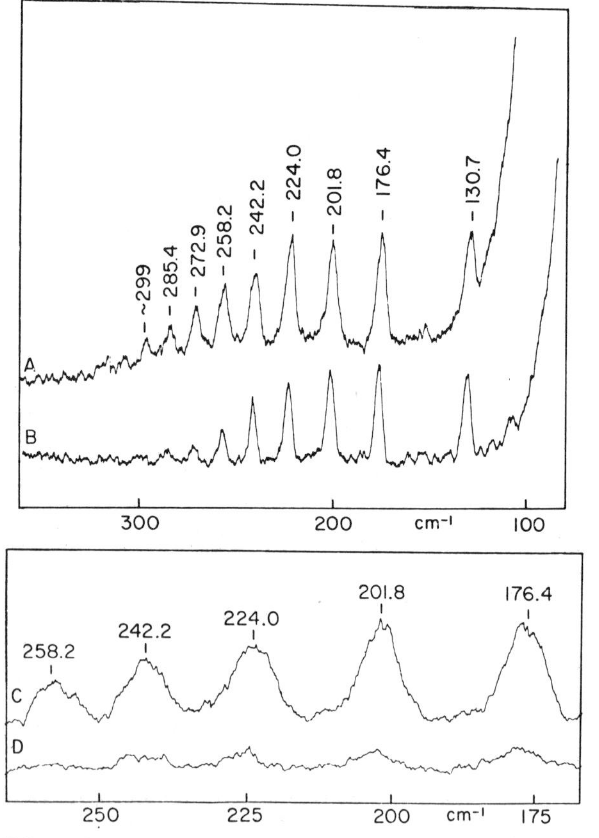

FIGURE 7 Vapor phase Raman spectrum of trimethylene oxide. (Reproduced by permission from Kiefer *et al.*[26])

$n = 0 \rightarrow 2$ transition, the 176.4 cm^{-1} peak, the $n = 1 \rightarrow 3$, etc. The resolution is somewhat poorer as a rule than that available from far-infrared instruments, but Raman studies have several important advantages. First, for symmetric molecules such as cyclobutane the ring-puckering mode is infrared inactive, but can be observed in the Raman spectrum. Second, since different selection rules are observed in the Raman spectrum, this technique serves as a very useful check on the assignment of infrared spectra. Given an infrared spectrum of the complexity shown in Figure 6, it is easy to see why such a check would be useful.

One systematic error that is usually ignored in both infrared and Raman studies should be pointed out here. Normally, the frequency of the peak of an observed Q branch is taken as the vibrational frequency. The Q branch of a C-type asymmetric rotor band is not perfectly sharp; moreover, since the observed infrared transitions take place between different vibrational levels, the rotational constants of the upper and lower states in the transition will differ. This results in an asymmetrically shaped Q branch where the maximum of the Q branch will not usually fall at the band origin. Borgers and Strauss[33] have calculated Q branch shapes in the trimethylene sulfide infrared spectrum and obtained shifts of the maximum from the origin as large as 1 cm^{-1}. Moreover, the lowest pairs of levels with $n = 0, 1$ and $n = 2, 3$ are close enough to undergo resonance that drastically changes the Q-branch shapes. This effect will be pronounced for transitions taking place from below the barrier to above it, since the change in the rotational constants is large in this case. Even when rotational constants have been measured, it is difficult to account quantitatively for Q-branch contours. The shift in the maximum of the Q branch from that of the true band origin will be more pronounced in Raman studies since $\Delta n = 2$. A large portion of the experimental measurement error in such studies is due to this effect.

Mid-Infrared Combination–Difference Bands

Recently, the discovery has been made that at long path lengths and high pressures combination–difference bands involving the ring-puckering mode and some other infrared active mode in the molecule can be observed.[34, 35] These bands have selection rules similar to those observed for the far-infrared transitions. The transitions involving $\Delta n = +1$, +2, etc., are observed on the high frequency side of the band, while those with $\Delta n = -1, -2$, etc., are observed on the lower frequency side of the band. The combination–difference transitions observed with $\Delta n \geqslant +1$ very closely approximate the spectrum obtained in the far-infrared region, while the lower frequency bands are almost a mirror image of those observed on the high frequency side. Such bands are most commonly observed in combination with a CH-stretching frequency, which generally falls in the 2900 cm^{-1} region, or with the CH_2-scissoring vibration falling around 1450 cm^{-1}. An example of such transitions is seen in Figure 8 in the spectra of trimethylene sulfide, trimethylene selenide, and trimethylene selenide-d_6 observed by Harvey, Durig, and Morrissey.[34] Although the combination–difference band Q branches are not distinct for the deuterated compound, these series are quite clear for the normal isotopic species and can be readily assigned.

The value of the observation of such transitions is obvious. If the puckering mode is inactive or too weak to be observed directly in the far-infrared region or by Raman spectroscopy the barrier height can be determined from these transitions if they are detectable. Even if the puckering vibration can be studied directly, these transitions provide a check to the assignment of the directly observed transitions. Further-

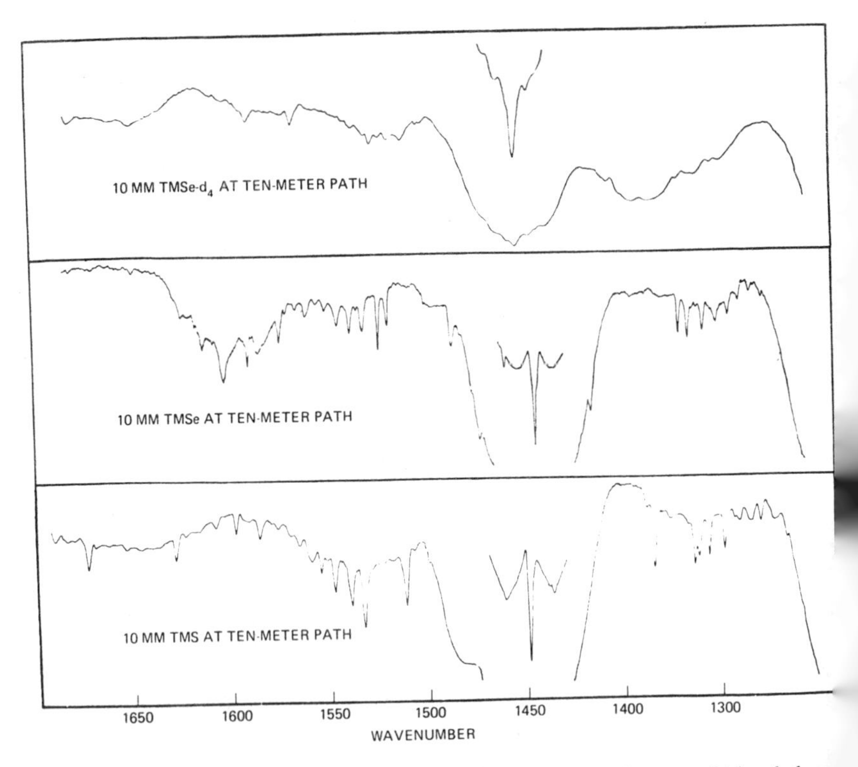

FIGURE 8 Example of vibration puckering bands in the mid-infrared spectra of trimethylene selenide (TMSe), trimethylene selenide-d_4 (TMSe-d_4), and trimethylene sulfide (TMS). (Reproduced by permission from Harvey *et al.*[34])

more, in some cases a more accurate measurement of the puckering levels can be obtained since the resolution obtainable in the mid-infrared region of the spectrum is better than that available in most far-infrared or Raman spectrometers.

Many workers have estimated ring-puckering transitions simply by subtracting the frequency of the fundamental vibration with which the puckering mode is combined and averaging the frequencies obtained from the high and low frequency side of the band. This is a fairly dangerous method. Stone and Mills[18] have shown that there is a small but significant change in the barrier between the upper state combination levels and the ground-state levels in cyclobutane. Furthermore, there is also a chance of an anharmonic resonance by one of the upper state combination levels with another vibration. Such crossings are especially likely in the C–H-stretching region. Ueda and Shimanouchi[35] have noted and explained such a resonance in the C–H-stretching region of trimethylene oxide. Such resonances can shift the combination–difference band origins and lead to erroneous results.

If both the low-frequency and high-frequency progressions have been observed, such difficulties can be avoided by using the standard spectroscopic technique of taking combination differences. This approach has been used by Stone and Mills.[18] Lower state combination differences can be formed by subtracting the frequencies of two transitions with a common upper state. The difference is simply the energy difference between the two ground-state levels involved in the transition. Upper state combination differences can be obtained in the same manner by subtracting transitions with a common lower state. Since these combinations involve only one state, they can be directly fit to obtain separate parameters for the ground and excited vibrational state. If the upper state is perturbed, energy level differences can still be obtained for the unperturbed lower state. Because of the chance of resonance and vibrationally dependent barrier height, barrier heights determined where only a low-frequency or a high-frequency band progression has been observed should be used with caution.

SUMMARY

In the early days of ring-puckering studies, far-infrared instrumentation was in its infancy. As a result many of the earlier infrared studies led to false conclusions because of incomplete data and wrong band assignments. Indirect but potent microwave techniques were necessary for the correct interpretation of the available data. The field is indebted indeed to the theoretical methods developed at that time by the Berkeley school of microwave spectroscopists.

At this point in time, however, far-infrared and Raman instrumentation have been greatly improved, and almost routine studies are possible. Since microwave studies are usually very time-consuming, it would now appear most efficient to use optical methods for the study of these molecules. Band assignments in many cases are not altogether clear; however, cross checks are now available through Raman and infrared studies, and once a molecule has been subjected to such checks, one can now accept optical results with some confidence.

During the course of writing this paper, the author had the benefit of much advice. He is particularly indebted to T. B. Malloy for many helpful conversations. He would also like to thank R. C. Lord, L. A. Carreira, J. Laane, J. R. Durig, J. K. G. Watson, L. H. Scharpen, J. T. Hougen, H. Wieser, and V. W. Laurie for their comments and suggestions.

REFERENCES

1. R. P. Bell, *Proc. R. Soc. (London)*, **A183**, 328 (1945).
2. G. W. Rathjens, N. K. Freeman, W. D. Gwinn, and K. S. Pitzer, *J. Am. Chem. Soc.*, **75**, 5634 (1953).
3. J. Laane, in *Advances in Vibrational Spectroscopy*, ed. by J. R. Durig, Marcell Dekker, N.Y. (1971).
4. A. Danti, W. J. Lafferty, and R. C. Lord, *J. Chem. Phys.*, **33**, 294 (1960).
5. W. D. Gwinn, *Discuss. Faraday Soc.*, **19**, 43 (1955).
6. S. I. Chan, J. Zinn, and W. D. Gwinn, *J. Chem. Phys.*, **33**, 294 (1960).
7. S. I. Chan, J. Zinn, J. Fernandez, and W. D. Gwinn, *J. Chem. Phys.*, **33**, 1643 (1960).
8. S. I. Chan, T. R. Borgers, J. W. Russell, H. L. Strauss, and W. D. Gwinn, *J. Chem. Phys.*, **44**, 1103 (1966).
9. C. S. Blackwell and R. C. Lord, in *Advances in Vibrational Spectroscopy*, ed. by J. R. Durig, Marcell Dekker, N.Y. (1971).
10. J. Laane and R. C. Lord, *J. Chem. Phys.*, **48**, 1508 (1968).
11. W. C. Pringle, *J. Chem. Phys.*, **54**, 4979 (1971).
12. J. Laane, *Appl. Spectrosc.*, **24**, 73 (1970).
13. T. Ueda and T. Shimanouchi, *J. Chem. Phys.*, **47**, 4042 (1967).
14. L. A. Carreira and R. C. Lord, *J. Chem. Phys.*, **51**, 2735 (1969).
15. S. I. Chan and D. Stelman, *J. Mol. Spectrosc.*, **10**, 278 (1963).
16. C. E. Reid, *J. Mol. Spectrosc.*, **36**, 183 (1970).
17. W. J. Lafferty, *J. Mol. Spectrosc.*, **36**, 84 (1970).
18. J. M. R. Stone and I. M. Mills, *Mol. Phys.*, **18**, 631 (1970).
19. F. A. Miller and R. J. Capwell, *Spectrochim. Acta*, **27A**, 947 (1971).
20. T. B. Malloy, *J. Mol. Spectrosc.*, **44**, 504 (1972).
21. L. A. Carreira, I. M. Mills, and W. B. Person, *J. Chem. Phys.*, **56**, 1444 (1972).
22. D. O. Harris, H. W. Harrington, A. C. Luntz, and W. D. Gwinn, *J. Chem. Phys.*, **44**, 3467 (1966).
23. D. O. Harris, G. G. Engerholm, C. A. Tolman, A. C. Luntz, R. A. Keller, H. Kim, and W. D. Gwinn, *J. Chem. Phys.*, **50**, 2438 (1969).

24. H. Wieser, M. Danyluk, and R. A. Kydd, *J. Mol. Spectrosc.*, **43**, 382 (1972).
25. R. A. Kydd, H. Wieser, and M. Danyluk, *J. Mol. Spectrosc.*, **44**, 14 (1972).
26. W. Kiefer, H. J. Bernstein, H. Wieser, and M. Danyluk, *J. Mol. Spectrosc.*, **43**, 393 (1972).
27. A. S. Esbitt and E. B. Wilson, *Rev. Sci. Instr.*, **34**, 901 (1963).
28. H. W. Harrington, *J. Chem. Phys.*, **44**, 3481 (1966).
29. L. H. Scharpen and V. W. Laurie, *J. Chem. Phys.*, **49**, 3041 (1968).
30. L. H. Scharpen, *J. Chem. Phys.*, **48**, 3552 (1968).
31. M. G. Petit, J. S. Gibson, and D. O. Harris, *J. Chem. Phys.*, **53**, 3408 (1970).
32. S. S. Butcher and C. C. Costain, *J. Mol. Spectrosc.*, **15**, 40 (1965).
33. T. R. Borgers and H. L. Strauss, *J. Chem. Phys.*, **45**, 947 (1966)
34. A. B. Harvey, J. R. Durig, and A. C. Morrissey, *J. Chem. Phys.*, **50**, 4949 (1969).
35. T. Ueda and T. Shimanouchi, *J. Chem. Phys.*, **47**, 5018 (1967).
36. J. Laane and R. C. Lord, *J. Chem. Phys.*, **47**, 4941 (1967).
37. J. R. Durig, A. C. Shing, L. A. Carreira, and Y. S. Li, *J. Chem. Phys.*, **57**, 4398 (1972).

Discussion Leader: ROBERT C. LIVINGSTON
Reporters: DAVID O. HARRIS
IRA W. LEVIN

Discussion

Commenting on Dreizler's paper concerning determination of barriers to internal rotation, LIDE emphasized that a combination of data from both microwave and far-infrared sources often gives a better value of the barrier height than either method alone. The quantity determined most accurately from the measurement of splittings in the microwave spectrum is s which is roughly proportional to the product of V_3 and I_α:

$$s \propto V_3 I_\alpha \, .$$

The measurement of the torsional fundamental gives an accurate value of

$$\nu_{0\to 1} \propto (V_3 / I_\alpha)^{1/2}.$$

Thus, by using both pieces of data, one can separate V_3 from I_α; if the microwave data alone are used, the uncertainty in I_α is directly reflected in V_3.

Following the papers presented by Rush and Lafferty, several comments were directed to the difficulties of interpreting frequencies of large-amplitude motions in the condensed phases. RAMSAY noted that studies of the near-ultraviolet spectrum of pyridine indicate that the molecule behaves as a quasiplanar molecule in the excited state. The out-of-plane vibration, which has a frequency of 403 cm^{-1} in the ground state, gives rise to energy levels at 60, 138, 239, 346, and 464 cm^{-1} in the excited state. These energy levels can be fitted within ± 2 cm^{-1} by a potential of the type

$$V(x) = ax^2 + bx^4 . \quad a < 0, b > 0$$

The height of the barrier is only 4 cm^{-1}. In the gas phase the 2–0 transition is at ν_{00} + 138 cm^{-1}. In the solid phase no band has been seen in this region, and, indeed, the corresponding transition has not been identified.

LORD continued in connection with the effects of the solid state on the energy levels of ring-puckering modes. His group has studied the absorption spectrum of 2,5-dihydrofuran in a matrix made of hydrocarbon wax. The effect of the matrix was to increase the quadratic term in the mixed quadratic–quartic potential sufficiently to cause the widely spaced gas-phase transitions to coalesce into a single relatively narrow band of about 20 cm^{-1} width. MILLER pointed out further that in cyclobutane vapor the ring-puckering mode gives a series of 4–6 Raman bands below 200 cm^{-1}. In the liquid these collapse into a single, broad, somewhat asymmetric band. Thus, there clearly is a considerable condensed-state effect.

MILLER also remarked that it is common practice to measure methyl torsion frequencies in solids, often at low temperatures and then to use these frequencies to calculate barrier heights. These barriers certainly should not be mixed with gas-phase barriers. The presence of nearby molecules must contribute to the potential function for the torsion and thus affect the deduced barrier. This is expecially clear in the case of those molecules for which the torsion is forbidden for the gas, but in which it becomes observable in the solid. The appearance in the solid surely is due to the influence of neighboring molecules. The potential no longer has the symmetry of the isolated molecule, or else the band would not be observed. Some day, perhaps, these "barriers" from the solid state may give us useful information about intermolecular interactions. In the meantime, there is a real hazard in putting them in the literature without adequate caveats. In molecules for which torsional frequencies have been measured for both gas and solid, it is common to find that they differ by 10 percent, which implies a difference of 20 percent in the barriers. Discrepancies as large as 100 percent are known (e.g., in benzaldehyde).

With regard to Lafferty's discussion of the tilting of the methylene hydrogens during the ring puckering motion, SNYDER remarked that, when the geometry of cyclobutane is studied by NMR in liquid crystals and without correction for vibrations of large or small amplitude, the dihedral angle is found to be 27° and the methylene group to be tilted back by 3.9°. *Ab initio* SCF calculations by Ditchfield in a 4-31G basis set, devised by Pople and co-workers, give a dihedral angle of 21.6° and a methylene tilt of 3.8° when the energy is minimized with respect to geometry. Both studies suggest that the ring-puckering motion will be coupled to the methylene tilt.

A general comment was offered by FUNG on studies of the O–H and O–D vibrations of some hydrogen-bonded systems. It was found that the double-minimum asymmetric quartic oscillator potential seemed to work well. The results were not published because of some uncertainties in the assignments.

ZERBI questioned Rush as to whether the neutron diffraction results on torsional vibrations in solid *o*-xylene and durene indicate dispersion of these mode frequencies in the crystal. RUSH replied that, as far as he could ascertain, the observed torsional peak widths at low temperature were very close to the resolution limit. This observation suggests little dispersion of mode frequencies as a result of intermolecular coupling.

VII

Electronic Charge Distribution by Resonance Methods

Bruce R. McGarvey

Determination of Electron Distribution from Hyperfine Interaction

MEASUREMENT OF THE HYPERFINE INTERACTION

The magnetic interaction between the electron spin **S** and nuclear spin **I** is called the hyperfine interaction. Since these are vector quantities, the general hyperfine interaction must be represented by a second-rank tensor

$$\mathbf{S} \cdot A \cdot \mathbf{I}.$$

This tensor is symmetric so that it is necessary in the general case to determine six separate terms in the hyperfine interaction. These terms are customarily reported in units of frequency or wave number (MHz or cm^{-1}), and sometimes in gauss, but this latter unit is best avoided due to factors that will be discussed later. For a rapidly tumbling molecule or ion in solution the hyperfine interaction takes on the simpler form

$$a\,\mathbf{S} \cdot \mathbf{I}$$

in which only one term is needed to characterize the interaction.

The majority of hyperfine interactions in molecules and ions have been determined by one of three methods: electron spin resonance (ESR), electron nuclear double resonance (ENDOR), and paramagnetic shifts in nuclear magnetic resonance (NMR). Other methods have been employed such as electron double resonance (ELDOR), but they have been employed only in a limited number of cases and will not be discussed here. These three methods are not always applicable to the same system nor do they measure the hyperfine values with the same precision. In comparing the general features of the three methods, we shall consider the case of a rapidly tumbling molecule in solution with one unpaired electron and one nuclear spin **I**. The general spin Hamiltonian for such a system is

$$\mathcal{H} = g\beta \mathbf{S}\cdot\mathbf{H} - g_N\beta_N \mathbf{I}\cdot\mathbf{H} + a\mathbf{I}\cdot\mathbf{S}, \tag{1}$$

where β and β_N are the Bohr and nuclear magnetons, respectively. In Eq. (1) we are assuming any orbital angular momentum of the electron has been quenched by molecular fields and that any second-order spin-orbit effects are incorporated in the value of g, which is different from that of the free electron. Similarly, we assume that any chemical shift associated with diamagnetic shielding of the nucleus by electrons is incorporated into g_N, which is therefore different from that of the bare nucleus.

For ESR, ENDOR, and NMR measurements, the magnetic field is generally large enough to make the interaction between the electron spin and the magnetic field the largest term in Eq. (1). In this case the electron and nuclear spins are quantized along the magnetic field and to a first order approximation the energy of the various spin states can be given by

$$E(M_s, M_I) = g\beta H M_s - g_N\beta_N H M_I + aM_sM_I, \tag{2}$$

where M_s and M_I are the magnetic quantum numbers for the electron and nucleus. It should be emphasized here that Eq. (2) is only approximate and leads to errors in the determination of a when a is large or when high precision measurements are made. This problem is discussed later in "Second-Order Effects." For the normal arrangement of an oscillating magnetic field perpendicular to the main magnetic field, the allowed transitions are the ESR transitions of $\Delta M_s = \pm 1$, $\Delta M_I = 0$, and the NMR transitions of $\Delta M_I = \pm 1$, $\Delta M_s = 0$. In Figure 1 are shown the various energy levels and allowed transitions for a molecule with $I = 1$. We have assumed in Figure 1 that $a > g_N\beta_N H$. In many instances the reverse is true.

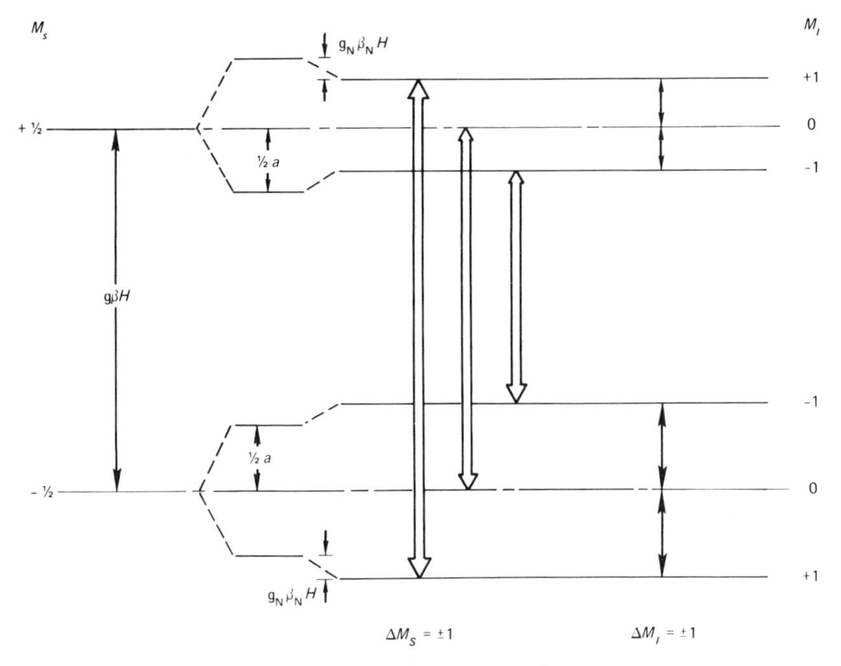

FIGURE 1 ESR and NMR transitions in the $S = \frac{1}{2}$, $I = 1$ spin system.

ESR

In the ESR method the $M_s = \pm 1$ transitions are detected. At constant magnetic field the $2I + 1$ transitions occur at frequencies given by

$$h\,\nu(M_I) = g\beta H + aM_I. \tag{3}$$

Therefore, the absorption lines are separated by a/h in frequency. The precision of measurement of a is therefore determined by the width of each absorption line. In transition metal complexes line widths are almost always greater than 3 MHz while in favorable cases line widths a hundred-fold smaller can be observed in free radicals. Thus, in practice, hyperfine interactions can at best be determined to ±0.1 MHz by ESR.

In most ESR spectrometers the frequency is constant and the magnetic field is varied instead. The fields at which the ESR lines are observed is obtained from Eq. (3)

$$H(M_I) = (h\nu_0/g\beta) - (a/g\beta)M_I, \tag{4}$$

where ν_0 is the constant frequency at which the experiments are done. The equation also predicts $2I + 1$ lines separated by $(a/g\beta)$. Many researchers have taken to reporting the values of $(a/g\beta)$ rather than a, which is a poor practice since these are not directly comparable between systems with different g values. The too common practice of reporting values of $(a/g\beta)$ without giving the experimental values of g is particularly deplorable. The need to convert the experimentally measured splitting term in gauss to a, using the experimentally determined g value, has led to incorrect values of a being reported in the literature. Several papers have been found in the literature in which the authors have used a standard value of $g = 2$ instead of the experimental g value in the conversion of gauss units into frequency units. Since in most papers not enough data are given to check on the method of conversion, it is difficult to assess how common this mistake is in the literature. For free radicals little error would be incurred since g is close to 2.00, but for transition and rare earth metal complexes the errors can be significant since g's often vary by 10–20 percent or more from the value of 2. There are also problems in determining g accurately when the hyperfine interaction is large. These will be discussed later in "Second-Order Effects."

It should be noted here that, since there is no way to determine the sign of M_I for each hyperfine line, the ESR method gives only the absolute magnitude of a and not the sign of a.

ENDOR

The $\Delta M_I = \pm 1$ or NMR transitions are not normally detected directly because the lines are too broad in concentrated solutions or pure solids due to interactions with neighboring electron spins. At dilute concentrations the intensity is generally too low for detection even though the lines are narrower. The ENDOR method is a double resonance method that uses the electron to detect an NMR transition. Use of the electron as a detector provides a large "enhancement" of the signal over normal NMR detection and allows studies of systems at low concentrations.

In ENDOR an ESR transition is irradiated at sufficient power to produce saturation effects in the observed ESR signal. A second radio frequency is introduced capable of causing $\Delta M_I = \pm 1$ transitions. If the radio frequency power is high enough to cause a rate of change in the nuclear spin states comparable with the rate of electron spin relaxation, the saturation conditions are upset causing a change in the ESR signal. Normally, in ENDOR only those $\Delta M_I = \pm 1$ transitions are detected that have one energy level in common with the ESR transition being saturated.

Examination of Figure 1 shows that, no matter which ESR line is saturated, only two ENDOR transitions are observed, at frequencies

$$h\nu = |(1/2)a \pm g_N \beta_N H|. \tag{5}$$

If $g_N \beta_N H > (1/2)a$, we would expect to see two lines split in frequency by a and centered at frequency $g_N \beta_N H/h$. In ENDOR, frequency is always varied so that the units of a are normally reported in frequency units. It should be noted from Eq. (5) that ENDOR like ESR can only determine absolute magnitudes of a in most systems.

A major advantage of the ENDOR method is the smaller line widths of the NMR transitions. Line widths as small as 10 kHz are observed making it possible to determine a to ±0.01 MHz. Hyperfine interactions too small to be resolved in the ESR spectrum have been readily detected in ENDOR measurements. Another advantage occurs in those systems that have more than one type of nucleus giving a hyperfine interaction. The presence of the $g_N \beta_N H$ term in Eq. (5) makes it an easy matter to decide which hyperfine term belongs to which nucleus. In ESR the $g_N \beta_N H$ term does not appear, and assigning hyperfine splittings to different nuclei often requires isotopic substitution to remove ambiguities.

A major disadvantage is the greater sophistication required in instrumentation. Our use of a liquid solution model to demonstrate the basic principles of ENDOR is perhaps misleading, as most ENDOR studies are done on solid samples at very low temperatures where long relaxation times allow ENDOR to be done at microwave and radio frequency powers that are easily obtained. ENDOR studies have been done on free radicals in solution,[1-5] but the high radio frequency power levels needed require special instrumentation.

NMR

Although NMR transitions are normally not directly detectable, a situation occurs in some systems that does allow for normal NMR detection. When there is an extremely rapid relaxation among the electron spin states the broadening effects of neighboring spins are effectively destroyed and line widths for NMR transitions become narrow even in concentrated solutions. In this situation we observe one NMR transition that has been averaged over all electron spin states. This averaging fails to remove completely the effect of a on the NMR frequency because the two spin states have unequal populations.

If we assume a Boltzmann distribution among the two electron spin states, the average energy for a given M_I spin state is obtained by averaging Eq. (3) over M_s.

$$\overline{E}(M_I) = -g_N\beta_N HM_I + \tfrac{1}{2}(g\beta H + aM_I)\,[1 - e^{(g\beta H/kT)}]/[1 + e^{(g\beta H/kT)}] \quad (6)$$

$$\simeq -g_N\beta_N HM_I - (g\beta H + a\,M_I)(g\beta H/4kT), \quad (7)$$

where Eq. (7) has been obtained by expanding exp $(g\beta H/kT)$ in a power series and keeping only terms linear in H since at normal temperatures $g\beta H \ll kT$. From Eq. (7) we find the frequency of a $\Delta M_I = \pm 1$ transition to be given by

$$\overline{\nu} = g_N\beta_N H/h + (g\beta H/4kTh)a. \quad (8)$$

If we let $\nu^0 = g_N\beta_N H/h$ then

$$\Delta\nu/\nu^0 = (\nu - \nu^0)/\nu^0 = (g\beta/4kTg_N\beta_N)a. \quad (9)$$

Thus, if the magnetic field is kept constant, there is a shift in frequency that is proportional to a. This is called the paramagnetic shift since it is proportional to the paramagnetic susceptibility resulting from the unpaired spin. If we keep frequency constant and vary the magnetic field, Eq. (8) gives

$$(H - H_0)/H \approx \Delta H/H_0 = -(g\beta/4kTg_N\beta_N)a \quad (10)$$

$$H_0 = h\nu_0/g_N\beta_N,$$

where ν_0 is the constant frequency of the spectrometer and H_0 is the resonant field in the absence of any hyperfine coupling.

Examination of Eq. (9) and (10) shows that the sign of a is readily determined from the direction of the shift. The magnitude is less accurately determined because it requires a knowledge of g that must be determined from ESR studies. Since the rapid electron spin relaxation that makes this NMR method feasible makes ESR detection extremely difficult, g values are not often known. In a few cases the ESR has been determined by going to low temperatures (generally 4 K) where the relaxation has become slow enough to allow detection of the ESR spectrum. Most NMR studies have been done on systems in which $S > 1/2$, or if $S = 1/2$, there are excited states close in energy to the ground state. For these systems the proportionality factor between $\Delta H/H_0$ and a is not given by Eq. (10) but is more complex and often not well known.[6] Therefore, the NMR method is good for determining the sign of a but much less reliable for determining absolute magnitudes. Since the proportionality constant between a and $\Delta H/H_0$ is the same for all nuclei in the same molecule, relative values of a can be accurately determined for different nuclei in the same molecule.

Thus we see that the NMR method is complementary to ESR and ENDOR methods in that it can be used on systems for which ESR and ENDOR studies are not possible. Its sensitivity is equal to or better than that of ENDOR. For example, Eq. (9) would predict a 10-Hz shift at 60 MHz for a proton in which $a = 0.006$ MHz at 300 K. Thus values of a as small as 0.001 MHz can be detected in some cases if high enough NMR frequencies are used.

Second-Order Effects

We have seen in the discussion of ESR that, to the extent Eq. (2) is correct, the value of $(a/g\beta)$ is obtained from an ESR spectrum by simply measuring the different in the magnetic field for two adjacent hyperfine lines. The value of g needed to convert $(a/g\beta)$ to a is then found by taking the average of the magnetic fields of all hyperfine lines, which is $(h\nu_0/g\beta)$. Similarly, for ENDOR a is easily found by measuring either the frequency difference in the two ENDOR lines when $g_N\beta_N H > (1/2)a$ or the average frequency when $(1/2)a > g_N\beta_N H$. It is the purpose of this section to point out that these commonly used methods of determining a are based on an approximate solution of Eq. (1), which assumes $a \ll g\beta H$. If a is large or if high precision of measurement is possible (as in ENDOR), use of these methods to calculate a can lead to serious errors.

An exact solution of Eq. (1) for $S = 1/2$ is

$$E(M_s, M_I) = -(1/4)a - (M_I + M_s)g_N\beta_N H + M_s\,[(g\beta H + g_N\beta_N H)^2 + 2(M_I + M_s)(g\beta H + g_N\beta_N H)a + (I + 1/2)^2 a^2]^{1/2}. \quad (11)$$

A useful approximate form of this equation good to second order is

$$E(M_s, M_I) \simeq g\beta H M_s - g_N\beta_N H M_I + a\,M_I M_s + (a^2/2g\beta H)\left\{[I(I+1) - M_I^2]\,M_s - (1/2)M_I\right\}. \quad (12)$$

The frequency for an ESR transition is then

$$h\nu\,(M_I) = g\beta H + a\,M_I + (a^2/2g\beta H)\,[I(I+1) - M_I^2] \quad (13)$$

or for constant frequency the magnetic field for the M_I transition is

$$H(M_I) = (h\nu_0/g\beta) - (a/g\beta)M_I - (a^2/2g\,h\beta\nu_0)\,[I(I+1) - M_I^2]\,. \qquad (14)$$

An examination of Eq. (13) and (14) reveals that the $2I + 1$ hyperfine lines are not evenly spaced; therefore, $(a/g\beta)$ cannot be determined from a simple measurement of the line spacings to an accuracy better than $\pm(a^2/g\beta h\nu_0)$. A value of $(a/g\beta)$ good to second order can be easily obtained by measuring the spacing between hyperfine lines with the same value of M_I^2 since the second-order term is the same for both lines, but the second-order terms must be considered when computing g since the average field for all hyperfine lines is

$$\overline{H(M_I)} = (h\nu_0/g\beta) + (a^2/3g\beta h\nu_0)I(I+1)(2I+1) \qquad (15)$$

rather than $(h\nu_0/g\beta)$.

For the two ENDOR transitions Eq. (12) predicts the frequencies for the M_I to $(M_I + 1)$ transition to be

$$h\,\nu = |g_N\beta_N H \mp (1/2)a \pm (a^2/4g\beta H)\,(2M_I + 1 \pm 1)|. \qquad (16)$$

Thus for $g_N\beta_N H > (1/2)a$, the frequency separation for the two ENDOR lines is

$$\Delta\nu = (a/h) - (a^2/2g\beta Hh)\,(2M_I + 1). \qquad (17)$$

Thus second-order corrections are important in ENDOR studies. For example, if $(a/h) = 10$ MHz and the ESR spectrometer operates at 9.5 GHz, then the second-order term is 0.1 percent of the value of a in magnitude so that a precision of ±0.01 MHz in the determination of a requires attention to second-order corrections.

When a is very large, second-order corrections are not enough, and one must go to higher order corrections or use exact equations such as Eq. (11). For example, in the hydrogen atom, $a = 1420.40573$ MHz and $g = 2.002256$.[7,8] At a frequency of 9.500 GHz, Eq. (11) predicts a separation in the two hyperfine lines of 509.70 gauss while Eq. (14) predicts a separation of $(a/g\beta) = 506.86$ gauss, a difference of 2.85 gauss. When evaluating literature reports on the hyperfine interaction, it is a useful practice to compute $(a^2/g\beta h\nu_0)$ or $(a^2/h\nu_0)$ and compare it with the quoted error in measurement. If the second-order term is larger than the reported error and no mention is made of making second-order corrections, then it is wise to assume the corrections were not made and the true error to be of the magnitude of the second-order term.

In the preceding discussion we have not mentioned the NMR shift method of measuring a because second-order terms are not present in this case. The second-order terms in ENDOR and ESR arise because the aS_xI_x and aS_yI_y terms in Eq. (1) cause mixing of spin terms and destroy the purity of the spin states. In the NMR method we deal with averaged spin operators $\bar{S}_z$, $\bar{S}_x$, and $\bar{S}_y$ and over a period of time $\bar{S}_x$ and $\bar{S}_y$ are zero if z is along the magnetic field. Therefore, the averaging process removes any effect of such second-order terms.

In free radicals we often deal with hyperfine spectra that are degenerate. For example, the splitting caused by a rotating methyl group appears as a simple four-line spectrum similar to that produced by a nucleus of $I = 3/2$. The $M_I = \pm 1/2$ lines, however, are three times as intense as the $\pm 3/2$ lines because they are really a superposition of three separate transitions that have the same frequency to first order. The second-order terms remove this degeneracy so that in favorable cases of high resolution the ESR spectrum will show a splitting of these degenerate ESR transitions.[9, 10]

Measurement of Hyperfine Interaction in Rigid Media

In the introductory remarks it was pointed out that the hyperfine interaction in rigid media is represented by a tensor with six different terms. It is always possible to choose a coordinate system in which the tensor is diagonal and the hyperfine interaction can be written as

$$A_xI_xS_x + A_yI_yS_y + A_zI_zS_z, \tag{18}$$

where x, y, and z refer to the "principal axes" in which the tensor is diagonal. The g tensor also can be diagonalized and at higher symmetries the principal axes of both the g and A tensors are the same. For low symmetry molecules the two tensors may have different sets of principal axes. In rigid media the experimental problem is to determine A_x, A_y, and A_z and, if possible, the three angles that specify the orientation of the principal axes to the laboratory frame of reference. In solution the hyperfine constant a is the average

$$a = \tfrac{1}{3}(A_x + A_y + A_z). \tag{19}$$

Two types of rigid media have been employed in hyperfine studies: supercooled liquid solutions or powders and single crystals. The anisotropy in the g and A tensors produces anisotropic line shapes in the ESR of supercooled solutions that can be analyzed to deduce the spin Hamiltonian parameters that will produce such shapes. In practice, line shapes are calculated either analytically or by computer simulation for various choices of spin Hamiltonian parameters until the best fit between the calculated and experimental lines is obtained. Many assumptions as to line shape and line intensity for each transition as a function of orientation have to be made so that one should not expect hyperfine values obtained this way to be accurate to more than a few percent. Very often the best computed line shape still differs significantly from the experimental shape. Narayana and Sastry[11] recently analyzed the single-crystal spectrum of VO^{2+} in an alum crystal and then powdered the crystal to obtain the powder spectrum. They found that the best fit between the computed and experimental spectrum required a value for A_x 10 percent different from the single-crystal value and a value of g_x 1 percent different. The values of A_z and g_z agreed to better than 1 percent for the two methods. Although less reliable than single-crystal methods, supercooled solutions and powders have been used in ESR often to obtain hyperfine data because it is often difficult to obtain single crystals of the systems that are to be studied. One disadvantage of the method is that no information can be obtained on the orientation of the principal axes relative to the coordinate axes of the molecule.

Supercooled liquid solutions can also give meaningful ENDOR results,[12] but most ENDOR has been done on single crystals. NMR studies can also be done, but there are few reports in the literature where any analysis has been done on an asymmetric NMR line.

ESR and ENDOR single-crystal studies are done with diamagnetic crystals having a small percentage of lattice sites occupied by a paramagnetic species to be studied. This arrangement is necessary to keep spin–spin interactions of electrons to a low value as otherwise they broaden the resonance lines. NMR single-crystal studies are almost always done with pure crystals of the compound being investigated, because here electron spin–spin broadening is absent due to the rapid spin relaxation. It is necessary in NMR to use pure materials to get sufficient intensity because of the large line widths (several kilohertz) still present due to nuclear spin–spin interactions that cannot be reduced by dilution.

In all single-crystal studies, the variation in resonance frequency or magnetic field is studied as a function of the orientation of the crystal in the magnetic field. A spin Hamiltonian of appropriate form is then solved and the parameters adjusted to fit the calculated variation with the experimental data. Most errors in doing this occur because approximate solutions of spin Hamiltonians are used for systems for which the approximations are not justified. Second-order effects are often very important in analyzing single-crystal ESR and ENDOR measurements.

If $g\beta H >> A >> g_N \beta_N H$, a first-order solution of the spin Hamiltonian

$$\mathcal{H} = \mathbf{S} \cdot g \cdot \mathbf{H} + \mathbf{S} \cdot A \cdot \mathbf{I} - g_N \beta_N \mathbf{H} \cdot \mathbf{I} \tag{20}$$

for $S = 1/2$ is

$$E(M_s, M_I) = g\beta H M_s + K M_I M_s - g_N \beta_N H F M_I \tag{21}$$

$$g = (g_x^2 \sin^2\theta \cos^2\phi + g_y^2 \sin^2\theta \sin^2\phi + g_z^2 \cos^2\theta)^{1/2} \tag{22}$$

$$K = (g_x^2 A_x^2 \sin^2\theta \cos^2\phi + g_y^2 A_y^2 \sin^2\theta \sin^2\phi + g_z^2 \cos^2\theta)^{1/2}/g \tag{23}$$

$$F = (g_x A_x \sin^2\theta \cos^2\phi + g_y A_y \sin^2\theta \sin^2\phi + g_z A_z \cos^2\theta)/gK, \tag{24}$$

where θ and ϕ are the polar angles of the magnetic field in the principal axes of g and A. It is assumed in deriving Eq. (21) that g and A tensors have the same principal axes. Thus to first order the ESR transitions occur at the frequencies

$$h\nu(M_I) = g\beta H + K M_I \tag{25}$$

and the ENDOR transitions at

$$h\nu = (1/2)K \pm g_N \beta_N H F. \tag{26}$$

Inclusion of second-order terms of magnitude $(A^2/g\beta H)$ are often required. Bleaney[13] has calculated these for the case where $A_x = A_y$.

It is important to recognize that Eq. (21) is not correct when $g_N \beta_N H$ is similar in magnitude to or greater than A_x, A_y, or A_z. Also, when both terms are comparable in magnitude, more than $2I + 1$ resonance lines are often observed due to a scrambling of nuclear spin states that allows transitions in which $\Delta M_I \neq 0$ to occur. There are many cases in the literature where the hyperfine spectrum has been incorrectly assigned because the authors failed to realize that they were dealing with a situation in which $g_N \beta_N H$ was too large in magnitude. Poole and Farach[14] have recently analyzed the effect of this nuclear Zeeman term on the spectra of irradiated free radicals and have found that many reported hyperfine values found in the literature may be in error by several percent because this term was not properly taken into account. The best procedure when faced with large nuclear Zeeman terms is to solve the spin Hamiltonian exactly with a computer program and adjust all parameters to fit experimental data.

Due to large widths of the resonance lines in the solid state, the NMR method has only been used on systems with large hyperfine interactions and these interactions can only be determined to a few megahertz at best. For a crystal with isotropic bulk susceptibility

$$\Delta H/H = 1 - \left\{ [A_x(\chi/g\beta g_N\beta_N) + 1]^2 \sin^2\theta \cos^2\phi + [A_y(\chi/g\beta g_N\beta_N) + 1]^2 \sin^2\theta \sin^2\phi + [A_z(\chi/g\beta g_N\beta_N) + 1]^2 \cos^2\theta \right\}^{1/2}, \quad (27)$$

where χ is the paramagnetic susceptibility per molecule or ion. In the very common case of $A(\chi/g\beta g_N\beta_N) \ll 1$, Eq. (27) becomes

$$\Delta H/H = -(A_x \sin^2\theta \cos^2\phi + A_y \sin^2\theta \sin^2\phi + A_z \cos^2\theta)(\chi/g\beta g_N\beta_N), \quad (28)$$

which is the form usually found in the literature. If the crystal is not isotropic or if exchange effects are important the equation for $\Delta H/H$ becomes much more complex.[15] In some of the systems that lend themselves to NMR studies, there are excited states with energies comparable with kT. In this situation the NMR shift measures some sort of complex average of the hyperfine constants in the several electronic states.[6,15]

Quadrupole Interactions

Nuclei with $I > 1/2$ have electric quadrupole moments that interact with electric field gradients in the molecule or ion. The presence of this interaction tends to orient the nuclei in the direction of the electric field gradient and hence will have major effects on the rigid-media spectra when it is comparable in magnitude with the hyperfine interaction. It has little effect on solution spectra except on line widths because the field gradient is averaged out by the rapid tumbling of the molecule.

The spin Hamiltonian for the quadrupole interaction is

$$\mathcal{H}_Q = [e^2qQ/4I(2I-1)]\left\{[3I_z^2 - I(I+1)] + \eta(I_x^2 - I_y^2)\right\}, \quad (29)$$

where Q is the quadrupole moment of the nucleus, q is the electric field gradient and η is an asymmetry parameter. Since $\mathcal{H}_Q$ depends on only the nuclear spin operators, it does not affect the ESR transition frequencies. It does, however, mix up the nuclear spin states giving rise to $\Delta M_s = \pm 1, \Delta M_I = \pm 1$ transitions. It does affect the ENDOR transitions directly and ENDOR studies can be used to measure the quadrupole interactions as well as the hyperfine interactions. The quadrupole interaction should affect the NMR shift in the solid state but most of the work reported in the literature has been done with nuclei for which $I = 1/2$.

$S>1/2$ *Systems*

Ions or molecules with more than one unpaired spin have large spin–spin interactions that in the presence of symmetries lower than octahedral or tetrahedral lead to splittings of the spin states even in the absence of a magnetic field. When these zero-field interactions are large they compete with the magnetic field in orienting the electron spin and make the predictions of Eq. (2) and (21) invalid. Since most studies of spin distribution by ESR or ENDOR have been done on $S = 1/2$ systems, we will not consider the problems of determining the hyperfine interaction in $S>1/2$ systems. The majority of systems studied by NMR, however, have $S>1/2$. Kurland and McGarvey[6] have considered the problem of determining the proportionality factor between $\Delta H/H$ and a for such systems.

THEORY OF THE HYPERFINE INTERACTION

Electron Spin–Nuclear Spin Interaction

It has been customary to use the electron spin–nuclear spin operator in its two limiting forms

$$\mathcal{H}_{SI} = 2.0023 g_N \beta\beta_N \sum_i [3(\mathbf{s}_i \cdot \mathbf{r}_i)(\mathbf{I} \cdot \mathbf{r}_i) - r_i^2(\mathbf{s}_i \cdot \mathbf{I})] r_i^{-5} + (8\pi/3)(2.0023) g_N \beta\beta_N \sum_i \delta(r_i) \mathbf{I} \cdot \mathbf{s}_i, \quad (30)$$

where $\mathbf{r}_i$ is the vector connecting the nucleus to the ith electron and $\delta(r_i)$ is the Dirac delta function. The first term in Eq. (30) is the limiting form of the operator when $r_i \gg$ radius of the nucleus and is the form to use when the wave function goes rapidly to zero (p,d,f orbitals) in the region of the nucleus. This term automatically gives zero integrals for functions of spherical symmetry as long as the limits of integration do not get too close to $r_i = 0$. The second term is the limiting form of the operator when r_i is close to the radius of the nucleus and the wave function has spherical symmetry in this region. The first term is commonly called the "dipolar interaction" and is applicable to p,d,f electrons while the second term is called the "Fermi contact interaction" and is applicable to s electrons. This terminology has left many scientists with the mistaken impression that there are two distinct hyperfine interactions while in fact they are two limits of the same interaction.

The dipolar term in Eq. (30) always gives a traceless hyperfine tensor so that its contribution is averaged out in solution. In the absence of orbital angular momentum for the electron, the Fermi contact term in

Eq. (30) gives an isotropic hyperfine tensor. It is a common technique to divide the hyperfine tensor into an isotropic tensor with principal value A_s and a traceless tensor with components $2A_d$, $-(1-\epsilon)A_d$, and $-(1+\epsilon)A_d$ such that the hyperfine components can be written

$$\begin{aligned} A_x &= A_s - (1-\epsilon)A_d \\ A_y &= A_s - (1+\epsilon)A_d \\ A_z &= A_s + 2A_d \\ 0 &\leqslant \epsilon \leqslant 1, \end{aligned} \tag{31}$$

where A_s is the average hyperfine constant measured in solution. If no other interactions contribute to the hyperfine interaction, A_s can be attributed to the Fermi contact term and A_d and ϵ to the dipole term in Eq. (30). We shall see in the next section that such a simple separation cannot be done if the molecule has unquenched orbital momentum for the unpaired electrons or has an appreciable spin–orbit interaction.

Spin–Orbit Effects

Although we have treated the hyperfine interaction as an interaction between the nuclear spin **I** and the intrinsic spin s of the electron, it is in reality the interaction of **I** with the total angular momentum of the electron. Most lanthanide and actinide ions and some transition metal ions in high symmetry do not have their orbital momentum quenched by the crystal fields. In these cases a complete treatment of the hyperfine interaction must include the following term

$$\mathscr{H}_{LI} = 2g_N\beta\beta_N \sum_i \langle r_i^{-3} \rangle \, l_i \cdot \mathbf{I} \tag{32}$$

where l_i is the orbital momentum operator of the ith electron centered at the nucleus. Although this appears to be a dipolar term since it depends on r^{-3}, it does not lead to a traceless tensor and hence contributes to both the isotropic and anisotropic parts of the hyperfine tensor.

Most transition metal ions have the orbital motion quenched to first order by the crystal field, but the presence of a large spin–orbit interaction coupled with Eq. (32) will cause second-order orbital terms in the hyperfine interaction of the magnitude

$$(\lambda/\Delta E)\,(2g_N\beta\beta_N)\,\langle r^{-3}\rangle,$$

where λ is the spin–orbit coupling constant, ΔE is the energy difference between the ground state and an excited state that is coupled to the ground state by the spin–orbit interaction, and $\langle r^{-3}\rangle$ is the average of r^{-3} for a d orbital. Since deviations of g from 2.0023 are proportional to $\lambda/\Delta E$, it should come as no surprise that this spin–orbit term makes a contribution to the isotropic constant A_s of

$$(\bar{g} - 2.0023)\,(2g_N\beta_N)\langle r^{-3}\rangle, \tag{33}$$

where $\bar{g} = (1/3)(g_x + g_y + g_z)$. Similar but more complex terms appear in the anisotropic terms the exact nature of which are dependent on what the ground state is. Eq. (33) applies to the transition metal complex. A related equation can be derived for ligand nuclei that has proved important in the analysis of NMR shifts in solution since it gives rise to what has come to be called the "pseudo contact shift."[16]

Thus the prevalent notion that the isotropic component of the hyperfine interaction rises solely from the Fermi contact term is really true only for free radicals in which the orbital momentum is strongly quenched and the spin–orbit terms are small. Not only do dipolar terms contribute to the isotropic term when unquenched orbital momentum is present but it has been shown recently[15, p. 583] that in the case of $Fe(CN)_6{}^{3-}$ there is a mechanism whereby much of the anisotropic component in the ^{13}C hyperfine interaction can result from the Fermi contact term. Thus when orbital momentum is present, there is no clear-cut separation into dipolar and Fermi contact terms in the hyperfine interaction.

Electron Spin Density from Hyperfine Constants

Examination of Eq. (30) and (32) shows the hyperfine interaction to be strongly dependent on the distribution of unpaired electrons in the vicinity of the nucleus and as such provides a sensitive testing ground for theoretical wave functions. Researchers working with free radicals often talk about determining the spin density from the hyperfine interaction but this is not strictly correct. A true spin density ρ_s would be computed from the molecular wave function Ψ in the following manner

$$\rho_s(x, y, z) = 2\sum_i^N \iint \ldots \int \Psi^* S_{zi}\Psi \, d\tau_1 \ldots d\tau_{i-1}\, d\tau_{i+1} \ldots d\tau_N,$$

where N is the number of electrons in the molecule. This function would map out the density of spin as a function of the spatial coordinates and could have both positive and negative values in different regions of space. Integration of $\rho_s(x, y, z)$ over all space would give the difference between the number of electrons with spin up and spin down. A mapping of spin density of this type would require much more data than are given by the few hyperfine constants obtainable from one molecule, so that the spin density referred to in the literature is a much more restricted quantity.

Most approaches to obtaining electron densities from hyperfine interactions have started with the assumption that the wave function of the molecule or ion could be approximated by a Slater determinant made up of molecular orbitals that are linear combinations of atomic orbitals.

$$\psi_{\mathrm{MO}} = \sum_n b_n \phi_n, \tag{34}$$

where ϕ_n is an atomic orbital centered on the nth atom. If overlap is ignored, the density of the electron on atom n is taken to be

$$\rho_n = b_n^2. \tag{35}$$

Thus this model maps density over atoms rather than spatial coordinates. If overlap is included some other definition of charge density such as Mulliken's[17] may be employed. Eq. (30) and (31) are then used with this wave function to calculate the hyperfine constants as a function of the ρ_n's. If symmetry is high enough, there will be enough hyperfine constants to determine all the ρ_n's, otherwise additional approximations may be necessary. For transition metal complexes, where spin–orbit effects are appreciable, it is necessary to include admixtures of excited-state configurations that are mixed with the ground state by the spin–orbit operator. To determine the extent of admixture, we must know the value of the spin–orbit constant λ and the energy of the excited states. It is customary to obtain these from spectroscopic studies.

A serious drawback to this model is its inability to account for observed Fermi contact interactions. The usual Slater determinant uses the same molecular orbital for spin up as for spin down and will therefore yield a Fermi contact contribution only if the orbital with the unpaired electron contains the s orbital of the atom concerned. Symmetry restrictions, however, prevent the presence of s orbitals in the molecular orbital for aromatic free radicals and many transition metal complexes; nevertheless, the large isotropic terms observed in these systems require an extensive contribution from the Fermi contact term. This is explained by assuming the unpaired electrons polarize the inner-filled orbitals having s character to produce a small net unpairing of spin. Very small polarizations will produce large Fermi terms due to the large density of s orbitals in the vicinity of the nucleus. Theoretically, this problem is handled in one of two ways: The MO coefficients b_n in Eq. (34) are allowed to be different for spin up than for spin down, or small admixtures are made to the ground state of excited states in which an inner electron is promoted to an empty orbital with s character.

Free Radicals The bulk of spin density work has been done using isotropic hyperfine constants of aromatic free radicals. In these free radicals the unpaired electron is in a π orbital that has a node in the plane that includes the carbon atoms of the ring and the atoms attached to those carbon atoms. Thus, the Fermi contact interaction that gives rise to the isotropic hyperfine constant must come from the polarization mechanism discussed earlier. The model that is now used was first developed for the ^1H splittings observed for those hydrogen atoms attached to the aromatic ring. It is assumed in these molecules, with some theoretical justification, that the polarization effect occurs through the C–H bond and that the hyperfine constant for the hydrogen attached to the nth carbon atom is proportional to ρ_n as defined by Eq. (35). That is

$$a_n = Q\rho_n. \tag{36}$$

It is further assumed that the proportionality constant Q is the same for all aromatic molecules. Q is then determined from experimental measurements on high-symmetry molecules where ρ_n is determined from symmetry alone. This empirical approach has worked surprisingly well, and values of ρ_n determined experimentally agree reasonably well with values obtained from Hückel theory. Since this approach is somewhat empirical, it should not come as a surprise that different values of Q are proposed for aromatic molecules with different charge. Equation (36) is also found to work for ^1H splittings observed for CH_3 groups attached to aromatic rings. In this case Q is of opposite sign.

To explain the presence of hyperfine constants for atoms in which Hückel theory predicts $\rho_n = 0$, the approach outlined above has been modified by assuming that the filled π orbitals are polarized by the unpaired spin resulting in some unpaired spin in these orbitals. A modified Hückel theory[18] has been developed, which works quite well, to include this polarization. In this theory, ρ_n in Eq. (36) is reinterpreted to be a spin density on the nth carbon atom

$$\rho_n = \sum_i (b_{ni}^+)^2 - \sum_j (b_{bj}^-)^2, \tag{37}$$

where b_{ni}^+ is the MO coefficient for atom n for the ith molecular orbital with positive spin. It is found that ρ_n is negative for those carbon atoms in which b_n^+ is zero for the unfilled molecular orbital.

This empirical approach has been extended to other nuclei but it requires postulating a whole series of polarization constants. For ^{13}C splittings in the aromatic ring, the following equation has proved successful[19]

$$a_n^C = Q_n^C \rho_n + \sum_j Q_{X_jC}^C \rho_j, \tag{38}$$

where ρ_j is the spin density on atoms bonded to the carbon atom and $Q_{X_jC}^C$ is the polarization constant for bond X_jC. Q_n^C is also dependent on the atoms carbon is bonded to and is written as

$$Q_n^C = S^C + \sum_j Q_{CX_j}^C, \tag{39}$$

where S^C is polarization of 1s electrons of carbon atom and $Q_{CX_j}^C$ is polarization constant for electrons in the CX bond. Similar approaches have been found necessary for other nuclei. The large number of empirical constants needed for nuclei other than hydrogen have made this procedure less satisfactory and open to disputes about the best values of some constants.

Transition Metal Complexes Attempts have been made to use the Fermi contact portion of the metal ion's hyperfine interaction to determine electron distribution in transition metal complexes, but these attempts have not been very successful or useful. Most of the useful data have been obtained from the anisotropic portion of the hyperfine constant as determined from single-crystal or supercooled solution studies. For ligand nuclei both the isotropic and anisotropic parts of the hyperfine interaction have yielded useful results on electron distribution. Although the most useful data still come from the anisotropic portion of the hyperfine interaction, there is need for more work in this area in that the available data are rather sparse for many important types of complexes.

A large collection of data is available in the literature for isotropic hyperfine interactions of ligand nuclei from the study of NMR shifts in solution. Although the magnitudes of the hyperfine constants are not well determined, the relative values within one complex are known. The relative values have been used to establish electron spin distributions within many aromatic ligands. Two serious defects in the analysis of these NMR shifts limit the validity of some of the conclusions deduced.

The first defect is in ignoring or overlooking the fact that orbital effects, discussed under "Electron Spin–Nuclear Spin Interaction," can make major contributions to the observed shifts and must be accounted for in order to get the correct Fermi contact contributions. Many researchers have assumed the total isotropic term is from the Fermi contact interaction.

The second defect occurs in the prevailing model used to interpret the Fermi contact shifts. This model assumes that an electron is transferred to the metal ion to pair up with its unpaired electron. This leaves a ligand free radical similar to that created by removing one electron from the highest-energy π orbital of the free ligand. The hyperfine constants are then treated using the theory developed for aromatic free radicals, as discussed in "Free Radicals." In some cases where the relative hyperfine constants do not fit the expected pattern, the possibility is considered of an electron from the metal ion going into the first unfilled π orbital of the free radical. If this does not work, a third model is proposed in which an electron is donated from a σ orbital and discussion is couched in terms of transmittal of spin density through the σ framework.

That this approach is crude and unsatisfactory is obvious. For example, proposing that electrons be back-donated from a positive metal ion into an empty π orbital seems intuitively bad since this would increase the positive charge on the metal ion. The difficulty is in the implied assumption of only one unpaired spin on the metal ion when, in fact, most systems studied by NMR have several unpaired spins in different d orbitals.

To understand how this arises, let us consider the hypothetical complex with two unpaired spins pictured in Figure 2. Y=Y represents a bidentate aromatic ligand containing the nuclei whose hyperfine interactions we measure. The symmetry of this complex is D_{2h} and we fur-

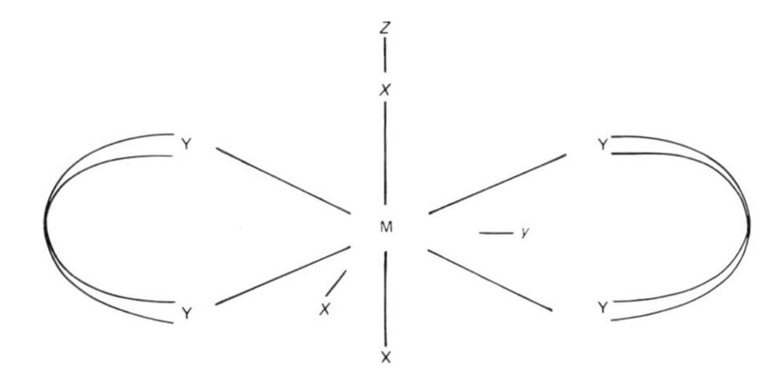

FIGURE 2 Hypothetical complex used as an example to discuss problems in determining spin distribution in complexes having more than one unpaired spin.

ther assume in our example that the ground state is a $^3B_{1g}$ state $d_{xz}{}^+d_{yz}{}^+$. The π molecular orbitals of the ligand can be classified as symmetric or antisymmetric with respect to a twofold rotation about the y axis. The molecular orbital associated with d_{xz} must be symmetric for rotation about the y axis and therefore only symmetric π orbitals mix with d_{xz}, while antisymmetric π orbitals mix with d_{yz}. The resulting ground-state configuration can only be viewed as a state in which unpaired spin is found in both symmetric and antisymmetric π ligand orbitals in nearly equal amounts. Thus, discussions in the literature that assume all unpaired spin must be in only one π molecular orbital for any given complex are making a fundamental error. Many of the cases in which researchers have stated that the spin delocalization must occur through the σ system because they could not fit their results to any π molecular orbital of the ligand may in reality involve spin delocalized in more than one π molecular orbital simultaneously. This whole problem requires a more careful analysis than it has yet received in the literature.

REMARKS

The careful reader will have noticed that a complete set of references to the literature has not been provided. The purpose of this paper is to acquaint the reader with the advantages and disadvantages of the various methods that have been used to measure the hyperfine interaction and with the common errors that have been made in its measurement and use to determine electron distributions. For a more complete treatment of ESR and ENDOR plus appropriate literature references, the reader is referred to the excellent books by Abragam and Bleaney[20] and by Wertz and Bolton.[21] For a treatment of NMR shifts in paramagnetic molecules, the book edited by LaMar, Horrocks, and Holm[15] should be consulted.

REFERENCES

1. J. S. Hyde and A. H. Maki, *J. Chem. Phys.*, **40**, 3117 (1964).
2. J. S. Hyde, *J. Chem. Phys.*, **43**, 1806 (1965).
3. J. H. Freed, *J. Chem. Phys.*, **43**, 2312 (1965); *J. Phys. Chem.*, **71**, 38 (1967).
4. J. H. Freed, D. S. Leniert, and J. S. Hyde, *J. Chem. Phys.*, **47**, 2762 (1967).
5. R. D. Allendoerfer and A. H. Maki, *J. Magn. Resonance*, **3**, 396 (1970).
6. R. J. Kurland and B. R. McGarvey, *J. Magn. Resonance*, **2**, 286 (1970).
7. R. Beringer and M. A. Heald, *Phys. Rev.*, **95**, 1474 (1954).
8. P. Kusch, *Phys. Rev.*, **100**, 1188 (1955).
9. R. W. Fessenden, *J. Chem. Phys.*, **37**, 747 (1962).
10. R. W. Fessenden and R. H. Schuler, *J. Chem. Phys.*, **43**, 2704 (1965).

11. P. A. Narayana and K. V. L. N. Sastry, *J. Chem. Phys.*, **57**, 1805 (1972).
12. G. H. Rist and J. S. Hyde, *J. Chem. Phys.*, **52**, 4633 (1970).
13. B. Bleaney, *Philos. Mag.*, **42**, 441 (1951).
14. C. P. Poole, Jr., and H. A. Farach, *J. Magn. Resonance*, **4**, 312 (1971).
15. G. N. LaMar, W. D. Horrocks, Jr., and R. H. Holm, ed., *NMR in Paramagnetic Molecules: Principles and Applications*, Academic Press, N.Y. (1973).
16. H. M. McConnell and R. E. Robertson, *J. Chem. Phys.*, **29**, 1361 (1958).
17. R. S. Mulliken, *J. Chem. Phys.*, **23**, 1833 (1955).
18. A. D. McLachlan, *Mol. Phys.*, **3**, 233 (1960).
19. M. Karplus and G. K. Fraenkel, *J. Chem. Phys.*, **35**, 1312 (1961).
20. A. Abragam and B. Bleaney, *Electron Paramagnetic Resonance of Transition Ions*, Oxford University Press, London (1970).
21. J. E. Wertz and J. R. Bolton, *Electron Spin Resonance*, McGraw-Hill, N.Y. (1972).

B. M. Fung

Nuclear Quadrupole Coupling Constants

Nuclei can be classified into three types according to their magnetic properties: those with no magnetic moment ($I = 0$); those with a nuclear spin equal to one half ($I = 1/2$) but no quadrupole moment; and those with a spin of one or larger ($I > 1/2$) and a quadrupole moment.

The existence of a nuclear quadrupole moment (eQ) is due to a distribution of nuclear charge density (ρ) with less than cubic symmetry. The nuclear charge distribution can be described by an electric quadrupole moment tensor, the components of which are related to a scalar quantity eQ, which is defined[17] by

$$eQ = \int \langle I, I|\rho(X)|I, I\rangle (3z^2 - r^2)\mathrm{d}^3X. \tag{1}$$

In atoms, ions, and molecules a nucleus is surrounded by electrons. The spatial distribution of the electronic charge can be described by the electrostatic field gradient tensor, V_{ij}:

$$V_{ij} = \partial^2 V/\partial x_i \partial x_j, \tag{2}$$

where V is the electric field potential. V_{ij} is a symmetric tensor and can be diagonalized in a suitable axis system. Furthermore, the tensor V_{ij} is traceless. If the axis system is chosen such that the z axis lies along the direction of maximum field gradient, the field gradient tensor is conveniently expressed in terms of the two quantities

$$eq = V_{zz} = (\partial^2 V/\partial z^2)_0, \tag{3}$$

and

$$\eta = (V_{xx} - V_{yy})/V_{zz}. \tag{4}$$

The parameter q is called the electric field gradient, and η is called the asymmetry parameter.

The electric field gradient has contributions from other nuclei and all electrons,

$$q = \sum_K Z_K(3\cos^2\theta_K - 1)/R_K{}^3 - \sum_k \langle\psi|(3\cos^2\theta_k - 1)/r_k{}^3|\psi\rangle, \tag{5}$$

where Z is the nuclear charge, R and r are the distances of another nucleus and the electron, respectively, from the nucleus concerned, θ is the angle between R or r and the z axis, and ψ is the molecular wave function. The electric field gradient can be calculated theoretically, and its accuracy depends on that of the geometrical parameters (θ_K and R_K) and the wave function.

The electrostatic interaction between the nucleus and its environment is defined by the quadrupole Hamiltonian

$$\mathcal{H}_Q = (1/6)\sum_{ij} Q_{ij} V_{ij}. \tag{6}$$

The nuclear quadrupole interaction often produces measurable effects in electronic, rotational, atomic and molecular beam resonance, nuclear magnetic resonance, and Mössbauer spectra. In many cases it can be directly studied by nuclear quadrupole resonance (NQR). The magnitude of interaction is expressed by the nuclear quadrupole coupling constant, e^2qQ/h, which is the product of Eq. (1) and (3) expressed as a frequency.

Since the nuclear quadrupole moments for almost all nuclei are known, the study of the nuclear quadrupole coupling constant and the asymmetry parameter furnishes valuable information on molecular structure and symmetry.

General aspects of nuclear quadrupole interaction, with many examples, have been discussed in several excellent reviews.[17,19,29] Here we will concentrate on the discussion of deuteron quadrupole interaction because of the abundance and importance of compounds containing hydrogen, the particularly small magnitudes of the quadrupole coupling constants (which distinguish deuterium from other nuclei in covalent compounds), and the appearance of a large amount of data on deuteron quadrupole coupling constants in recent years that is not covered by previous reviews.

EXPERIMENTAL DETERMINATION OF DEUTERON QUADRUPOLE COUPLING CONSTANTS

Deuterium has a nuclear spin of unity ($I = 1$) and a small nuclear quadrupole moment, $Q = 2.7965 \times 10^{-27}$ cm^2.[35] In compounds containing deuterium, the molecular environment of the deuterium nucleus usually has a symmetry less than T_d or O_h, with only a few exceptions, e.g., solid LiD. Therefore, in most cases, the deuteron quadrupole coupling constant is non-zero. However, its value is quite small because both the nuclear quadrupole moment and the electric field gradient have small values. The reported values of e^2qQ/h for deuterium are always less than 400 kHz.

For small molecules in the gaseous state, quadrupole coupling constants for deuterium and other nuclei can be determined from microwave spectroscopy[54,56] and molecular beam resonance.[43] For systems that behave differently in isolated molecules and in the condensed states, these techniques give unique information. Examples are lithium deuteride,[61] which has an NaCl structure and therefore a zero field gradient for deuterium in the solid state, and deuterium fluoride,[36] which forms strong hydrogen bonds in both the liquid and the solid state.

In microwave spectroscopy, pure rotational transitions are studied. If one or more nuclei in the molecule has a nuclear quadrupole moment, the quadrupole Hamiltonian [Eq. (6)] has to be included in the quantum mechanical treatment because the field gradient q [Eq. (5)] is dependent on the rotational wave function. The nuclear quadrupole interaction, which causes the rotational transitions to split into hyperfine structure, can usually be treated as a perturbation to the rotational Hamiltonian. The general cases of symmetric and asymmetric tops have been reviewed.[5] For deuterium, the nuclear quadrupole interaction is small and the hyperfine structure is not always resolved, so that the deuteron quadrupole coupling constant must be obtained by curve-fitting. The accuracy of the data diminishes as the size and the asymmetry of the molecule increases.

Molecular beam resonance can be performed in a magnetic field or an electric field. In molecular beam magnetic resonance, when the Zeeman interaction is much larger than the quadrupole interaction, as it is for deuterium in a high magnetic field, transitions between nuclear spin levels are detected with splittings caused by quadrupole interactions.

The resulting spectrum represents a weighted average over the rotational states and a careful analysis of it yields the nuclear quadrupole coupling constant. Molecular beam electric resonance is complementary to pure rotational spectroscopy since transitions between the Stark levels of the rotational states ($\Delta J = 0$) are observed (sometimes $\Delta J = \pm 1$ transitions are also studied). Specially constructed maser spectrometers[55] that can detect transitions of rotationally selected molecules have been used to determine very small coupling constants, such as those for deuterium compounds. Again, molecular beam resonance is currently limited to the study of small molecules.

Pure NQR in the absence of a magnetic field is a simple theoretical problem.[20,29] Only the quadrupole Hamiltonian [Eq. (6)] needs to be considered. For a spin $I = 1$, the energy levels are

$$\begin{aligned} E_{-1} &= e^2qQ(1-\eta)/4, \\ E_0 &= -e^2qQ/2, \\ E_{+1} &= e^2qQ(1+\eta)/4, \end{aligned} \tag{7}$$

with corresponding quadrupole transitions

$$\nu = \left|\frac{3e^2qQ}{4h}\right|\left(1 \pm \frac{\eta}{3}\right). \tag{8}$$

Although the study of NQR in the megahertz range is a standard technique for determining quadrupole coupling constants, a direct measurement of the NQR for deuterium by the conventional method is not possible because of the small values of its coupling constants. To observe pure quadrupole transitions for deuterium, double resonance methods can be used. In the double resonance experiments, the effect of the deuterium nuclear quadrupole transitions on the NMR signal[50] or the NQR signal[42] of another nucleus is observed. For example, an ingenious method of detecting pure quadrupole transitions by nuclear double resonance was suggested by Schwab and Hahn.[50] In their experiment, the quadrupole transitions of deuterium in a 12 percent deuterium-enriched sample of *p*-dichlorobenzene were excited, and the effect on the proton resonance signal through dipolar cross relaxation was detected. The ^{35}Cl spins served to establish and monitor an ordered metastable spin state of the protons.

Direct or indirect observation of pure quadrupole transitions yields the most accurate values for quadrupole coupling constants and asymmetry parameters in the solid state. Polycrystalline samples can be used because the transition frequencies [Eq. (8)] have no geometrical dependence. Hence, the crystal structure of the sample need not be known. The limitation of the double resonance methods for deuterium lies in the fact that only special systems with suitable relaxation characteristics can be studied.

Nuclear magnetic resonance (NMR) is perhaps the simplest technique for obtaining deuterium quadrupole coupling constants in solids or in liquid crystalline solutions. In ordinary NMR experiments with a magnetic field $H_0 \gtrsim 10^4$ gauss, the nuclear quadrupole interaction [Eq. (6)] for deuterium is much smaller than the Zeeman interaction and can be treated as a perturbation to the Hamiltonian

$$\mathcal{H}_M = -\gamma\hbar H_0 I_Z \tag{9}$$

where γ is the gyromagnetic ratio and Z is the direction of the magnetic field. If the polar angles of the Z axis with respect to the molecular frame (in which the tensor V_{ij} is diagonal) are θ and ϕ, the first-order energies for Eq. (6) and (9) are

$$E_m = -\gamma\hbar H_0 m + \frac{e^2qQ[3m^2 - I(I+1)]}{8I(2I-1)}(3\cos^2\theta - 1 + \eta\sin^2\theta\cos 2\phi). \tag{10}$$

The transition frequencies for $I = 1$ are then

$$\nu = (\gamma H_0/2\pi) \pm 3e^2qQ(3\cos^2\theta - 1 + \eta\sin^2\theta\cos 2\phi)/8h\,. \tag{11}$$

For single crystals, a direct measurement of the angular dependence of the resonance frequencies followed by a transformation of the coordinate system[57] provides the values of e^2qQ/h and η. Their accuracies are only limited by the stability and resolution of the spectrometer. In rare cases, the dipolar splitting can also be resolved (Figure 1), and the expressions for the energy levels and transition frequencies are more complicated.[15]

For polycrystalline samples, the parentheses in Eq. (11) are replaced by a distribution function. The line shape of the "powder spectrum" has been discussed for spin $I = 3/2$[17] and $I = 1$[14] nuclei with various values of η. In general, for $I = 1$ it shows singularities at $(\gamma H_0/2\pi) \pm 3e^2qQ(1-\eta)/8h$, shoulders or distribution edges at $(\gamma H_0/2\pi) \pm 3e^2qQ(1+\eta)/8h$, and steps or distribution edges at $(\gamma H_0/2\pi) \pm 3e^2qQ/4h$ (Figure 2). However, because of dipolar broadening, the powder spectrum is not always well resolved. Then, the data have to be obtained from curve-fitting, which may result in larger uncertainties in the values of e^2qQ/h and η. Special care must be taken in analyzing the powder spectra of samples undergoing rapid internal rotation.[4]

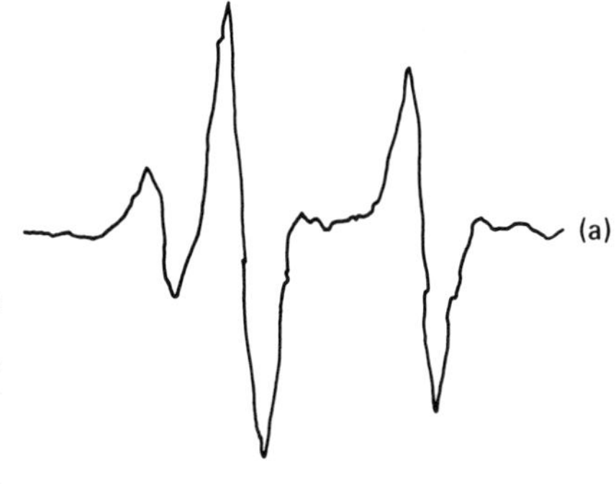

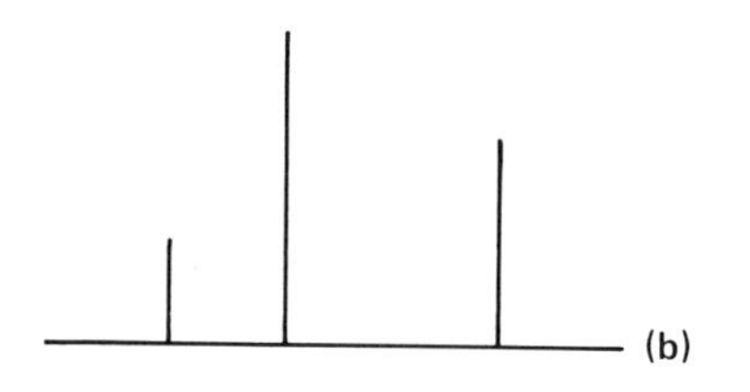

FIGURE 1 Deuterium NMR spectrum of a single crystal of $Ba(ClO_3)_2 \cdot D_2O$ showing dipolar and quadrupole splittings.[15] (a) Half of the experimental spectrum; the other half is antisymmetrical with respect to ω_0 and not shown here. (b) Theoretical pattern.

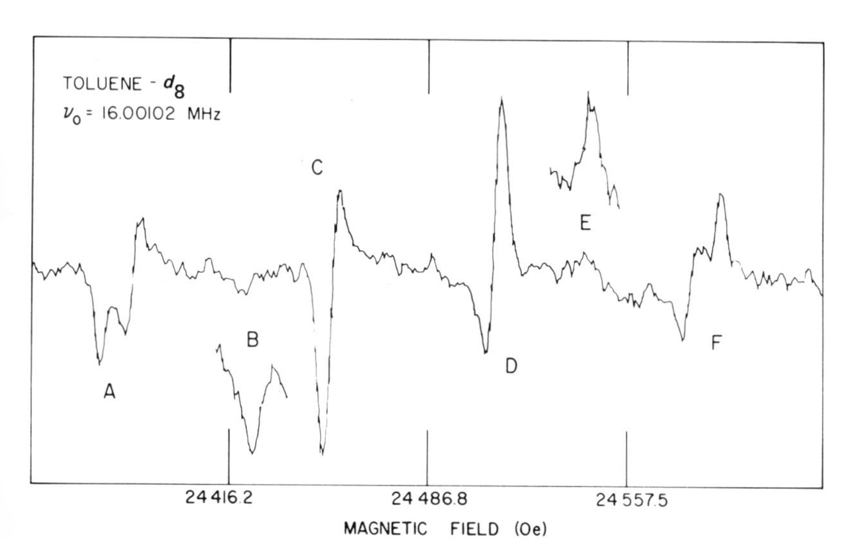

FIGURE 2 Deuterium NMR spectrum (derivative mode) of polycrystalline toluene-d_8 at 4.2 K. The methyl group shows singularities at C and D and steps at B and E. The ring deuterons show singularities at A and F ($\eta \neq 0$ with no internal rotation), and the steps lie outside the region shown.[5]

For liquid crystals or liquid crystalline solutions, the parentheses in Eq. (11) are replaced by a factor $S_{zz}/2$, where S_{zz} is the orientation factor of the z-axis. The value of S_{zz} can be obtained from the proton resonance spectrum provided that the molecular geometry is known. If the z axis reorients rapidly about a major axis with an angle θ between them, then instead of $S_{zz}/2$ the factor $(3 \cos^2\theta - 1 - \eta \sin^2\theta) \cdot S_{33}$ has to be used,[32] where S_{33} is the orientation factor of the major axis. In any event, the nuclear quadrupole coupling constant and the asymmetry factor cannot both be determined from a single liquid crystalline solution. For many covalent deuterium compounds, η is very small (<0.10), and is neglected so that e^2qQ/h can be calculated from the spectrum. This procedure inevitably brings in a systematic error, amounting to about 5 percent or more in many cases. To avoid this difficulty, the value of η obtained from the solid state of similar compounds can be used in the calculation. In some (but not all) cases, either the composition or the temperature of the liquid crystalline solution can be varied to obtain a dependence of the quadrupole splitting on the orientation factor, so that both e^2qQ/h and η can be determined.[48]

If there is a thermal motion, the quadrupole splitting is slightly modified. A vibration of the X–D bond about its equilibrium position with an angular displacement of $\pm\alpha$ introduces a factor of $(\cos^2\alpha + \cos\alpha$ into the angular part of the splitting.[62]

In the liquid state, the nuclear quadrupole coupling constant is difficult to determine. In the extreme narrowing case ($\omega^2\tau_c^2 << 1$), where ω is the Larmor frequency and τ_c is the rotational correlation time, the spin-lattice relaxation time (T_Q) of the quadrupole nucleus in a spherical diamagnetic molecule can be expressed[1] by:

$$\frac{1}{T_Q} = \frac{3}{40} \frac{2I+3}{I^2(2I-1)} \left(\frac{e^2qQ}{\hbar}\right)^2 \left(1 + \frac{\eta^2}{3}\right) \cdot \tau_c . \tag{12}$$

However, because of the difficulty in measuring the correlation time, the quadrupole coupling constant determined from T_Q in the liquid state is not always reliable even when η is truly negligible. In special instances, τ can be determined from Raman line shapes[26] or the T_Q of another quadrupole nucleus in the same molecule whose nuclear quadrupole coupling constant is accurately known.[24]

Deuterium quadrupole coupling constants can also be obtained from electron nuclear double resonance (ENDOR).[19,30] An observation of the hyperfine structure caused by quadrupole coupling in the electron paramagnetic resonance (EPR) spectrum, as for many lanthanide complexes, has not been reported for deuterium. The determination of nuclear quadrupole coupling constants from Mössbauer spectroscopy is not applicable to the deuterium nucleus.

CHARACTERISTICS OF DEUTERON QUADRUPOLE COUPLING CONSTANTS

Since the immediate electronic environment of the deuterium nucleus in most compounds is relatively simple, deuterium quadrupole coupling constants show very systematic variations with molecular structure. In covalent compounds, except those with hydrogen bonds or multicentered bonds (as in many boron deuterides), deuterium is bonded to only one atom. The bonding electrons have essentially cylindrical symmetry, and the only asymmetry in the electronic environment of deuterium comes from indirectly bonded atoms. Therefore, in many cases, η is very small (< 0.1).

Although many experimental methods cannot determine the sign of the deuterium quadrupole coupling constant, theoretical calculations and available experimental data indicate that it is always positive. Unlike the quadrupole coupling constants for ^{35}Cl and ^{14}N, which are often about 10 percent larger in the gas phase than in the solid,[34,64] deuterium quadrupole coupling constants seem to have little dependence on the physical state of the molecule (Table 1) unless hydrogen bonding is involved (see below). This may be related to the fact that deuterium does not have p electrons in the valence shell. The value for CH_3D in the gaseous state, 191.48 ± 0.77 kHz,[63] was not included in Table 1 because there are no corresponding measurements in other states, and it was felt that the comparison for other alkanes with at least two carbon atoms is more appropriate.

The deuterium quadrupole coupling constant bears an interesting relationship to the force constant ($k = \partial^2 W/\partial z^2$, where W = electro-

TABLE 1 Deuterium Quadrupole Coupling Constants (in kilohertz) for Several Compounds in Different Phases

	CD_3R	C_6D_6	RC≡CD	CD_3CN	DCN
Gas	176 ± 15^a		200 ± 10^b 208 ± 10^c	168 ± 4^d	194 ± 2^e
Liquid crystal	167 ± 12^f	183 ± 10^g	198 ± 7^h 199 ± 2^i	171 ± 17^g	199 ± 3^g
Solid	165^j 168 ± 3^k	187 ± 2^l	215 ± 5^m		

[a] R = C≡CH.[49]
[b] R = D.[44]
[c] R = CH_3.[60]
[d] From Kukolich *et al.*[28]
[e] From De Lucia and Gordy.[21]
[f] R = CH_3.[32]
[g] From Millett and Dailey.[32]
[h] R = D.[11]
[i] R = CH_3.[32]
[j] R = C_6H_5.[45] Uncertainty not given.
[k] R = CH_3.[9]
[l] From Pyykko.[40]
[m] R = C_6H_5.[37]

static energy) for an X–D bond. Salem[47] first investigated this relationship and pointed out that the approximation $k = q$ (in atomic units) should hold for small diatomic molecules. We have collected data for a large variety of compounds, most of them polyatomic, and found that e^2qQ/h bears a fairly good linear relation with k,[22,31,59] irrespective of whether the other valence electron in the X–D bond is an s, p, or d electron. The available data are presented in Table 2 and Figure 3. A least-squares fit of the data yields the relation (in atomic units)

$$k = 0.02 + 1.07\,q. \tag{13}$$

In Figure 3, quadrupole coupling constants for deuterium bonded to Mo, W, and Al exhibit the largest percentage deviations from Eq. (13). The experimental data for those elements were obtained from compounds with uncommon structures: $\pi\text{–}(C_5H_5)_2MoD_2$, $\pi\text{–}(C_5H_5)_2WD_2$, and $LiAlD_4$, which does not have a truly isolated AlD_4^- unit and Al–D bonds.

Recently, Parr and co-workers derived an exact theoretical relation between k and q.[3] A more detailed study and comparison of that relation with experimental data should result in a better understanding of molecular structure and wave functions.

If the X atom in the X–D bond can have different forms of hybridization, the force constant and the deuterium quadrupole coupling constants vary accordingly. The trend $sp^3 < sp^2 < sp$ for deuterocarbons has been established[32,37,46] (Table 1). Non-electron-deficient boron deuterides behave similarly.[62] A molecular orbital calculation[37] showed that the change of the quadrupole coupling constant is mainly due to the shorten-

TABLE 2 Representative Values of Deuterium Quadrupole Coupling Constants (in kilohertz) for Covalent Compounds Containing the X–D Bond

H(225)[a]	Li(33)[a]	Be(70)[a]		B(105)[a]	C(167)[b]	N(230)[a]	O(314)[a]	F(354)[c]
				Al(72)[d]	Si(95)[d]	P(115)[d]	S(155)[d]	Cl(190)[e]
						As(94)[f]	Se(123)[g]	Br(153)[e]
			Mo(52)[h]					I(113)[e]
			W(54)[h]					

[a] See Merchant and Fung[31] for complete reference. The datum for Be is theoretical.[12]
[b] Table 1.
[c] From Muenter and Klemperer.[33]
[d] See Fung and Wei[22] for complete reference.
[e] From Genin *et al.*[23]
[f] From Burnett and Zeltmann.[10]
[g] From Chandra and Dymanus.[13]
[h] From Wei and Fung.[59]

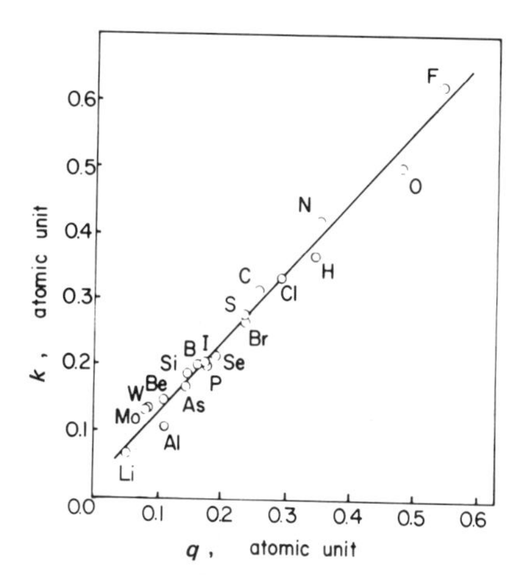

FIGURE 3 Relation between the force constant for the X–D bond and the electric field gradient at the deuterium nucleus. Values of k for X = H, Li, and Be are taken from G. Herzberg, *Spectra of Diatomic Molecules*, p. 458 (D. Van Nostrand Co., Princeton, N.J., 1950); the values for B, Al, Mo, and W are obtained from the stretching frequencies of 2360 cm^{-1}, 1670 cm^{-1}, 1847 cm^{-1}, and 1896 cm^{-1}, respectively, for the corresponding hydrogen compounds by utilizing the isolated harmonic oscillators approximation; k values for the other atoms are taken from E. B. Wilson, Jr., J. C. Decius, and P. C. Cross, *Molecular Vibrations*, p. 175 (McGraw-Hill, New York, 1955). The q's are taken from Table 2 with a conversion factor of 1/657.[37]

ing of the C–D bond with the increase in the s character of carbon, rather than a systematic change of the charge distribution around deuterium with the change in hybridization of carbon. The molecular orbital calculation also indicated that an increase in the electronegativity of the substituent on carbon would cause a reduction in the deuterium quadrupole coupling constant. This has been borne out by experimental results (Table 3). Unusually large values of deuterium quadrupole coupling constants for formic acid were reported.[27] However, the validity

TABLE 3 Effect of Substituents on Deuterium Quadrupole Coupling Constants (in kilohertz) in Deuterocarbons[a]

$CDCl_3$ 167.6 ± 6.8	CD_2Cl_2 169.6 ± 1.1	CD_2ClCD_2Cl 171.7 ± 0.8	CD_3CCl_3 174.3 ± 1.6	$CD_3CCl_2CD_3$ 174.7 ± 1.9	(*b*)
CD_3CN 171 ± 17	CD_3Br 177 ± 18	CD_3I 189 ± 19			(*c*)
C_6D_6 183 ± 10	$C_6D_5NO_2$ 181 ± 10	*m*-$C_6D_4(NO_2)_2$ 176 ± 10	*s*-$C_6D_3(NO_2)_3$ 171 ± 10		(*d*)
C_6D_6 183 ± 10	D_2CO 170 ± 2	DCOOH 161 ± 2			(*e*)

[a] A more complete list is available in Millett and Dailey.[32]
[b] From Ragle and Sherk.[42]
[c] From Caspary *et al.*[11]
[d] Recalculated from Wei and Fung[58] by assuming $\eta = 0.06$ instead of $\eta = 0.0$.
[e] From Adriaensseus and Bjorkstam.[2]

of these data must be doubted because they were obtained from an assumed geometry and are in gross disagreement with data on similar molecules.[2,30,52]

In hydrogen bonded systems, the deuterium quadrupole coupling constant decreases steadily with increase in the strength of the hydrogen bond.[6,7,16,53] It has been suggested that the deuterium quadrupole coupling constant is a more sensitive quantity than the O–D stretching frequency in evaluating the strength of the hydrogen bond.[6,7] An empirical relation between the O-------O distance in an O–D-------O system and the deuterium quadrupole coupling constant has been proposed.[53] Molecular orbital calculations have also been made for hydrogen-bonded systems.[38] It was shown that the four-electron model of Coulson and Danielsson[18] is inadequate in explaining the deuterium quadrupole

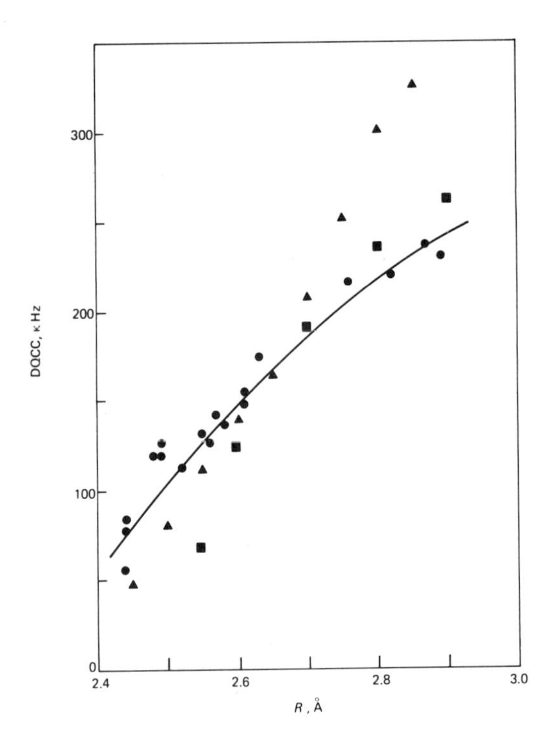

FIGURE 4 Deuterium quadrupole coupling constant of hydrogen-bonded crystals as a function of O—O distance. The circles are experimental values, with a line drawn to show the general trend. The triangles and squares are theoretical values for two different sets of O–D distances in a molecular orbital calculation.[38]

coupling constant, and a 13-electron or a 17-electron system reproduces the experimental data with only moderate success (Figure 4).

Phase transitions of ferroelectric materials[8,39] can also be studied through the variation of the deuterium quadrupole coupling constant with temperature.

Theoretical calculation of the quadrupole coupling constant for deuterium is generally simpler than that for other nuclei. The accuracy of the result depends on a correct choice of Q, the molecular geometry, and the molecular wave function [Eq. (5)]. In many cases, fair agreement with experimental data has been obtained.[25,37,38,41,51]

REFERENCES

1. A. Abragam, *The Principles of Nuclear Magnetism*, Oxford University Press, London (1961).
2. G. J. Adriaensseus and J. L. Bjorkstam, *J. Chem. Phys.*, **56**, 1223 (1972).
3. A. B. Anderson, N. C. Handy, and R. G. Parr, *J. Chem. Phys.*, **50**, 3634 (1969).
4. R. G. Barnes and J. W. Bloom, *Mol. Phys.*, **25**, 493 (1973); R. G. Barnes, *Proceedings of NQR Symposium*, Heyden & Sons, London (in press).
5. R. G. Barnes and J. W. Bloom, *J. Chem. Phys.*, **57**, 3082 (1972).
6. R. Blinc and D. Hadzi, *Nature*, **212**, 1307 (1966).
7. R. Blinc, *Adv. Magn. Resonance*, **3**, 141 (1968).
8. J. L. Bjorkstam, *Phys. Rev.*, **155**, 599 (1967).
9. L. J. Burnett and B. H. Muller, *J. Chem. Phys.*, **55**, 5829 (1971).
10. L. J. Burnett and A. H. Zeltmann, *J. Chem. Phys.*, **56**, 4695 (1972).
11. W. J. Caspary, F. Millett, M. Reichbach, and B. P. Dailey, *J. Chem. Phys.*, **51**, 623 (1969).
12. A. C. H. Chan and E. R. Davidson, *J. Chem. Phys.*, **49**,727 (1968).
13. S. Chandra and A. Dymanus, *Chem. Phys. Lett.*, **13**, 105 (1972).
14. T. Chiba, *J. Chem. Phys.*, **36**, 1122 (1962).
15. T. Chiba, *J. Chem. Phys.*, **39**, 947 (1963).
16. T. Chiba, *J. Chem. Phys.*, **41**, 1352 (1964).
17. M. H. Cohen and R. Reif, *Solid State Phys.*, **5**, 322 (1957).
18. C. A. Coulson and U. Danielsson, *Askiv Fys.*, **8**, 245 (1954).
19. L. R. Dalton and A. L. Kwiram, *J. Am. Chem. Soc.*, **94**, 6930 (1972).
20. T. P. Das and E. L. Hahan, *Solid State Phys.*, **Suppl. 1** (1958).
21. F. C. De Lucia and W. Gordy, *Phys. Rev.*, **187**, 58 (1969).
22. B. M. Fung and I. Y. Wei, *J. Am. Chem. Soc.*, **92**, 1497 (1970).
23. D. J. Genin, D. E. O'Reilly, E. M. Peterson, and T. Tsang, *J. Chem. Phys.*, **58**, 4525 (1968).
24. K. T. Gillen and J. H. Noggle, *J. Chem. Phys.*, **52**, 4905 (1970).
25. J. Harrison, *J. Chem. Phys.*, **48**, 2379 (1968).
26. E. F. Johnson and R. S. Drago, *J. Am. Chem. Soc.*, **95**, 1391 (1973).
27. S. G. Kukolich, *J. Chem. Phys.*, **51**, 358 (1969).
28. S. G. Kukolich, D. J. Ruben, J. H. S. Wang, and J. R. Williams, *J. Chem. Phys.*, **58**, 3155 (1973).
29. E. A. C. Lucker, *Nuclear Quadrupole Coupling Constants*, Academic Press, London (1969).

30. R. C. McCalley and A. L. Kwiram, *Phys. Rev. Lett.*, **24**, 1279 (1970).
31. S. Z. Merchant and B. M. Fung, *J. Chem. Phys.*, **50**, 2265 (1969).
32. F. S. Millett and B. P. Dailey, *J. Chem. Phys.*, **56**, 3249 (1972).
33. J. S. Muenter and W. Klemperer, *J. Chem. Phys.*, **52**, 6033 (1970).
34. W. B. Moniz and H. S. Gutowsky, *J. Chem. Phys.*, **38**, 1155 (1963).
35. H. Narumi and J. Watanabe, *Progr. Theor. Phys. (Kyoto)*, **35**, 1154 (1966).
36. H. M. Nelson, J. A. Leavitt, M. R. Baker, and N. F. Ramsey, *Phys. Rev.*, **122**, 856 (1961).
37. P. L. Olympia, Jr., I. Y. Wei, and B. M. Fung, *J. Chem. Phys.*, **51**, 1610 (1969).
38. P. L. Olympia, Jr., and B. M. Fung, *J. Chem. Phys.*, **51**, 2976 (1969).
39. D. E. O'Reilly and T. Tsang, *J. Chem. Phys.*, **46**, 1291 (1967).
40. P. Pyykko, *Ann. Univ. Turku Ser. A*, **88**, 93 (1966).
41. P. Pyykko, *Proc. Phys. Soc. London*, **92**, 841 (1967).
42. J. L. Ragle and K. L. Sherk, *J. Chem. Phys.*, **50**, 3553 (1969).
43. N. F. Ramsey, *Molecular Beams*, Oxford University Press, London (1956).
44. N. F. Ramsey, *Am. Sci.*, **49**, 509 (1961).
45. J. Rowell, W. D. Phillips, L. Melby, and M. Panar, *J. Chem. Phys.*, **43**, 3442 (1965).
46. J. Royston and J. A. S. Smith, *Trans. Faraday Soc.*, **66**, 1039 (1970).
47. L. Salem, *J. Chem. Phys.*, **38**, 1227 (1963).
48. E. T. Samulski and H. J. C. Berendsen, *J. Chem. Phys.*, **56**, 3920 (1972).
49. R. L. Schoemaker and W. H. Flygare, *J. Am. Chem. Soc.*, **91**, 5417 (1969).
50. M. Schwab and E. L. Hahn, *J. Chem. Phys.*, **52**, 3152 (1970); **54**, 1431 (1971).
51. L. C. Snyder and H. Basch, *Molecular Wave Functions and Properties*, Wiley Interscience, New York (1972).
52. G. Soda and T. Chiba, *J. Chem. Phys.*, **48**, 4328 (1968).
53. G. Soda and T. Chiba, *J. Chem. Phys.*, **50**, 439 (1969).
54. T. M. Sugden and C. N. Kenney, *Microwave Spectroscopy of Gases*, Van Nostrand, London (1965).
55. P. Thaddeus and L. C. Krisher, *Rev. Sci. Instrum.*, **32**, 1083 (1961).
56. C. H. Townes and A. L. Schawlon, *Microwave Spectroscopy*, McGraw-Hill, New York (1955).
57. G. M. Volkoff, H. E. Petch, and D. W. L. Smellie, *Can. J. Phys.*, **30**, 270 (1952).
58. I. Y. Wei and B. M. Fung, *J. Chem. Phys.*, **52**, 4917 (1970).
59. I. Y. Wei and B. M. Fung, *J. Chem. Phys.*, **55**, 1486 (1971).
60. V. W. Weiss and W. H. Flygare, *J. Chem. Phys.*, **45**, 8 (1966).
61. L. Wharton, L. P. Gold, and W. Klemperer, *J. Chem. Phys.*, **37**, 2149 (1962).
62. J. Witschel, Jr., and B. M. Fung, *J. Chem. Phys.*, **56**, 5417 (1972).
63. S. C. Wofsy, J. S. Muenter, and W. Klemperer, *J. Chem. Phys.*, **53**, 4005 (1972).
64. P. N. Wolfe, *J. Chem. Phys.*, **25**, 976 (1956).

W. H. Flygare

Magnetic Interactions and the Electronic Structure of Diamagnetic Molecules

In this paper we outline the theory necessary to understand the important magnetic interactions in a molecule that has zero electronic spin and orbital angular momenta in the ground electronic state. We concentrate on a description of the magnetic interactions that are normally measured by microwave resonance techniques.[1,2]

The necessary theory for a proper interpretation of the molecular Zeeman effect in diamagnetic molecules is described in the first section and involves first a description of the coupling of the electronic and rotational motions leading to the molecular g-value tensor, $\mathcal{G}$. We also describe there the magnetic-field-induced effects leading to the magnetic susceptibility tensor, $\mathcal{X}$, and show the connection between the $\mathcal{G}$ and $\mathcal{X}$ tensors. The diagonal elements in the $\mathcal{G}$ and $\mathcal{X}$ tensors can also be combined with the moment of inertia tensor $\mathcal{I}$ to give the molecular quadrupole moments in the principal inertial axis system. The molecular structure can be combined with the above information to yield the electric dipole moment (and sign), the diagonal elements in the paramagnetic susceptibility tensor, and the anisotropies in the second moments of the electronic charge distribution. Adding the bulk magnetic susceptibility to the above numbers yields the diagonal elements in the total and the diamagnetic susceptibility tensor elements. The diagonal elements in the $\mathcal{X}$ and $\mathcal{G}$ tensors can be measured by various spectroscopic methods. These data obtained from the molecular Zeeman effect are given for a number of molecules in order to illustrate the methods, pitfalls, and accuracy of the results.

The nuclear magnetic interactions that give rise to nuclear magnetic shielding, σ, and the nuclear spin–rotation interaction, $\mathcal{M}$, are described in the second section. Nuclear magnetic shielding arises from an external magnetic field–molecule perturbation similar to the interaction leading to the field-induced moment and magnetic susceptibility. The spin–rotation interaction arises from a rotational-induced field at the nucleus in a way similar to the rotational-induced magnetic moment. The parallels between the $\mathcal{G}$ and $\mathcal{X}$ tensors with molecular center of mass origins and the $\boldsymbol{\sigma}$ and $\mathcal{M}$ tensors with nuclear origins are very strong. For instance, the diagonal elements in the spin–rotation interaction tensor, can be used to calculate the diagonal elements in the paramagnetic shielding tensor, a relation analogous to that between the molecular g-values and the paramagnetic susceptibility. If the total nuclear shielding is known through magnetic resonance measurements, the spin–rotation interaction constants can be used to extract the diagonal elements in the diamagnetic shielding tensor. We also review briefly the rotational molecular Zeeman effect in the presence of strong nuclear-rotational coupling.

Finally, in the third section of this paper, we describe a semiempirical atom dipole model that allows a reliable prediction of molecular electric dipole and quadrupole moments, diamagnetic susceptibilities, and diamagnetic nuclear shieldings. A set of localized bond and atom values are developed for the individual diagonal elements in the total molecular magnetic susceptibility tensor.

INTRAMOLECULAR ELECTRONIC-ROTATIONAL INTERACTIONS, MAGNETIC FIELD DEPENDENT INTERACTIONS, AND THE MOLECULAR ZEEMAN EFFECT

Discussion here is limited to nonvibrating molecules that have zero electronic spin and orbital angular momentum in the ground electronic state. In that the majority of molecules satisfy this criterion, this limitation is not severe. We also ignore the effects of molecular vibrations.

The zero-field kinetic energy for a system of rigid nonmagnetic nuclei and corresponding electrons in a molecule is given by[2]

$$T = \frac{1}{2}(\mathbf{J}-\mathbf{L})\cdot \mathcal{I}_n^{-1}\cdot(\mathbf{J}-\mathbf{L}) + \frac{1}{2m}\sum_i p_i^2$$

$$= \frac{1}{2}\mathbf{J}\cdot \mathcal{I}_n^{-1}\cdot \mathbf{J} - \mathbf{J}\cdot \mathcal{I}_n^{-1}\cdot \mathbf{L} + \frac{1}{2}\mathbf{L}\cdot \mathcal{I}_n^{-1}\cdot \mathbf{L} + \frac{1}{2m}\sum_i p_i^2$$

$$= \frac{1}{2}\sum_g \frac{J_g^2}{I_{ggn}} - \sum_g \frac{J_g L_g}{I_{ggn}} + \frac{1}{2}\sum_g \frac{L_g^2}{I_{ggn}} + \frac{1}{2m}\sum_i p_i^2, \tag{1}$$

in which **J** is the total angular momentum of the molecule and

$$\mathbf{L} = \sum_i \mathbf{r}_i \times \mathbf{p}_i \tag{2}$$

is the intrinsic electronic angular momentum or the electronic angular momentum in the rotating rigid nuclear frame. The sum over i is over all electrons and $\mathbf{r}_i$ and $\mathbf{p}_i$ are the center-of-mass position and linear momentum vectors for the electrons in the molecule.

$$\mathcal{I}_n^{-1} = \begin{pmatrix} \frac{1}{I_{xxn}} & 0 & 0 \\ 0 & \frac{1}{I_{yyn}} & 0 \\ 0 & 0 & \frac{1}{I_{zzn}} \end{pmatrix}$$

$$\mathcal{I}_n^{-1} \cdot \mathcal{I}_n = \mathbf{1}$$

$$\mathcal{I}_n = \sum_\alpha M_\alpha (r_\alpha^2 \mathbf{1} - \mathbf{r}_\alpha \mathbf{r}_\alpha) \tag{3}$$

$\mathcal{I}_n$ is the *principal* inertial tensor of the rigid nuclei and $\mathcal{I}_n^{-1}$ is the corresponding inverse nuclear moment tensor. M_α is the mass of the αth nucleus in the molecule, **1** is the unit 3 × 3 tensor and $\mathbf{r}_\alpha$ is the center-of-mass nuclear coordinate. The sums over g in Eq. (1) are over the three principal inertial axes. The last term in Eq. (1) is the electronic kinetic energy contribution to the zero-order electronic energy Hamiltonian and m is the electronic mass. Adding the electronic–electronic and nuclear–electronic potential energies to $\frac{1}{2m}\sum_i p_i^2$ leads to the zero-order electronic states given by

$$\psi^0, \psi^1, \psi^2, \ldots \psi^k, \tag{4}$$

where ψ^0 is the ground electronic state, and ψ^k are all the excited electronic states. We are assuming that the ground electronic state, ψ^0, does not possess any electronic angular momentum. However, the excited states, ψ^k, may possess electronic angular momentum.

The corrections to the electronic kinetic energy in the presence of an external magnetic field are derived by starting with the Lorentz force for a charged particle in a field or, alternately, by defining the correct Lagrangian function and using standard classical equations of motion to give the Hamiltonian function. The field dependent perturbation Hamiltonian is given by[2]

$$\mathcal{H} = \frac{e}{2mc}\,\mathbf{H}\cdot\mathbf{L} + \frac{e^2}{8mc^2}\,\mathbf{H}\cdot\left(\sum_i r_i^2\,\mathbf{1} - \mathbf{r}_i\mathbf{r}_i\right)\cdot\mathbf{H} - \mathbf{H}\cdot\left[\frac{e}{2c}\sum_\alpha Z_\alpha(r_\alpha^2\,\mathbf{1} - \mathbf{r}_\alpha\mathbf{r}_\alpha)\right]\cdot$$

$$\mathcal{I}_n^{-1}\cdot(\mathbf{J}-\mathbf{L}) + \frac{e^2}{8Mc^2}\mathbf{H}\cdot\sum_\alpha Z_\alpha(r_\alpha^2\,\mathbf{1} - \mathbf{r}_\alpha\mathbf{r}_\alpha)\cdot\mathbf{H}. \quad (5)$$

Here, $\mathbf{H}$ is the external magnetic field, e is the electronic charge, M is the proton mass, c is the speed of light, and Z_α is the atomic number of the αth nucleus. Combining Eq. (1) and (5) and dropping the $\frac{1}{2m}\sum_i p_i^2$ term gives the complete perturbation Hamiltonian (for noninteracting nuclei) describing the rotational energy in the presence of the electronic–rotational interaction and in the presence of a magnetic field.

$$\mathcal{H}' = \frac{1}{2}\mathbf{J}\cdot\mathcal{I}_n^{-1}\cdot\mathbf{J} - \mathbf{L}\cdot\omega' + \frac{1}{2}\mathbf{L}\cdot\mathcal{I}_n^{-1}\cdot\mathbf{L} + \frac{e^2}{8c^2m}\mathbf{H}\cdot\sum_i(r_i^2\,\mathbf{1} - \mathbf{r}_i\mathbf{r}_i)\cdot\mathbf{H}$$

$$-\frac{e}{2c}\sum_\alpha Z_\alpha\mathbf{H}\cdot(r_\alpha^2\,\mathbf{1} - \mathbf{r}_\alpha\mathbf{r}_\alpha)\cdot\mathcal{I}_n^{-1}\cdot\mathbf{J} + \frac{e^2}{8c^2M}\mathbf{H}\cdot\sum_\alpha Z_\alpha(r_\alpha^2\,\mathbf{1} - \mathbf{r}_\alpha\mathbf{r}_\alpha)\cdot\mathbf{H}$$

$$\omega' = \mathcal{I}_n^{-1}\cdot\mathbf{J} - \frac{e}{2mc}\mathbf{H} - \frac{e}{2c}\,\mathcal{I}_n^{-1}\cdot\sum_\alpha Z_\alpha(r_\alpha^2\,\mathbf{1} - \mathbf{r}_\alpha\mathbf{r}_\alpha)\cdot\mathbf{H}. \quad (6)$$

The $\frac{e}{2mc}\mathbf{H}$ contribution to $\boldsymbol{\omega}'$ is the well-known Larmor frequency.

We now employ perturbation theory to first calculate the electronic corrections to the zero-order energy due to the Hamiltonian in Eq. (6). Using the zero-order electronic functions in Eq. (4) leads to the following *rotational* Hamiltonian corrected to several orders in electronic states (the higher order terms are dropped)[2] :

$$\mathcal{H} = \frac{1}{2}\mathbf{J}\cdot\mathcal{I}_{\text{eff}}^{-1}\cdot\mathbf{J} - \frac{\mu_0}{\hbar}\mathbf{H}\cdot\mathcal{G}\cdot\mathbf{J} - \frac{1}{2}\mathbf{H}\cdot\mathcal{X}\cdot\mathbf{H}$$

$$\mathcal{G} = \mathcal{G}_n + \mathcal{G}_e$$

$$\mathcal{G}_n = M \sum_\alpha Z_\alpha (r_\alpha{}^2 \mathbf{1} - \mathbf{r}_\alpha \mathbf{r}_\alpha) \cdot \mathcal{I}^{-1}_{\text{eff}}$$

$$\mathcal{G}_e = \frac{2M}{m} \mathcal{I}^{-1}_{\text{eff}} \cdot \sum_{k>0} \frac{\langle 0|\mathbf{L}|k\rangle\langle k|\mathbf{L}|0\rangle}{E_0 - E_k}$$

$$\mathcal{X} = \mathcal{X}^{\text{d}} + \mathcal{X}^{\text{p}}$$

$$\mathcal{X}^{\text{d}} = -\frac{e^2}{4c^2 m} \langle 0|\sum_i (r_i{}^2 \mathbf{1} - \mathbf{r}_i \mathbf{r}_i)|0\rangle$$

$$\mathcal{X}^{\text{p}} = -\frac{e^2}{2m^2c^2} \sum_{k>0} \frac{\langle 0|\mathbf{L}|k\rangle\langle k|\mathbf{L}|0\rangle}{E_0 - E_k} \tag{7}$$

where $\mu_0 = \frac{\hbar e}{2Mc}$ is the nuclear magneton, m and M, respectively, the electron and proton masses, and $\hbar$ Planck's constant divided by 2π. $\frac{1}{2}\mathbf{J} \cdot \mathcal{I}^{-1}_{\text{eff}} \cdot \mathbf{J}$ is the rigid rotor term and $\mathcal{I}^{-1}_{\text{eff}}$ is the measured inverse moment of inertia at zero field. $\mathcal{G}_n$ and $\mathcal{G}_e$ are the nuclear and electronic contributions to the molecular g-value tensor, $\mathcal{G}$. It is evident that the molecular magnetic moment, $\frac{\mu_0}{\hbar} \mathcal{G} \cdot \mathbf{J}$, is a sum of positive nuclear and negative ($E_0 < E_k$) electronic terms. If the molecule had no excited electronic states that possessed electronic angular momentum, the rotational magnetic moment would be due entirely to the nuclei. The mixing of excited electronic states reduces the pure nuclear moment and sometimes the electronic contribution exceeds the nuclear value. The individual diagonal elements in the g-value tensor are given by

$$g_{xx} = \frac{M}{I_{xx}} \sum_\alpha Z_\alpha (r_\alpha{}^2 - x_\alpha{}^2) + \frac{2M}{mI_{xx}} \sum_{k>0} \frac{|\langle 0|L_x|k\rangle|^2}{E_0 - E_k} \tag{8}$$

and cyclic permutations for g_{yy} and g_{zz}.

$\mathcal{X}^{\text{d}}$, $\mathcal{X}^{\text{p}}$, and $\mathcal{X}$ are the diamagnetic, paramagnetic, and total magnetic susceptibilities, respectively, and the individual elements are given by

$$\chi_{xx} = \chi^{\text{d}}_{xx} + \chi^{\text{p}}_{xx}$$

$$\chi_{xx} = -\frac{e^2}{4c^2 m} \langle 0|\sum_i (y_i{}^2 + z_i{}^2)|0\rangle - \frac{e^2}{2c^2 m^2} \sum_{k>0} \frac{|\langle 0|L_x|k\rangle|^2}{E_0 - E_k} \tag{9}$$

and cyclic permutations for χ_{yy} and χ_{zz}. χ^{d}_{xx} and χ^{p}_{xx} are the diamagnetic and paramagnetic components of the susceptibility, respectively. χ^{d}_{xx} is always negative and depends only on the ground-state distribution of electrons. χ^{p}_{xx} is generally positive as $E_k > E_0$ and depends on a sum over all excited electronic states.

Eq. (9) shows that the magnetic-field-induced electronic moment, $\frac{1}{2}\mathbf{H}\cdot\mathfrak{X}$, is a sum of negative and positive terms. The negative terms arise through the diamagnetic response of the molecule's electrons, which gives a moment that opposes the field. The second and positive contribution to the electronic magnetic moment arises through a paramagnetic response of the molecule's electrons; the paramagnetic moment complements the applied magnetic field.

It is clear that the second terms in g_{xx} [Eq. (8)] and χ_{xx} [Eq. (9)] have the same dependence on the sum over all excited molecular electronic states. Therefore, if the total g_{xx} can be measured and if the nuclear component of g_{xx} can be computed from the known molecular structure, the numerical value for the paramagnetic dependence in the susceptibility can be obtained. Substituting the $\sum_{k>0} \frac{|\langle 0|L_x|k\rangle|^2}{E_0 - E_k}$ dependence in g_{xx} into χ_{xx} gives

$$\chi_{xx} = \chi^{d}_{xx} + \chi^{p}_{xx}$$

$$= -\frac{e^2}{4c^2 m}\langle 0|\sum_i (y_i^2 + z_i^2)|0\rangle - \frac{e^2}{4c^2 m}\left\{\frac{g_{xx}I_{xx}}{M} - \sum_\alpha Z_\alpha(y_\alpha^2 + z_\alpha^2)\right\} \quad (10)$$

and cyclic permutations for χ_{yy} and χ_{zz}.

Now, if the elements in the total magnetic susceptibility tensor can also be measured, the numerical value of χ^{d}_{xx} can be extracted. χ^{d}_{xx} depends only on the ground electronic state electron distribution and gives some notion of the outer electronic shape of the molecule.

The diagonal elements of the molecular quadrupole moment tensor Q are related to the diagonal elements of the magnetic susceptibility tensor in Eq. (10) by[3]

$$Q_{zz} = \frac{e}{2}\sum_\alpha Z_\alpha(3z_\alpha^2 - r_\alpha^2) - \frac{e}{2}\langle 0|\sum_i (3z_i^2 - r_i^2)|0\rangle$$

$$= \frac{2mc^2}{e}(\chi_{xx} + \chi_{yy} - 2\chi_{zz}) + \frac{e}{2M}(g_{xx}I_{xx} + g_{yy}I_{yy} - 2g_{zz}I_{zz}) \quad (11)$$

and cyclic permutations for Q_{yy} and Q_{xx}. The molecular quadrupole moments of several molecules have been determined by measuring the diagonal elements in the $\mathcal{G}$ and $\mathcal{I}$ tensors and the magnetic susceptibility anisotropies $(\chi_{xx} + \chi_{yy} - 2\chi_{zz})$; the results will be discussed later. Equation (11) simplifies for a linear molecule to

$$Q_{zz} = e\sum_{\alpha} Z_\alpha z_\alpha{}^2 - e\langle 0|\sum_i (z_i{}^2 - x_i{}^2)|0\rangle$$

$$= \frac{e}{M} g_{xx} I_{xx} + \frac{4mc^2}{e}(\chi_{xx} - \chi_{zz}). \qquad (12)$$

A similar equation is evident for symmetric tops.

$$Q_{zz} = \frac{e}{M}(g_{zz}I_{zz} - g_{xx}I_{xx}) + \frac{4mc^2}{e}(\chi_{xx} - \chi_{zz}), \qquad (13)$$

where z is the symmetry axis for both Eq. (12) and (13).

We now recall that only the value of the first non-zero electric multipole moment is independent of the origin. Consider a molecular quadrupole moment along the internuclear z axis in a linear molecule at two different origins along the z axis separated by Z. According to Eq. (12)

$$Q'_{zz} = e\sum_{\alpha} Z_\alpha (z'_\alpha)^2 - e\langle 0|\sum_i [(z'_i)^2 - (x'_i)^2]|0\rangle. \qquad (14)$$

In a linear molecule, $z' = z - Z$ and $x' = x$, which gives

$$Q'_{zz} = e\sum_{\alpha} Z_\alpha (z_\alpha - Z)^2 - e\langle 0|\sum_i [(z_i - Z)^2 - x_i{}^2]|0\rangle$$

$$= Q_{zz} - 2Z\,D_z + Z^2 M_0. \qquad (15)$$

Here D_z is the electric dipole moment and M_0 is the molecular monopole moment. If $M_0 = 0$ (neutral molecule) and $D_z = 0$, $Q_{zz} = Q'_{zz}$. If the molecule is neutral and if the molecule has a non-zero electric dipole moment,

$$Q'_{zz} - Q_{zz} = -2Z\,D_z = \frac{e}{M}(g'_{xx}I'_{xx} - g_{xx}I_{xx}), \qquad (16)$$

where the last step is obtained by Q'_{zz} and Q_{zz} from Eq. (12) in terms of $g_{xx}I_{xx}$ and $\chi_{xx} - \chi_{zz}$. (Remember that the magnetic susceptibility anisotropy $(\chi_{xx} - \chi_{zz})$ is independent of the origin of the measurement.) These equations are easily generalized to nonlinear molecules.

The anisotropies in the second moment of the electronic charge distributions are also available from the above information and the molecular structure:

$$\langle y^2\rangle - \langle x^2\rangle = \langle 0|\sum_i y_i^2|0\rangle - \langle 0|\sum_i x_i^2|0\rangle = \sum_n Z_n(y_n^2 - x_n^2)$$

$$+\frac{1}{M}[g_{yy}I_{yy} - g_{xx}I_{xx}] + \frac{4mc^2}{3e^2}[\chi_{yy} - \chi_{xx}], \tag{17}$$

where y_n and x_n are the nuclear center of mass coordinates obtained from the molecular structure.

Finally, if the bulk magnetic susceptibility is known, the individual tensor elements in $\mathcal{X}^{\mathrm{d}}$ and $\mathcal{X}$ can be determined. The bulk or average magnetic susceptibility is given by

$$\chi = \frac{1}{3}(\chi_{xx} + \chi_{yy} + \chi_{zz}). \tag{18}$$

The individual second moments of the electronic charge distributions can also be determined by

$$\langle x^2\rangle = -\frac{2mc^2}{e^2}\left(\chi^{\mathrm{d}}_{yy} + \chi^{\mathrm{d}}_{zz} - \chi^{\mathrm{d}}_{xx}\right)$$

$$= -\frac{2mc^2}{e^2}\left[(\chi_{yy} + \chi_{zz} - \chi_{xx}) - (\chi^{\mathrm{p}}_{yy} + \chi^{\mathrm{p}}_{zz} - \chi^{\mathrm{p}}_{xx})\right]. \tag{19}$$

A summary of the interconnections between the magnetic parameters is shown in Figure 1. The parameters in the top row are measured directly by the molecular Zeeman effect.

To calculate the rotational energy due to the Hamiltonian in Eq. (7), we write $\mathcal{G}$ and $\mathcal{X}$ in the principal *inertial* axis system with the direction cosine transformation, $\mathbf{\Phi}$, leading to

$$\mathcal{H} = \mathcal{H}_{rr} - \frac{\mu_0}{\hbar}\mathbf{H}\cdot[\mathbf{\Phi}\mathcal{G}]\cdot\mathbf{J} - \frac{1}{2}\mathbf{H}\cdot[\tilde{\mathbf{\Phi}}\mathcal{X}\mathbf{\Phi}]\cdot\mathbf{H}, \tag{20}$$

where we now use the a, b, and c axes to denote the principal inertial axis system. Flygare and Benson[1] have discussed the errors inherent in truncating Eq. (20) at quadratic terms in the magnetic field. Flygare[2] has discussed the errors obtained by truncating the perturbation expansion leading to Eq. (7). In virtually all cases the Hamiltonian in Eq. (20)

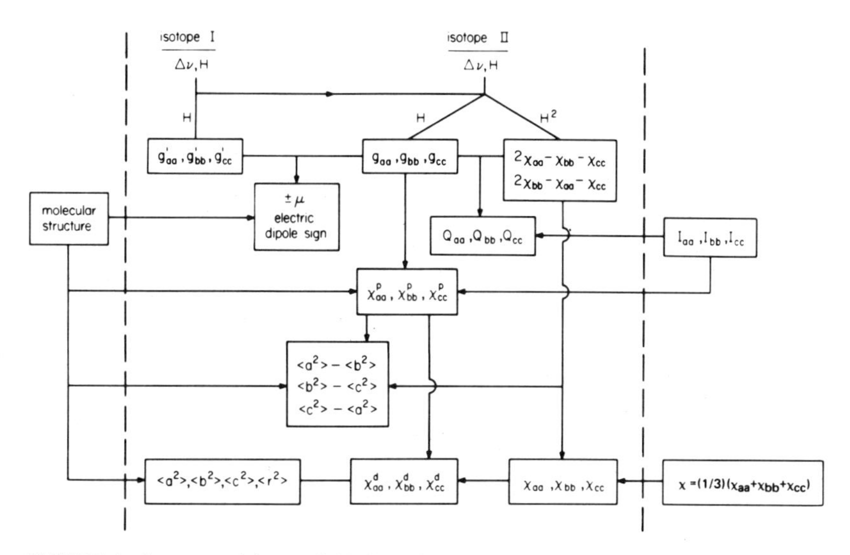

FIGURE 1 Summary of data available from the measurement of the magnetic field (H), giving rise to the frequency shifts ($\Delta\nu$) in the molecular Zeeman effect. The information between the dotted lines must be added to the non-Zeeman information outside of the dotted lines when indicated.

at fields up to 25 kGs will give g-values accurate to about 0.1 percent and χ anisotropies to better than 1 percent.[1]

Huttner and Flygare[4] have discussed in detail the rotational average of the Hamiltonian in Eq. (20) in both the nuclear coupled (discussed later) and uncoupled cases. In the absence of nuclear spin, the rotational energy for a molecule is computed by first obtaining the eigenfunctions of $\mathcal{H}_{rr} = \frac{1}{2}\,\mathbf{J}\cdot\mathcal{I}_{\text{eff}}^{-1}\cdot\mathbf{J}$ from Eq. (7) at zero field. These rigid-rotor and J^2 eigenfunctions are then used to evaluate the first-order corrections to the rotational energy due to the $\mathcal{G}$ and $\mathcal{X}$ terms in Eq. (20). The rotational energy for a general molecule in the uncoupled basis first order in J is given in the absence of nuclear spin as[4]

$$E(J,\tau,M) = E^0(J,\tau,M) - \frac{\mu_0 MH}{J(J+1)}\sum_{\alpha} g_{\alpha\alpha} <J\tau||J_\alpha{}^2||J\tau> - \frac{1}{2}H^2\chi$$

$$-\frac{H^2}{J(J+1)}\left[\frac{3M^2 - J(J+1)}{(2J-1)(2J+3)}\right]\sum_{\alpha}(\chi_{\alpha\alpha} - \chi) <J\tau||J_\alpha{}^2||J\tau>. \quad (21)$$

$\chi = \frac{1}{3}(\chi_{aa} + \chi_{bb} + \chi_{cc})$ is the average magnetic susceptibility, $E^0(J,\tau,M)$

is the zero-field rotational energy, and $\langle J_\tau \,\|\, J_\alpha{}^2 \,\|\, J\tau\rangle$ is the reduced matrix element of $J_\alpha{}^2$ in the general asymmetric top basis indicated by quantum number τ. The $\langle J\tau \,\|\, J_\alpha{}^2 \,\|\, J\tau\rangle$ matrix elements are easily computed from the experimental principal moments of inertia by transforming the symmetric top $\langle JK \,\|\, J_\alpha{}^2 \,\|\, J'K'\rangle$ matrix to the asymmetric top basis.[4] The sums over α are over the three principal inertial axes a, b, and c. Thus, we note that the three independent diagonal elements in the molecular g-value tensor (g_{aa}, g_{bb}, and g_{cc}) can generally be measured along with two independent magnetic susceptibility anisotropies, $2\chi_{aa}-\chi_{bb}-\chi_{cc}$ and $2\chi_{bb}-\chi_{aa}-\chi_{cc}$. *Thus, in general, the Zeeman perturbation provides five parameters.* In a symmetric top there are only two independent g values and a single independent anisotropy. In a linear molecule there is a single independent g value and a single anisotropy.

Equation (21) reduces to a simpler form for symmetric top and linear molecules, where for a prolate top, with the symmetry axis along a, we write

$$\langle J_c^2\rangle = \frac{1}{2}\,[J(J+1) - K^2] = \langle J_b^2\rangle,$$

$$\langle J_a^2\rangle = K^2. \tag{22}$$

Substituting into Eq. (21) gives (where a is the rotational symmetry axis):

$$E(JKM) = E^0(JKM) - \mu_0 MH\left[g_{bb} + (g_{aa} - g_{bb})\frac{K^2}{J(J+1)}\right] - \frac{1}{2}H^2\chi$$
$$-\frac{H^2}{3}\left[\frac{3M^2 - J(J+1)}{(2J-1)(2J+3)}\right]\left[(\chi_{bb}-\chi_{aa}) - \frac{3K^2}{J(J+1)}(\chi_{bb}-\chi_{aa})\right]. \tag{23}$$

In the case of a linear molecule, $K = 0$ giving (where a is the internuclear axis):

$$E(JM) = E^0(JM) - H\mu_0 M g_{bb} - \frac{H^2}{3}\left[\frac{3M^2 - J(J+1)}{(2J-1)(2J+3)}\right](\chi_{bb}-\chi_{aa}) - \frac{1}{2}H^2\chi. \tag{24}$$

Eq. (21), (23), and (24) have been used extensively for the determination of the g values and magnetic susceptibility anisotropies in rotating molecules in the absence of nuclear spin.[1] Molecular beam techniques have been employed to extract g_{bb} and $\chi_{bb}-\chi_{aa}$ values from several linear molecules.[5]* It is difficult to extend the molecular beam method to more complex molecules, however, and most of the recent work has been generated by using high-resolution microwave spectroscopy with high magnetic fields[1]; some of the recently obtained data on linear and symmetric top molecules are listed in Table 1.

Figure 2 illustrates three cases of relative H-dependent and H^2-dependent spectra in asymmetric top molecules. In light- or medium-sized molecules the linear Zeeman effect dominates the spectra leading to symmetric looking spectra and very accurate measurements of the molecular g values. In heavy molecules the field-induced magnetic moments, $\frac{1}{2}\,\mathbf{H}\cdot\mathcal{X}$, are larger than the rotationally induced moments leading to more accurate measurements of $2\chi_{aa}-\chi_{bb}-\chi_{cc}$ and $2\chi_{bb}-\chi_{aa}-\chi_{cc}$ than the molecular g values. The progression as described above is shown in Figure 2 starting with $Hg > H^2\chi$ in SO_2 to $Hg \approx H^2\chi$ in ethylene sulfide to $Hg < H^2\chi$ in fluorobenzene. The same transition is shown in all cases where the increasing asymmetry is evident. An extreme case in the $Hg < H^2\chi$ region is found in tropone where the magnetic susceptibility anisotropies were obtained from the Zeeman splittings, in spite of the fact that the g values could not be determined.[6]

Table 2 lists some typical molecular g values, magnetic susceptibility anisotropies, and molecular quadrupole moments for planar nonring asymmetric top, three-membered ring, four-membered ring, and other planar ring molecules. The numbers presented in these tables were obtained by least squares fitting the observed splittings on 40–100 transitions for each molecule to the energy expressions in Eq. (21). The uncertainties are single standard deviations. The results listed in Tables 1 and 2 give only a small, but representative, sample of the data that have been obtained by microwave spectroscopic methods over the past few years. The uncertainties are also representative of the averages of the 40–100 transitions analyzed as mentioned above. Additional examples and references to the original references are available.[1,2]

The uncertainties in the molecular quadrupole moments in Tables 1 and 2 calculated with Eq. (11), (12), or (13) depend in most cases on the uncertainties in the magnetic susceptibility anisotropies as the g values are normally the more accurate Zeeman parameter. It is evident from Table 1 that the determination of the sign of the electric dipole moment is a marginal experiment as the standard deviations are normally as large as the measured moments. Nevertheless, most of the di-

* A current summary of linear molecule g values and $\chi_{bb} - \chi_{aa}$ values determined by molecular beams is given in Flygare and Benson.[1]

TABLE 1 Molecular g Values, Electric Dipole Moments (and Signs), Magnetic Susceptibility Anisotropies, and Molecular Quadrupole Moments of a Number of Linear and Symmetric Top Molecules as Measured by Microwave Spectroscopy

	$g_{\perp}$	μ^{a}	$g_{\parallel}$	$\chi_{\perp}-\chi_{\parallel}{}^{b}$	$Q_{\parallel}{}^{c}$	Reference
$-^{16}O^{12}C^{32}S+$	−0.028711 ± 0.00004	+0.75 ± 0.30	–	9.27 ± 0.10	−0.88 ± 0.15	*d*
$^{15}N^{15}N^{16}O$	−0.07606 ± 0.0001	–	–	10.15 ± 0.15	−3.65 ± 0.25	*d*
$-^{16}O^{12}C^{80}Se+$	−0.01952 ± 0.0002	+0.9 ± 1.9	–	10.06 ± 0.18	−0.32 ± 0.24	*d*
$-^{12}C^{32}S+$	−0.2702 ± 0.0004	+2.7 ± 1.0	–	24.2 ± 1.2	0.8 ± 1.4	McGurk *et al.*[23]
$^{12}C^{80}Se$	−0.2431 ± 0.0016	–	–	27.8 ± 1.4	−2.6 ± 1.6	McGurk *et al.*[23]
$HC^{15}N$	−0.0904 ± 0.0003	–	–	7.2 ± 0.4	3.1 ± 0.6	Hartford *et al.*[21]
$+CH_3C^{15}N-$	−0.0317 ± 0.0003	−3.5 ± 1.5	–	10.5 ± 0.5	−1.8 ± 1.2	*d*
$FC^{15}N$	−0.0504 ± 0.0008	–	–	7.2 ± 0.8	−3.7 ± 1.0	Rock *et al.*[24]
$+^{35}ClC^{15}N-$	−0.0384 ± 0.0003	−2.4 ± 3.6	–	10.8 ± 0.8	−3.9 ± 1.0	*d*
$^{79}BrC^{15}N$	−0.0325 ± 0.0010	–	–	11.8 ± 1.0	−6.0 ± 1.1	*d*
FCCH	−0.0077 ± 0.0002	–	–	5.2 ± 0.2	4.0 ± 0.2	*d*
$-^{35}ClCCH+$	−0.00630 ± 0.00014	−0.3 ± 0.2	–	9.3 ± 0.5	8.8 ± 0.4	*d*
$^{79}BrCCH$	−0.00395 ± 0.00032	–	–	9.5 ± 0.9	8.5 ± 1.1	Hartford *et al.*[21]
HCP	−0.0430 ± 0.0010	–	–	8.4 ± 0.9	4.4 ± 1.2	Hartford *et al.*[21]
$+CH_3CCH-$	+0.0035 ± 0.0002	−0.7 ± 0.2	0.312 ± 0.0002	7.7 ± 0.2	4.8 ± 0.3	*d*
CH_3CCCCH	0 ± 0.0005	–	(0.310)*d*	13.1 ± 0.2	9.9 ± 0.8	*d*
$CH_3{}^{14}NC$	−0.0317 ± 0.003	–	(0.310)*d*	13.5 ± 1.7	−2.7 ± 1.6	*d*
$+CH_3C^{15}N-$	−0.0338 ± 0.0008	+3.5 ± 1.5	(0.310)*d*	10.2 ± 1.0	1.8 ± 1.2	*d*
CH_3F	−0.0612 ± 0.002	–	(0.310)*d*	8.2 ± 0.8	−1.4 ± 1.1	*d*
$+CH_3{}^{35}Cl-$	−0.0165 ± 0.0003	+2.3 ± 2.0	(0.305)*d*	8.0 ± 0.5	1.2 ± 0.8	*d*
$CH_3{}^{79}Br$	−0.0057 ± 0.0003	–	+0.294 ± 0.016	8.5 ± 0.4	3.6 ± 0.8	*d*
$CH_3{}^{127}I$	−0.0068 ± 0.0004	–	+0.310 ± 0.016	11.0 ± 0.5	5.4 ± 0.9	*d*
$-SiH_3CH_3+$	−0.03583 ± 0.0001	0.96 ± 0.40	+0.0182 ± 0.0069	2.4 ± 0.2	−6.3 ± 0.5	Shoemaker and Flygare[25]

^{a} Units of 10^{-18} esu cm.
^{b} Units of 10^{-6} erg/Gs^{2} mol.
^{c} Units of 10^{-26} esu cm^{2}.
^{d} See Flygare and Benson[1] for original literature references.

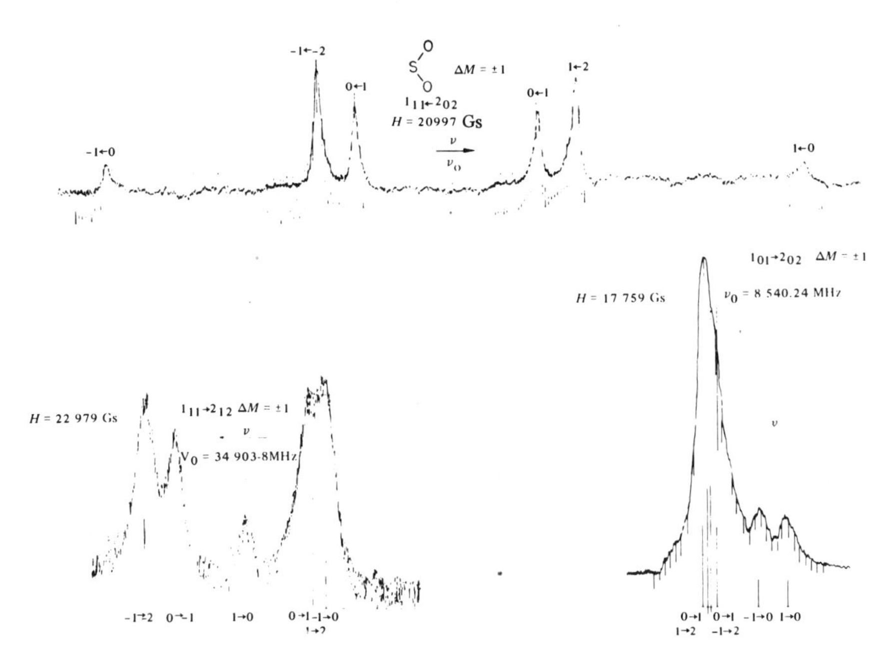

FIGURE 2 A series of spectra showing different relative contributions to the linear and quadratic Zeeman effect in the same $J=1_{11}\rightarrow 2_{02}$ transitions in different molecules. The top spectrum for SO_2 shows a nearly symmetric linear field Zeeman effect.[42] The lower left shows an intermediate $Hg \cong H^2\chi$ spectrum for the same $1_{11}\rightarrow 2_{02}$ transition in ethylene sulfide.[43] The lower right diagram shows the $1_{11}\rightarrow 2_{02}$ transition in fluorobenzene where $Hg < H^2\chi$.[44]

poles are in agreement with the more accurately determined magnitudes as obtained by the Stark effect; and in most cases the signs are the expected values. The major exception is the recent work on FCl where the molecular Zeeman effect gives a dipole moment sign that disagrees within a single standard deviation with the *ab initio* value.[7] An experimental uncertainty of two standard deviations will allow agreement with the calculated sign. Higher magnetic fields and higher resolution will be necessary to bring the ClF dipole moment signs into agreement.

We refer again to Figure 1 that shows the relation between the measurements, which are the field-dependent frequency shifts and the corresponding magnetic fields, and the remaining magnetic parameters. The parameters g_{aa}, g_{bb}, g_{cc}, $2\chi_{aa}-\chi_{bb}-\chi_{cc}$, and $2\chi_{bb}-\chi_{aa}-\chi_{cc}$ are measured directly and all others are determined from these numbers, the moments of inertia, the molecular structure, and the value of the bulk magnetic susceptibility. The sources of error in these measurements are in frequencies and fields. Of course, for very large Zeeman splittings, the accuracy of the magnetic field becomes critical. The

TABLE 2 Molecular g Values, Magnetic Susceptibility Anisotropies, and Molecular Quadrupole Moments in Some Nonring Planar Asymmetric Tops and Planar Ring Compounds

y ↑ → x	g_{xx} g_{yy} g_{zz}	$2\chi_{xx}-\chi_{yy}-\chi_{zz}$ $-\chi_{xx}+2\chi_{yy}-\chi_{zz}$	Q_{xx} Q_{yy} Q_{zz}
C–C(=C)–C=C	−0.0621 ± 0.0013 −0.0339 ± 0.0016 +0.0080 ± 0.0016	16.7 ± 1.2 19.2 ± 1.0	+1.7 ± 2.2 +3.3 ± 2.3 −5.0 ± 3.2
C=C–C=O	−0.5512 ± 0.0019 −0.0567 ± 0.0010 −0.0080 ± 0.0010	24.1 ± 0.9 17.1 ± 1.5	−2.5 ± 1.1 +3.3 ± 1.7 −0.8 ± 2.2
C=C–C	−0.0789 ± 0.0006 −0.0424 ± 0.0004 +0.0107 ± 0.0005	−0.7 ± 0.3 13.4 ± 0.5	+0.6 ± 0.3 +2.9 ± 0.5 −3.5 ± 0.7
+ H_2C=O −	−2.9017 ± 0.0008 −0.2243 ± 0.0001 −0.0994 ± 0.0001	25.5 ± 0.5 −3.9 ± 0.3	−0.1 ± 0.3 +0.2 ± 0.2 −0.1 ± 0.5
H_2C=S[a]	−5.6202 ± 0.0068 −0.1337 ± 0.0004 −0.0239 ± 0.0004	52.3 ± 1.1 −5.1 ± 0.7	+3.0 ± 0.7 −2.4 ± 0.5 −0.6 ± 1.1
O–S–O	−0.6037 ± 0.0005 −0.1161 ± 0.0002 −0.0882 ± 0.0004	+6.4 ± 0.5 +3.1 ± 0.3	−5.3 ± 0.4 +1.3 ± 0.3 +4.0 ± 0.6
F–O–F	−0.213 ± 0.005 −0.058 ± 0.002 −0.068 ± 0.002	−8.8 ± 1.4 −4.4 ± 0.7	−1.6 ± 1.4 +2.1 ± 1.1 −0.5 ± 1.9
H–O–F[b]	+0.642 ± 0.001 −0.119 ± 0.001 −0.061 ± 0.001	−19.6 ± 0.6 +12.8 ± 1.2	0.2 ± 0.4 1.9 ± 0.8 −2.1 ± 1.1
+ H_2O −	+0.718 ± 0.007 +0.657 ± 0.001 +0.645 ± 0.006	−0.199 ± 0.048 +0.464 ± 0.024	−0.13 ± 0.03 +2.63 ± 0.02 −2.50 ± 0.02
▷=	−0.0672 ± 0.0007 −0.0231 ± 0.0004 +0.0244 ± 0.0004	18.3 ± 0.5 14.9 ± 0.6	−0.7 ± 0.5 +0.9 ± 0.6 −0.2 ± 0.9
−▷+	−0.0897 ± 0.0009 −0.1492 ± 0.0002 +0.0536 ± 0.0002	7.1 ± 0.6 26.8 ± 0.4	−0.4 ± 0.4 +2.4 ± 0.3 −2.0 ± 0.6

TABLE 2 (*Continued*)

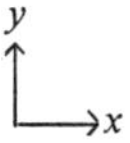 y ↑ → x	g_{xx} g_{yy} g_{zz}	$2\chi_{xx}-\chi_{yy}-\chi_{zz}$ $-\chi_{xx}+2\chi_{yy}-\chi_{zz}$	Q_{xx} Q_{yy} Q_{zz}
 N	+0.0229 ± 0.0009 −0.0422 ± 0.0008 +0.0539 ± 0.0010	4.6 ± 0.8 16.5 ± 0.7	−2.6 ± 0.6 +1.3 ± 0.6 +1.3 ± 0.6
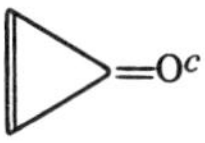 =O[c]	−0.2900 ± 0.0013 −0.0963 ± 0.0004 −0.0121 ± 0.0004	13.6 ± 1.0 22.0 ± 0.8	−3.0 ± 0.9 4.0 ± 0.7 −1.0 ± 1.3
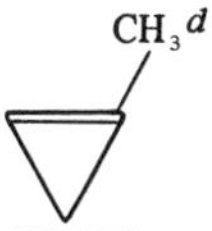 CH_3[d]	−0.0813 ± 0.007 −0.0261 ± 0.004 +0.0166 ± 0.003	13.9 ± 0.3 16.4 ± 0.6	+0.6 ± 0.4 −0.3 ± 0.6 −0.3 ± 0.8
	−0.0516 ± 0.0007 −0.0663 ± 0.0007 −0.0219 ± 0.0006	−0.9 ± 0.5 +5.0 ± 0.7	−0.3 ± 0.6 +1.6 ± 0.7 −1.3 ± 1.0
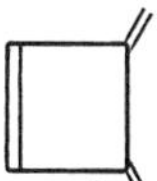 [e]	−0.0532 ± 0.0007 −0.0703 ± 0.0007 +0.0023 ± 0.0007	21.2 ± 0.6 22.1 ± 0.7	4.0 ± 1.2 4.0 ± 1.2 −8.0 ± 1.8
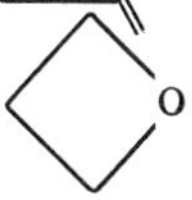 O	−0.0073 ± 0.0005 −0.0429 ± 0.0007 −0.0747 ± 0.0005	−20.1 ± 0.5 −13.5 ± 0.8	−4.9 ± 0.5 +2.3 ± 0.7 +2.6 ± 1.0
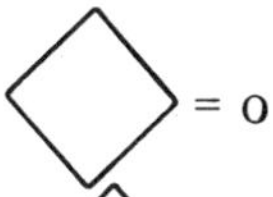 = O	−0.0740 ± 0.0020 −0.0325 ± 0.0004 −0.0279 ± 0.0004	14.8 ± 0.9 −10.6 ± 1.0	−9.4 ± 1.2 +4.6 ± 1.1 +4.8 ± 1.7
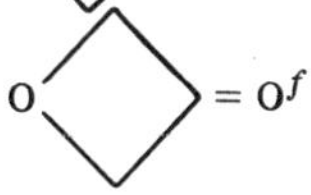 O, = O[f]	−0.1059 ± 0.0008 −0.0581 ± 0.0004 −0.0437 ± 0.0004	9.6 ± 0.5 −7.8 ± 0.6	−12.8 ± 0.8 7.9 ± 0.8 4.9 ± 0.8
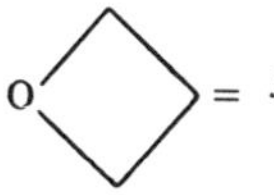 O, = [f]	−0.0510 ± 0.002 −0.0435 ± 0.001 −0.0313 ± 0.001	−10.9 ± 4.7 2.3 ± 0.9	−5.4 ± 1.0 5.1 ± 1.2 0.3 ± 1.5
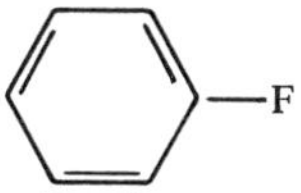 —F	−0.0670 ± 0.0008 −0.0397 ± 0.0015 +0.0266 ± 0.0017	52.9 ± 0.8 63.6 ± 1.5	−1.9 ± 0.8 +5.1 ± 1.0 −3.2 ± 1.0
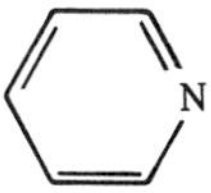 N	−0.0770 ± 0.0005 −0.1010 ± 0.0008 +0.0428 ± 0.0004	54.3 ± 0.6 60.5 ± 0.8	−3.5 ± 0.9 +9.7 ± 1.1 −6.2 ± 1.5
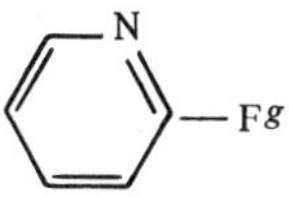 N, — F[g]	−0.0880 ± 0.0007 −0.0405 ± 0.0006 +0.0233 ± 0.0006	50.5 ± 1.6 53.7 ± 1.1	4.6 ± 1.8 2.8 ± 1.7 −7.4 ± 2.7

TABLE 2 (*Continued*)

Molecule (y ↑, → x)	g_{xx} / g_{yy} / g_{zz}	$2\chi_{xx}-\chi_{yy}-\chi_{zz}$ / $-\chi_{xx}+2\chi_{yy}-\chi_{zz}$	Q_{xx} / Q_{yy} / Q_{zz}
[cyclohexadiene]	−0.0433 ± 0.0011 −0.0400 ± 0.0024 −0.0062 ± 0.0009	9.1 ± 2.2 5.7 ± 1.6	+3.2 ± 2.8 +2.7 ± 2.8 −5.9 ± 3.8
[cyclopentene]	−0.0827 ± 0.0003 −0.0700 ± 0.0003 +0.0385 ± 0.0002	37.8 ± 0.3 30.7 ± 0.3	+3.7 ± 0.4 +1.4 ± 0.4 −5.1 ± 0.5
[cyclopentadienone, =O] [h]	−0.0914 ± 0.001 −0.0503 ± 0.001 −0.0122 ± 0.001	25.3 ± 1.0 8.3 ± 1.1	−7.1 ± 1.6 8.1 ± 1.8 −1.0 ± 2.4
[fulvene, =] [e]	−0.1059 ± 0.0014 0.0482 ± 0.0007 +0.0219 ± 0.0007	35.9 ± 0.7 38.1 ± 1.1	5.8 ± 1.4 3.6 ± 1.6 −9.4 ± 2.1

Note: The dipole moment signs are also listed when appropriate. The units are the same as in Table 1. References are from Flygare and Benson[1] unless indicated. More data are given in Ref. 1 and 2.

[a] From Rock and Flygare.[26]
[b] From Rock *et al.*[27]
[c] From Benson *et al.*[28]
[d] From Benson and Flygare.[29]
[e] From Benson and Flygare.[30]
[f] From Norris *et al.*[31]
[g] From Sutter.[33]
[h] From Norris *et al.*[34]

complete molecular Zeeman results with uncertainties for the other molecules listed in Tables 1 and 2 can be found in the original references.

NUCLEAR MAGNETIC SHIELDING, NUCLEAR SPIN-ROTATION INTERACTIONS, AND THE MOLECULAR ZEEMAN EFFECT IN THE PRESENCE OF NUCLEAR INTERACTIONS

We have just examined the rotational–electronic interaction and the external field–electronic interaction with the resultant rotational Hamiltonian in Eq. (6), which is valid in the absence of *nuclear*–electronic interactions. We now add in the effects of the nuclear–electronic interactions that give rise to the spin–rotation interaction and nuclear magnetic shielding.

The energy of the interaction of the magnetic dipole moment of the kth nucleus, $\boldsymbol{\mu}^k$, with the total external and internal field at the kth nucleus, $\mathbf{H}^k$, is given by[2]

$$\mathcal{H}_n = -\mu^k \cdot \mathbf{H}^k = -\gamma_k \mathbf{I}_k \cdot \mathbf{H}^k = -\frac{\mu_0}{\hbar} g_k \mathbf{I}_k \cdot \mathbf{H}^k$$

$$= -\gamma_k \mathbf{I}_k \cdot \mathbf{H} + \frac{e}{mc} \gamma_k \mathbf{I}_k \cdot \sum_i \frac{\mathbf{l}_{ki}}{r_{ki}^3} + \frac{e^2}{2mc^2} \gamma_k \mathbf{I}_k \cdot \sum_i \frac{(r_{ki}^2 \, \mathbf{1} - \mathbf{r}_{ki}\mathbf{r}_{ki})}{r_{ki}^3} \cdot \mathbf{H}$$

$$-\frac{e}{c} \gamma_k \mathbf{I}_k \cdot \sum_\alpha{}' \frac{Z_\alpha}{r_{k\alpha}^3} (r_{k\alpha}^2 \mathbf{1} - \mathbf{r}_{k\alpha}\mathbf{r}_{k\alpha}) \cdot \mathcal{I}_n^{-1} \cdot (\mathbf{J} - \mathbf{L}). \quad (25)$$

$\gamma_k = \frac{\mu_0}{\hbar} g_I$ is the gyromagnetic ratio of the kth nucleus, $\mu_0 = \frac{\hbar e}{2Mc}$ is the nuclear magneton, g_k is the nuclear g value, and $\mathbf{I}_k$ is the nuclear angular momentum of the kth nucleus. Combining the nuclear Hamiltonian in Eq. (25) with the results in Eq. (6) and correcting for the electronic effects with the zero-order basis in Eq. (4) gives the following terms that arise *in addition to* the results in Eq. (7):

$$\mathcal{H} = -\gamma_k \mathbf{I}_k \cdot (\mathbf{1} - \sigma) \cdot \mathbf{H}$$

$$-\frac{1}{\hbar^2} \mathbf{I}_k \cdot \mathcal{M} \cdot \mathbf{J}$$

$$\sigma = \sigma^d + \sigma^p$$

$$\sigma^d = \frac{e^2}{2mc^2} \langle 0 | \sum_i \frac{r_{ki}^2 \mathbf{1} - \mathbf{r}_{ki}\mathbf{r}_{ki}}{r_{ki}^3} | 0 \rangle$$

$$\sigma^p = \frac{e^2}{m^2c^2} \sum_{k>0} \frac{\langle 0 | \sum_i \frac{\mathbf{l}_{ki}}{r_{ki}^3} | k \rangle \langle k | \mathbf{L} | 0 \rangle}{E_0 - E_k}$$

$$\mathcal{M} = \mathcal{M}_n + \mathcal{M}_e$$

$$= \frac{e}{c} \gamma_k \hbar^2 \sum_\alpha{}' \frac{Z_\alpha}{r_{k\alpha}^3} (r_{k\alpha}^2 \mathbf{1} - \mathbf{r}_{k\alpha}\mathbf{r}_{k\alpha}) \cdot \mathcal{I}_{\text{eff}}^{-1}$$

$$+ 2 \frac{e}{mc} \gamma_k \hbar^2 \sum_{k>0} \left[\frac{\langle 0 | \sum_i \frac{\mathbf{l}_{ki}}{r_{ki}^3} | k \rangle \langle k | \mathbf{L} | 0 \rangle}{E_0 - E_k} \right] \cdot \mathcal{I}_{\text{eff}}^{-1}, \quad (26)$$

where σ^d and σ^p are the diamagnetic and paramagnetic nuclear magnetic shielding tensors with individual diagonal elements given by[8]

$$\sigma_{xx} = \sigma^{\text{d}}_{xx} + \sigma^{\text{p}}_{xx} = \frac{e^2}{2mc^2} <0|\sum_i \frac{(y_{ki}^2 + z_{ki}^2)}{r_{ki}^3}|0> + \frac{e^2}{2m^2c^2} \sum_{k>0}$$

$$\left[\frac{<0|\sum_i \frac{(l_{ki})_x}{r_{ki}^3}|k><k|L_x|0> + <0|L_x|k><k|\sum_i \frac{(l_{ki})_x}{r_i^3}|0>}{E_0 - E_k} \right] \quad (27)$$

and cyclic permutations for σ_{yy} and σ_{zz}.

It is evident that σ^{d}_{xx} is the diamagnetic and σ^{p}_{xx} is the paramagnetic shielding. The diamagnetic shielding is always positive and decreases the net field at the nucleus. σ^{p}_{xx} is normally negative, as $E_0 < E_k$, leading to an enhanced magnetic field at the nucleus. $\mathcal{M}$ is the spin–rotation constant with individual diagonal elements given by

$$M^k_{xx} = \frac{\hbar e g_k \mu_0}{c I_{xx}} \sum_\alpha{}' \frac{Z_\alpha}{r^3_{k\alpha}} (r^2_{k\alpha} - x^2_{k\alpha})$$

$$= \frac{\hbar e g_k \mu_0}{m c I_{xx}} \sum_{k>0} \left[\frac{<0| \sum_i \frac{(l_{ki})_x}{r^3_{ki}} |k> <k|L_x|0> + <0|L_x|k><k| \sum_i \frac{(l_{ki})_x}{r^3_{ik}} |0>}{E_0 - E_k} \right]$$

(28)

and cyclic permutations for M^k_{yy} and M^k_{zz}. The spin–rotation tensor constants are a sum of a positive pure nuclear term and a negative ($E_0 < E_k$ electronic term. We note the similarities in concept between the rotational g values in Eq. (8) and the spin–rotation constants in Eq. (28). In both cases, the bare nuclei contribute and the ground-state electronic distribution of electrons do not contribute. In both constants, the electronic states that possess electronic orbital angular momentum contribute to the electronic term. It is quite evident that the second-order dependences in Eq. (27) and (28) are identical, and the diagonal elemen in the magnetic shielding tensor can be expressed in terms of the spin–rotation tensor elements by[9]

$$\sigma_{xx} = \sigma^{\text{d}}_{xx} + \sigma^{\text{p}}_{xx} = \frac{e^2}{2mc^2} <0|\sum_i \frac{(y_i^2 + z_i^2)}{r_i^3}|0>$$

$$+ \frac{e^2}{2mc^2} \left[\frac{M_{xx} I_{xx} c}{\mu_0 e \hbar g_k} - \sum_\alpha{}' \frac{Z_\alpha}{r^3_\alpha} (y^2_\alpha + z^2_\alpha) \right]. \quad (29)$$

All distances are from the nucleus in question, and the sum over α omits this nucleus. Normally, only the average magnetic shielding is measured by nuclear magnetic resonance experiments. The average shielding is given by[9]

$$\sigma_{\mathrm{av}} = \sigma^{\mathrm{d}}_{\mathrm{av}} + \sigma^{\mathrm{p}}_{\mathrm{av}} = \frac{1}{3}(\sigma_{xx} + \sigma_{yy} + \sigma_{zz}) = \frac{e^2}{3mc^2} \langle 0| \sum_i \frac{1}{r_i} |0\rangle$$

$$+ \frac{e^2}{6mc^2}\left(\frac{M_{xx}I_{xx}c}{\mu_0 e\hbar g_k} + \frac{M_{yy}I_{yy}c}{\mu_0 e\hbar g_k} + \frac{M_{zz}I_{zz}c}{\mu_0 e\hbar g_k} - 2\sum_\alpha{}' \frac{Z_\alpha}{r_\alpha}\right). \quad (30)$$

This expression further simplifies for high symmetry molecules. Figure 3 illustrates the relationships between σ, $\mathcal{M}$, measurement, molecular structure, and other nuclear parameters.

The spin–rotation constants can be measured at zero field according to Eq. (26), where we write the $\mathcal{M}$ tensor in terms of the molecular-fixed principal inertial values by using the direction cosines

$$\mathcal{H} = \frac{1}{\hbar^2}\mathbf{J}\cdot(\mathcal{M}\,\Phi)\cdot\mathbf{I}. \quad (31)$$

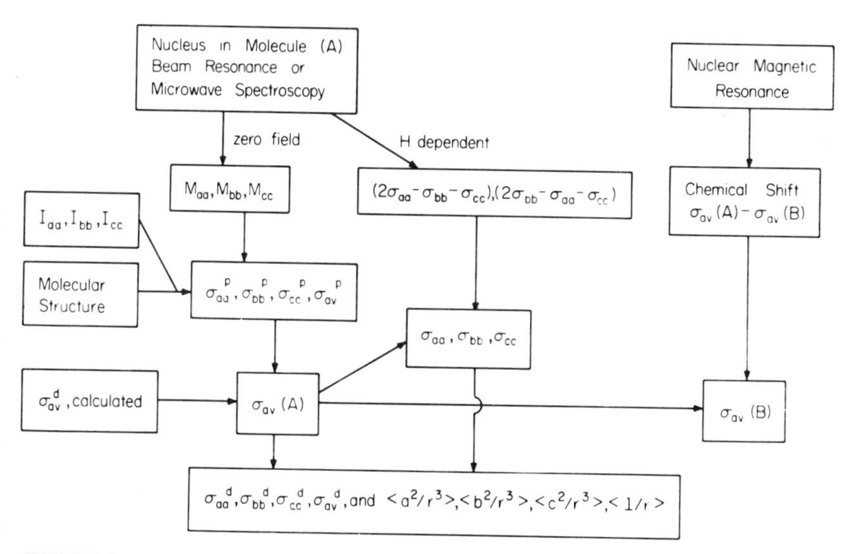

FIGURE 3. Diagram showing the relation between the diagonal elements in the spin–rotation and magnetic shielding tensor elements. $\sigma^{\mathrm{d}}_{\mathrm{av}} = \frac{1}{3}(\sigma^{\mathrm{d}}_{aa} + \sigma^{\mathrm{d}}_{bb} + \sigma^{\mathrm{d}}_{cc})$, $\sigma^{\mathrm{p}}_{av} = \frac{1}{3}(\sigma^{\mathrm{p}}_{aa} + \sigma^{\mathrm{p}}_{bb} + \sigma^{\mathrm{p}}_{cc})$, and $\langle a^2/r^3\rangle = \langle 0|\sum_i \frac{a_i^2}{r_i^3}|0\rangle$.

This Hamiltonian is similar to the linear-field term in Eq. (20), and Hüttner and Flygare[4] have developed the equations necessary for the solution. The resultant energy is given (in the coupled **F** = **J** + **I** basis) by

$$E(I, J, \tau, F) = -\frac{[F(F+1) - J(J+1) - I(I+1)]}{2J(J+1)} \sum_{\alpha} M_{\alpha\alpha} \langle J\tau \| J_{\alpha}^{2} \| J\tau \rangle , \quad (32)$$

where the reduced matrix elements over J_α^2 are defined following Eq. (21). [See, for instance, Eq. (23) for a symmetric top.] It is evident from Eq. (32) that the diagonal elements in the spin–rotation interaction tensor, $\mathcal{M}$, are measured by the rotational perturbation at zero field. Several spin–rotation constants in linear molecules are listed in Table 3. Combining Eq. (26) and (27) with the *nuclear* quadrupole coupling at a single nucleus where we relate the molecular fixed quantities to the laboratory values with the direction cosines gives

$$\mathcal{H} = \mathcal{H}_{rr} - \frac{\mu_0}{\hbar} \mathbf{H} \cdot [\tilde{\Phi}\ \mathcal{G}\,] \cdot \mathbf{J} - \frac{1}{2} \mathbf{H} \cdot [\tilde{\Phi} \mathcal{X} \Phi] \cdot \mathbf{H}$$

$$- \gamma_I\, \mathbf{I} \cdot (1 - \tilde{\Phi} \sigma \Phi) \cdot \mathbf{H} - \frac{1}{\hbar^2} \mathbf{J} \cdot [\mathcal{M}\, \Phi] \cdot \mathbf{I}$$

$$+ \frac{1}{2} Q_{zz}\, [\tilde{\Phi}\ \mathcal{G}\ \Phi]_{zz} \,. \quad (33)$$

The $\mathcal{G}$, $\mathcal{X}$, σ, and $\mathcal{M}$ tensors are all defined in the principal inertial axes systems. Q_{zz} is the scalar quadrupole moment of the nucleus [defined by the convention in Eq. (11)] and $\mathcal{G}$ is the field-gradient tensor at the nucleus described again in the principal inertial axes systems. All other terms have been defined previously.

Hüttner and Flygare[4] have discussed in detail the matrix representation of Eq. (33) in both the *coupled* and *uncoupled* bases. In the coupled basis, the matrix representations of the I^2, J^2, F^2, and F_z operators are all diagonal. I is the angular momentum of the coupled nucleus, J the rotational angular momentum, F the vector sum of I and J, and F_z and z component of F. The quantum numbers labeling the coupled representation are $|IJFM_F\rangle$. In the uncoupled representation the matrices of the z components of I and J, I_z and J_z, are each diagonal, as are I^2 and J^2 since the nuclear angular momentum is independent of the rotational angular momentum. The appropriate quantum numbers labeling the uncoupled representations are $|IM_IJM_J\rangle$. One may use either representation to find diagonal and off-diagonal matrix elements of the interaction of the magnetic field with the molecule. We will refer to the original literature for the details in the rotational dependence of

TABLE 3 Spin-Rotation, $M_\perp$, and the Rotational Field, $R_\perp$, Constants for Several Linear Molecules

Molecule	Nucleus	g_k	$M_\perp/h$ (Hz)	$M_\perp/\mu_0 g_k = R_\perp$ (Gs)	r (10^{-8} cm)	$I_\perp$ (10^{-40} g cm^2)	$\sigma^p_{av} \times 10^6$ [Eq. (30) or (34)]	$\sigma^d_{av} \times 10^6$	$\sigma_{av} \times 10^6$
H–H	H	5.854	$+112.734 \times 10^{3}$ [a]	25.2	0.7416	0.467	−6	32	26
H–^{19}F	H	5.854	$+71.0 \times 10^{3}$ [a]	16.1	0.917	1.34	−80	110	30
H–^{19}F	^{19}F	5.2546	-284.0×10^{3} [b]	−71.0	0.917	1.34	−63	483	420(410) [b]
^{6}Li–^{19}F	^{19}F	5.2546	-37.3×10^{3} [c]	−9.3	1.525	10.61	−73	489	416
^{7}Li–^{19}F	^{19}F	5.2546	-32.9×10^{3} [c]	−8.2	1.525	11.91	−73	489	416
^{19}F–^{19}F	^{19}F	5.2546	-157.0×10^{3} [d]	−39.0	1.418	32.0	−750	530	−230
^{15}N–^{15}N	^{15}N	−0.5660	$+22.0 \times 10^{3}$ [e]	−51.0	1.094	15.0	−485	386	−99
^{13}C–^{16}O	^{13}C	1.4042	-32.6×10^{3} [f]	−30.5	1.128	15.16	−323	327	+4
^{12}C–^{17}O	^{17}O	−0.7572	$+29.0 \times 10^{3}$ [g]	−50.3	1.128	14.9	−460	445	−15
H–^{12}C≡^{12}C–H	H	5.854	3.58×10^{3} [a]	0.80	d_{HC}=1.060 d_{CC}=1.207	23.8	−70	98	28

Note: g_k is the nuclear g value, and $I_\perp$ is the moment of inertia. σ^d_{av} for H_2 is an accurate calculation. σ^d_{av} for the remaining molecules is estimated from Eq. (35). σ^p_{av} is computed with $I_\perp$, r, and Eq. (30) or (34). $\sigma_{av} = \sigma^d_{av} + \sigma^p_{av}$.

[a] From Ramsay.[35]
[b] Corrected for vibrational effects.[36]
[c] See Townes and Schawlow.[37]
[d] From Baker *et al.*[38]
[e] From Kern and Lipscomb.[39]
[f] From Ozier *et al.*[40]
[g] From Flygare and Weiss.[41]

Eq. (33).[4] It should be clear that the diagonal elements in each of the $\mathcal{G}$, $\mathcal{X}$, σ, and $\mathcal{M}$ tensors are available from the molecular Zeeman data.

We now return to a discussion of the parameters in Table 3 starting with the spin–rotation constants, which are related to the average shielding in a diatomic molecule by [see Eq. (30)]:

$$\sigma_{av}\ (\text{diatomic}) = \sigma_{av}^{d} + \sigma_{av}^{p}$$

$$= \frac{e^2}{3mc^2} <0| \sum_i \frac{1}{r_i} |0> + \frac{e^2}{6mc^2}\left(\frac{M_\perp Ic}{\mu_0 e\hbar g_k} - \frac{2Z}{r}\right), \tag{34}$$

where Z is the atomic number of the other nucleus in the diatomic and r is the internuclear distance. The paramagnetic shielding can be computed directly from the molecular spin–rotation constants (or rotational field constants) and the molecular structure. The results for several diatomics are listed in Table 3 under σ_{av}^{p}. The diamagnetic contributions to the total average magnetic-shielding values must be obtained by calculation from the ground-state molecular wave functions. However, as σ_{av}^{d} is a local atomic property that weights most heavily the inner electrons, we can approximate σ_{av}^{d} in molecules by a sum over atomic properties[13]

$$\sigma_{av}^{d}\ (\text{nucleus } k \text{ in a molecule}) = \sigma_{av}^{d}\ (\text{free } k \text{ atom}) + \frac{e^2}{3mc^2} \sum_\alpha{}' \frac{Z_\alpha}{r_\alpha}. \tag{35}$$

Reliable calculated free atom diamagnetic shieldings are available in the literature.[14] Eq. (35) implies that other atom corrections may be added by assuming the other atom's charges are distributed at the nuclear points throughout the molecule. The values of σ_{av}^{d} from Eq. (35) in several molecules are listed in Table 3. The values of σ_{av}^{d} in Table 3 are all within 1 percent of the values calculated with the best available molecule wave functions. Thus, Eq. (35) is an extremely good approximation. If Eq. (35) is valid, we can further simplify Eq. (30). Substituting Eq. (35) into Eq. (30) shows that the pure nuclear term vanishes giving[13, 15]

$$\sigma_{av}(\text{linear}) = \sigma_{av}^{d}\ (\text{atom}) + \left(\frac{e}{3m\hbar c}\right) R_\perp I, \tag{36}$$

where $R_\perp$ is defined in Table 3 as $R_\perp = M_\perp/\mu_0 g_k$.

Returning to Table 3, we note that the net shielding at O in CO,

N in N_2, and F in F_2 is negative (paramagnetic). Thus, the actual fields in these nuclei are larger than the external fields. The other examples in Table 3 indicate a net shielding at the nuclei.

The average shielding, σ_{av}, for a nucleus in a specific molecule allows a standard of comparison for the shielding at this nucleus in other molecules. The average magnetic shielding cannot be measured directly by nuclear magnetic resonance. Only the chemical shift—the *difference* in the magnetic shielding—is obtained. Thus, to obtain the average shielding in any molecule, the shifts must be related to the actual shieldings in Table 3. The standard shielding is obtained by the link with the spin-rotation interaction and the calculated diamagnetic shielding. A summary of the interconnections between the spin–rotation and shielding tensor elements is shown in Figure 3. This figure is similar in design to Figure 1 showing the relations between molecular *g* values and magnetic susceptibilities.

A great deal of interpretation has been applied to the observed chemical shifts in molecules. The major share of these theories assume that the chemical shifts arise primarily from the change in the paramagnetic terms. However, it is evident from Table 3 and Eq. (35) that the diamagnetic shielding can also vary considerably at the same nucleus in two different molecules.

SEMIEMPIRICAL ATOM DIPOLE APPROACH TO PREDICTING MOLECULAR PROPERTIES

A localized model for the interpretation and prediction of molecular properties obtained from the magnetic studies described previously has been developed and is called the *atom dipole model.*[16] The model replaces the ground electronic state average values of several molecular properties into sums over the corresponding free atom values plus corrections that depend on the atom dipoles. The *semiempirical* atom dipoles are obtained by fitting a set of molecular dipole and quadrupole moments to the expressions relating the experimental moments to the atom dipole dependent expressions.[16] These atom dipole moments are useful in predicting molecular dipole and quadrupole moments in new molecules. In addition, we outline the use of the atom-dipole moments to evaluate the molecular diamagnetic susceptibilities and nuclear diamagnetic shielding in a molecule.[17] Finally, we use the localized picture of electron distribution to obtain a semiempirical set of atom and bond magnetic susceptibility tensor elements.

In the *atom-dipole* approach we assume that molecules are constructed of atoms where the small electron density differences between molecules and free atoms (with the same molecular structure) is described by polarized atoms. The expressions for the dipole moment, quadrupole moment, diamagnetic susceptibility, and second moment of the charge distribution are given by[16]

$$D_x = -e\sum_n \langle x\rangle_n$$

$$Q_{xx} = -e\sum_n (x_n \langle x\rangle_n - \mathbf{r}_n \cdot \langle \boldsymbol{\rho}\rangle_n)$$

$$\chi^{\mathrm{d}}_{xx} = -\frac{e^2N}{4mc^2}\left[\sum_n Z_n(y_n^2 + z_n^2) + 2\sum_n (y_n\langle y\rangle_n + z_n\langle z\rangle_n) + \sum_n (\langle y^2\rangle_n + \langle z^2\rangle_n)\right]$$

$$\langle 0|\sum_i x_i^2|0\rangle = \sum_n Z_n x_n^2 + \sum_n 2x_n\langle x\rangle_n + \sum_n \langle x^2\rangle_n, \qquad (37)$$

where $\langle \rho\rangle_n$ and $\langle x\rangle_n$ are the vector- and component-reduced atom dipoles that are listed for a limited number of atoms in Table 4. Additional semiempirical values are found for other atoms in Gierke *et al.*[16] The values of $\langle x^2\rangle_n$, $\langle y^2\rangle_n$, and $\langle z^2\rangle_n$ are also needed to evaluate χ^{d}_{xx} and $\langle 0|\sum_i x_i^2|0\rangle$ for a molecule by this method. We will use free atom values for $\langle x^2\rangle_n = \langle r^2/3\rangle_n$.* This approximation is quite reasonable, and close inspection of electron density maps in bonded atoms reveals that changes in density with respect to the free-atom distribution are relatively large only when the absolute electron density is relatively small. In other words, it appears that the charge distribution in atoms is not dramatically changed by covalent bonding. Using the reduced atom dipoles, $\langle x\rangle_n$, and values of $\langle r^2/3\rangle_n$ gives the calculated results in Table 5 for $\mathbf{D}$, $\mathcal{X}$, $\langle x^2\rangle$, and $\mathcal{X}^{\mathrm{d}}$. These calculated values are also compared with the experimental results. It is evident that the results are quite good. The calculated results in Table 5 for χ^{d}_{xx} and $\langle x^2\rangle$ are better than the results on molecular moments, since uncertainties in atom dipoles will introduce relatively small errors in the second moments of

* The values of $\langle r^2/3\rangle_n$ for the free atoms needed here are obtained from the free-atom diamagnetic susceptibilities $\chi^{\mathrm{d}}_{\mathrm{av}}$ (atom).[18]

TABLE 4 Reduced Atom Dipole Moments[a]

Hydrogen	
H–C<	+0.20
H–C=C	+0.15
H–C≡C	+0.06
H–C≡N	+0.06
H–C=O	+0.05
H–O	+0.30
H–X(halogen)	+0.31
H–N	+0.25
H–C(aromatic)	+0.16
Fluorine	
F–C<	−0.15
F–C=C	−0.13
F–C≡C, F–C≡N	−0.10
F–H	−0.07
F–C(aromatic)	−0.13
F–C=O	(−0.20 to −0.29)
sp^3 Carbon	
C–H	0
C–F	0.0 to +0.03
C–C=C	0
C–C=O	0
C–O	0.06 to 0.08
C–C	0
sp^2 Carbon	
C=C	+0.02 to +0.05
=C–H, =C–C, =C–	0
C=O	+0.15
C (aromatic)	−0.16
=C–F	+0.02
=C–O	≈0

[a] Plus sign means that the electrons in the *n*th atom are polarized away from the atom in the bond. These values along with Eq. (37) give the calculated values in Table 5. See Gierke *et al.*[16] for more details.

TABLE 5 Calculated Molecular Dipole Moments, Molecular Quadrupole Moments, Second Moments of the Electronic Charge Distribution, and Diamagnetic Susceptibilities[a]

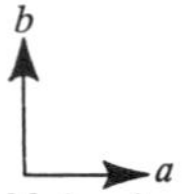 Molecule (axes b, a)	$\left.\begin{matrix}D_a\\D_b\\D_c\end{matrix}\right\} \times 10^{18}$ Experimental	Calculated	$\left.\begin{matrix}Q_{aa}\\Q_{bb}\\Q_{cc}\end{matrix}\right\} \times 10^{26}$ Experimental	Calculated	$\sum_n Z_n a_n^2$, $\sum_n Z_n b_n^2$, $\sum_n Z_n c_n^2$	$\left.\begin{matrix}\langle a^2\rangle\\\langle b^2\rangle\\\langle c^2\rangle\end{matrix}\right\} \times 10^{16}$ Experimental	Calculated	$\left.\begin{matrix}\chi_{aa}^d\\\chi_{bb}^d\\\chi_{cc}^d\end{matrix}\right\} \times 10^{6}$ Experimental	Calculated
NH_3 (N—H, H, H; axes a, c; − +)	0	0	+1.2 ± 0.1	+1.3	1.34	2.6 ± 0.4	2.5	−19.3 ± 1.4	−19.2
	0	0	+1.2 ± 0.1	+1.3	1.34	2.6 ± 0.4	2.5	−19.3 ± 1.4	−19.2
	|1.47|	+1.44	−2.4 ± 0.1	−2.6	0.31	1.9 ± 0.4	2.0	−22.0 ± 1.4	−21.5
[illegible] (+ −)	[illegible]	[illegible]	[illegible]	−1.0	30.77	34.7 ± 0.4	34.6	−44.2 ± 0.5	−44.5

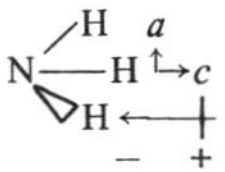

	0	0	+0.4 ± 0.2	+0.5	0	4.5 ± 0.6	5.0	[illegible]	[illegible]
	0	0	+0.4 ± 0.2	+0.5	0	4.5 ± 0.6	5.0	−215.5 ± 2.0	−218.0
	0	0	+0.4 ± 0.2	+0.5	0	4.5 ± 0.6	5.0	−215.5 ± 2.0	−218.0
O=C=O	0	0	−4.4 ± 0.2	−4.4	21.64	25.5 ± 1.0	25.5	−24.9 ± 3.0	−27.6
	0	0	+2.2 ± 0.2	+2.2	0	2.9 ± 1.0	3.2	−120.6 ± 2.0	−122.0
	0	0	+2.2 ± 0.2	+2.2	0	2.9 ± 1.0	3.2	−128.6 ± 2.0	−122.0
Cyclobutanone (ring structure, C=O)	|2.98|	−3.23	−9.4 ± 1.2	−8.8	55.3	63.2 ± 1.8	62.4	−185.0 ± 3.9	−182.4
	0	0	+4.5 ± 1.1	3.6	26.9	32.8 ± 1.8	32.6	−314.1 ± 5.2	−309.1
	0	0	+4.9 ± 1.7	5.2	4.9	10.8 ± 1.8	10.4	−407.5 ± 5.6	−403.4
H_3C–C(H)=O	|2.55|	−2.54	−1.2 ± 1.5	−2.9	28.9	32.3 ± 0.5	32.5	−64.1 ± 1.4	−62.5
	|0.87|	+0.93	+1.0 ± 0.9	+1.9	5.5	9.6 ± 0.4	9.3	−160.4 ± 1.8	−160.1
	0	0	+0.2 ± 1.8	+1.0	1.5	5.6 ± 0.5	5.4	−177.4 ± 1.8	−177.0
H–C≡C–F	|0.75|	−0.77	+4.0 ± 0.2	+3.8	28.34	31.5 ± 0.6	31.0	31.0	−31.0
	0	0	−2.0 ± 0.2	−1.9	0	3.5 ± 0.6	3.5	147.9	−147.9
	0	0	−2.0 ± 0.2	−1.9	0	3.5 ± 0.6	3.5	147.9	−147.9

[a] Eq. (37) using the atom dipoles and the known molecular structures. The experimental results are also shown. The numbers are from Gierke *et al.*,[16] where more values are listed.

electronic charge. Deviations in second moments calculated with Eq. (37) are estimated to be 0.1–1.0 Å^2, which will produce deviations in the diamagnetic susceptibility of about 0–8 × 10^{-6} erg/Gs^2 mol).

Gierke and Flygare[17] have also applied the atom-dipole method to the evaluation of diamagnetic shieldings. Starting with Eq. (27), the resultant atom dipole expressions are

$$\sigma^{d}_{av}(A) = \sigma^{d}_{atom}(A) + \frac{e^2}{3mc^2} \sum_n{}' \frac{Z_n}{r_n} - \frac{e^2}{3mc^2} \sum_n{}' \frac{r_n \cdot \langle\rho\rangle_n}{r_n^3}. \tag{38}$$

The corresponding result for the individual diagonal tensor elements is

$$\begin{aligned}\sigma^{d}_{xx}(A) = \sigma^{d}_{atom}(A) &+ \frac{e^2}{2mc^2} \sum_n{}' \frac{Z_n}{r_n^3}(y_n^2 + z_n^2) \\ &+ \frac{e^2}{2mc^2} \sum_n{}' \left[(2r_n^{-3})(y_n\langle y\rangle_n + z_n\langle z\rangle_n) - (3r_n^{-5})(y_n^2 + z_n^2)(\mathbf{r}_n \cdot \langle\rho\rangle_n)\right] \\ &+ \frac{e^2}{2mc^2} \sum_n{}' \left[(2r_n^{-3})\langle\rho^2/3\rangle_n - 3(r_n^{-5})(y_n^2 + z_n^2)\langle\rho^2/3\rangle_n\right].\end{aligned} \tag{39}$$

The interpretation of the various terms in these equations are given by Gierke and Flygare.[17]

Table 6 lists the values of σ^{d}_{xx}, σ^{d}_{yy}, σ^{d}_{zz}, and σ^{d}_{av} calculated from Eq. (38) and (39) from Gierke and Flygare.[17] Of course, the molecular structure must be known also in order to make these calculations. The first column in Table 6 lists the diamagnetic susceptibilities; the numbers in parentheses are from reliable *ab initio* calculations. It is evident that the values of σ^{d}_{av} are in excellent agreement. The individual diagonal elements in σ^{d} are also in quite good agreement in most cases. The second column lists the values of σ^{p}_{xx}, σ^{p}_{yy}, σ^{p}_{zz}, and σ^{p}_{av}, which are calculated with Eq. (29) and (30), the known moments of inertia, the molecular structure, and the spin–rotation constants. The total values of the magnetic shielding, $\sigma = \sigma^{d} + \sigma^{p}$, are also listed along with any available experimental data. These data are used to fix the magnetic shielding scales.

We now turn to a final analysis that uses a localized approach to magnetic parameters: the concept of localized magnetic susceptibilities. The individual elements in the magnetic susceptibilities can be obtained

TABLE 6 σ Tensor Elements in Several Molecules[a]

Molecule (y ↑, → x)	Atom	σ^d_{xx} σ^d_{yy} σ^d_{zz} σ^d_{av}	σ^p_{xx} σ^p_{yy} σ^p_{zz} σ^p_{av}	σ_{xx} σ_{yy} σ_{zz} σ_{av}	σ_{xx}(exp) σ_{yy} σ_{zz} σ_{av}
H–H	^{1}H	35.7	0.0	34.7	
		30.4	–8.4	22.0	
		30.4	–8.4	22.0	
		32.9(32.0)	–5.6	26.3	26.6
H–F	^{1}H	54	0	54	
		137	–119	18	
		137	–119	18	
		109(109)	–80	29	29.2
	^{19}F	480	0	480	
		487	–94	393	
		487	–94	393	
		484(482)	–63	421	415
H–C≡N	^{14}N	348	0	348	
		398	–627	–229	
		398	–627	–229	
		381	–418	–37	
	^{1}H	48	0	49	
		121	–109	12	
		121	–109	12	
		97	–71	26	28
N≡N	^{15}N	349(338)	0	349	
		402(407)	–729	–327	
		402(407)	–729	–327	
		384(385)	–486	–102	–101
C≡O	^{13}C	280(271)	0	280	
		349(354)	–484	–135	
		349(354)	–484	–135	
		326(326)	–323	+3	
	^{17}O	419(410)	0	419	
		456(461)	–690	–234	
		456(461)	–690	–234	
		444(444)	–460	–16	
O=C=S	^{13}C	299	0	299	
		483	–579	–96	
		483	–579	–96	
		421	–386	35	
	^{17}O	421	0	421	
		538	–978	–440	
		538	–978	–440	
		499	–652	–153	
	^{33}S	1060	0	1060	
		1141	–810	331	
		1141	–810	331	
		1114	–540	574	

TABLE 6 (continued)

Molecule	Atom	σ^d_{xx} σ^d_{yy} σ^d_{zz} σ^d_{av}	σ^p_{xx} σ^p_{yy} σ^p_{zz} σ^p_{av}	σ_{xx} σ_{yy} σ_{zz} σ_{av}	σ_{xx}(exp) σ_{yy} σ_{zz} σ_{av}
H–C≡C–H	^{1}H	50	0	50	
		120	−99	21	
		120	−99	21	
		97	−66	31	
$H_2C=O$	^{1}H	87(93)	−69	18	
		94(95)	−70	24	
		137(147)	−120	17	
		106(111)	−86	20	21.3
	^{17}O	418(415)	−1600	−1182	
		468(462)	−870	−402	
		470(475)	−10	460	
		452(452)	−827	−375	

^{a} The first column gives the diamagnetic shieldings from Eq. (38) and (39) with the *ab initio* results in parentheses. The second column gives σ^p evaluated from the molecular structure, moments of inertia, and spin-rotation constants. The final columns give the total calculated and experimental values of σ. The results are from Gierke and Flygare.[17]

by combining the bulk values, χ, with the anisotropies as discussed previously (see also Figure 1).

Starting with Eq. (9), which defines the diamagnetic, χ^d_{xx}, and paramagnetic, χ^p_{xx}, tensor elements, we partition the sum over all molecular electrons into sums over each free-atom number of electrons. Using again this atom-dipole approximation leads to the following values for the paramagnetic susceptibility [the diamagnetic component is given in Eq. (37)]:

$$\chi^p_{xx} = \frac{e^2}{4mc^2}\left[\sum_n Z_n(x_n^2 + y_n^2) + 2\sum_n (x_n \langle x\rangle_n + y_n \langle y\rangle_n)\right]$$

$$- \frac{e^2}{2m^2c^2}\sum_n\sum_{k>0}\left[\frac{\langle 0|\sum_{i_n=1}^{Z_n}(l_i^n)_z|0\rangle\langle k|\sum_{i_n=1}^{Z_n}(l_i^n)_z|0\rangle + \text{c.c.}}{E_0 - E_k}\right]. \quad (40)$$

The vector $l_i^n = -i\hbar(\rho_{i_n} \times \nabla_i)$ is the electronic angular momentum of the ith electron about the nth nucleus, and c.c. stands for the conjugate complex expression.

Considering $\chi_{xx} = \chi_{xx}^{\mathrm{p}} + \chi_{xx}^{\mathrm{d}}$ from Eq. (39) and (40), we note that the first two terms in $\mathcal{X}^{\mathrm{d}}$ and $\mathcal{X}^{\mathrm{p}}$ cancel removing all explicit dependences on molecular coordinates. Hence, $\mathcal{X}$ is "gage invariant" while $\mathcal{X}^{\mathrm{d}}$ and $\mathcal{X}^{\mathrm{p}}$ are not. What remains is

$$\chi_{xx} = \sum_n \left[\frac{e^2}{4mc^2} (\langle x^2 \rangle_n + \langle y^2 \rangle_n) \right]$$

$$- \left(\frac{e^2}{2m^2c^2}\right) \sum_{k>0} \left[\frac{\langle 0| \sum_{i_n=1}^{Z_n} (1_i^n)_z |k\rangle\langle k| \sum_{i_n=1}^{Z_n} (1_i^n)_z |0\rangle + \text{c.c.}}{E_0 - E_k} \right]$$

$$= \sum_n (\mathcal{X}_n^{\mathrm{d}} + \mathcal{X}_n^{\mathrm{p}})_{zz} \,. \tag{41}$$

The bulk value is given by

$$\chi = \frac{1}{3} (\chi_{xx} + \chi_{yy} + \chi_{zz}) = \sum_n (\chi_n^{\mathrm{d}} + \chi_n^{\mathrm{p}}) = \sum_n \chi_n \,. \tag{42}$$

The total molecular susceptibility has now been expressed as a sum over *operators* localized on the various atomic nuclei. But they operate on wave functions that extend over the whole molecule. If the average values of these atomic operators are not greatly dependent on parts of the wave function far removed from the nucleus in question and if the relevant properties of the electron distribution around each nucleus are not much different for a given type of atom in different molecules, the terms within each sum over n in Eq. (41) will be independent and constant. They will, in short, be additive atomic susceptibilities that can be evaluated from measured molecules and used to predict the susceptibility of any desired molecule. We have already demonstrated the additivity of the diamagnetic susceptibilities [Eq. (37)].

It is hard to show theoretically that the average values of the atomic angular momentum operators in $\mathcal{X}_n^{\mathrm{p}}$ of Eq. (41) are localized. Physically, this requirement can be understood to imply that there be no long-range circulation of electrons; in other words, each electron circulation is confined to localized orbitals. However, the usual model for aromatic compounds that involves a molecular ring current would indicate that this class of molecules cannot be treated by localized theories. There is also question as to whether small strained rings meet this localization criterion. In conclusion, there is reason to believe that an attempt to construct a system of local rules for the diagonal components of the susceptibility tensor might succeed for nonstrained, nonaromatic molecules. In fact, we are extending the well-known Pascal's rules[20] for the bulk magnetic susceptibility to the individual diagonal elements in the total magnetic susceptibility tensor. We list the experimental values (obtained by microwave methods[1,2]) for the individual components of the susceptibility tensor in a number of nonstrained molecules in Table 7. These molecules should all satisfy the criterion necessary to determine a set of localized atomic values.

For convenience in application we have obtained both atom and bond values. In the atom approach we have assumed that an atom in a particular bonding situation (particular hybridization) will always contribute the same amount to the molecular susceptibility. This contribution consists of the three principal components as shown in Table 8 under atom susceptibilities. To evaluate the molecular susceptibility, the atom or bond values in Table 8, which are principal values, are rotated into the principal inertial axis system (a, b, and c) of the molecule. The atom and bond susceptibilities were determined by least squares fitting the experimental molecular susceptibility components of the 14 common nonstrained, nonaromatic molecules shown in Table 7.

It is evident from the results in Table 7 that the molecular susceptibilities obtained from the local values in Table 8 are in good agreement with experimental results. These local values can now be used to predict the molecular susceptibility of nonstrained, nonaromatic compounds on which measurements are not available. In conclusion we have extended the well-known Pascal's rules for bulk susceptibility to the individual diagonal elements in the molecular magnetic susceptibility tensor. Of course, the diagonal sums of the values given in Table 8 also allow a prediction of the bulk susceptibilities.

Since these local group susceptibility values are derived under the explicit assumption that all electron motions are localized, the difference between the observed magnetic susceptibility and that calculated from local group values should provide a quantitative measure of electron delocalization and perhaps aromaticity.[19] There are also a variety of other applications for the local susceptibilities given in Table 8. For instance, molecular g values can be measured in a number of small molecules by molecular beam techniques.[1] In many of these molecules an accurate measurement of the susceptibility anisotropy is difficult. Magnetic susceptibilities calculated by the methods of this paper can thus be combined with g values obtained by other methods to extract quadrupole moments. Alternatively, calculated susceptibility anisotropies

TABLE 7 Experimental and Semiempirically Calculated Values of χ_{aa}, χ_{bb}, and χ_{cc}[a]

Molecule	χ_{aa} χ_{bb} χ_{cc} Experimental	Calculated Atom	Calculated Bond
Methyl formate	−28.3 ± 0.5	−28.6	−28.1
	−30.9 ± 0.6	−29.8	−30.2
	−36.7 ± 0.8	−36.7	−36.4
Acetaldehyde	−20.0 ± 0.9	−21.3	−20.6
	−19.5 ± 0.9	−19.1	−18.0
	−28.6 ± 1.5	−29.2	−28.9
Propene	−30.9 ± 1.0	−28.5	−28.0
	−26.2 ± 1.0	−26.6	−25.9
	−34.9 ± 1.1	−34.9	−35.9
Acrolein	−16.0 ± 3.0	−18.1	−20.3
	−18.3 ± 3.0	−18.5	−18.5
	−37.7 ± 4.0	−37.7	−39.4
Maleic anhydride	−25.7 ± 1.5	−25.1	−27.0
	−28.2 ± 1.5	−27.5	−29.4
	−53.5 ± 1.7	−54.3	−54.0
Cyclopent-2-ene-1-one	−33.1 ± 4.0	−37.0	−36.2
	−39.6 ± 4.0	−38.0	−37.6
	−55.4 ± 6.0	−55.7	−52.8
Cyclopent-3-ene-1-one	−34.3 ± 4.5	−39.1	−38.1
	−40.0 ± 4.0	−35.9	−35.7
	−53.9 ± 6.0	−55.7	−52.8
Vinylene carbonate	−35.3 ± 4.0	−32.0	−34.1
	−30.5 ± 4.0	−32.0	−34.8
	−47.4 ± 5.0	−48.1	−48.2
2-Pyrone	−36.1 ± 4.0	−32.9	−35.2
	−34.4 ± 4.0	−34.5	−36.3
	−60.0 ± 6.0	−60.0	−61.0
4-Pyrone	−31.2 ± 4.0	−32.6	−37.4
	−31.9 ± 4.0	−34.8	−34.8
	−54.5 ± 6.0	−60.0	−61.0
Isoprene	−35.4 ± 4.0	−36.9	−36.1
	−34.5 ± 4.0	−39.7	−38.8
	−52.8 ± 5.0	−54.3	−56.1
Formic acid	−18.8 ± 1.5	−16.9	−17.9
	−16.8 ± 1.5	−16.3	−15.7
	−24.2 ± 2.0	−25.8	−25.4
Glycolaldehyde	−27.6 ± 4.0	−28.9	−29.4
	−23.7 ± 4.0	−27.8	−26.5
	−38.6 ± 4.5	−37.1	−35.1
Water	−12.2 ± 2.0	−13.0	−11.3
	−13.4 ± 2.0	−13.6	−12.7
	−13.4 ± 2.5	−12.1	−11.0

[a] The calculated values are from the atom and bond values in Table 8.

TABLE 8 Local Atom and Bond Susceptibilities[a]

Atom Susceptibilities (axes: x, y; z out-of-plane)

Atom	x	y	z
H^b	−2.0	−2.3	−2.3
H_3C^b (tetrahedral, three H)	−9.5	−6.7	−6.7
H_2C^b (two H)	−6.5	−8.0	−7.1
O^b (two single bonds)	−9.2	−8.8	−7.5
$>C^b=$	−2.9	−3.5	−7.4
$O^b=$	−0.1	−0.1	−6.3
$-C^c\equiv$	−9.9	−7.4	−7.4
$N^c\equiv$	−9.5	−4.5	−4.5
$-N^c<$ (planar)	−11.3	−14.8	−2.5
$S^c<$	−18.7	−15.7	−16.8
$S^c=$	+4.7	−13.1	−23.0
$P^c\equiv$	−24.1	−17.9	−17.9

Bond Susceptibilities (axes: x, y; z out-of-plane)

Bond	x	y	z
C–H[a]	−5.6	−3.1	−3.1
C–C[a]	−7.9	−0.2	−0.2
C=C[a]	−0.8	+4.0	−13.8
C–O[a]	−7.2	−6.7	−3.8
C=O[a]	−1.3	+2.2	−13.0
O–H[a]	−4.5	−7.5	−5.5
C–S[b]	−12.6	−11.3	−8.9
C=S[b]	+5.8	−11.0	−28.8
C≡C[b]	−13.7	−14.2	−14.2
C≡N[b]	−16.5	−11.6	−11.6
C≡P[b]	−30.4	−24.5	−24.5

[a] From Schmalz, Norris, and Flygare.[19]
[b] Determined from least squares fit of molecules in Table 7.
[c] Determined from a limited number of molecules assuming the least squares values for the other parameters.

can be combined with calculated quadrupole moments to yield the g values. For instance, Hartford *et al.*[21] recently used a series of substituted acetylenes and substituted cyanides to obtain the magnetic susceptibility of H–C ≡ C–H. Their result was $\chi_\perp - \chi_\parallel = (4.5 \pm 0.5) \times 10^{-6}$ erg/Gs^2 mol, which is in excellent agreement with the values of $\chi_\perp - \chi_\parallel$ for acetylene obtained for Table 8; the acetylene values are 4.4 from the atom values and 4.5 from the bond values. Using $\chi_\perp - \chi_\parallel = 4.5$ and a reliable calculated quadrupole moment in HC≡CH, Hartford *et al.*[21] were able to show that the *g* value in this molecule must be positive rather than negative as reported in the literature.[22]

Supported in part by the National Science Foundation Grant GP-12382X4.

REFERENCES

1. W. H. Flygare and R. C. Benson, *Mol. Phys.*, **20**, 225 (1971).
2. W. H. Flygare, *Chem. Rev.* (in press).
3. W. Hüttner, M. L. Lo, and W. H. Flygare, *J. Chem. Phys.*, **48**, 1206 (1968).
4. W. Hüttner and W. H. Flygare, *J. Chem. Phys.*, **47**, 4137 (1967).
5. N. F. Ramsey, *Molecular Beams*, Oxford Press, New York (1956).
6. C. L. Norris, R. C. Benson, P. Beak, and W. H. Flygare, *J. Am. Chem. Soc.*, **95**, 2766 (1973).
7. J. C. McGurk, C. L. Norris, H. L. Tigelaar, and W. H. Flygare, *J. Chem. Phys.*, **58**, 3118 (1973).
8. N. F. Ramsey, *Phys. Rev.*, **78**, 699 (1950).
9. W. H. Flygare, *J. Chem. Phys.*, **41**, 793 (1964).
10. W. Hüttner and W. H. Flygare, *J. Chem. Phys.*, **49**, 1912 (1968).
11. D. VanderHart and W. H. Flygare, *Mol. Phys.*, **18**, 77 (1970).
12. J. J. Ewing, H. L. Tigelaar, and W. H. Flygare, *J. Chem. Phys.*, **56**, 1957 (1972).
13. W. H. Flygare and J. Goodisman, *J. Chem. Phys.*, **49**, 3122 (1968).
14. G. Malhi and C. Froese, *Int. J. Quantum Chem.*, **15**, 95 (1967).
15. H. L. Tigelaar and W. H. Flygare, *Chem. Phys. Lett.*, **7**, 254 (1970).
16. T. D. Gierke, H. L. Tigelaar, and W. H. Flygare, *J. Am. Chem. Soc.*, **94**, 330 (1972).
17. T. D. Gierke and W. H. Flygare, *J. Am. Chem. Soc.*, **94**, 7277 (1972).
18. G. Malli and S. Fraga, *Theor. Chim. Acta*, **5**, 284 (1960).
19. T. G. Schmalz, C. L. Norris, and W. H. Flygare, *J. Am. Chem. Soc.*, **95**, 7961 (1973).
20. J. A Pople, W. G. Schneider, and H. J. Bernstein, *High Resolution Nuclear Magnetic Resonance*, McGraw-Hill, New York (1959).
21. S. L. Hartford, W. C. Allen, C. L. Norris, E. F. Pearson, and W. H. Flygare, *Chem. Phys. Lett.*, **18**, 153 (1973).
22. C. W. Cederberg, C. H. Anderson, and N. F. Ramsey, *Phys. Rev.*, **136**, A960 (1964).
23. J. McGurk, H. L. Tigelaar, S. L. Rock, C. L. Norris, and W. H. Flygare, *J. Chem. Phys.*, **58**, 1420 (1973).
24. S. L. Rock, J. C. McGurk, and W. H. Flygare, *Chem. Phys. Lett.*, **19**, 153 (1973).
25. R. L. Shoemaker and W. H. Flygare, *J. Am. Chem. Soc.*, **94**, 684 (1972).
26. S. L. Rock and W. H. Flygare, *J. Chem. Phys.*, **56**, 4723 (1972).
27. S. L. Rock, E. F. Pearson, E. Appleman, C. L. Norris, and W. H. Flygare, *J. Chem. Phys.*, **59**, 3940 (1973).
28. R. C. Benson, W. H. Flygare, M. Oda, and R. Breslow, *J. Am. Chem. Soc.*, **95**, 2772 (1973).

29. R. C. Benson and W. H. Flygare, *J. Chem. Phys.*, **58**, 2651 (1973).
30. R. C. Benson and W. H. Flygare, *J. Chem. Phys.*, **58**, 2366 (1973).
31. C. L. Norris, H. L. Tigelaar, and W. H. Flygare, *Chem. Phys.*, **1**, 1 (1973).
32. H. L. Tigelaar, T. D. Gierke, and W. H. Flygare, *J. Chem. Phys.*, **56**, 1966 (1972).
33. D. Sutter, *Z. Naturforsch.*, **26a**, 1644 (1971).
34. C. L. Norris, R. C. Benson, and W. H. Flygare, *Chem. Phys. Lett.*, **10**, 75 (1971).
35. N. F. Ramsey, *Am. Sci.*, **49**, 509 (1961).
36. D. K. Hinderman and C. D. Cornwell, *J. Chem. Phys.*, **48**, 4148 (1968).
37. C. H. Townes and A. L. Schawlow, *Microwave Spectroscopy*, McGraw-Hill, New York (1955).
38. M. R. Baker, C. H. Anderson, and N. F. Ramsey, *Phys. Rev. A.*, **133**, 1533 (1964).
39. C. W. Kern and W. N. Lipscomb, *J. Chem. Phys.*, **37**, 260 (1962).
40. I. Ozier, L. M. Crapo, and N. F. Ramsey, *J. Chem. Phys.*, **49**, 2315 (1968).
41. W. H. Flygare and V. W. Weiss, *J. Chem. Phys.*, **45**, 2785 (1966).
42. J. M. Pochan, R. G. Stone, and W. H. Flygare, *J. Chem. Phys.*, **51**, 4278 (1969).
43. D. H. Sutter and W. H. Flygare, *Mol. Phys.*, **16**, 153 (1969).
44. W. Hüttner and W. H. Flygare, *J. Chem. Phys.*, **50**, 2863 (1969).

Bernard R. Appleman and Benjamin P. Dailey

Anisotropies of Chemical Shifts and Their Relation to Magnetic Susceptibility

The nuclear magnetic resonance (NMR) spectrometer has become in recent years one of the most popular and useful tools available to the chemist for structural studies. In the usual high-resolution NMR spectrum of a liquid the scalar parameter called the "chemical shift" is readily obtained and may be interpreted in an approximate and qualitative fashion to establish features of the overall structure. The task of placing the interpretation of the "chemical shift" on a more quantitative and rigorous basis is not an easy one but considerable progress has been made in this direction. The potential rewards of improved understanding in this area seem most attractive.

The magnetic field experienced by a nucleus differs from the external field due to the screening effect of the electrons as expressed in Eq. (1)

$$\mathbf{H}_{\mathrm{eff}} = \mathbf{H}_0(1 - \sigma). \tag{1}$$

In general, the direction of $\mathbf{H}_{\mathrm{eff}}$ is not the same as the initially applied field $\mathbf{H}_0$; therefore, the shielding $\boldsymbol{\sigma}$ is a tensor with a possibility of nine independent components.*

* Recently, Buckingham and Malm[3] have pointed out that, in general, the shielding tensor is asymmetric. For example in a C_{2v} system for a nucleus off the unique axis (z), there are five independent components, namely, σ_{xx}, σ_{yy}, σ_{zz}, σ_{xy}, and σ_{yx}.

In principle, however, it is always possible to diagonalize the shielding tensor to its three principal components. Unless stated otherwise, it will be assumed that the shielding tensor is diagonal. The isotropic shielding is defined as one-third the trace of the shielding tensor. The anisotropy can be defined in several ways. Probably the most general definition is

$$(\sigma_{\alpha\alpha})_{\text{anisotropic}} = \sigma_{\alpha\alpha} - \sigma_{\text{iso}} = \frac{1}{3}(2\sigma_{\alpha\alpha} - \sigma_{\beta\beta} - \sigma_{\gamma\gamma}), \tag{2}$$

where α, β, and γ are the principal axes. Since the sum of the three anisotropies is zero, there can be at most two independent anisotropies.

One of the crucial arguments for studying tensor components or anisotropies is that they provide much information that is lost when only the isotropic chemical shift is reported. The linear and symmetric-top molecules have been strongly emphasized in the study of anisotropies. For these types of molecules there are two independent shielding components, normally designated as $\sigma_{\|}$ and $\sigma_{\perp}$ to indicate the shielding parallel or perpendicular to the symmetry axis. The anisotropy is then uniquely defined as $\Delta\sigma = \sigma_{\|} - \sigma_{\perp}$. Additionally, it is noted that from a knowledge of the anisotropic and isotropic shift, the complete shielding tensor is known. This feature has been very useful for comparing and evaluating shielding tensors.

The numerical values of the chemical shift tensor elements depend, of course, on the axis system used to define the tensor. This fact has led to considerable difficulties in the analysis and intercomparison of the experimental data. The transformation of tensor elements from one axis system to another may require assumptions as to bond symmetry or geometry that are not necessarily justified. The axis system that yields tensor elements that are easiest to compare from one molecule to another has an axis parallel to the bond axis of the nucleus under study and an axis perpendicular to the bond. Unfortunately, in many highly symmetric molecules the tensor experimentally determined in the molecular symmetry axis system cannot be transformed to the bond axis system unambiguously and some hard-to-justify assumptions frequently must be made. Some of the difficulties of this sort that arise from an intercomparison of shielding data from liquid crystals and the coherent averaging of solids for aromatic fluorine compounds have been discussed by Mehring *et al.*[1]

For the so-called asymmetric tops, the situation is not nearly so convenient since there are often four or five independent tensor components. For theoretical determinations the problem can be formally solved by diagonalizing the tensor. For many molecules, however, the three principal axes so determined do not correspond to readily defined molecular axes, and in some cases it is questionable whether the additional information provided is of any real use. To avoid such matters Ditchfield, Miller, and Pople[2] report only the isotropic shift.

The experimentalist dealing with these asymmetric molecules often has a more serious problem, namely, "What do the data mean?" In several of the methods described later one obtains experimental anisotropies or components that are defined with respect to a particular set of axes, but the relationship between these and the molecular axes can only be assumed. In some cases one can only extract a quantity such as[4]

$$\tfrac{2}{3}(\sigma_{AA} - \sigma_{BB})[-\tfrac{1}{2}S_{33} + \tfrac{1}{2}(S_{11} - S_{22})\cos 2\epsilon] + \tfrac{2}{3}(\sigma_{CC} - \sigma_{BB})S_{33} - \tfrac{1}{3}(\sigma_{AB} + \sigma_{BA})\sin 2\epsilon(S_{11} - S_{22}),$$

which may be of comparatively little value. (See below for definition of S matrix.)

Nevertheless, there are a reasonable number of cases in which these quantities are sufficiently unambiguous to permit the acquisition of useful, interpretable data, often with the aid of theoretical quantities.

EXPERIMENTAL METHODS: SOURCES OF DATA

In principle, the anisotropy could be experimentally determined by observing the response of a molecule in magnetic fields of different directions. However, in liquids and gases the rapid molecular reorientation allows one to observe only the rotationally averaged isotropic shielding constants. In solids the molecular motion is greatly reduced, and on an NMR time scale one observes a spectrum that is a superposition rather than an average of the shielding tensor elements. The solid-state NMR spectra are normally very broad due to the dipolar coupling terms, which are not averaged to zero as in the liquid phase NMR.

Some early attempts were made to determine anisotropies from second-moment analysis in which use is made of the anisotropic contribution to the observed second moment.[5,6] Limitations of this method are the need for large shifts and fields and the often-required assumption of axial symmetry of the shielding tensor. In 1968 a method was reported that uses coherent averaging techniques[7] to effectively narrow the dipolar-broadened lines of powdered solids. The observed spectrum is curve-fitted using a computer-broadened theoretical chemical shift distribution to give the principal components of the shielding tensor.

This method has been used by Mehring *et al.*[1] to study fluorine shielding.

Another relatively new method[8] allows one to observe magnetically dilute nuclei, e.g., ^{13}C, in a solid by cross-polarizing them with the abundant spins, e.g., ^{1}H. The basic steps follow[8]: (1) Polarize the abundant nuclei I in high field. This step can be performed at low temperature to increase the Boltzmann population difference between the spin states and, hence, the total magnetization. (2) Cool I to a low spin temperature in the rotating frame. In this step the I magnetization initially directed along the steady field is brought into the rotating frame by applying a small radio frequency field at resonance. It is further made to point along the radio frequency field direction to attain the spin-locked position. The I spins that now retain the initial magnetization in the weak H_1 field are thus at a lowered spin temperature. (3) Establish I–S contact. The rare nuclei S are then brought into simultaneous resonance with the I spins by applying a second radio frequency field adjusted such that

$$\gamma_S H_{1S} = \gamma_I H_{1I}. \tag{3}$$

The exchange of energy (cross relaxation) that occurs between the two spin systems (repeated N_I/N_S times) enhances the population difference and hence the "free induction decay" of the rare nucleus, which is the signal observed in this version of the double resonance experiment. This method has been used for natural abundance spectra of ^{13}C,[9] ^{15}N,[10] and ^{29}Si.[11]

Another classification of methods for measuring anisotropies is that of partially oriented molecules, the most important examples of which are the liquid crystals. The utility of this method lies in the fact that many molecules can be dissolved in liquid crystals without destroying the liquid crystal phase.[12–15] The solute molecules are able to translate relatively freely in one direction—along the optic axis. As a result of this motion the intermolecular dipole–dipole interactions are averaged to zero but the intramolecular interactions are not. Also since the solute molecules cannot rotate equally in all directions, there will be an anisotropic contribution to the observed shielding.

The orientation of the solute molecule is described by the ordering matrix **S**,[13] which is defined by Eq. (4)

$$S_{ij} = \tfrac{1}{2}\langle 3\cos\theta_i \cos\theta_j - \delta_{ij}\rangle, \tag{4}$$

where θ_i and θ_j are the angles between the axes of the solute molecule and the magnetic field. The brackets indicate that one requires a value averaged over the molecular motion. The diagonal elements of the **S** tensor are called the degrees of order.

The observed chemical shift in the nematic phase depends on the isotropic shift, the ordering parameters, and the elements of σ.[14]

$$\sigma_{\text{Nem}} = \sigma_{\text{Iso}} + (2/3) \sum_{\alpha\beta} S_{\alpha\beta}\sigma_{\alpha\beta}. \tag{5}$$

This relation is simplified by molecular symmetry. The splitting of the Zeeman term above is determined by the indirect couplings J_{ij} and the direct couplings B_{ij}, which are not averaged to zero in the nematic phase. The average Hamiltonian for a symmetric-top molecule is given in Eq. (6)

$$\langle H \rangle_{\text{Av}} = \frac{-H_0}{2\pi} \sum_i \gamma_i (1 - \sigma_i - \tfrac{2}{3} S_{zz} \Delta\sigma_i) I_{zi} + \sum_{i<j} J_{ij} I_i \cdot I_j$$

$$+ \tfrac{1}{2} \sum_{i<j} (B_{ij} + \Delta J_{ij}) S_{zz} (3 I_{zi} I_{zj} - I_i \cdot I_j). \tag{6}$$

The J_{ij} term can be measured from the isotropic spectrum. The usual procedure in determining S_{zz} from the splittings is to calculate B_{ij} from the molecular geometry and neglect ΔJ_{ij}. Using the relation

$$\sigma_{\text{Nem}} = \sigma_{\text{Iso}} + \tfrac{2}{3}(\sigma_{\parallel} - \sigma_{\perp}) S_{zz}, \tag{7}$$

the anisotropy can be obtained as the slope of the plot of σ_{Nem} vs. $(2/3)S$. If one assumes that medium effects can be neglected, it is also possible to obtain $\Delta\sigma$ from the nematic-to-isotropic phase shift at the transition temperature. For ^{13}C these two methods have been found to give entirely equivalent results (P. K. Bhattacharyya and B. P. Dailey, unpublished data). For ^{19}F and ^{1}H shielding, there are usually large medium effects[16,17] and the single temperature method is less reliable.

Other experimental methods for obtaining chemical shift anisotropies are the spin-lattice relaxation, molecular beam, and clathrate methods. The spin-lattice relaxation time, T_1, measures how soon a nuclear spin or assemblage of spins returns to its equilibrium position by releasing energy to the total system. Any mechanism that provides a fluctuating magnetic field can, in principle, contain the proper Fourier components to permit this relaxation to take place.[18] The experimentally measured spin-lattice relaxation rate $1/T_1$ is composed of several contributing terms, one of which is that due to chemical shift anisotropy.[19]

$$\frac{1}{T_1(\Delta\sigma)} = (8/15)\pi^2(\sigma_{||} - \sigma_{\perp})^2 H_0{}^2 \tau, \tag{8}$$

where τ is the correlation time of molecular reorientation and H_0 the applied field. This method, however, is limited due to the difficulty of detecting this particular relaxation term among other terms and the inadequacy of the methods for determining τ.[20]

The molecular beam methods provide spin–rotation constants that are directly related to the paramagnetic shielding terms[20,21] as shown in Eq. (9)

$$\sigma_{xx}^{\mathrm{p}} = \frac{-e^2}{2mc^2} \sum_l Z_l \left(\frac{z_l{}^2 + y_l{}^2}{r_l{}^3}\right) + \frac{1}{2}\left(\frac{M_{\mathrm{p}}}{mg_I}\,\frac{C_{xx}}{G_{xx}}\right) \tag{9}$$

Z_l is the nuclear charge, z_l, y_l, and r_l are internuclear distances, M_{p} is the proton mass, g_I the nuclear g-factor, G_{xx} the rotational constant, and C_{xx} the spin–rotation constant. To derive shielding anisotropies and tensors, it is necessary to combine the paramagnetic shielding components with diamagnetic components. The latter have been available from accurate *ab initio* theoretical calculations[20,22] for many small molecules and from reliable semiempirical calculations[23] for larger molecules. Originally used primarily for diatomics this method has become an important source of shielding anisotropies and absolute shielding constants for many types of molecules.[20,23]

The clathrate method[24] consists of trapping small gaseous molecules in the cavities of large organic clathrates. The nearly 100 percent alignment that results enables the anisotropy to be determined directly from the measured spectra at 0 and 90 degrees orientation angles.

An additional source of chemical shielding anisotropies is that of *ab initio* theoretical calculations.[20,25] There has been considerable progress in this area of molecular quantum mechanics, particularly with the use of gauge-invariant atomic orbitals within the framework of self-consistent-field (SCF) perturbation theory.[26] In many cases the theoretical quantities have been extremely accurate and have served not only as a corroboration of experimental quantities but also as a reliable source of new data for molecules of second-row atoms (i.e., Li through F).

DISCUSSION OF SHIELDING ANISOTROPIES

Carbon-13 Anisotropies

The correlation between theoretical and experimental shielding anisotropies has been most important for ^{13}C shielding results. Several previously discussed experimental techniques including the liquid crystal and cross-polarization techniques have recently supplied a large amount of data. The theoretical approach, in progressing from diatomics to small polyatomics has also placed strong emphasis on carbon-containing molecules.

First it should be pointed out that the agreement between R. Ditchfield's theoretical results (private communication and ref. 25, 26) and the experimental results has been extremely good for the isotropic chemical shift.[27] Normally, both the experimental and theoretical shifts are taken with respect to a reference compound, such as CH_4, and the relative shifts compared. It is also possible—through the use of experimental spin–rotation constants or anisotropies in conjunction with accurate theoretical diamagnetic quantities—to place the experimental quantities on an absolute scale (B. R. Appleman, C. W. Kern, T. Tokuhiro, and G. Fraenkel, unpublished data, and ref. 20) and compare them directly with the theoretical quantities.

In the complete series of fluoroethylenes and fluoroacetylenes, Ditchfield and Ellis[27] found the theoretical shifts relative to methane varied from the experimental shifts by an average of only several parts per million. For the substituted methanes (R. Ditchfield, private communication) the agreement was fairly good, although there were some slight discrepancies.

The comparison of some anisotropies is shown in Table 1, which also gives the absolute isotropic shifts and tensor components. The agreement is again very good, almost entirely within experimental error, indicating that both theoretical and experimental methods are extremely reliable. The experimental values of $\sigma_{\|}$ and $\sigma_{\perp}$ are determined from the primary experimental quantities $-\Delta\sigma$ and σ_{Iso}.

The correlation between experimental methods of obtaining ^{13}C anisotropies is demonstrated in Table 2. In each instance the methods compared agree within experimental error. It is particularly noteworthy that the ^{13}C-shielding anisotropy does not appear to depend on the state of matter, e.g., liquid crystal, solid, or gaseous phases.

One of the interesting trends noticed from studying shielding components is the constancy of the shielding along the parallel axis of a CH_3– system (B. R. Appleman *et al.*, unpublished data, and ref. 8, 20). Thus, for the molecules presented in Table 3 the values of $\sigma_{\|}$ lie

TABLE 1 Theoretical vs. Experimental ^{13}C Shielding Anisotropies

	$\sigma_{\parallel}{}^{a}$	$\sigma_{\perp}{}^{a}$	$\sigma_{Iso}{}^{b}$	$\Delta\sigma^{a}$	Method
CO	271.2	−118.4	11.5	389.5	Theory[c]
	268 ± 10	−116 ± 10	12 ± 10[b]	384 ± 10	Molecular beam[d] + σ^d (theory)[c]
HCN	278.4	−17.3	81.3	295.7	Theory[e]
	264 ± 15	−18 ± 15	76 ± 10[b]	282 ± 20	Liquid crystal[f]
	287.3	−31.7	74.7	319.0	Molecular beam[g] + σ^d (SE)[h]
C_2H_2	279.2	54.8	129.6	224.4	Theory[e]
	283 ± 20	38 ± 20	120 ± 10[b]	245 ± 20	Liquid crystal[i]
CH_3F	196.5	112.4	140.4	84.1	Theory[e]
	167 ± 15	99 ± 15	122 ± 15[b]	68 ± 15	Liquid crystal
CH_3OH[l]	201.2	138.3	159.3	62.9	Theory[e]
	193 ± 20	123 ± 20	146 ± 10[b]	70 ± 20	Liquid crystal[j]
	188 ± 10	125 ± 10	146 ± 10	63 ± 10	Cross relaxation[k]

[a] $\sigma_{\parallel}$ is shielding along symmetry axis; $\sigma_{\perp}$ perpendicular to axis; $\Delta\sigma \equiv \sigma_{\parallel} - \sigma_{\perp}$. Units are ppm.

[b] Absolute shielding scales based on weighted average of five experimental points. See Appleman and Dailey.[20]

[c] Extended-basis perturbed Hartree–Fock calculation.[58]

[d] From I. Ozier *et al.*[81]

[e] SCF perturbation theory with gage-invariant atomic orbitals by R. Ditchfield (private communication, 1972).

[f] From Millett and Dailey.[59]

[g] From Garvey and de Lucia.[60]

[h] Semiempirical calculation of diamagnetic shielding.[23]

[i] S. Mohanty (private communication, 1972).

[j] P. Bhattacharyya (private communication, 1972).

[k] From Pines *et al.*[9]

[l] Axial symmetry assumed.

entirely within the range from 167 to 199 ppm. This trend holds even for molecules that do not possess a true threefold axis, and it holds for molecules in the gaseous, liquid, liquid crystal, and solid phases. Theoretical results (R. Ditchfield, personal communication, and B. R. Appleman *et al.*, unpublished data) confirm this trend, which has been attributed to the fact that the carbon's nearest neighbors (besides the protons) contribute very little to $\sigma_{\parallel}$. The variations in $\sigma_{\parallel}$ are primarily due to long-range and distortion effects that are usually quite small. A similar trend has been observed for carbon atoms in linear or near-linear environments.[20]

TABLE 2 Experimental ^{13}C-Shielding Anisotropies: Correlation of Methods

	$\Delta\sigma$	Method
CS_2	438 ± 44	T_1 relaxation[a]
	425 ± 16	Powder pattern[b]
CH_3OH	70 ± 20	Liquid crystal[c]
	63 ± 10	Cross relaxation[c]
C_6H_6	180 ± 10	Cross relaxation[d]
	190 ± 4	Liquid crystal[e]
$CaCO_3$	75 ± 5	Single crystal[f]
	76 ± 5	Powder pattern[b]
HCN	285 ± 20	Liquid crystal[g]
	319 ± 10	Molecular beam[g]

[a] From Spiess *et al.*[61]

[b] From Pines and co-workers.[62]

[c] See footnotes *b* and *j* of Table 1.

[d] From Yannoni and Bleich.[63]

[e] From Englert.[64]

[f] From Lauterbur.[65]

[g] See footnotes *f, g, h* of Table 1.

Fluorine Anisotropies

In the area of fluorine shielding, in contrast to that of carbon-13 shielding, there has been a great deal of conflicting experimental data and interpretation of results. The most successful theoretical models, moreover, have not yet been applied to several of the most controversial fluorine compounds. The task of comparing fluorine tensors and anisotropies is complicated by the fact that there are usually several sets of axes for which these quantities can be defined, since the fluorine nuclei often do not lie on the molecular symmetry axis.[20]

There has been considerable interest in the fluorine shielding of the fluoromethanes. In particular, the results have cast serious doubt upon the reliability of liquid crystal fluorine anisotropies. Bernheim[16] obtained a value of $\Delta\sigma = -159$ ppm for CH_3F that disagreed sharply with both the clathrate[24] and molecular beam[28] results of -66 ± 8 and -61 ± 15 ppm, respectively. For CF_2H_2 and CF_3H there are also large discrepancies between liquid crystal anisotropies and those of other methods, as indicated in Table 4.

The shielding components are defined with respect to the C–F bond

TABLE 3 Anisotropies and Parallel Shielding in CH_3-Containing Systems[a]

	$\Delta\sigma^b$	$\sigma_{\parallel}{}^b$	Method
CH_4	0	196 ± 10	isotropic[c]
CH_3F	68 ± 15	167	LC[d]
CH_3Cl	28 ± 10	189	LC
CH_3Br	0 ± 5	185	LC
CH_3I	−75 ± 20	169	LC
CH_3OH	66 ± 10	190	LC, CR[e]
CH_3CN	5 ± 10	196	LC
$CH_3C{\equiv}C{-}CH_3$	14 ± 10	199	CR
$C_6(CH_3)_6$	0 ± 5	173	CR
CH_3CHO	46 ± 10	193	CR
CH_3CH_2OH	19 ± 10	189	CR
$(CH_3)_2CO$	50 ± 10	197	CR
$(CH_3)_2SO$	37 ± 10	177	CR
$(CH_3CH_2)_2O$	73 ± 10	178	CR
$(CH_3CO)_2O$	36 ± 10	197	CR
$(CH_3O)_2CO$	68 ± 10	190	CR
$(CH_3S)_2$	27 ± 10	193	CR
CH_3CO_2H	35 ± 10	197	CR
$C_6H_5CH_3$	22 ± 10	188	CR

[a] See Tables 8 and 9 of reference given in footnote *b* of Table 1 of this article.

[b] $\sigma_{\parallel}$ is shielding along C_3 axis of CH_3 group. $\Delta\sigma = \sigma_{\parallel} - \sigma_{\perp}$.

[c] $\sigma_{CH_4} = \sigma_{CS_2} + 196$; $\sigma_{CS_2} = \sigma_{CO} - 12$; $\sigma_{CO} = 12$ ppm. See Ettinger *et al.*[66] and Emsley *et al.*[67]

[d] Liquid crystal measurements.

[e] Cross-relaxation technique.

axes in each case. For some of these molecules the experimentally determined anisotropies were reported in terms of the molecular axes and were transformed to the C–F axes according to the approximate transformation formula,[7]

$$\Delta\sigma(\text{C-F bond}) = \Delta\sigma\,(C_3\ \text{axis})/P_2 \cos\theta, \tag{10}$$

where θ is the angle between the axis of symmetry and the set of C–F bonds. For $CDFCl_2$ the analysis[29] assumed that rapid molecular motion about the C–F bond gave the shielding tensor approximately axial symmetry.

A proposed explanation of the discrepancy of results between the liquid crystal and other methods is the existence of a temperature dependence of the chemical shift in the isotropic and nematic phase (P. K. Bhattacharyya, private communication). Thus, Eq. (7) is modified so that the difference $(\sigma_{Nem} - \sigma_{Iso})$ is explicitly a function of the

TABLE 4 Fluorine Shielding Anisotropies in Fluoromethanes

	$\sigma_{\parallel}$(C–F)[a]	$\sigma_{\perp}$(C–F)[a]	σ_{Iso}[b]	$\Delta\sigma$ (C–F)[a]	Method
CH_3F	438	509	486	−71	Theory–GIAO[c]
	424	490	468 ± 15	−66 ± 8	Clathrate[d]
				−61 ± 15	Molecular beam[e]
				−157 ± 10	Liquid crystal (LC)[f]
				−73 ± 10	LC corrected[g]
CH_2F_2	447	288[h]	341 ± 10	159 ± 1	LC[f]
	415	229[i]	341 ± 10	110 ± 15	Molecular beam[j]
		380			
CHF_3	344	239	274 ± 10	105 ± 20	Clathrate[k]
	424	199	274 ± 10	225 ± 15	LC[f]
				160 ± 15	LC corrected[l]
CF_3Cl	274	196	222 ± 10	78 ± 6	LC[l]
CF_3I	210	186	194 ± 10	24 ± 9	LC[m]
$CDFCl_2$	130	340	270 ± 10	−210 ± 30	Broad-line NMR[h, n]
	410	200	270 ± 10	+210 ± 30[o]	
CF_3CH_3	383	191	255 ± 10	192 ± 15	LC[p]
CF_3CCl_3	423	196	271 ± 10	228 ± 9	LC[m]

[a] Shielding defined with respect to C–F bond axis. In some cases experimental shielding data were transformed according to Eq. (10) of text.

[b] Absolute shielding scale is from D. K. Hindermann and C. D. Cornwell.[68] See also Section IIIB of Appleman and Dailey.[20]

[c] R. Ditchfield (private communication, 1972).

[d] From Hunt and H. Meyer.[24]

[e] From Wofsky and co-workers.[28]

[f] From Bernheim and co-workers.[16]

[g] P. Bhattacharyya (private communication, 1972); see text.

[h] Axial symmetry about C–F bond assumed.

[i] Two values of $\sigma_{\perp}$ due to lack of C_3 symmetry.

[j] From Kukolich and A. C. Nelson.[69]

[k] From Harris and co-workers.[70]

[l] N. Zumbulyadis (private communication, 1972).

[m] From Yannoni and co-workers.[71]

[n] From MacLean and E. L. Mackor.[29]

[o] Positive value for $\Delta\sigma$ assumed.

[p] From Silverman and Dailey.[72]

temperature as well as implicitly via the temperature dependence of S.

$$\sigma_{Nem} = \sigma_{Iso} + \alpha T + (2/3)S\Delta\sigma. \tag{11}$$

P. K. Bhattacharyya (private communication) and N. Zumbulyadis (private communication) have used such an equation to reinvestigate the liquid crystal results of CH_3F and CHF_3. By using several points in the nematic range they were able to obtain values for both α and $\Delta\sigma$ as follows:

$CH_3F: \Delta\sigma = -73 \pm 10$ ppm $\qquad \alpha = 20 \times 10^{-3}$ ppm/°C
$CHF_3: \Delta\sigma = +160 \pm 15$ ppm $\qquad \alpha = 9.1 \times 10^{-3}$ ppm/°C

which are in better agreement with the molecular beam and clathrate results.

These studies are being extended to determine the effect of the temperature-dependent shift on other liquid crystal results. It should be pointed out that for other fluorine compounds, such as CF_3CO_2H, $CF_3CO_2^-Ag^+$, and C_6F_6, there has been good agreement between anisotropies[20] of liquid crystals and other methods. In addition, for PF_3 the temperature dependence of the fluorine shielding was found to be very slight (N. Zumbulyadis, private communication) and had only a minimal effect on the experimental anisotropy.

There have been several liquid crystal studies of CF_3-containing systems,[20] some of which are included in Table 4. An interesting aspect of these systems is that the fluorine-shielding perpendicular to the C–F bond is quite constant (CF_3Cl, 196; CF_3I, 186; CF_3CH_3, 191; CF_3CCl_3, 196) whereas that along the C–F bond is much less regular. $CDFCl_2$ also follows this pattern[29] with $\sigma_\perp = 200$ if the anisotropy is assumed to be positive. Exceptions to this trend are CH_3F and CH_2F_2 for which the carbon substituents are primarily hydrogens rather than halogens.

Proton Anisotropies

Because of the proton's chemical importance and its favorable characteristics for NMR detection, the overwhelming bulk of experimental investigation and correlation of chemical shifts has been centered on the proton. However, the situation regarding the determination of proton shielding anisotropies has been unsatisfactory in many respects. Similarly, the field of *ab initio* theoretical calculations of proton shielding tensors has, until very recently, enjoyed little success. Both these factors are related to the fact that the proton chemical shifts are comparatively small and hence influenced strongly by secondary factors such as neighboring-atom electron distribution and medium effects.

The theoretical magnetic shielding is a property of the second-order energy that depends critically on the quality of the wavefunction. In particular, the wave function must provide a good description of the effect of the perturbations in the region near the nucleus. For almost all polyatomic molecules it is in practice necessary to limit the total number of basis functions used to construct the molecular wave function as a determinant of LCAO-MO's. Because of the incompleteness of the basis set, the calculated shielding is not an invariant quantity as would be expected for an exact calculation; rather it has a value that varies with the origin chosen for the magnetic vector potential.[30–32] Usually this gauge origin is chosen to coincide with some physically important molecular site, principally the resonant nucleus itself or the nuclear center of mass.[2]

The theoretical proton shielding is quite strongly dependent on the gauge origin: It has been found that choosing the proton as gauge origin frequently gives results[33–35] in very poor agreement with experiment. In several cases it has been found that the best agreement is obtained using the nearest heavy atom as gauge. However, there are still large discrepancies, more so than for other nuclei. The effects of transforming the gauge origin, which are often comparatively small (though not insignificant), for the heavy nuclei (e.g., C, N, O, F) become dominant for proton shielding, thereby greatly reducing the reliability of these quantities.[35]

Detailed analyses (B. R. Appleman and C. W. Kern, unpublished data, and ref. 36) of the constitution of the proton shielding indicate that (again, unlike other nuclei) the major part of the shielding is from contributions of atomic orbitals centered on atoms other than the protons. To improve the calculations, it would probably be necessary to place many additional basis functions on the heavy atoms nearest the proton as well as some additional functions–e.g., 2s, 2p on the proton itself. However, even such greatly extended basis sets have failed to eliminate the gauge dependence or to give very good agreement with experiment. A more successful approach[37] has been to allow the atomic orbitals to depend explicitly on the magnetic vector potential. This procedure has the effect of completely eliminating the gauge dependence of the shielding. It has been extremely successful in calculating accurate proton shielding constants using only moderately extended wave functions (Ditchfield, private communication, and ref. 25).

The most widely used method for obtaining experimental proton anisotropies has been that of liquid crystal NMR measurements. This method has several inherent sources of experimental error, which, in the case of proton shielding, often overshadow the resulting anisotropy. Many of the analyses of liquid crystal proton spectra have been based on the observed shift between the nematic and isotropic phases. Such a procedure relies on the assumption that the solvent effects are the same in both the ordered and isotropic phases.[3]

Another difficulty in interpreting the spectrum is that it is usually necessary to use some other proton peak as a reference. The use of an external reference has been found generally unsatisfactory.[38] Care must also be exercised when using internal references, since experiments have shown that even such symmetric molecules as tetramethylsilane and CH_4 show solvent and phase-dependent shifts.[3] From the observed isotropic–nematic shift of the reference, Ceasar *et al.*[39] determined "corrections" to the proton anisotropies of methyl halides, which in most cases yielded significantly different answers and very large uncertainties.

The problem of shifts due to isotropic–nematic transitions can be eliminated by using only nematic phase measurements and obtaining $\Delta\sigma$ from the slope of σ_{Nem} vs. S. However, there still exists[40] the possibility of specific solvent–solute interactions that change with the orientation parameter and/or the temperature, which are different for the solute and the reference. For molecules such as HCN and C_2H_2 (H. Spiesecke, private communication, and S. Mohanty, private communication) the direction of orientation has been observed to change with different liquid crystals, suggesting some sort of interaction that does not depend solely on the magnetic susceptibility of the solute. The linear dependence of the nematic phase shift on temperature can also cause serious errors in the resulting $\Delta\sigma$ if not properly accounted for as discussed in the section on fluorine anisotropies.

Several investigators have gone to great length to insure that the anisotropies be independent of the medium. Hayamizu and Yamamoto[41] measured the proton anisotropy of trihalobenzenes in two liquid crystals using the slope method and found virtually identical results. Mohanty (private communication) studied the molecule acetylene in four liquid crystals at various pressures using several references. The results were a fairly reproducible anisotropy of 22 ± 2 ppm.

A still unresolved question is whether the anisotropies observed in the nematic phase are equivalent to those in the gas, liquid, or solid states. For acetylene the spin–rotation measurements[42] along with *ab initio* calculations of σ^d (B. R. Appleman and C. W. Kern, unpublished data) yield a value for $\Delta\sigma$ of 18 ± 1 ppm that is not quite within the experimental uncertainty of the liquid crystal value. It still appears, however, that, for small or moderate-sized proton anisotropies, the large uncertainties associated with the use of liquid crystal measurements make this method one of doubtful reliability.

RELATION BETWEEN SHIELDING AND MAGNETIC SUSCEPTIBILITY ANISOTROPY

The magnetic susceptibility is a molecular property that is often considered in relation to the nuclear magnetic shielding. In many respects there are great similarities in the mathematical formulations, physical explanations, and the observed trends of these quantities. There are, additionally, expressions that give the proton shielding in terms of magnetic anisotropies. Since there are now available large amounts of data on magnetic shielding and susceptibility anisotropies, it is possible to compare these quantities directly for many molecules in order to examine their trends and correlations.

A well-known characteristic of the magnetic susceptibility is its facile subdivision into localized contributions of atomic, bond, or group susceptibilities.[43,44] However, such additivity schemes have had little success in correlating or predicting values of chemical shifts from localized shielding terms. The shielding is an intensive property of the nuclei, whereas the susceptibility is an extensive or bulk property of the molecular or submolecular units. Mathematically, this difference is attributed to the appearance of a $1/r^3$ factor in the shielding, which localizes this term primarily within the region near the nucleus.[20,45]

The isoelectronic molecules, Ne, HF, H_2O, NH_3, and CH_4, clearly demonstrate the differences between shielding and susceptibility. In Table 5 it is noted that as one progresses from Ne to CH_4 the susceptibility increases while the shielding decreases, although the total electronic charge is the same for each of these molecules. These trends indicate that the susceptibility depends more on the charge displaced from the nucleus or center of mass whereas the shielding depends most strongly on the electron density nearest the nucleus. The anisotropies of both shielding and susceptibility are comparatively small in accord with the approximate tetrahedral distribution of electron pairs; for HF, however, the slight distortion caused by the H–F bond is enough

TABLE 5 Magnetic Susceptibility and Central-Atom Shielding of an Isoelectronic Series[a]

	χ_{av}^{b}	$\Delta\chi$	σ_{Iso}	$\Delta\sigma$
Ne	-7.46	0	552.3	0
HF	-10.3^{c}	0.3	411	104
H_2O	-13.1	-0.1^{d}	328	-29^{d}
NH_3	-16.3	0.4	266	-39
CH_4	-19.4^{e}	0	196	0

[a] See Table 26 in Appleman and B. P. Dailey.[20]

[b] Units of χ are 10^{-6} erg gauss^{-2} mole^{-1}.

[c] Based on experimental molecular g value and theoretical diamagnetic susceptibility.[73]

[d] Parallel component refers to out-of-plane axis, perpendicular to the average of components along in-plane axes.

[e] Based on experimental molecular g value and theoretical diamagnetic susceptibility.[74,75]

to give a paramagnetic shielding of 106 ppm in the direction perpendicular to the bond.

In Table 6 the methyl and hydrogen halides are compared. First, it is pointed out that although both σ_{Iso} and χ_{av} of the methyl halides increase in going from CH_3F to CH_3I, the trends are probably caused by different factors—the nature of the substituent for the shielding and the molecular size for the susceptibility. This interpretation is supported by observing that the shielding anisotropies decrease greatly along the series, while the susceptibility anisotropies remain approximately constant with a slight increase in magnitude for CH_3I. The behavior of the proton shielding and average susceptibilities of the hydrogen halides reproduces the trends of the methyl halides. It should be added that since the quantities $\sigma_{\parallel}$ of ^{13}C of CH_3X and of ^1H of HX are constant, an increase in σ_{Iso} automatically ensures a decrease in $\Delta\sigma$. The anisotropic and isotropic shielding of the halogens in the hydrogen halides are quite regular indicating that the HX bond has a strong effect on the innermost halogen electrons for the larger halo-

TABLE 6 Shielding[a] and Susceptibility in Some Methyl and Hydrogen Halides

	Nucleus	σ_{Iso}	$\Delta\sigma$	χ_{av}	$\Delta\chi$
CH_3F	C	122[a]	68 ± 15	−17.8[c]	−8.2 ± 0.8[d]
CH_3Cl	C	170	28 ± 10[b]	−32.0	−7.9 ± 0.5[f]
CH_3Br	C	185	0 ± 5[b]	−42.8	−8.5 ± 0.4[f]
CH_3I	C	219	−75 ± 20[g]	−57.2	−11.0 ± 0.5[f]
HF	H	27.9[h]	36[i]	−8.6[c]	
	F	411[j]	104[j]	(−10.3)[j]	
HCl	H	30.8[h]	33[i]	−22.1[c]	
	Cl	952[i]	292[i]		
HBr	H	34.8[h]	25[i]	−32.9[c]	
	Br	2617[i]	760[i]		
HI	H	43.7[h]	15[i]	−47.7[c]	
	I	4510[i]	1488[i]		

[a] Absolute ^{13}C-shielding scale based upon $\sigma(CO) = 12 \pm 10$ ppm. See also footnote *b* of Table 1.
[b] P. Bhattacharyya (private communication, 1972).
[c] Liquid-phase value given in Barter *et al.*[76]
[d] From Flygare and Benson.[77]
[e] From Pascal.[78]
[f] From Vander Hart and Flygare.[79]
[g] From Morishima *et al.*[80]
[h] From Emsley *et al.*,[67] Vol. I, p. 125.
[i] From Gierke and Flygare.[23]
[j] See Table 5.

gens as well as fluorine, and its effects are not limited to the valence orbitals.

The success of the concept of group and atom average susceptibilities has led to investigations of the use of group susceptibility anisotropies. McConnell,[46] Pople,[47] and McWeeny[48] developed a theory that describes the effect on the isotropic proton shift of the magnetic anisotropy of benzene rings. This ring-current model has given rise to a large number of moderately successful semiempirical correlations between σ^H and $\Delta\chi$ in spite of several shortcomings.[49-52] Among these are the neglect of local factors and the invalidity of the point-dipole model at short distances.

The application of the magnetic anisotropy effect to functional groups other than benzene has been considerably less successful, at least partially because of inadequate or improper assessment of the local effects.[20-26] In addition to giving conflicting values for the group anisotropies, the use of these NMR-derived quantities[26] gave erroneous values for molecular susceptibility anisotropies.[20] Ap Simon *et al.*[53,54] have tried to correct some of the deficiencies of the original models by taking explicit account of the variations among molecules in bond lengths and angles and the finite distances from and lengths of the induced dipoles. They have determined principal bond susceptibilities for C–C, C–H, C=C, and C=O bonds, each in a variety of environments by attributing differences in shielding to the total number of bonds that must be broken or created to form the new molecule.

It still remains to be determined whether or not the group susceptibility anisotropies and principal susceptibilities are transferable to any large extent among different molecules. If the concept of bond susceptibilities is valid, it should be possible to establish sets of these quantities, which, when combined, are able to reproduce the molecular susceptibility components and anisotropies.[55] Of course, it would be necessary to classify the bonds according to the environment, as ApSimon *et al.*[53,54] did.

Benson and Flygare[56,57] have used a similar approach to investigate the possibility of nonlocal (i.e., ring-current) contributions to molecular anisotropies in a series of open-chained and conjugated and nonconjugated ring systems. They have shown that through the use of empirically derived quantities of $\chi_{\text{out-of-plane}} - (\chi_{\text{in-plane}})_{\text{av}}$, the total molecular anisotropies could be fairly closely reproduced for nonconjugated systems. However, the use of bond anisotropies rather than principal bond susceptibilities makes it difficult to extend the model to molecules with heavy atoms that do not form a plane.

As an example of the latter procedure, we have found that using the haloacetylenes and halocyanides (including HCN and HCCH) one can derive a consistent set of principal susceptibilities for the C≡C, C≡N, C–X, and C–H bonds. However, for nonlinear systems there arises the question of whether such bonds as C–H or C–F are still axially symmetric. Thus, it may be necessary to specify three susceptibilities for each bond in each change of environment, e.g., change of hybridization in neighbors. Our overall feeling is that, because of the large differences in bonding environments, the model will be best suited for series of limited variability.

REFERENCES

1. M. Mehring, R. G. Griffin, and J. S. Waugh, *J. Chem. Phys.*, **55**, 746 (1971).
2. R. Ditchfield, D. P. Miller, and J. A. Pople, *J. Chem. Phys.*, **54**, 4186 (1971).
3. A. D. Buckingham and S. M. Malm, *Mol. Phys.*, **22**, 1127 (1971).
4. A. D. Buckingham, E. E. Burnell, and C. A. de Lange, *Mol. Phys.*, **16**, 299 (1969).
5. D. L. Vander Hart and H. S. Gutowsky, *J. Chem. Phys.*, **49**, 261 (1968).
6. N. Bloembergen and T. J. Rowland, *Phys. Rev.*, **97**, 1679 (1955).
7. U. Haeberlen and J. S. Waugh, *Phys. Rev.*, **175**, 453 (1968).
8. A. Pines, M. G. Gibby, and J. S. Waugh, *J. Chem. Phys.*, **56**, 1776 (1972).
9. A. Pines, M. G. Gibby, and J. S. Waugh, *Chem. Phys. Lett.*, **15**, 373 (1972).
10. M. G. Gibby, R. G. Griffin, A. Pines, and J. S. Waugh, *Chem. Phys. Lett.*, **17**, 80 (1972).
11. M. G. Gibby, A. Pines, and J. S. Waugh, *J. Am. Chem. Soc.*, **94**, 6231 (1972).
12. A. D. Buckingham and K. A. McLauchlan, *Progr. NMR Spectrosc.*, **2**, 63 (1967).
13. A. Saupe, *Z. Naturforsch.*, **19a**, 161 (1964).
14. G. Englert and A. Saupe, *Mol. Crys.*, **1**, 503 (1966).
15. L. C. Snyder, *J. Chem. Phys.*, **43**, 4041 (1965).
16. R. A. Bernheim, D. J. Hoy, T. R. Krugh, and B. J. Lavery, *J. Chem. Phys.*, **50**, 1350 (1969).
17. A. D. Buckingham, E. E. Burnell, and C. A. de Lange, *J. Am. Chem. Soc.*, **90**, 2972 (1968).
18. A. Abragam, *The Principles of Nuclear Magnetism*, Ch. VIII, Oxford University Press, London (1961).
19. W. T. Huntress, Jr., *J. Chem. Phys.*, **48**, 3524 (1968).
20. B. R. Appleman and B. P. Dailey, *Adv. Magn. Resonance*, **7**, (1974).
21. W. Hüttner and W. H. Flygare, *J. Chem. Phys.*, **47**, 4137 (1967).
22. W. N. Lipscomb, *Adv. Magn. Resonance*, **2**, 137 (1966).
23. T. D. Gierke and W. H. Flygare, *J. Am. Chem. Soc.*, **94**, 7277 (1972).
24. E. Hunt and H. Meyer, *J. Chem. Phys.*, **41**, 353 (1964).
25. R. Ditchfield, *Chem. Phys. Lett.*, **15**, 203 (1972).
26. R. Ditchfield, in *MTP International Reviews of Science: Physical Chemistry*, A. Allen, ed., Medical Technical Press, New York (1972).
27. R. Ditchfield and P. D. Ellis, *Chem. Phys. Lett.*, **17**, 342 (1972).
28. S. C. Wofsky, J. S. Muenter, and W. Klemperer, *J. Chem. Phys.*, **55**, 2014 (1971).
29. C. MacLean and E. L. Mackor, *Proc. Phys. Soc. (London)*, **88**, 341 (1966).

30. G. P. Arrighini, M. Maestro, and R. Moccia, *Chem. Phys. Lett.*, **7**, 351 (1970).
31. T. Tokuhiro, B. R. Appleman, G. Fraenkel, C. W. Kern, and P. K. Pearson, *J. Chem. Phys.*, **57**, 20 (1972).
32. B. R. Appleman, T. Tokuhiro, G. Fraenkel, and C. W. Kern, *J. Chem. Phys.*, **58**, 400 (1973).
33. G. P. Arrighini, M. Maestro, R. Moccia, *J. Chem. Phys.*, **52**, 6411 (1970).
34. H. J. Kolker and M. Karplus, *J. Chem. Phys.*, **41**, 1259 (1964).
35. C. W. Kern and W. N. Lipscomb, *J. Chem. Phys.*, **37**, 260 (1962).
36. H. Kato, *J. Chem. Phys.*, **52**, 3723 (1970).
37. R. Ditchfield, *J. Chem. Phys.*, **56**, 5688 (1972).
38. A. D. Buckingham and E. E. Burnell, *J. Am. Chem. Soc.*, **89**, 3341 (1967).
39. G. P. Ceasar, C. S. Yannoni, and B. P. Dailey, *J. Chem. Phys.*, **50**, 373 (1969).
40. A. D. Buckingham, E. E. Burnell, and C. A. de Lange, *J. Chem. Phys.*, **54**, 3242 (1971).
41. K. Hayamizu and O. Yamamoto, *J. Chem. Phys.*, **51**, 1676 (1969); **54**, 3243 (1971); *J. Magn. Resonance*, **2**, 377 (1970).
42. J. W. Cederberg, C. H. Anderson, and N. F. Ramsey, *Phys. Rev.*, **136**, A960 (1964).
43. A. Pacault, *Rev. Sci.*, **86**, 38 (1948).
44. P. W. Selwood, *Magnetochemistry*, Sect. III 6, Interscience, New York (1943).
45. A. Saika and C. P. Slichter, *J. Chem. Phys.*, **22**, 26 (1954).
46. H. M. McConnell, *J. Chem. Phys.*, **27**, 226 (1957).
47. J. A. Pople, *J. Chem. Phys.*, **24**, 1111 (1956).
48. R. McWeeny, *Mol. Phys.*, **1**, 311 (1958).
49. C. E. Johnson and F. A. Bovey, *J. Chem. Phys.*, **29**, 1012 (1958).
50. F. A. Bovey, *Nuclear Magnetic Resonance Spectroscopy*, Academic Press, New York (1969).
51. B. P. Dailey, *J. Chem. Phys.*, **41**, 2304 (1964).
52. L. M. Jackman, *Nuclear Magnetic Resonance Spectroscopy*, Pergamon Press, London (1959).
53. J. W. ApSimon, W. G. Craig, P. V. DeMarco, D. W. Mathieson, L. Saunders, and W. B. Whalley, *Tetrahedron*, **23**, 2339, 2357, 2375 (1967); *Chem. Commun.*, 359 (1966).
54. J. W. ApSimon, W. G. Craig, P. V. DeMarco, D. W. Mathieson, A. K. G. Nasser, L. Saunders, and W. B. Whalley, *Chem. Commun.*, 754 (1966).
55. J. A. Pople, *J. Chem. Phys.*, **37**, 60 (1962).
56. R. C. Benson, C. L. Norris, W. H. Flygare, and P. A. Beak, *J. Am. Chem. Soc.*, **93**, 5591 (1971).
57. R. C. Benson and W. H. Flygare, *J. Chem. Phys.*, **53**, 4470 (1970).
58. R. M. Stevens and M. Karplus, *J. Chem. Phys.*, **49**, 1094 (1968).
59. F. Millett and B. P. Dailey, *J. Chem. Phys.*, **54**, 5434 (1971).
60. R. M. Garvey and F. C. de Lucia, *Bull. Am. Phys. Soc. II*, **18**, 574 (1973).
61. H. W. Spiess, D. Schweitzer, U. Haeberlen, K. H. Hausser, *J. Magn. Resonance*, **5**, 101 (1971).
62. A. Pines, W. K. Rhim, and J. S. Waugh, *J. Chem. Phys.*, **54**, 5438 (1971).
63. C. S. Yannoni and H. E. Bleich, *J. Chem. Phys.*, **55**, 5406 (1971).
64. G. Englert, *Z. Naturforsch.*, **27**, 715 (1972).
65. P. C. Lauterbur, *Phys. Rev. Lett.*, **1**, 343 (1958).
66. R. Ettinger, P. Blume, A. Patterson, Jr., and P. C. Lauterbur, *J. Chem. Phys.*, **33**, 1597 (1960).

67. J. W. Emsley, J. Feeney, and L. H. Sutcliffe, *High Resolution Nuclear Magnetic Resonance Spectroscopy*, Vol. II, Sect. 12.2, Pergamon Press, London (1966).
68. D. K. Hindermann and C. D. Cornwell, *J. Chem. Phys.*, **48**, 4148 (1968).
69. S. G. Kukolich and A. C. Nelson, *J. Chem. Phys.*, **56**, 4446 (1972).
70. A. B. Harris, E. Hunt, and H. Meyer, *J. Chem. Phys.*, **42**, 2851 (1965).
71. C. S. Yannoni, B. P. Dailey, and G. P. Ceasar, *J. Chem. Phys.*, **54**, 4020 (1971).
72. D. N. Silverman and B. P. Dailey, *J. Chem. Phys.*, **51**, 655 (1969).
73. R. M. Stevens and W. N. Lipscomb, *J. Chem. Phys.*, **41**, 184 (1964).
74. C. H. Anderson and N. F. Ramsey, *Phys. Rev.*, **149**, 14 (1966).
75. G. P. Arrighini, M. Maestro, and R. Moccia, *J. Chem. Phys.*, **49**, 882 (1968).
76. C. Barter, R. G. Meisenheimer, and D. P. Stevenson, *J. Chem. Phys.*, **64**, 1312 (1960).
77. W. H. Flygare and R. C. Benson, *Mol. Phys.*, **20**, 225 (1971).
78. M. P. Pascal, *Ann. Chim. Phys.*, **19**, 5 (1910).
79. D. L. Vander Hart and W. H. Flygare, *Mol. Phys.*, **18**, 77 (1970).
80. I. Morishima, A. Mizuno, and T. Yonezawa, *Chem. Phys. Lett.*, **7**, 633 (1970).
81. I. Ozier, L. M. Crapo, and N. F. Ramsey, *J. Chem. Phys.*, **49**, 2314 (1968).

Discussion Leader: STEPHEN G. KUKOLICH

Reporters: MARLIN D. HARMONY
E. JEAN JACOB

Discussion

BUCKINGHAM suggested that one-electron molecular orbitals do not give an adequate description for calculation of spin densities and that many-electron wave functions are needed. He also did not see why the hyperfine tensor should be symmetric. McGARVEY replied that this is only an approximation and that in practice the antisymmetric terms are usually not detectable since they enter in higher order.

McCALLEY commented on the exhortation to report hyperfine coupling constants in energy units (such as megahertz) rather than in gauss. This implicitly assumes that the hyperfine "constant" is independent of the effective g value as the latter varies with orientation. However, the *anisotropic* component of the hyperfine tensor is well known to be proportional to g_{eff}. (This causes the recently important pseudocontact hyperfine effect in solution complexes of lanthanides.) Thus, it appears that at least *this* component should be expressed in gauss in order that the angular dependence follow the tensor relationship precisely. On the other hand, most derivations of the Fermi contact hyperfine component suggest that it is due to only the spin part of the electron magnetic moment, so that it should not vary with g_{eff}. McCALLEY stated that he did not believe that this possible deviation from the standard $\mathbf{S} \cdot \mathbf{A} \cdot \mathbf{I}$ hyperfine Hamiltonian had been investigated experimentally. McGARVEY replied that in practice it has not been possible to separate these effects. In addition, for the equations he used, it has been assumed that the **g** tensor and the **A** tensor have the same principal axis system. There are a number of examples in the literature where this is not the case.

FLYGARE mentioned that studies of deuterium quadrupole coupling constants in substituted acetylenes had suggested the use of this constant as a probe of the electron density on the adjacent carbon atom. However, measurements on fluorobenzene have shown that the coupling constants of deuterons at the *ortho, meta*, and *para* positions are essentially the same. Thus, the quadrupole coupling constant seems to be an extremely insensitive probe of electron density at an adjacent atom.

There was a discussion by FUNG, SNYDER, KUKOLICH, and McCALLEY on the relative importance of hybridization and electronegativity in determining deuterium quadrupole coupling constants. The consensus was that hybridization is more important; both experimental and theoretical evidence suggests a grouping according to sp, sp^2, and sp^3 hybridization. However, this point is somewhat obscured by an uncertainty in the correct value for CH_3D. It was also noted that such comparisons should be restricted to measurements in the same physical state, unless suitable corrections are made. In particular, quadrupole coupling constants measured in liquid crystals tend to be low. The influence of vibrational averaging effects on quadrupole coupling constants should not be overlooked, especially when hydrogen atoms are involved.

BUCKINGHAM pointed out that the relation given by Flygare between molecular g values and the paramagnetic susceptibility tensor depends on the rigid-rotor approximation. He noted that calculations for the hydrogen molecule indicated deviations of a few percent from the rigid-rotor calculation. The deviations should be particularly large for this case, but it is difficult to estimate the error introduced by this approximation in other molecules.

SCHAEFER pointed out that *ab initio* calculations generally give better values for quadrupole moments than for dipole moments. An exception is ozone, where the calculated quadrupole moment is in poor agreement with experiment. However, FLYGARE noted that the experimental uncertainty in the ozone measurement is very large.

BUCKINGHAM raised a question concerning the interaction between nuclear and electron spin through the Fermi contact Hamiltonian. When this is carried to second order, it gives an energy term quadratic in I, which depends on the square of the magnetic moment. This leads to two problems: a theoretical problem because of the infinite self-coupling and a practical one because the term transforms exactly like the quadrupole coupling term. Bleaney had previously pointed out the difficulty in separating this magnetic pseudoquadrupole effect from the true quadrupole coupling. FLYGARE added that nuclear polarization would also contribute a term of this form. The sense of the discussion was that these effects are probably small; if significant, they would have been detected through comparison of measured quadrupole coupling constants for different isotopes. However, no one was able to give a quantitative estimate of their importance.

DITCHFIELD indicated that the theoretical value for the proton-shielding anisotropy in acetylene was 14 ppm. DAILEY reported a value of 22 ppm from his measurements and a molecular beam value of 18 ppm; however, correction for temperature dependence may bring the experimental values closer to the theoretical.

VIII

Electronic Charge Distribution by Other Methods

A. D. Buckingham

Electric Moments and Polarizabilities of Molecules

The distribution of electric charge in a molecule is intimately related to its structure and reactivity. Knowledge of the distribution gives us a "feeling" for the physical and chemical properties of the molecule and provides a valuable assessment of the accuracy of approximate molecular wavefunctions. The charge distribution in the nth stationary state is determined by the many-electron wave function $\Psi^{(n)}$ of the free molecule. If the molecule interacts with an external electric perturbation **E**, the wave function $\Psi^{(n)}(\mathbf{E})$ determines the distortion and, hence, the various molecular polarizabilities.

Full knowledge of the charge distribution of a molecule requires specification of the charge density at all points. For some purposes the charge density provides excess information; thus, the potential outside a sodium ion is independent of the distribution of the electrons, and the interaction of a molecule with a uniform external field is determined by its dipole moment and dipole polarizabilities. The *electric multipole moments* characterize the charge distribution; the first three are defined as follows:

$$\text{the monopole or charge } q = \sum_i e_i$$

$$\text{the first moment or dipole } \mu = \sum_i e_i \mathbf{r}_i$$

the second moment or quadrupole $\boldsymbol{\Theta} = \frac{1}{2} \sum_i e_i (3\mathbf{r}_i\mathbf{r}_i - r_i^2 \mathbf{1})$,

where $\mathbf{1}$ is the unit second-rank tensor and $\mathbf{r}_i$ is the vector from the origin 0 to the element of charge e_i. Actually, these are the multipole moment operators and, if the molecule is in the state $\Psi^{(n)}$, its dipole moment is $\boldsymbol{\mu}^{(n)} = \langle \Psi^{(n)} | \boldsymbol{\mu} | \Psi^{(n)} \rangle / \langle \Psi^{(n)} | \Psi^{(n)} \rangle$. The lth order electric moment operator $\xi^{(l)}$ is defined, in cartesian tensor notation, as

$$\xi^{(l)}_{\alpha\beta\gamma\ldots\lambda} = (-1)^l \, (l!)^{-1} \sum_i e_i r_i^{2l+1} \frac{\partial}{\partial r_{i\alpha}} \frac{\partial}{\partial r_{i\beta}} \frac{\partial}{\partial r_{i\gamma}} \cdots \frac{\partial}{\partial r_{i\lambda}} (r_i^{-1})$$

or, in terms of spherical tensors, by

$$\xi_{l,m} = \left(\frac{4\pi}{2l+1}\right)^{\frac{1}{2}} \sum_i e_i r_i^l \, Y_{l,m} (\theta_i, \varphi_i),$$

where $Y_{l,m}(\theta,\varphi)$ is the normalized spherical harmonic of order $l (l = 0, 1, 2, \ldots; -l \leqslant m \leqslant l)$. Hence

$$\xi_{l,0} = \xi^{(l)}_{zzz\ldots z}.$$

Since the transformation properties of spherical harmonics are well known, the spherical-tensor notation has some advantages, particularly in the derivation of general theorems; however, the reality of cartesian tensors also has its attractions, especially for small values of l. Normally, the moments of a particular three-dimensional molecule are most conveniently given in an x,y,z frame.

Only the first nonvanishing multipole moment is independent of the choice of origin. For example, if the dipole moment of HF is 1.8262 ± 0.0003 D = 6.0915 ± 0.0010 × 10^{-30} C m[1,2] with the sense H^+F^-[3], and if the dipole of DF is the same as that of HF (actually $\mu_{HF} - \mu_{DF}$ = 0.00772 ± 0.00007 D = 0.0258 ± 0.0002 × 10^{-30} C m),[1] the quadrupole moments of HF and DF relative to their centers of mass differ by $\Theta_{HF} - \Theta_{DF} = 0.15 \times 10^{-26}$ esu = 0.51×10^{-40} C m^2.

STARK EFFECT

Spectroscopic studies of the Stark effect lead, through application of quantum-mechanical perturbation theory, to values for the dipole moments of molecules in particular stationary states.[4] Very careful work, such as the microwave Stark studies of Scharpen, Muenter, and Laurie[5] on OCS, NNO, and $CD_3-C{\equiv}C-H$, and the molecular beam electric resonance observations of Muenter[2] on HF, also provides values for static polarizability anisotropies $\alpha_{\parallel}-\alpha_{\perp}$. Because the energy differences arising from the polarizability are generally very small, it is necessary to incorporate into the effective energy expression a term arising from vibration/rotation interaction related to the centrifugal stretching constant and proportional to the ratio $r(\partial\mu/\partial r)/\mu$, where r is an internuclear distance.[6] For a $^1\Sigma$ diatomic in the vth vibrational state, the appropriate expression is[2,6]

$$W_{vJM} = \frac{J^2 + J - 3M^2}{(2J-1)(2J+3)} \left[\frac{\mu_v^2 E^2}{2J(J+1)hcB_v} + \frac{\mu_v^2 E^2 D_v}{3hcB_v{}^2} \left(1 + \frac{r_v}{\mu_v} \frac{\mathrm{d}\mu}{\mathrm{d}r} \right) - \tfrac{1}{3}(\alpha_{\parallel} - \alpha_{\perp})_v E^2 \right] + \frac{\mu_v^4 E^4}{8h^3 c^3 B_v^3} F(J,M),$$

where the subscript v denotes an expectation value for the nonrotating state of the vth vibrational state, B_v and D_v are the rotational and centrifugal distortion constants (in wave numbers), and $F(J,M)$ is a function of J and M giving the fourth-order Stark energy.[5] The polarizability anisotropy $\alpha_{\parallel}-\alpha_{\perp}$ incorporates the contributions arising from distortion of the electrons relative to the nuclei (the electronic polarizability anisotropy) and distortion of the nuclei relative to one another [the atomic polarizability anisotropy, which is equal to $2r_v^2\,(\mathrm{d}\mu/\mathrm{d}r)^2 B_v/hc\omega_v^2$]. For HF with $J = 1$, the three terms in E^2 in W_{vJM} are calculated, using published constants,[1,2,7] to contribute to the energy difference between the $M = \pm 1$ and $M = 0$ levels in the ratio $1 : 2.5 \times 10^{-4} : -3.6 \times 10^{-4}$. Hence, in this case, the vibration/rotation interaction term is of magnitude comparable to the polarizability contribution to the Stark splitting of the $J = 1$ levels; however, it is proportional to $\mu(\mathrm{d}\mu/\mathrm{d}r)$ and is, therefore, less important when μ is small.

The energy change induced by a field E (in V m^{-1}) acting on a dipole of 1 D = 3.33564×10^{-30} C m is $5E$ kHz. In high-resolution spectroscopy, weak fields ($E < 10^4$ V m^{-1}) may be used but in optical spectroscopy shifts of at least 0.1 cm^{-1} are desirable and these are normally obtained only in strong fields ($> 10^6$ V m^{-1}). In electronic spectroscopy, the Stark splittings yield dipole moments of molecules in excited states[4]; the magnitude and sign is obtained relative to that in the ground state. In some instances there are two possible solutions for the unknown dipole moment, but this ambiguity can be eliminated by studying different rotational lines in the band.[8] The presence of spin-splittings, hyperfine interactions, and nonrigidity may complicate the interpretation of a spectrum.[4] In optical spectroscopy, it is normally necessary to work with an envelope rather than with resolved transitions involving molecules having particular M values.

The dipole moments of formaldehyde and propynal have been measured in both singlet and triplet states arising from $\pi^* \leftarrow n$ excitation. The results are shown in Table 1. As expected, $\pi^* \leftarrow n$ excitation may be considered to move an electron away from the oxygen atom (for the sense of the dipole moment of formaldehyde in its ground state is $H_2{}^+CO^-$).[14] The smaller moment of the triplets is in accord with the generalization that relaxation of the electron clouds to a lower energy distribution is normally in the direction of reducing the dipole moment.[10,15] Another illustration is the excessive dipole moment generally computed using the Hartree–Fock self-consistent-field wave function[16]; relaxation due to electron correlation reduces both the energy and the dipole moment.

NONRIGIDITY

The nonrigidity of a molecule causes the dipole moment and other properties to differ in different vibrational and rotational states. Within the Born–Oppenheimer approximation the property may be expanded as a Taylor series in the normal coordinates and expectation values taken for particular vibration/rotation states. For a diatomic

$$\mu(\xi) = \mu_e + \mu_e'\xi + \tfrac{1}{2}\mu_e''\xi^2 + \ldots,$$

TABLE 1 Dipole Moments of Formaldehyde and Propynal in Their Ground and Excited States[a]

Formaldehyde	μ	Ref.	Propynal	μ	Ref.
$\tilde{X}\,{}^1A_1$	2.33 ± 0.02	(9)	$\tilde{X}\,{}^1A'$	2.39 ± 0.04	(11)
$\tilde{A}\,{}^1A_2$	1.56 ± 0.07	(8)	$\tilde{A}\,{}^1A''$	{0.7 ± 0.2	(12)
$\tilde{a}\,{}^3A_2$	1.29 ± 0.03	(10)		{0.97 ± 0.11	(13)
			$\tilde{a}\,{}^3A''$	0.5 ± 0.1	(13a)

[a] The values are for μ_a, the component along the a inertial axis (the "symmetric-top" axis), which is the total moment of formaldehyde in its planar ground state. All values are in debyes.

where $\xi = (r - r_e)/r_e$ is the displacement from the equilibrium internuclear distance r_e, and[4,17]

$$\mu_{v,J} = \mu_e + (v+\tfrac{1}{2})(B_e/\omega_e)(\mu_e'' - 3a\mu_e') + 4(J^2+J)(B_e/\omega_e)^2\mu_e' + \ldots,$$

where a is the ratio of the coefficient of the term in ξ^3 to that in ξ^2 in the potential energy function. Normally, μ_e, μ_e', and μ_e'' are of comparable magnitude, a is approximately -2, and B_e/ω_e a small number of the order of 10^{-2} to 10^{-3}. Hence, the relative change in dipole moment with v (or on deuteration of HX) is $\sim 10^{-2}$, and that with J much less except for very high J.

For polyatomic molecules the dependence of μ on J and K is barely detectable, except for extremely nonrigid molecules like NH_3[18] and HNCO and HN_3.[19] Symmetrical nonlinear molecules like CH_4 and BF_3 may exhibit dipoles in excited vibrational[20] or rotational states.[21–25]

BREAKDOWN OF THE BORN-OPPENHEIMER APPROXIMATION

When very accurate dipole moments are deduced, it is proper to query the significance of a breakdown of the Born–Oppenheimer approximation. This approximation justifies the assignment of molecular property tensors, such as dipole moments and polarizabilities, to specific directions in a molecule-fixed frame and supports the use of a property function or surface representing the variation of the property with nuclear position. The dipole moment of HD (5.85×10^{-4} D)[26] arises solely from the breakdown of the approximation and may have the sense H^+D^-.[27–29] In HCl and DCl, there is an isotope effect on the dipole moment that has been attributed to a violation of the Born–Oppenheimer approximation[30]; there is an apparent difference of 0.0010 ± 0.0002 D between the dipole functions of HCl and DCl, with HCl having the bigger moment. This result is in accord with a recent theoretical analysis by Bunker.[31]

ELECTRIC PERMITTIVITY

Measurement of the electric permittivity, or dielectric constant, ϵ of a gas over a range of temperature leads—via Debye's[32] classical equation

$$\epsilon - 1 = 4\pi N(4\pi\epsilon_0)^{-1}\left(\alpha + \frac{\mu^2}{3kT}\right),$$

where N is the number density and ϵ_0 the permittivity of free space ($4\pi\epsilon_0 = 1.112650 \times 10^{-10}$ C V^{-1} m^{-1} = 1 esu)—to values of the dipole moment μ and mean polarizability α. The values obtained are averages over all thermally populated states. If the dipole moment is approximately the same in these states and if the energy difference between adjacent rotational states is small compared with kT, Debye's equation can be used to deduce μ through measurements of ϵ over a range of T.[33] At low temperatures quantization of the rotational motion must be considered; for a rigid diatomic rotor $\mu^2/3kT$ has to be multiplied by $f_1(T)$ where

$$f_1(T) = (kT/hcB_0)\left\{\sum_{J=0}^{\infty} (2J+1)\exp\left[-(J^2+J)hcB_0/kT\right]\right\}^{-1}$$

$$= 1 - \tfrac{1}{3}(hcB_0/kT) + \tfrac{2}{45}(hcB_0/kT)^2 - \ldots .$$

For HF at 300 K, $f_1(T) = 0.967$, but for most molecules the correction is insignificant.

MOLECULAR QUADRUPOLE MOMENTS

Energy changes due to the interaction of a molecular quadrupole moment Θ with a laboratory field gradient are normally too small to lead to observable splittings. However, small frequency shifts of about 100 Hz were observed in a resonance experiment on Al atoms in the $^2P_{3/2}$ state in a beam in a field gradient of 4×10^8 V m^{-2}.[34] The second-order Stark effect, acting through the anisotropic polarizability, complicates the experiment but can be separated by exploiting the fact that the quadrupole interaction changes sign with the voltage, unlike the second-order Stark energy.

Molecular quadrupole moments may be obtained directly by observing the optical birefringence induced in a gas by an electric field-gradient $E'_{xx} = -E'_{yy}$.[35–37] The birefringence $n_x - n_y$, like the dielectric polarization in Debye's treatment, is comprised of a temperature-independent contribution due to distortion of the molecule and an orientation term proportional to T^{-1} [35]:

$$n_x - n_y = 2\pi N E'_{xx}(4\pi\epsilon_0)^{-1}\left[B + \frac{2\Theta(\alpha_{\parallel} - \alpha_{\perp})}{15kT}\right].$$

This approach leads to an average quadrupole moment for all the thermally populated states. Deduction of Θ is dependent on a knowledge of the anisotropy in the polarizability $\alpha_{\parallel} - \alpha_{\perp}$, and this may be obtained from relative intensity measurements of the light scattered by the gas.[38] An ambiguity in the sign of $\alpha_\parallel - \alpha_\perp$ exists but, in principle, this may be resolved by Kerr effect observations, and in practice knowledge of bond polarizabilities gives an indication of the sign. The accuracy in Θ is of the order of ±2 percent and is limited by (a) uncertainty in the temperature-independent distortion contribution B (for spherical molecules B may be obtained from the Verdet constant determining the Faraday rotation[39]), (b) uncertainty in $\alpha_\parallel - \alpha_\perp$, (c) experimental errors, and (d) representation of the molecule by a classical rigid rotor.

Quantum corrections to the above classical formula[40] are more important than in Debye's equation. For a rigid diatomic rotor, the term $2\Theta(\alpha_\parallel - \alpha_\perp)/15kT$ must be multiplied by

$$f_2(T) = 1 - (hcB_0/kT) + \tfrac{8}{15}(hcB_0/kT)^2 - \ldots,$$

which is 0.76 for H_2, 0.87 for D_2, 0.90 for HF, and 0.99 for N_2 at 300 K.

In a dipolar molecule, the quadrupole moment is origin-dependent, and it is important to know the appropriate origin for Θ in a particular experiment. In spectroscopic experiments on isolated molecules[34,41] the center of mass is the natural choice, and one should obtain different values for say HF and DF. However, in a bulk equilibrium property like birefringence, the distribution of molecules is determined by their potential energy and is not influenced by their mass; thus gaseous HF and DF would exhibit the same birefringence in identical field gradients. The appropriate origin is "the effective quadrupole center," which is the point at which the polarizability anisotropy may be considered to be located[37,42]; this is the center in a centrosymmetric molecule like H_2 and is the same point in HD. In a polar molecule such as OCS, measurement of Θ relative to both the center of mass and the effective quadrupole center yields new information about the *distribution of polarizability* in the molecule; in $^{16}O^{12}C^{32}S$ the effective quadrupole center is -0.4×10^{-10} m from the center of mass in the direction of the dipole (from − to +), which has the sense −OCS+,[14] placing it $0.1_4 \times 10^{-10}$ m along the CS bond from the C atom.

MOLECULAR POLARIZABILITIES AND HYPERPOLARIZABILITIES

Mean molecular polarizabilities and anisotropies may be obtained from measurements of dielectric polarization, refractive indexes, Kerr constants, relative intensities of scattered light and from Stark effect observations. Higher order dipole polarizabilities such as hyperpolarizabilities, optical activity and Verdet constants may also be determined, but there are difficulties in isolating the higher multipole polarizabilities. The dipole polarizabilities and hyperpolarizabilities are independent of the choice of origin but higher multipole polarizabilities are not.[43,44]

REFERENCES

1. J. S. Muenter and W. Klemperer, *J. Chem. Phys.*, **52**, 6033 (1970).
2. J. S. Muenter, *J. Chem. Phys.*, **56**, 5409 (1972).
3. N. F. Ramsey, *Am. Sci.*, **49**, 509 (1961).
4. A. D. Buckingham, *MTP International Review of Physical Chemistry*, Series I, Vol. 3, ed. by D. A. Ramsay, p. 73–117, Medical Technical Publishing Co., New York, 1972.
5. L. H. Scharpen, J. S. Muenter, and V. W. Laurie, *J. Chem. Phys.*, **46**, 2431 (1967); **53**, 2513 (1970).
6. L. Wharton and W. Klemperer, *J. Chem. Phys.*, **39**, 1881 (1963).
7. D. U. Webb and K. N. Rao, *J. Mol. Spectrosc.*, **28**, 121 (1968).
8. D. E. Freeman and W. Klemperer, *J. Chem. Phys.*, **45**, 52 (1966).
9. R. D. Nelson, Jr., D. R. Lide, Jr., and A. A. Maryott, *Selected Values of Electric Dipole Moments for Molecules in the Gas Phase*, NSRDS–NBS 10, National Bureau of Standards, Washington (1967).
10. A. D. Buckingham, D. A. Ramsay, and J. Tyrrell, *Can. J. Phys.*, **48**, 1242 (1970).
11. J. A. Howe and J. H. Goldstein, *J. Chem. Phys.*, **23**, 1223 (1955).
12. D. E. Freeman, J. R. Lombardi, and W. Klemperer, *J. Chem. Phys.*, **45**, 58 (1966).
13. R. M. Conrad and D. A. Dows, *J. Mol. Spectrosc.*, **32**, 276 (1969).
13a. R. Y. Dong and D. A. Ramsay, *Can. J. Phys.*, **51**, 2444 (1973).
14. W. H. Flygare and R. C. Benson, *Mol. Phys.*, **20**, 225 (1971).
15. J. F. Liebman, *Mol. Phys.*, **21**, 563 (1971).
16. P. E. Cade and W. M. Huo, *J. Chem. Phys.*, **45**, 1063 (1966).
17. C. Schlier, *Fortschr. Phys.*, **9**, 455 (1961).
18. F. Shimizu, *J. Chem. Phys.*, **51**, 2754 (1969); **52**, 3572 (1970); **53**, 1149 (1970).
19. K. J. White and R. L. Cook, *J. Chem. Phys.*, **46**, 143 (1967).
20. K. Uehara, K. Sakurai, and K. Shimoda, *J. Phys. Soc. Japan*, **26**, 1018 (1969).
21. K. Fox, *Phys. Rev. Lett.*, **27**, 233 (1971).
22. J. K. G. Watson, *J. Mol. Spectrosc.*, **40**, 536 (1971).
23. I. Ozier, *Phys. Rev. Lett.*, **27**, 1329 (1971).
24. M. R. Aliev, *Soviet Phys. JETP Lett.*, **14**, 417 (1971).
25. A. Rosenberg and I. Ozier, *J. Chem. Phys.*, **58**, 5168 (1973).
26. M. Trefler and H. P. Gush, *Phys. Rev. Lett.*, **20**, 703 (1968).
27. S. M. Blinder, *J. Chem. Phys.*, **35**, 974 (1961).
28. W. Kołos and L. Wolniewicz, *J. Chem. Phys.*, **45**, 944 (1966).
29. P. R. Bunker, *J. Mol. Spectrosc.*, **46**, 119 (1973).
30. E. W. Kaiser, *J. Chem. Phys.*, **53**, 1686 (1970).
31. P. R. Bunker, *J. Mol. Spectrosc.*, **45**, 151 (1973).

32. P. Debye, *Phys. Z.*, **13**, 97 (1912).
33. J. H. Van Vleck, *The Theory of Electric and Magnetic Susceptibilities*, Ch. VII, Oxford University Press, New York (1932).
34. J. R. P. Angel, P. G. H. Sandars, and G. K. Woodgate, *J. Chem. Phys.*, **47**, 1552 (1967).
35. A. D. Buckingham, *J. Chem. Phys.*, **30**, 1580 (1959).
36. A. D. Buckingham and R. L. Disch, *Proc. R. Soc. A.*, **273**, 275 (1963).
37. A. D. Buckingham, R. L. Disch, and D. A. Dunmur, *J. Am. Chem. Soc.*, **90**, 3104 (1968).
38. N. J. Bridge and A. D. Buckingham, *Proc. R. Soc. A*, **295**, 334 (1966).
39. A. D. Buckingham and M. J. Jamieson, *Mol. Phys.*, **22**, 117 (1971).
40. A. D. Buckingham and M. Pariseau, *Trans. Faraday Soc.*, **62**, 1 (1966).
41. N. F. Ramsey, *Molecular Beams*, Oxford University Press, London (1956), p. 229.
42. A. D. Buckingham and H. C. Longuet-Higgins, *Mol. Phys.*, **14**, 63 (1968).
43. A. D. Buckingham, *Adv. Chem. Phys.*, **12**, 107 (1967).
44. A. D. McLean and M. Yoshimine, *J. Chem. Phys.*, **47**, 1927 (1967).

Rolfe H. Herber and Yehonathan Hazony

Chemical Structure Studies by Mössbauer Spectroscopy

Although more than 50 nuclei are known for which gamma-ray resonance fluorescence (the Mössbauer effect) has been observed,[1] the present contribution will concern itself primarily with data pertinent to experiments using the 14.4 keV radiation of ^{57}Fe and the 23.8 keV radiation of ^{119}Sn, since the vast majority of studies related to chemical structure problems have exploited these two nuclides.

Among the parameters derived from the data in Mössbauer spectroscopy, the isomer shift (IS) plays the most prominent role in chemical structure analysis. This property relates to the s-electron density at the Mössbauer nucleus and provides a measure of the average bonding interaction of the Mössbauer atom with the surrounding ligands. It has been shown to be closely related to the average ligand electronegativity.

The quadrupole splitting (QS) provides a measure of the anisotropy of the bonding electron density distribution and is used primarily as a "finger-printing" technique to identify gross features of the symmetry within a given valence state. It is very useful in identifying phase changes in solids. In bonding situations involving partially filled electronic shells, as in the case of iron chemistry, QS may provide a measure of crystal field parameters as well as estimates for $\langle r^{-3} \rangle$ for the particular electronic orbital.

The magnetic hyperfine interaction (H_{int}), which is frequently observed in the Mössbauer spectra of iron compounds, provides a measure of the coupling between the nuclear magnetic moment and the magnetic field arising from the electronic environment, as measured at the Mössbauer nucleus. Because of the different terms contributing to the magnetic interaction, it provides information concerning the unpaired s-spin density at the nucleus as well as symmetry and radial properties of the three-dimensional density distribution.

In the interpretation of the numerical results that can be extracted from Mössbauer spectroscopic data, it is necessary to recognize three sources of errors that can affect the accuracy of the data. These three contributions to the experimental error, which may not always be distinguishable from each other, can be identified as (a) statistical, (b) systematic, and (c) model-dependent errors. The statistical error, which arises from the fact that a finite number of observations are made in order to evaluate a given parameter, is the most readily estimated from the conditions of the experiment, provided that a Gaussian error distribution is assumed. Systematic errors are those that arise from factors influencing the absolute value of an experimental parameter but not necessarily the internal consistency of the data. Hence, such errors are the most difficult to diagnose and their evaluation commonly involves measurements by entirely independent experimental procedures. Finally, the model errors arise from the application of a theoretical model that may have only limited applicability in the interpretation of the experimental data. The errors introduced in this manner can often be estimated by a careful analysis of the fundamental assumptions incorporated in the theoretical treatment.

An experimental program designed to yield accurate Mössbauer data will provide for the independent estimation of the three types of errors referred to above, to the extent that this is possible, and should lead ultimately to a realistic assessment of the accuracy and precision of the reported results. The errors commonly encountered in Mössbauer effect experimentation will be illustrated by several examples in the following discussions.

COMMENTS ON METHODOLOGY

As in all other spectroscopic techniques, the numerical data extracted from the experiment can only be related to the solution of structural problems via results that have been cataloged for reference or model substances whose spectra are well understood in terms of their structural implications. To effect such comparisons in Mössbauer spectroscopy, the experimenter must concern himself *inter alia,* with the following experimental conditions:

- Velocity calibration
- Spectrometer linearity
- Isomer shift calibration
- Source quality and integrity
- Counting statistics and signal-to-noise ratio
- Temperature control of source and scatterer

as well as related variables.

Velocity calibration of the spectrometers can be accomplished by Mössbauer or by optical methods, depending on the resources of the experimenter and/or the demands imposed by the nature of the experiment. In the former, high purity (99.99 percent or better), thin (0.02-mm) natural iron foil is used as a reference absorber in conjunction with a narrow-line gamma-ray source. Since the ground-state (1/2 spin state) magnetic moment is accurately known from NMR measurements (equivalent to 3.9177 mm/s at 294 K),[2] spectrometer calibration can be effected by determining experimentally the position of the four inner resonance maxima of the resulting Mössbauer spectrum. Moreover, the separations of the 1-2, 2-3, 4-5, and 5-6 resonance peaks correspond to the excited state (3/2 spin state) magnetic moment that has now been determined[3] with high precision (equivalent to 2.237 ± 0.002 mm/s at 294 K). Using these data it is possible to evaluate the spectrometer linearity over the full velocity range covered by the metallic iron spectrum (approximately ±5.3 mm/s from the midpoint of the spectrum), which will encompass the overwhelming majority of the hyperfine effects encountered in ^{57}Fe and ^{119}Sn Mössbauer spectroscopies (Figure 1). A further advantage to this technique is that it also defines an appropriate isomer shift reference point[4] for iron Mössbauer spectroscopy, and such shifts are now generally reported with respect to the center of the metallic iron spectrum. In the case of ^{119}Sn, the reference point is usually taken as the center of a room temperature $BaSnO_3$, SnO_2, or SnTe spectrum.

Optical methods of spectrometer calibration include laser interferometric methods and Moiré fringe counting techniques. Since such methods depend on the accurately known wavelength of light (for example, 6328.1983 Å for a He–Ne laser at 293 K under standard conditions), they are independent of any assumptions made in the iron foil technique and, thus, are intrinsically more reliable. Moreover, such methods do not require the acquisition of ^{57}Co Mössbauer sources and reference materials for iron absorber studies and may thus be attractive

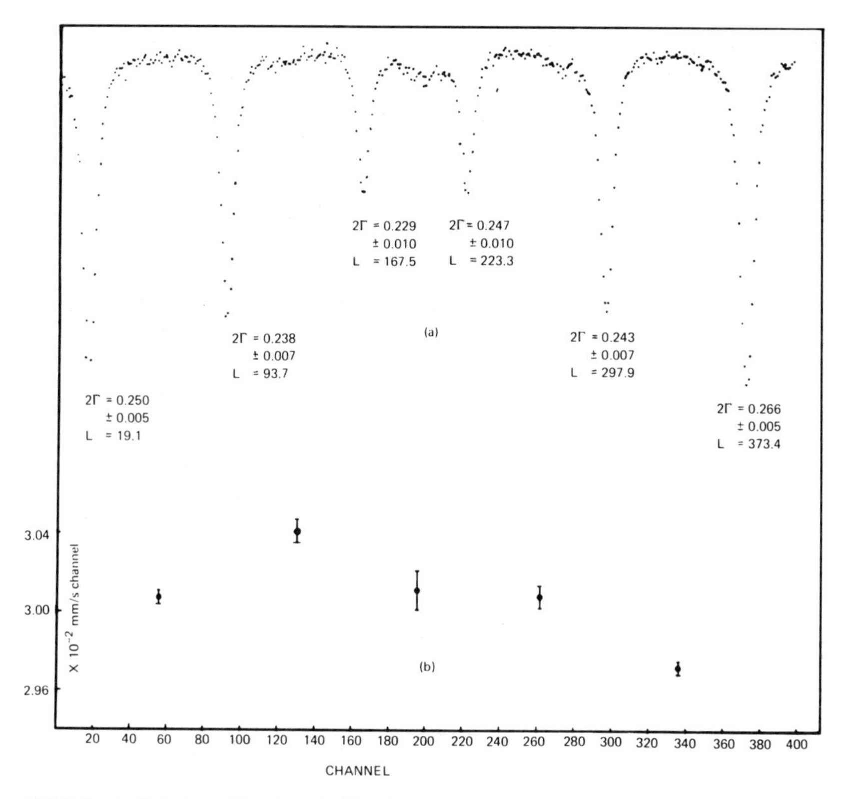

FIGURE 1 Velocity calibration of a Mössbauer spectrometer. The spectrum shown (a) is of metallic iron at room temperature. Line positions are given in channels and line widths in mm/s. The velocity calibration constant (b) is derived from the known energy differences between various components of the magnetic hyperfine spectrum. In the present data a differential nonlinearity of about 1 percent is observed. Such spectrometer nonlinearity may become a source for significant systematic errors in high-resolution experiments.

to spectroscopists working with other nuclides. Such factors may offset the initially higher costs of instrumentation for optical velocity calibration methods.

Although modern high-speed digital computer techniques have been of significant help in educting small signals from noisy (random) backgrounds, high precision Mössbauer effect parameters normally require a signal-to-(random) noise ratio on the order of 10 : 1. Since the latter is a consequence of the random decay statistics of a nuclear process (described by a Poisson distribution function), the standard deviation varies as the square root of the number of counts scaled for the data point in question. For a Mössbauer experiment in which the resonance effect magnitude (ϵ) is 1 percent, on the order of 10^6 counts will have to be scaled per channel in order to achieve a signal-to-noise ratio of 10. In transmission geometry, using a source-detector distance of 10 cm and standard proportional counter nuclear radiation detection techniques, a source strength of 2–10 mCi is needed to permit a data acquisition rate of 10^6 counts per channel in a 400-channel multichannel analyzer in a period of 1–20 h.

Since essentially all of the quantitative parameters that can be extracted from a Mössbauer experiment (isomer shift, quadrupole splitting, line width, effect magnitude, magnetic hyperfine splitting, etc.) are temperature-dependent, some effort must be made to control this variable, even in constant temperature experiments. Under normal conditions control to ±0.5 K in the range $4.2 < T < 300$ K and ±2 K in the range $300 < T < 1000$ K are practicable with routine instrumentation. Special requirements that demand control better by one or two orders of magnitude can be realized with existing components. In temperature-dependence measurements—which can be of extreme usefulness in elucidating questions of chemical structure—the precision of temperature control and stability, characteristic of all such studies, should be realized.

In addition to the above fundamental aspects of the technique, a variety of other experimental parameters can significantly affect the "quality" of the data extracted from Mössbauer effect experiments. The proper appreciation and understanding of the origin of such experimental variables can be used to extract additional, nontrivial information from the study, while its disregard can confuse the unwary and mask the subtleties of small effects that are often the key to a detailed understanding of the properties of the experimental sample. Among such parameters are effect of sample thickness (t factor) on line width, effect of optical geometry of the spectrometer on line width, crystal orientation effects on spectral shape (temperature-independent), recoil-free fraction anisotropy effects on spectral shape (temperature-dependent), crystalline solid packing forces, solvation effects, and influence of critical and noncritical scatterers.

The majority of Mössbauer experiments that are conceived to elucidate problems of chemical structure make use of velocity modulated spectrometers. Most frequently, these are of the constant acceleration type, in which equal time intervals are spent in equal velocity increments. The primary information storage device for such spectrometers is commonly a multichannel analyzer operated in the multiscaler mode. Since the minimum observable line widths ($2\Gamma_{nat} = h/\pi$) for ^{57}Fe and ^{119}Sn are 0.1946 and 0.6193 mm/s, respectively, and since on the order of at least 5–10 data points on each side of the resonance maximum are needed to adequately define the Lorentzian line shape of the peak, the spectrometer *resolution* should be on the order (for ^{57}Fe) of $\sim(2 \times 0.195)/10$ or 0.04 mm/s. On the other hand, the magnetic hyperfine spectrum of metallic iron, which is commonly employed to effect velocity calibrations in spectrometers used for ^{57}Fe and ^{119}Sn studies (see above), spans a range of 10.63 mm/s so that the number of channels used in the multiscalar mode of operation will be typically at least 200 channels. Thus, the data reduction for each spectrum will require the processing of approximately 200 data points and is most appropriately effected by high-speed digital computational methods. In this context, proper care must be taken with respect to the following points:

1. Data normalization to the nonresonant baseline;
2. Correction for baseline nonlinearity; and
3. Convergence of the iterative process to a true minimal value of the least mean square deviation.

The last point is especially important in the interpretation of complex spectra. Figure 2a represents a reconstruction of a case where the observed spectrum has been interpreted in terms of a theoretical model involving three sets of doublets. As is commonly done in reported literature, a smooth curve was passed through the data giving the impression of excellent agreement with the theoretical model, which has been optimized by a least mean squares fitting procedure by a digital computer. The superposition of the three multiplets displays a rather poor agreement with the data (Figure 2b), which may serve to demonstrate the consequences of the convergence of the iterative process to false minima. The superposition of the three components shown in Figure 2c displays a good fit to the experimental results as shown (broken curve). However, the poor quality of the data would not justify any degree of confidence in the physical significance of this fit either.[5]

Using the methodology outlined in the preceding discussion, good quality ^{57}Fe and ^{119}Sn Mössbauer effect spectroscopic data should yield isomer shift and quadrupole splitting parameters that are replicable to ±0.005 mm/s and have an experimental uncertainty not exceeding ±0.02 mm/s. State of the art experiments at present are capable of an order of magnitude improvement over these limits.

ISOMER SHIFT

While the state of the art in Mössbauer instrumentation permits the recording of high-precision, high-resolution spectra, the interpretation of

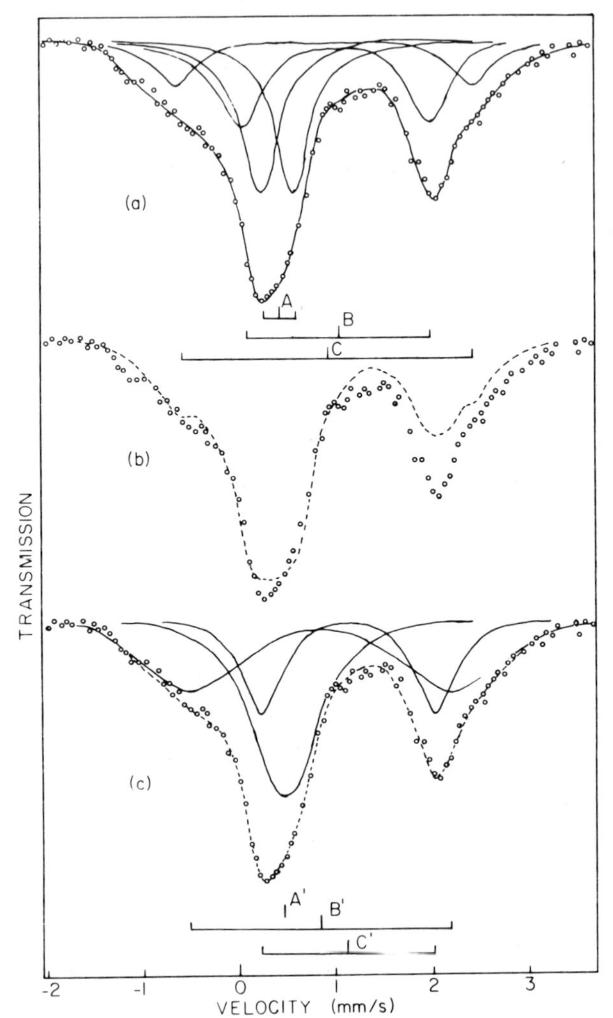

FIGURE 2 Example of a convergence to false minima of the least mean squares fit routine. (a) A fit produced by the computer of three sets of doublets (constrained to equal line widths and equal intensities of the components of each pair) to an experimental spectrum; (b) a comparison between the superposition of the six lines calculated from (a) and the experimental data illustrates a substantial misfit especially in the positive velocity region of the spectrum; (c) an alternate fit to the data (constrained to two doublets and a singlet of different width), giving a much better agreement between the calculated and experimental values. It should be noted, however, that the agreement obtained in (c) is not unique; hence, additional experimental information must be exploited to arrive at a reliable estimate of the spectroscopic parameters.

such spectra is subject to much greater uncertainties. In trying to take advantage of the availability of better measurements, one has to consider additional temperature-dependent contributions to the line position prior to assigning a precise value for the isomer shift.

The thermal shift δ_T of the Mössbauer spectrum is the sum of a contribution due to the second-order Doppler effect (δ_{SOD}) and a possible contribution due to an intrinsic dependence of the isomer shift (δ_I) on temperature. The second-order Doppler shift is proportional to the mean square velocity of the Mössbauer nucleus. For the purpose of a comparison with thermodynamic data, the SOD shift may be described in terms of the Debye approximation,[6]

$$\delta_{SOD} = -\frac{9}{16}\frac{E_0 k}{mc^2}\left[\theta_D + 8T\left(\frac{T}{\theta_D}\right)^3 \int_0^{\theta_D/T} \frac{x^3 dx}{e^x - 1}\right]. \quad (1)$$

The inadequacy of the Debye approximation in describing the details of the frequency distribution function in a real solid is well known. This results in noticeable disparities between Debye temperatures derived from the results of different experimental techniques used to elucidate this parameter on the same solid, or over different temperature ranges. Substantial discrepancies may be expected in solids containing two (or more) different atoms in the unit cell. This has been demonstrated by the Debye–Waller factors recorded for the two different Mössbauer nuclei in the case of SnI_4,[7] or when the Debye–Waller factor has been compared with the thermal shift results for the same Mössbauer nucleus in the iron cyanides.[8] The possible contribution due to an intrinsic thermal change of the isomer shift may be obscured by an improper assignment of an effective Debye temperature.

The thermal shift data[6] for $FeCl_3$ over the temperature range $4 < T < 350$ K are compared in Figure 3 with those for $FeCl_2$. The respective isomer shift scales (measured relative to metallic iron at room temperature) are shown on the right-hand side of the figure. They have been shifted vertically with respect to each other for a simultaneous display on the same SOD shift scale (left-hand side of the figure). The assignment of a relative isomer shift between the two compounds will be affected by the choice of effective Debye temperatures to account for the deviations from linearity of the experimental curves.

The analysis of the data for $FeCl_2$, in terms of the SOD contribution, results in the estimate of Θ_D (SOD) $\sim 200 \pm 20$ K. This value compares well with the effective Debye temperature derived from the Mössbauer–Debye–Waller (MDW) factor (Table 1). In contrast, the analysis of the $FeCl_3$ data yields Θ_D(SOD) $\sim 400 \pm 25$ K, which is nearly twice as large

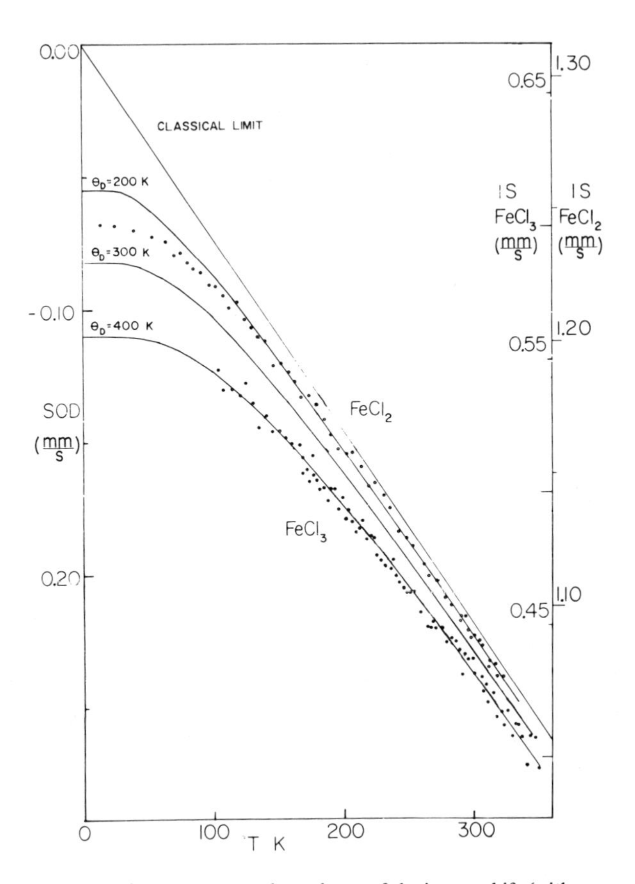

FIGURE 3 Temperature dependence of the isomer shift (with respect to metallic iron at room temperature) in iron chlorides. The corresponding isomer shift scales (right-hand side) have been shifted vertically with respect to each other for the purpose of a comparison on the same SOD scale (left-hand side).

as the corresponding Θ_D (MDW). Also given in the table are estimates for the Debye temperature derived from specific heat data. The large discrepancy between the different estimates of the Debye temperature serves to demonstrate the uncertainties involved in the detailed analysis of high-resolution thermal shift data. The large deviation of the Θ_D (SOD) for $FeCl_3$ has been shown[6] to be due to a significant contribution to the thermal shift from the intrinsic temperature dependence of the isomer shift. This effect is demonstrated in Figure 4, where the high-temperature thermal shift curves are compared[9] for FeF_2 and FeF_3. While the high-temperature slope of the thermal shift for FeF_2 is in

TABLE 1 Effective Debye Temperatures (K) for Iron Chlorides[a]

Compound	θ_D(MDW)	θ_D(SOD)	$\theta_D(C_p)$
$FeCl_2$	210 ± 25	200 ± 20	290
$FeCl_3$	225 ± 25	400 ± 25	360

[a] From Hazony.[6]

good agreement with the classic limit for the second-order Doppler shift, the additional contribution due to the intrinsic temperature dependence of the isomer shift is quite significant for FeF_3.

The study of these contributions either as a subject in itself or for the purpose of obtaining more accurate values for the isomer shift requires the combination of a high performance spectrometer and an extensive data reduction effort using a digital computer.

The results shown in Figures 3 and 4 demonstrate the extent of the

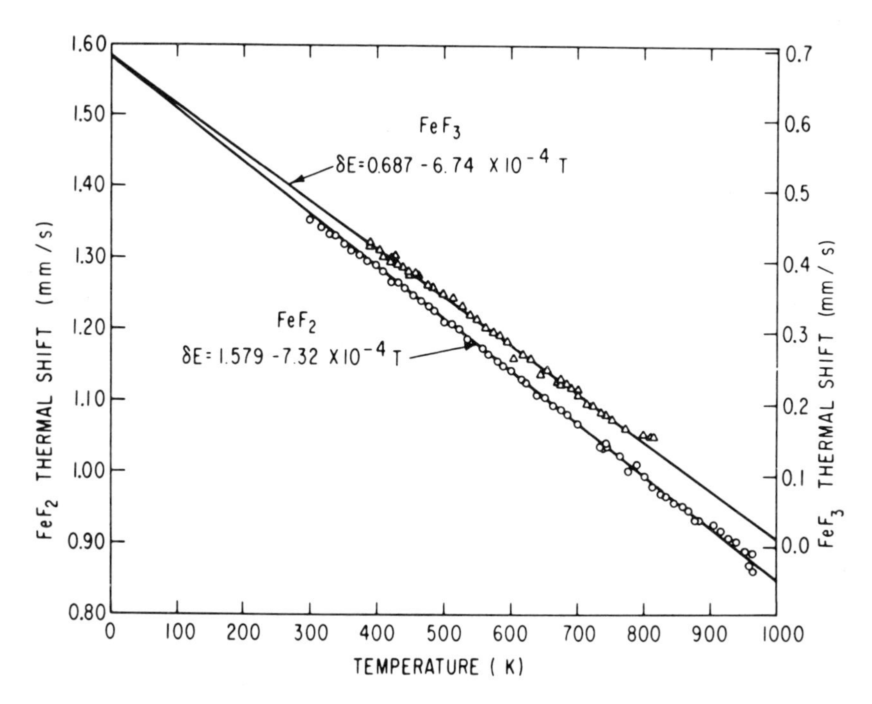

FIGURE 4 High-temperature slopes of the ^{57}Fe Mössbauer isomer shift in two iron fluorides. The data are given with respect to metallic iron at room temperature.

experimental effort required to ascertain the validity of the extrapolation procedures used to derive the isomer shift. As in other measurements of the thermophysical properties of solids, one has to be alert to the consequences of lattice anharmonicity as well as to structural phase transitions. This requires monitoring as many temperature points as are consistent with the resolution and statistics of the experiment. Such a detailed experiment typically extends over a period of several days and this dictates stringent specifications for the long-term stability of the spectrometer. Figure 5 presents a stability test performed during the development of a high-performance constant acceleration Mössbauer spectrometer.[10] It shows the location of the two resonance lines of sodium nitroprusside, with respect to a ^{57}Co source diffused in copper, as a function of time. The peak at channel 241.25, which is near zero velocity, was constant within ±0.001 mm/s over a period of 13 days. The peak at channel 57.5 shows a ±0.004 mm/s drift, which is clearly larger than the statistical fluctuations. The error flags correspond to one-sigma limits

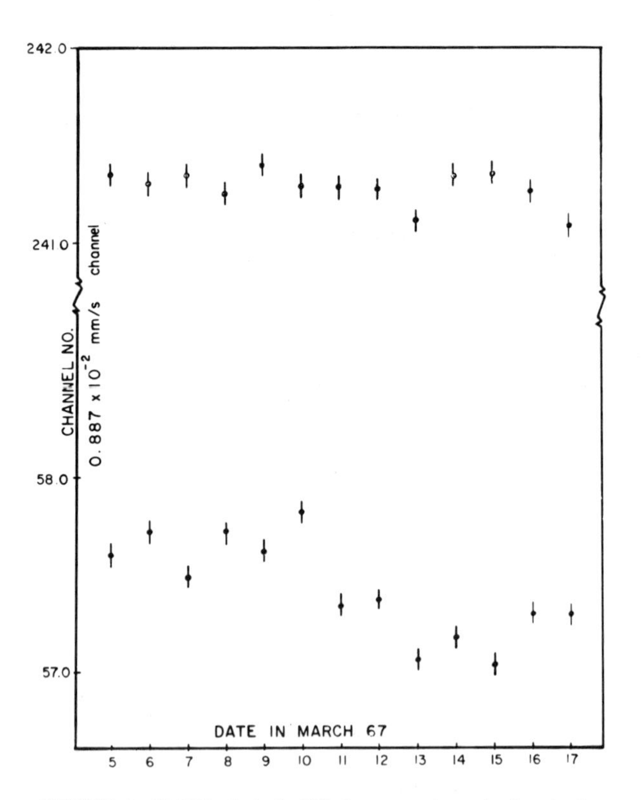

FIGURE 5 Stability test of a Mössbauer spectrometer (see text).

as determined by a least squares fit of the data to Lorentzian curves.

Extensive data of this type are necessary in order to verify the stability of a spectrometer and identify the potential sources of drift. The stability displayed in the figure is not sufficient for the thermal studies described above because the assignment of the isomer shift depends on the determination of the Debye temperature from the small changes in the slope of the thermal shift curve. The data presented in Figures 3, 4, and 5 have been monitored with a constant velocity spectrometer with a verified long-term stability of ±0.001 mm/s, which was integrated into a fully automated data-acquisition system.

The extrapolated values of the isomer shift are shown in Figure 6 for a well-characterized series of the hydrates of $FeCl_2$, with a confidence limit of ±0.005 mm/s. This represents a large improvement in the resolu-

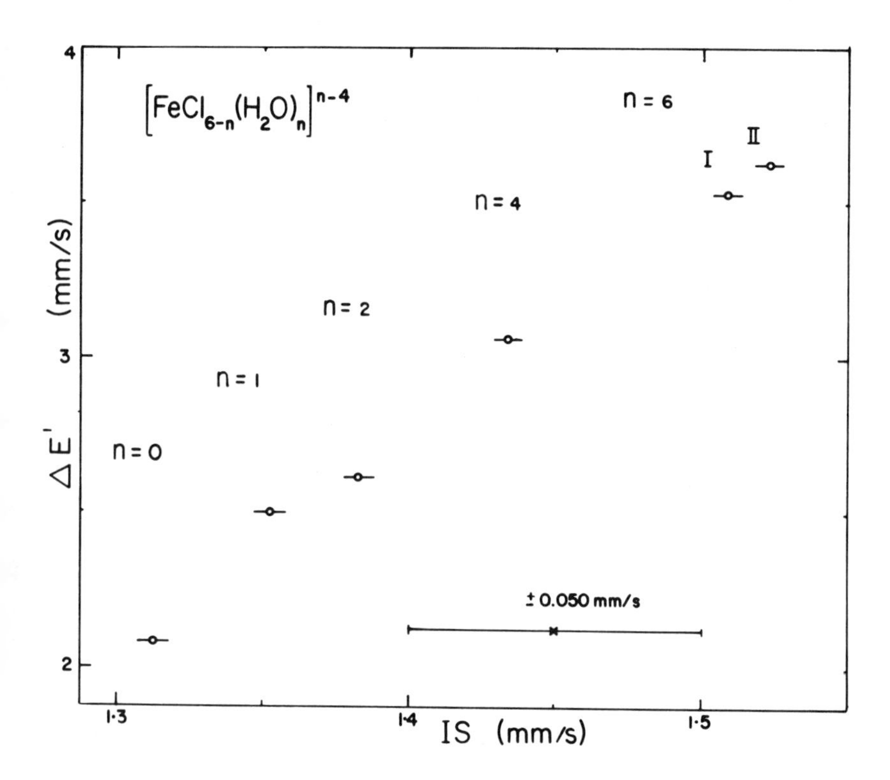

FIGURE 6 $\Delta E'$ vs. IS for the hydrates of $FeCl_2$. The IS values have been corrected for the effects of temperature, using the procedure described in the text, assuming no intrinsic temperature effects. The uncertainty limits shown correspond to ±0.005 mm/s. The effect of an order of magnitude increase in the experimental uncertainty is illustrated in the bottom of the figure.

tion of the isomer shift scale, compared with the commonly reported estimated errors. As a result of this improvement, dramatic and previously unobserved correlations have emerged between the isomer shift and quadrupole splitting within several well-defined series of iron compounds.[11]

A comparison between the data presented in Figure 3 and the results for other compounds, such as iron cyanides,[8] for example, demonstrates that errors as large as 0.05 mm/s may be introduced in the evaluation of the relative isomer shift if temperature effects are not considered. It is possible to measure the position of the centroid of a Mössbauer spectrum to two orders of magnitude better than the above-quoted error. However, the reliability of any substantial reduction of the magnitude of the confidence limit will depend on the validity of the explicit and implicit assumptions underlying the data reduction procedures. For example, it has been demonstrated[6] for the case of $FeCl_3$ that disregarding the contribution from the intrinsic temperature dependence of the isomer shift produced an error of about 0.025 mm/s in the extrapolated value of the IS.

Uncertainty limits of ±0.05 mm/s are shown in Figure 6 for comparison. It is apparent that such uncertainties will wipe out completely the chemical structural significance of the data. In the analysis of the above data it was implicitly assumed that there is no contribution from the intrinsic temperature dependence of the isomer shift. This assumption was recently confirmed[6] for several Fe^{2+} compounds including $FeCl_2$ and may very well be valid for the above group of compounds.

The situation is substantially different in the case of tin compounds.[12] The much larger line width, as well as the significantly smaller specific activity of the sources, make it more difficult to monitor spectra to the same degree of accuracy as in iron. However, the higher sensitivity of the isomer shift more than compensates for the reduced resolution. The relative isomer shift between FeF_2 and FeI_2 (Figure 7) is about 0.4 mm/s, compared with about 1.6 mm/s between SnF_6^{2-} and SnI_6^{2-} (Figure 8). Detailed studies of the thermal shift indicate a substantial reduction of the slope with respect to the expectation from the SOD effect,[13] probably due to the intrinsic thermal variation of the isomer shift. Figure 9 presents isomer shift data[14] for dimethyltin bis acetylacetonate, $(CH_3)_2Sn(acac)_2$, measured with a standard deviation of ±0.01 mm/s, which is typical for measurements with ^{119}Sn. It is apparent from the figure that an increase of one order of magnitude in the statistics and stability of the measurement will be needed for a meaningful study of the thermal shift. However, it demonstrates also that a single temperature measurement

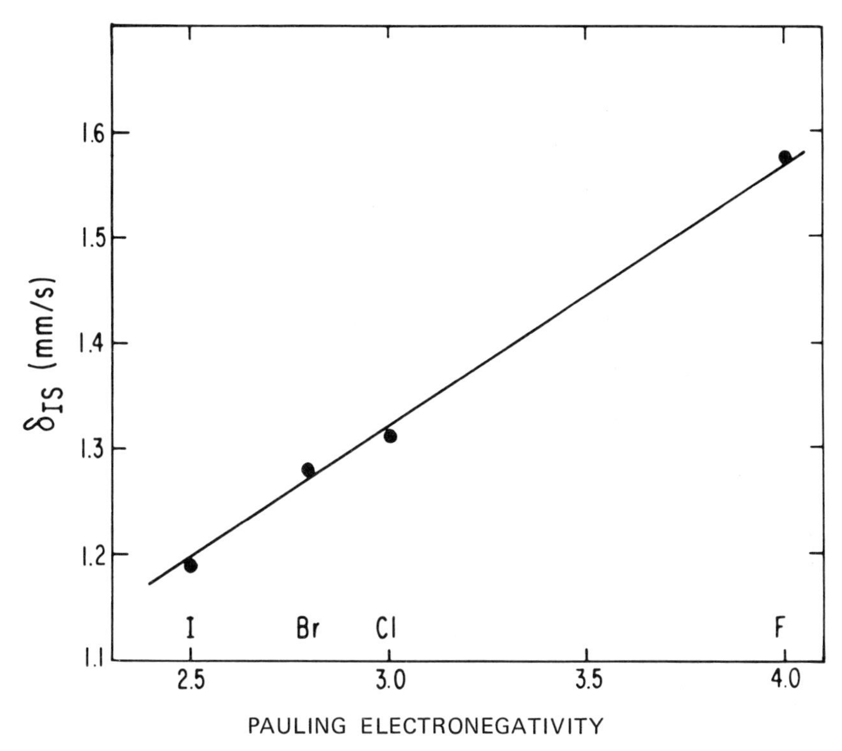

FIGURE 7 Isomer shift vs. Pauling electronegativity for anhydrous ferrous halides. The IS values are corrected for the effects of temperature, assuming no intrinsic thermal contribution. The uncertainty limits are represented by the size of the dots.

may be sufficient as long as an isomer shift resolution of ±0.02 mm/s is acceptable.

Although the concept of electronegativity has been used by chemists for many years in the qualitative description of bonding interactions between two dissimilar atoms, quantitative experimental techniques for elucidating this parameter have been limited in scope. The Mössbauer isomer shift has been demonstrated[15–18] to provide a quantitative measure of the relative electronegativity of a series of ligands in isostructural compounds. This has permitted, for the first time, an effective estimate[19] of the group electronegativity of polyatomic ligands such as OH, N_3, SCN, CH_3, and C_6H_5, *inter alia.* Such estimates are normally obtained by means of calibration data based on experimental isomer shift values for model compounds. Most frequently used are the halides, for which good estimates of the Pauling (or Mulliken–Jaffe–Whitehead) electronegativity are available.

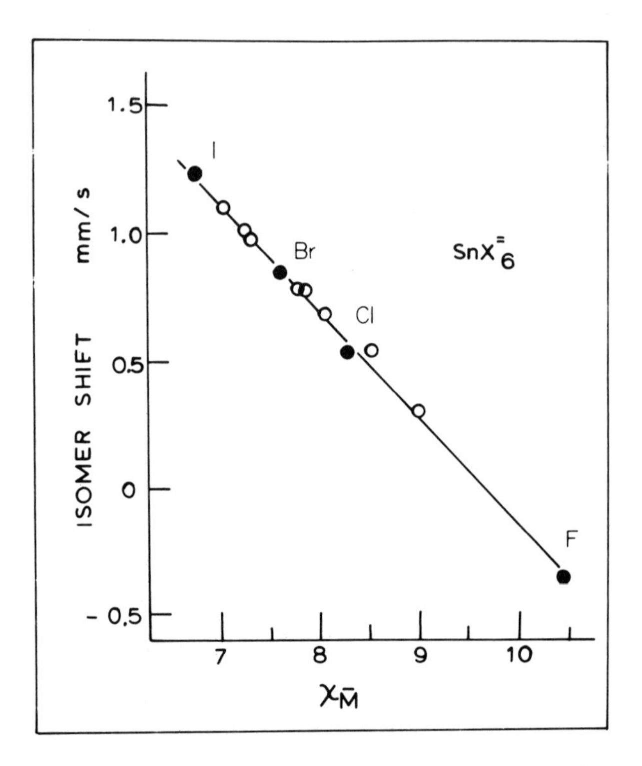

FIGURE 8 Isomer shift vs. Mulliken electronegativity for SnX_6^{-2} and $[SnX_4Y_2]^{-2}$ compounds (X, Y = F, Cl, Br, or I). The isomer shift (relative to $BaSnO_3$)–ligand electronegativity correlation is calibrated from the data for the octahedral complexes containing only a single kind of halogen atom (filled circles). The isomer shift for the mixed complexes of the type $[SnX_4Y_2]^{-2}$ (open circles) is plotted against the average Mulliken electronegativity of all of the halogen ligands bonded to the metal atom. The common cation for all of these Sn (IV) complexes is $[(CH_3)_4N]^+$.

A typical set of data[20] are summarized in Figure 7 for ^{57}Fe. The isomer shift for the series of octahedral ferrous halides (FeX_6^{-2}) is plotted against the Pauling electronegativity of the halogen. A similar correlation[15–18] for a series of octahedral tin compounds (of the type SnX_6^{-2}) is summarized in Figure 8, in which the abscissa is the average Mulliken electronegativity of the six halogen ligands bonded to the metal. Calibration of the linear relationship is afforded by the data for SnI_6^{-2}, $SnBr_6^{-2}$, $SnCl_6^{-2}$, and SnF_6^{-2} represented by the filled data points. The open circles represent the correlation between the observed isomer shift and calculated average ligand electronegativity for the mixed-halogen com-

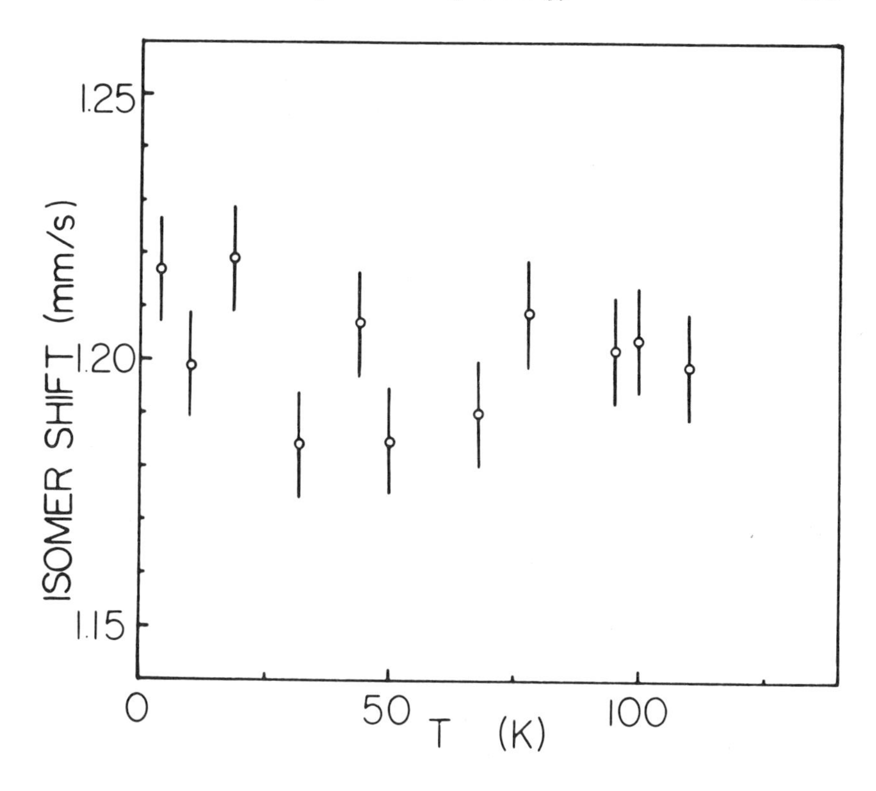

FIGURE 9 Temperature dependence of the ^{119}Sn Mössbauer isomer shift in $(CH_3)_2Sn(acac)_2$. Error-bars represent one standard deviation. The IS scale is with respect to a room temperature barium stannate absorber spectrum.

pounds of the type $(SnX_4Y_2)^{-2}$, where X and Y are two different halogen ligands.

THE QUADRUPOLE SPLITTING

Precision measurements of the temperature dependence of QS are not common in tin Mössbauer spectroscopy, primarily because the splittings vary little with temperature, and most chemical structural interpretations can be based on the QS value at a single temperature, typically 77 K. The theory needed for the interpretation of the QS parameter in tin compounds has not been developed to such a point as to make precision measurements worthwhile.[21–23] However, the thermal variation of the QS is commonly studied for two main reasons: for monitoring structural phase transitions and for the study of the motional anisotropy (the Gol'danskii–Karayagin effect) of the Mössbauer nucleus.[24–26] The latter provides an indirect measure of the anisotropy of the electron density around the Mössbauer atom.

The situation is different with iron compounds where significant thermal variation of the QS is observed. This effect is used for the study of crystal field parameters and the spin–orbit coupling interaction. The complexity of the theoretical analysis of the quadrupole splitting in high-spin ferrous or low-spin ferric compounds is reflected in the expression

$$\Delta E = (2/7)e^2 Q(l - R)\, \alpha_c^2 \langle r^{-3}\rangle_0 F(\Delta_1, \Delta_2, \alpha_s^2\lambda_0, T) + \text{lattice contribution} \qquad (2)$$

with the usually accepted notation.[27] From the simplicity of the typical curves for the temperature dependence of ΔE (Figure 10), it is apparent that such curves cannot be used for a unique determination of all the

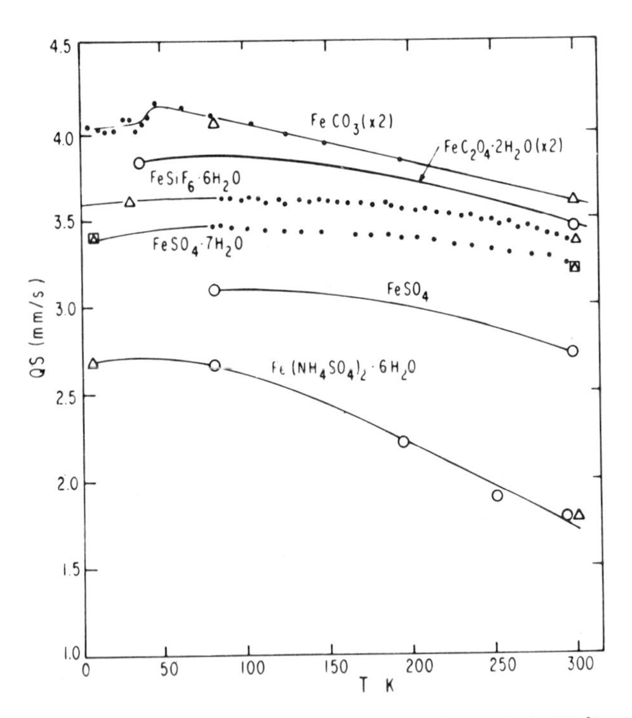

FIGURE 10 Typical quadrupole-splitting data for octahedral Fe^{2+} compounds. The results for $FeCO_3$ and $FeC_2O_4 \cdot 2H_2O$ are doubled (see text).

parameters involved. The most fruitful approach may be in the study of a series of related model compounds for which there are reasons to believe that the effects of some of the parameters in Eq. (2) are negligible. For example, if the distortion from octahedral symmetry is large enough so that the separations (Δ_1, Δ_2) of the electronic ground state from the upper two t_{2g} levels are significantly larger than the spin–orbit interaction ($\alpha_s^2\lambda_0$), but the distortion is small enough so that the lattice contribution to ΔE is relatively small, then Eq. (3) may be simplified

$$\Delta E = (\Delta E)_0 \alpha_c^2 F, \tag{3}$$

where $F \sim 1$ for a singlet ground state and ~ 0.5 for a doublet ($\Delta_1 = 0$) at the low-temperature limit. Possible variations of the antishielding parameter $(1 - R)$ with covalency are included in α_c^2, which may be considered as a measure of the changes of the charge density distribution parameter $\langle r^{-3}(t_{2g})\rangle$ due to the effects of covalency.

Equation (3) has been applied[28] to the data displayed in Figure 6 for $FeCl_2$ with $\Delta E' = 2\Delta E$ because of the doubly degenerate electronic ground state in this compound. A somewhat lower value has been reported for ΔE for $FeCl_6^{-4}$ with a singlet electronic ground state.[29] The observed linear correlation (Figure 6) demonstrates that, under well-specified conditions, it is possible to use the low-temperature saturation values of the quadrupole splitting as a gross measure of the effects of covalency on the mean charge density distribution of the t_{2g} electron.

Attempts to extract more information from the $\Delta E(T)$ curves involve determination of changes of curvature at different temperatures. This requires precision data points distributed in the more sensitive temperature regions. The evaluation of the spin–orbit coupling constant requires a detailed knowledge of the curvature in the region $0 < T < 150$ K. It is important to ascertain whether there is a maximum in the $\Delta E(T)$ curve and, if so, at which temperature. For a meaningful evaluation of the crystal field parameters and their possible variation with temperature, it is necessary to have data over a temperature range through which a substantial decrease of ΔE is observed.

MAGNETIC HYPERFINE INTERACTION

The magnetic hyperfine splitting observed in the spectra of many iron compounds is generally described[29] in terms of three components

$$H_{\text{int}} = H_c + H_L + H_D, \tag{4}$$

where H_c is the Fermi contact term, H_L is the contribution from the unquenched angular momentum, and H_D is the component due to the anisotropy of the unpaired spin density distribution. The interpretation of the magnetic hyperfine interaction for a specific compound is extremely sensitive to the details of the symmetry conditions, which are frequently very poorly understood. Here too, the most promising approach may be in the study of a series of model compounds, for which some of the uncertainties may be circumvented.

The derived values of H_c for the series $FeCO_3$, $FeTiO_3$, $FeCl_2$, $FeBr_2$, and FeI_2 are shown in Figure 11 as a function of $\langle r^{-3} \rangle$, as measured by the saturation values of the quadrupole splitting in the paramagnetic phase.[11,28] These compounds show a trigonal distortion from octahedral symmetry with a doublet electronic ground state. An outstanding feature of this ground state is the large value of the unquenched angular momentum operator along the z axis, which is a consequence of the trigonal crystal field. This simplifies the evaluation of H_L and H_D and the derivation of the contact term H_c in a systematic manner for all five compounds.

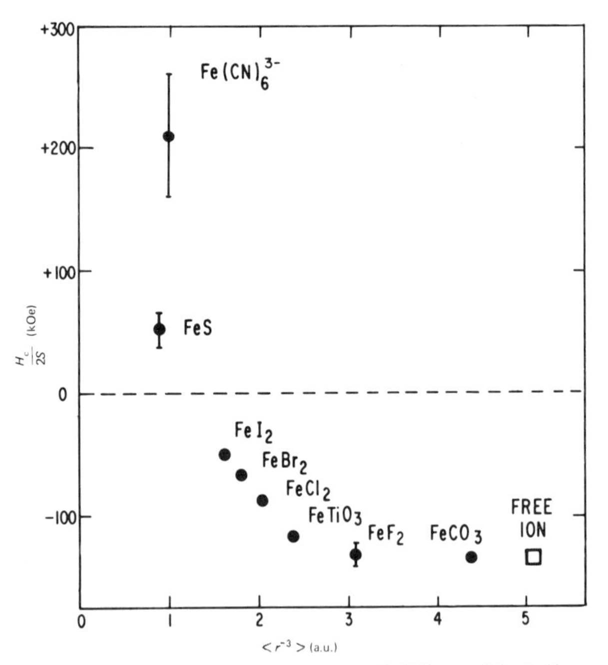

FIGURE 11 The magnetic contact term $H_c/2S$ vs. $\langle r^{-3}(t_{2g}) \rangle$ for some octahedral compounds of iron. $\langle r^{-3} \rangle$ values are derived from the quadrupole-splitting data.

The data point for FeF_2 is representative of a group of octahedral compounds displaying similar values of the quadrupole splitting at low temperature.[28] The data point for the $Fe(CN)_6^{-3}$ complex represents the range of values for a group of ferricyanides.[31] The available magnetic susceptibility data for this group indicate effective magnetic moments close to the "spin-only" values, which signifies the absence of a dominant contribution to H_{int} from H_L. The spread of H_{int} values observed within the group is reflected in the error bars. This may be attributed in part to the nonvanishing of the last two terms, as well as to anisotropic contribution to H_c due to anisotropy in the effective spin operator $\langle S_i \rangle$, produced by the nonvanishing unquenched angular momentum and the spin–orbit interaction. The positive sign of H_{int} for this complex has been determined experimentally (J. Sawicki, private communication, and ref. 32) and therefore for H_c.[11,33]

The results summarized in Figure 11 demonstrate the advantage of discussing the data for a well-characterized series of chemical compounds as a method to circumvent some of the complexity and uncertainty involved in the analysis of a single isolated compound. These results indicate that it may be possible to describe the effects of covalency on the electronic charge and spin distributions using a unified description over the full range of covalency between the extreme ionic and fully covalent compounds. This observation is corroborated by the results obtained by other experimental techniques.[34]

SUMMARY AND CONCLUSIONS

In the preceding discussion, some of the characteristic aspects of Mössbauer effect spectroscopy have been reviewed in light of the present status of the field. Mössbauer spectroscopy (specifically of ^{57}Fe and ^{119}Sn) is capable of yielding quantitative data related to electron densities at nuclei, the symmetry of electronic charge distribution, the magnitude of magnetic hyperfine fields at nuclei, the vibrational amplitudes of atoms (and molecules), and numerous other related parameters with an accuracy comparable to most other quantitative physicochemical measurements.

Being a spectroscopic technique, the interpretation of Mössbauer effect data is normally based on the calibration of a particular parameter by the use of model compounds and an interpolation or extrapolation of the available results to encompass the (new) experimental data point. In this context, reliable Mössbauer effect measurements should be reported in such a way that such comparisons can be reliably effected between different investigators and between different laboratories; in this manner a consistent and meaningful set of data can be cataloged in the literature that can then be exploited by the theoretician in arriving at a more meaningful description of bonding.

Finally, it is worth noting that Mössbauer effect spectroscopy can yield both unique and supportive data that find application in a wide range of areas, including structural and synthetic chemistry, solid state physics, biology, metallurgy, and nuclear structure systematics. Such questions as the structural integrity of a chemical compound in the real solid as compared with the frozen solution, the presence of phase transitions in the solid state, atomic and molecular motions, electrostatic and electromagnetic hyperfine fields, solvation effects, chemical consequences of nuclear transitions, and the relative strength of inter- and intramolecular bonding forces are examples among the wide range of chemical problems that have been elucidated by the Mössbauer technique.

REFERENCES

1. N. N. Greenwood and T. C. Gibb, *Mössbauer Spectroscopy*, Chapman Hall Ltd., London (1971); V. I. Gol'danskii and R. H. Herber, ed., *Chemical Applications of Mössbauer Spectroscopy*, Academic Press, New York (1968); G. M. Bancroft and R. H. Platt, in *Advances in Inorganic Chemistry and Radiochemistry*, ed. by H. J. Emeleus and A. G. Sharpe, Academic Press, New York (1972).
2. J. J. Spijkerman, J. R. deVoe, and J. C. Travis, Natl. Bur. Standards Spec. Publ., 260–20 (1970); R. H. Herber in *Mössbauer Effect Methodology*, Vol. 6, ed. by I. J. Gruverman, Plenum Press, New York (1971).
3. J. I. Budnick, L. J. Bruner, R. J. Blume, and B. L. Boyd, *J. Appl. Phys.*, **32**, 1205 (1961).
4. R. L. Cohen and G. M. Kalvius, *Nucl. Instrum. Methods*, **86**, 209 (1970).
5. S.L. Ruby, in *Mössbauer Effect Methodology*, Vol. 8, ed. by I. Gruverman, Plenum Press, New York (1973).
6. Y. Hazony, *Phys. Rev.*, **B7**, 3309 (1973).
7. S. Bukshpan and R. H. Herber, *J. Chem. Phys.*, **46**, 3375 (1967); Y. Hazony, *J. Chem. Phys.*, **49**, 159 (1968); **53**, 858(E) (1970).
8. Y. Hazony, *J. Chem. Phys.*, **45**, 2664 (1966).
9. H. K. Perkins and Y. Hazony, *Phys. Rev.*, **B5**, 7 (1972).
10. Y. Hazony, *Rev. Sci. Instrum.*, **38**, 1760 (1967).
11. Y. Hazony, in *Mössbauer Effect Methodology*, Vol. 7, ed. by I. Gruverman, Plenum Press, New York (1971).
12. J. J. Zuckerman, *Adv. Organomet. Chem.*, **9** (1970); M. C. Hayes, in *Chemical Applications of Mössbauer Spectroscopy*, ed. by V. I. Gol'danskii and R. A. Herber, Academic Press, New York (1968), Ch. 5.
13. G. M. Rothberg, S. Guimard, and N. Benczer-Koller, *Phys, Rev.*, **B1**, 136 (1970).
14. M. Leahy, R. H. Herber, and Y. Hazony, *J. Chem. Phys.* (in press).
15. R. H. Herber and S. C. Cheng, *Inorg. Chem.*, **8**, 2145 (1969).
16. C. Clausen and M. L. Good, *Inorg. Chem.*, **9**, 817 (1970).
17. J. E. Huheey and J. C. Watts, *Inorg. Chem.*, **10**, 1553 (1972).
18. J. E. Huheey, *J. Phys. Chem.*, **69**, 3284 (1965).

19. R. H. Herber, in *Mössbauer Spectroscopy and Its Applications,* International Atomic Energy Agency, Vienna (1972), p. 257.
20. R. C. Axtmann, Y. Hazony, and J. W. Hurley, Jr., *Chem. Phys. Lett.,* **2**, 673 (1968).
21. N. G. W. Debye and J. Y. Zuckerman, *Developments in Applied Spectroscopy,* Vol. 8, Plenum Press, New York (1970).
22. R. V. Parish and R. H. Platt, *J. Chem. Soc.,* **A**, 2145 (1969).
23. R. R. Berrett and B. W. Fitzsimmons, *Chem. Commun.,* **91** (1966); *J. Chem. Soc.,* **A**, 525 (1967).
24. V. I. Gol'danskii *et al., Dokl. Akad. Nauk SSSR,* **147**, 127 (1962).
25. S. V. Karayagin, *Dokl. Akad. Nauk SSSR,* **148**, 1102 (1963).
26. P. Flinn, S. L. Ruby, and W. L. Kehl, *Science,* **143**, 1434 (1964).
27. R. Ingalls, *Phys. Rev.,* **133**, A787 (1964).
28. Y. Hazony, *Phys. Rev.,* **B3**, 711 (1971).
29. G. R. Davidson, M. Eibschitz, D. E. Cox, and V. J. Minkiewicz, in *Magnetism and Magnetic Materials,* AIP Conference Proceedings, ed. by C. D. Graham, Jr., and J. J. Rhyne (1971), p. 436.
30. C. E. Johnson, in *The Mössbauer Effect,* Symposia of the Faraday Society, The Faraday Society, London (1967), No. 1, p. 7.
31. A. Z. Hrynkiewicz, B. Sawicka, and J. Sawicki, in *Proceedings of the Conference on the Application of the Mössbauer Effect (Tihany; 1969)*, ed. by I. Dezsi, Akademiai Kiado, Budapest (1971).
32. K. Ono, M. Shinohara, A. Ito, N. Sakai, and M. Suenaga, *Phys. Rev. Lett.,* **24**, 770 (1970).
33. Y. Hazony, *J. Phys. C., Solid State Phys.,* **5**, 2267 (1972).
34. J. A. Wilson, *Adv. Phys.,* **21**, 144 (1972).

Robert F. Stewart

Charge Densities from X-Ray Diffraction Data

The intensity of x rays coherently scattered by a molecule is proportional to $|F|^2$ where

$$F(\mathbf{S},\mathbf{Q}) = \int \rho(\mathbf{r},\mathbf{Q})\exp(i\mathbf{S}\cdot\mathbf{r})\,d\mathbf{r}. \tag{1}$$

Here $\mathbf{S} = 2\pi(\mathbf{k}-\mathbf{k}_0)/\lambda$, where $\mathbf{k}$ and $\mathbf{k}_0$ are unit vectors in the direction of scattered and incident radiation, respectively; $\mathbf{Q}$ is a vector of the nuclear coordinates; and $\rho(\mathbf{r},\mathbf{Q})$ is the one-electron density function for the molecule with a fixed $\mathbf{Q}$. Note that Eq. (1) is a Fourier transform of a static charge density function and is generally valid for a radiation energy that is much larger than core electron ionization energies. The actual, observed intensity is a statistical average over the states for $\mathbf{Q}$,

$$I(\mathbf{S})_{\text{coherent}} = \overline{F^*(\mathbf{S},\mathbf{Q}')\,F(\mathbf{S},\mathbf{Q})} \qquad (\mathbf{Q}' = \mathbf{Q}) \tag{2}$$

where the average is weighted by the Dirac density matrix,

$$t(\mathbf{Q}',\mathbf{Q};\beta) = \sum_n \psi_n(\mathbf{Q}')\exp(-\beta\mathcal{H})\psi_n(\mathbf{Q}). \tag{3}$$

ψ_n are state functions for the nuclei, $\beta = 1/kT$, and $\mathcal{H}$ is the Hamiltonian for the nuclei. The extension of Eq. (1), (2), (3) to a small crystal lattice does not introduce anything fundamentally different into the theory; translational symmetry comes into consideration and the dimensions of **Q** are dramatically increased. Equation (2) represents an essentially correct theory of "coherent" x-ray scattering developed by Waller, Hartree, Born, and others. An excellent summary can be found in a paper by Born.[1] It is crucial to remember that Eq. (2) is the ideal experimental result and it is an arduous, often speculative, path to work back to Eq. (1), the Fourier representation of the static charge density for a specified **Q** such as the equilibrium configuration. Perhaps Eq. (1) may be aptly described as the "Holy Grail" for the charge density buff who seeks such information from x-ray diffraction data.

For the diffracted intensities from real crystals, there is absorption as well as some interaction between incident and scattered radiation within the crystal. In this paper, attention will be focused on those crystals where the extinction phenomena are accounted for such that the experimental intensities can be reduced to the intensities of Eq. (2) with a systematic error no larger than 5 percent. These diffraction data are called accurate.

To render Eq. (2) tractable for a structure analysis, several approximations are made. The Born–Oppenheimer approximation is invoked, so that the potential energy term in Eq. (3) is just the total electronic energy. It is usually further assumed that the potential for the nuclei is quadratic. In this case it is easy to solve for $t(\mathbf{Q}, \mathbf{Q}'; \beta)$ in Eq. (3), but Eq. (2) cannot be easily evaluated without further assumptions of the charge density as a function of **Q**. The practice is to partition $\rho(\mathbf{r}, \mathbf{Q})$ into a sum of charge density functions that are centered on the several nuclei and to *further assume that each density function follows the motion of the nucleus on which it is centered.* This latter factorization has dubious validity, but nonetheless forms the basis of a working approximation for both structural and charge density analysis of x-ray diffraction data.

The above assumptions lead to an explicit form for Eq. (2), as developed by Born,[1]

$$\overline{F^*F} = \sum_{p,p'} I_p^* I_{p'} \prod_j \exp\left[-\tfrac{1}{2}(\mu_{pp'}^j)^2\, \overline{\epsilon}_j \omega_j^{-2}\right], \tag{4}$$

where $\overline{\epsilon}_j$ is the mean energy of a linear harmonic oscillator with angular frequency ω_j. I_p can be written as

$$I_p = f_p\,(\mathbf{S}) \exp\left[i\, \mathbf{S}\cdot\mathbf{r}_p\right], \tag{5}$$

where f_p (**S**) is the Fourier transform of the density functions on center p (nucleus p) and $\mathbf{r}_p$ is the position of the nucleus at the equilibrium configuration. The remaining important quantity $(\mu_{pp'}^j)^2$ is,

$$(\mu_{pp'}^j)^2 = |\mathbf{S}|^2 \left[\frac{(\mathrm{e}_{pq}^j)^2}{m_p} + \frac{(\mathrm{e}_{p'q}^j)^2}{m_{p'}} - \frac{2\,\mathrm{e}_{pq}^j\,\mathrm{e}_{p'q}^j}{\sqrt{m_p m_{p'}}}\right], \tag{6}$$

where e_{pq}^j is the component of $\mathbf{e}_p^j/\sqrt{m_p}$, the amplitude vector of nucleus p for the vibration j, along the direction of **S**. The first two terms in Eq. (6) give rise to the dominant intensity near a Bragg reflection and *the third term is responsible for thermal diffuse scattering.* With the neglect of the third term, it is then possible to factor Eq. (4) into the standard structure factor equation,

$$I = \sum_{p,p'} F_p^* F_{p'} \text{ and } F_p = I_p \exp\left[-\tfrac{1}{2}\mathbf{S}^t U_p \mathbf{S}\right], \tag{7}$$

where the components of the matrix U_p are,

$$U_{p,\alpha\beta} = \frac{1}{m_p} \sum_j \bar{\epsilon}_j \mathrm{e}_{p\alpha}^j \mathrm{e}_{p\beta}^j / \omega_j^2 \qquad \alpha,\beta = x, y, z. \tag{8}$$

When U_p is diagonalized, one has the three principal components of the mean square amplitude of vibration for nucleus p.

Thermal diffuse scattering [that is, the cross term in Eq. (6)] is the general phenomenon of phonon–photon momentum exchange that preserves the coherence of the scattered x rays such that they participate in the diffracted intensities.[2] It may well be the nemesis of charge density analysis of organic molecular crystals. Corrections for simple metals, where elastic constants are known, have been made; for organic compounds, however, only gross estimates are possible. The influence of thermal diffuse scattering on charge density studies is essentially unknown. The phenomenon can be suppressed by measuring diffraction intensities at very low temperatures; we must await results of accurate intensity measurements under these conditions.

It appears that present day x-ray diffraction measurements on well-selected organic molecular crystals have an accuracy of 5 percent or better (probably down to 0.5 percent in special cases). As a rule, these intensity data are not absolute and therefore are only proportional to Eq. (2) within a scale factor. There is indeed merit to Coppens's recent plea for efforts to *measure* the absolute scale of x-ray diffraction intensities from organic crystals.[3] Despite the fact that respectable accuracy can be achieved in the measurement of Eq. (2), the reduction of $I_{\text{coherent}}(\mathbf{S})$ to a structure factor is model dependent and most models used are fraught with uncertainties that probably outweigh the errors in the raw intensity data. One is faced with the standard problem of interpreting a thermodynamic measurement in terms of a microscopic model. Most charge density "results" are derived from the expressions in Eq. (7) and (8) as a representation for Eq. (2).

The discussions in this paper will proceed on the assumption that kinematic structure factors[4] can be properly obtained from raw diffraction data and that dynamical effects can be accounted for either by suitable corrections or by sheer neglect (such as thermal diffuse scattering). With this reservation or apology, some of the literature on accurate x-ray diffraction measurements will be reviewed with charge density information in mind.

FOURIER SYNTHESES OF ELECTRON DENSITY MAPS

The three-dimensional periodic electron density in a crystal can be represented with the three-dimensional Fourier series

$$\rho(x,y,z) = \frac{1}{V} \sum_{-\infty h}^{\infty} \sum_{-\infty k}^{\infty} \sum_{-\infty l}^{\infty} |F_{hkl}| \exp\left[-2\pi i(hx+ky+lz-\alpha_{hkl})\right], \tag{9}$$

where (h,k,l) are components of the Bragg vector (i.e., Miller indices of a reflection plane), (x,y,z) is a point in the cell where x, y, and z are in fractions of the unit cell axes, V is the volume of the unit cell, $|F_{hkl}|$ is the amplitude of the structure factor (which is derived from the measured diffraction intensities), and α_{hkl} is the phase angle of the structure factor. Aside from errors in $|F_{hkl}|$ or in the derived phases, α_{hkl}, Eq. (9) will produce a three-dimensional electron density map of the time-averaged one-electron density function for the unit cell in an infinite crystal. If $|F_{hkl}|$ and α_{hkl} are well determined, the obvious temptation is to compute Eq. (9) in order to have a look at the charge density in the crystallographic unit cell.

In practice, one is restricted to the Ewald sphere, since the Bragg vector cannot exceed $4\pi/\lambda$ in magnitude, where λ is the x-ray wavelength employed in the diffraction study. Such a restriction necessarily introduces a series termination error into the electron density map. For example, the charge density for a spherically symmetrical atom will be a convolution of the true time-average charge onto a first-order spherical Bessel function divided by its argument. That is,

$$\rho_{\substack{\text{series}\\\text{terminated}}}(R) = 4\pi S_0^3 \int [\rho(|\mathbf{R}+\mathbf{r}|) j_1(S_0 r)/(S_0 r)]\, d\mathbf{r}, \tag{10}$$

where $S_0 = (4\pi \sin\theta/\lambda)_{max}$ is the radius of the Ewald sphere. Note that $j_1(u)/u$ is an oscillatory function with the first several nodes at 4.49, 7.72, 10.90, 14.07, ... The effect of this convolution is to scramble the electron density about the nuclear position. At small S_0, such as 8.15 Å^{-1}, the limit of CuK_α radiation (λ=1.5418 Å), the maximum density is markedly reduced and spread out away from the nucleus. As S_0 is increased, the density about the nucleus becomes more sharply peaked; in addition, however, artificial oscillatory features are introduced. For example at S_0 = 15 Å^{-1}, a carbon atom with a root-mean-square amplitude of vibration of 0.17 Å will appear to have a node in the electron density at about one-half angstrom from the nucleus. This is just the region where one would hope to see chemical bonding effects. A rather dramatic example of series termination error on an extensive diffraction data set [$(\sin\theta/\lambda)_{max}$ = 1.0 Å^{-1} or S_0 = 12.6 Å^{-1}] for the molecular crystal of melamine is shown in Figure 1. The quality of the density map is somewhat impaired since it was taken from one half of a stereo pair slide, kindly supplied by Don Cromer. In the region of the bonds between carbon and nitrogen atoms, there is a definite pinching in of the electron density. This is not an experimental result for the quantum chemist or the quantum statistical mechanist to be concerned about. What we see is an old-fashioned optical effect that is easily explained in terms of a finite aperture. Thus, even if $|F_{hkl}|$ and α_{hkl} are virtually error free and the data are contained in an Ewald sphere sufficiently large to expect resolution of the charge density at a distance of ∿0.4 Å, a straightforward computation of Eq. (9) is not a practical method for construction of detailed charge density maps.

In a typical structure refinement of an organic molecular crystal, the structure factor is calculated by the expression,

$$F_{hkl} = \sum_{j=1}^{N} f_j(S) \exp[-\tfrac{1}{2}\mathbf{S}^t U_j \mathbf{S}] \exp[2\pi i \mathbf{H}^t \mathbf{X}_j]. \tag{11}$$

For Eq. (11) $\mathbf{S}$ is the Bragg vector $\mathbf{S} = 2\pi\mathbf{H}$, $\mathbf{H}^t$ is the row vector (h,k,l) and the scalar $S = |\mathbf{S}| = 4\pi \sin\theta/\lambda$. The index j covers the N atoms in the unit cell. The atomic scattering factor $f_j(S)$ is the Fourier–Bessel transform of the electronic, radial density function of the isolated atom. This density function is usually derived from a spin-restricted Hartree–Fock wave function for the atom in its ground state. The structure fac-

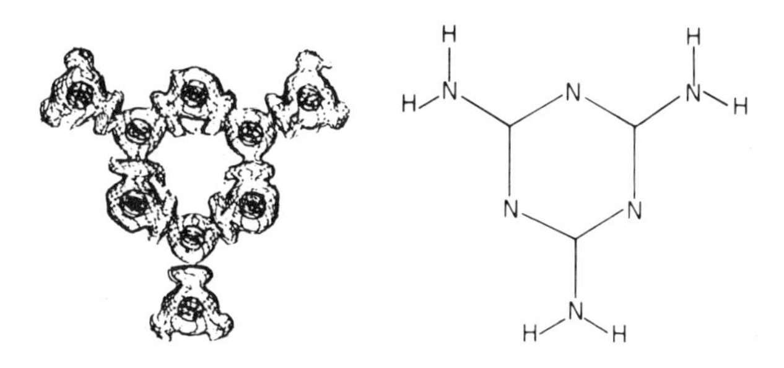

FIGURE 1 An F_0 Fourier synthesis for melamine that shows series termination error. Maximum $\sin\theta/\lambda$ is 1.0 Å^{-1}.

tor Eq. (11) is incorporated into a least squares function with parameter space spanned by $\mathbf{X}_j$ (the mean atomic positions) and U_j (the components of the symmetric tensor for the mean square amplitudes of atomic vibration). In well-determined organic molecular crystals, the least squares problem will typically include a number of "observations" (i.e., $|F_{\mathrm{hkl}}|$ or $|F_{\mathrm{hkl}}|^2$) that is at least ten times the number of atomic positions and thermal parameters.

It is crucial to point out here that Eq. (11) rests on the following assumptions: (a) the one-electron density function is a sum of spherically symmetrical Hartree–Fock atomic densities centered on the several nuclei; (b) the atomic charge densities perfectly follow the motions of the nuclei to which they are assigned; and (c) the nuclei execute harmonic motion. Departures from assumption (a) have been illustrated over the last several years with various types of Fourier difference maps. The effect of relaxation of the charge distribution during the displacement of nuclei on coherent x-ray scattering has not been studied to my knowledge. A breakdown of assumption (c) has been nicely illustrated with neutron diffraction data,[5] but the influence of anharmonic nuclear motion on charge density results has not been extensively explored. If the experimental intensities (*and their reduction to structure factors*) are sufficiently accurate to reflect significant departures from the rather simplified molecular physics in these assumptions, then one can expect the atomic positions (derived from the least squares analysis) to deviate from the true equilibrium positions; also, the thermal parameters will not represent a purely nuclear property of the crystal. The dependence of thermal parameters on atomic form factors has been reviewed elsewhere.[6] In this case, systematic errors in the mean square amplitude of vibration (particularly for hydrogen atoms) can be interpreted as due to chemical bonding. The charge density about the nucleus departs significantly from the Hartree–Fock atomic density.

When the application of Eq. (11) to a least squares analysis of x-ray structure factors has been completed, it is usual to calculate a Fourier synthesis of the difference between observed and calculated structure factors. The map is constructed by computation of Eq. (9), but now $|F_{hkl}|$ is replaced by $F^{o}_{hkl} - F^{c}_{hkl}$, where the phase of the calculated structure factor is assumed in the observed structure factor. In this case the series termination error is virtually too small to be observed. If the experimental errors are small and atomic parameters are accurate, the residual density map is a molecular bond density *convoluted onto the motion of the nuclear frame.* A molecular bond density is the difference between the true electron density and that of the isolated Hartree–Fock atoms placed at the mean nuclear positions. An extensive study of such residual density maps was reported in 1966.[7] From published crystallographic data of that period, the authors showed that peaking of electron density in the aromatic C–C bonds of five organic molecular crystals was systematic. The random error in the electron density maps was reduced by averaging over chemically equivalent bonds. The atomic parameters from the model Eq. (11), however, will refine by least squares to minimize residual densities in the unit cell.

O'Connell[8] has pointed out that residual density maps can be sharpened if the atomic parameters are based on a refinement of high order ($\sin\theta/\lambda > 0.60$ Å^{-1}) diffraction only. The rationale to this approach is the assumption that the core density [an SCF $(1s)^2$ product for first row atoms] is invariant to chemical bonding; the scattering by $(1s)^2$ for carbon is dominant in the high-angle region.[9] Difference density maps from low-temperature diffraction data reveal much larger peaks as well. An example is the study of cyanuric acid at liquid nitrogen temperature.[10] Coppens has investigated Fourier difference maps based on neutron diffraction data. In this case, the structure factor Eq. (11) is computed with atomic parameters from a least squares analysis of neutron diffraction intensities. [A scale factor parameter on the "observed" x-ray structure factors is also used to minimize $(kF^{o}_{hkl} - F^{N}_{hkl})^2$, where F^{N}_{hkl} is given by Eq. (11) with $\mathbf{X}_j$ and $\mathbf{U}_j$ derived from neutron data.] An X–N difference map is shown in Figure 2 after Coppens and Vos.[11] These authors interpret the lobes from the oxygen atom as due to lone pairs. The X–N maps show a larger contrast than corresponding maps based on x-ray data above. In the neutron case thermal parameters cannot accommodate bonding density features. The negative densities near the time average nuclear positions presumably reflect a migration of charge from the isolated atom into the bond. A similar story is reported from the neutron and x-ray diffraction measurements of tetracyanoethylene oxide.[12] The maps lend support to a bent bond picture for a

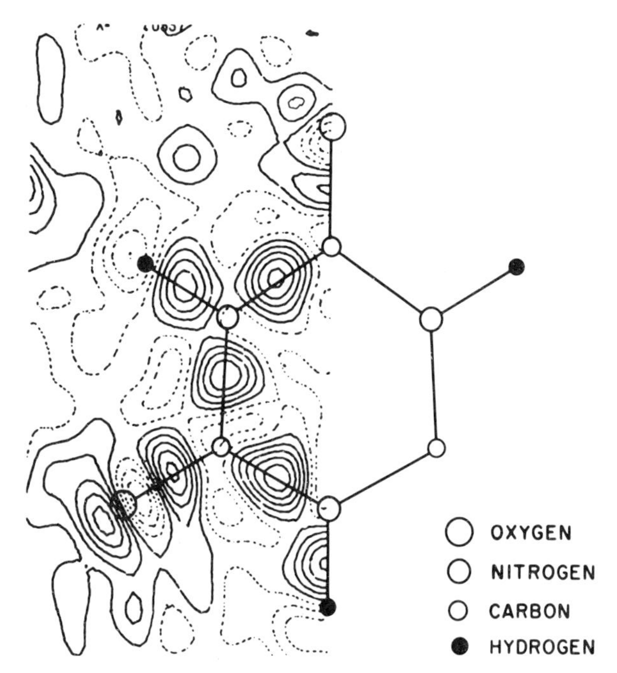

FIGURE 2 X–N map of cyanuric acid at $x = 0.25$. Contours are in units of 0.1 eÅ^{-3}; zero and negative contours are dotted lines (after Coppens and Vos[11]).

three-membered ring. A recent paper on sucrose[13] compares residual maps based on high-order ($\sin\theta/\lambda \geqslant 0.64$ Å^{-1}) x-ray data to those derived from neutron data.[14] The salient features are similar, although more pronounced in the X–N (neutron) case. Since sucrose ($C_{12}H_{22}O_{11}$) is rather rich in oxygen, I suspect these workers would find much closer agreement had they restricted the data for refinement to $\sin\theta/\lambda > 0.75$ Å^{-1}; this value is based on the L-shell-scattering amplitude of oxygen.[9]

In an effort to emphasize the valence structure of chemical bonds, valence electron density maps have been constructed.[9] In these studies the core electron density (the spin restricted Hartree–Fock ls orbital product for a first row atom) is assumed invariant to chemical bonding and is the basis of the scattering factor that is incorporated in Eq. (11). A valence density map of uracil is shown in Figure 3. Notice that contrast between O and N is clearly evident, but the density about the protons and C nuclei does not appear as a distinct maximum. For data that extend to ~ 1.0 Å^{-1} in $\sin\theta/\lambda$ and for root-mean-square amplitudes of vibration larger than 0.14 Å, these valence density maps should be free from series termination error.[15]

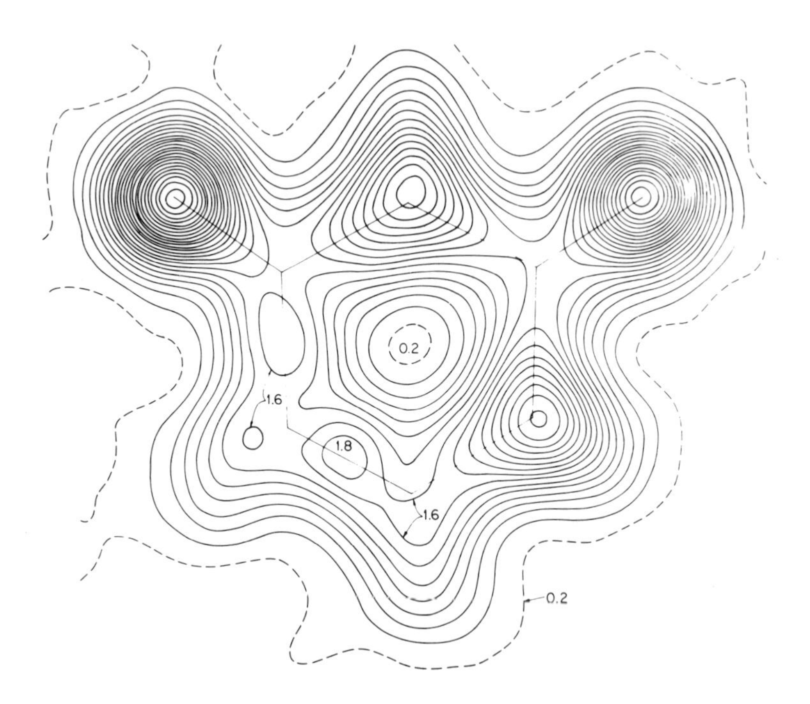

FIGURE 3 Valence Fourier difference map in the least squares plane of uracil. Contours in units of 0.2 e $Å^{-3}$. Dotted line is 0.2 e $Å^{-3}$ (after Stewart[9]).

Up to this point, I have not given a critical evaluation of charge density results. The kinds of residual density map are submitted as evidence to suggest that chemical bonding effects are measurable by present-day diffraction methods. The maps, however, are not amenable to quantitative analysis. Although the series termination error is virtually eliminated, one must consider errors introduced into the F^{c}_{hkl} values. Suppose Eq. (11) is a valid model so far as assumptions (b) and (c) are concerned, so that we seek possible departure from the isolated atomic density assumption (a). Errors in the atomic parameters, $\mathbf{X}_j$ and $\boldsymbol{U}_j$, can produce large distortions in the maps. For example, a small error in the atomic thermal parameters can lead to a very large error in the electron density map near the time-average nuclear positions.[15] But a similar error has only a marginal effect in the bonding density region. In summary then, difference density maps are useful for a qualitative exploration of chemical features, but are not useful for quantitative study. What is observed in these maps is most simply explained by a breakdown of assumption (a) in the structure factor model [Eq. (11)].

GENERALIZED X-RAY-SCATTERING FACTORS

In the above section I pointed out that the evidence from Fourier syntheses of residual electron density maps can be interpreted as a deficiency in the model for a static charge density of a molecule in a crystal. The easiest path to follow is to relax the constraints placed on an atomic scattering factor. If this route is chosen, there are several ground rules. One is that the representation for $f_j(\mathbf{S})$ [notice the use of **S** rather than S] in Eq. (11) must be rotationally invariant. The charge density function should obey classic laws of electrostatics. It is desirable, though not necessary, to use quantum chemical experience in the choice of bases to span $f_j(\mathbf{S})$.

Dawson[16] analyzed the measured structure factors for powdered diamond diffraction data with an atom deformation approach. The site symmetry of diamond is $\bar{4}3m$ (T_d) so that an atomic octapole moment is allowed. Dawson showed that the 222 reflection could arise from this third-order harmonic and pointed out that this deformation term built up charge density along the tetrahedral bonds to nearest neighbors. A fourth-order harmonic (a hexadecapole term) could account for second nearest neighbor bonds along the face diagonal of the cubic unit cell. Dawson chose to represent the monopole term as an isolated Hartree–Fock atom. Moreover, the radial function for the deformation terms was a single spherical Gaussian-type function, $r^2 \exp(-\alpha r^2)$, which *includes* the nuclear mean square amplitude of vibration, where α was an empirical fit to the data. The bonding terms are therefore a time-average representation of the charge density and are not suitable for the computation of one-electron expectation values.

Kurki-Suonio[17] developed a scattering model with greater flexibility than Dawson's. In this model surface harmonics span the angular components, and three-dimensional harmonic oscillator wave functions were proposed for a proper set of radial functions. In more recent work, he and Ruuskanen[18] define the radial function with a finite sphere of radius R, which is determined from the structure factors. The finite sphere model is used as a correction to a theoretical atomic scattering factor, such as a Hartree–Fock form factor. This proposal is rather interesting and has some merit for monatomic crystals like diamond or various simple salts. For organic molecular crystals, however, the model seems cumbersome. In addition physical properties, dependent on the true electrostatic nature of the crystal, are not easy to interpret from such artificial spheres of charge.

It has been proposed that generalized x-ray-scattering factors be used as a basis for electron population analysis.[19] In this model the core electron density function (a Hartree–Fock ls atomic orbital product for first-row atoms) is treated as an invariant to chemical bonding.[9] The valence density is then represented by an expansion of 2s and 2p atomic orbital products with the population coefficients to be determined from experimental diffraction data. The atomic orbitals are similar to those used by quantum chemists in spanning many-electron wave functions. The idea of the approach was to investigate the possible use of functions that span the first-order density matrix in a variational calculation as bases for the one-electron density function that can be extracted, perhaps, from an accurate set of diffraction data for a molecular crystal. This method has several pitfalls that have either been overlooked or ignored by workers who have attempted to apply the model to several organic molecular crystals.[20–22] I will outline these problems below.

The valence density model represents the one-electron density function for a molecule in a crystal as

$$\rho(\mathbf{r}) = 2 \sum_a [\chi_{1s}(r_a)]^2 + \sum P_{\mu\nu}\chi_\mu\chi_\nu, \tag{12}$$

where χ_μ and χ_ν are restricted to 2s and 2p atomic orbitals of first-row atoms and to ls for H atoms. χ_{1s} is a spin-restricted Hartree–Fock atomic orbital for a first-row atom. The $P_{\mu\nu}$ coefficients in Eq. (12) are adjustable parameters in a least squares treatment of x-ray diffraction data. It is to be emphasized that the orbital products $\chi_\mu\chi_\nu$ are to serve as possible basis functions that span a one-electron density function. The $P_{\mu\nu}$ are *not* elements of a first-order density matrix. The actual basis functions in an x-ray analysis are the generalized x-ray-scattering factors

$$X_{\mu\nu}(\mathbf{S},\mathbf{R}) = \int \chi_\mu\chi_\nu \exp(i\mathbf{S}\cdot\mathbf{r})d\mathbf{r}, \tag{13}$$

where the Bragg vector $|\mathbf{S}| = 4\pi \sin\theta/\lambda$ and $\mathbf{R}$ is an internuclear vector. The incorporation of Eq. (13) into the structure factor [Eq. (11)] presents several new problems. First, we need to know if the several $X_{\mu\nu}$ are linearly dependent or nearly so. Second, we must choose a set $\{X_{\mu\nu}\}$ that renders Eq. (12) rotationally invariant and satisfies the site symmetry of the atom. For the two-center case ($|\mathbf{R}| \neq 0$), our second assumption (the perfectly following density function) may have to be modified.

The one-center orbital products among the 2s and 2p atomic orbitals are

$$
\begin{aligned}
(2s)^2 &= \mathcal{R}_{ss}(r)P_0^0(\cos\theta)\\
(2p_x)^2 &= \mathcal{R}_{pp}(r)\,[(1/2)P_2^2\cos 2\phi - P_2^0 + P_0^0]\\
2p_x2p_y &= \mathcal{R}_{pp}(r)\,[(1/2)P_2^2\sin 2\phi]\\
2p_x2p_z &= \mathcal{R}_{pp}(r)P_2^1\cos\phi\\
(2p_y)^2 &= \mathcal{R}_{pp}(r)\,[-(1/2)P_2^2\cos 2\phi - P_2^0 + P_0^0]\\
2p_y2p_z &= \mathcal{R}_{pp}(r)P_2^1\sin\phi\\
(2p_z)^2 &= \mathcal{R}_{pp}(r)\,(2P_2^0 + P_0^0)\\
2s2p_x &= \mathcal{R}_{sp}(r)\sqrt{3}P_1^1\cos\phi\\
2s2p_y &= \mathcal{R}_{sp}(r)\sqrt{3}P_1^1\sin\phi\\
2s2p_z &= \mathcal{R}_{sp}(r)\sqrt{3}P_1^0\\
\text{and } \mathcal{R}_{\mu\nu} &= R_\mu(r)R_\nu(r)/4\pi
\end{aligned}
\tag{14}
$$

The radial functions in (r) are either normalized Slater-type orbitals or atomic Hartree–Fock orbitals. Note that I have explicitly written these functions in terms of surface harmonics, $P_l^m(\cos\theta)\begin{Bmatrix}\cos m\phi\\ \sin m\phi\end{Bmatrix}$, because they form a basis for the irreducible representation of the full-rotation group. Moreover, these functions are transformed by Eq. (13) into the corresponding surface harmonics of the angular components of the Bragg vector. The surface harmonics in Eq. (14) have a simple interpretation in terms of classic electrostatics: The P_0^0 is a monopole, P_1^m are components of a dipole, and P_2^m are quadrupole terms. A systematic way to apply these functions to charge density analyses is to take them in groups of their order. For an atom that has a site symmetry of 1, all three P_1^m and/or all five P_2^m must be included to ensure rotational invariance. This is assuredly not the case in Matthews *et al.* Tables 1–3.[21] These results are hopelessly dependent on the choice of the authors' coordinate system and do not allow one to critically evaluate the possible charge density information inherent in the diffraction data. For many aromatic ring compounds the atom has an approximate $\bar{6}m2$ (D_{3h}) site symmetry. In this example one would expect all dipole deformation terms to have negligible population coefficients, and only one dominant local quadrupole term. To establish this with diffraction data, one must include all three sp type products in the analysis. If the local dipole is essentially zero, one can neglect these deformation terms and analyze the data for local quadrupole deformations, but all five functions must be included. The final orientation of the local quadrupole can be a key factor in appraising the diffraction data for valence structure information.

If population coefficients are assigned to each multipole term, rather than orbital products,[23] the expansion is now in terms of orthogonal functions. [The functions centered on different centers are not orthogonal to each other.] For this case correlation coefficients should be very small. Except for $(2p_x)^2$, $(2p_y)^2$, $(2p_z)^2$, and $(2s)^2$, the orbital product population coefficient is proportional to the one assigned to a multipole. If we assign P_v to $P_0^0(\cos\theta)$, Q_1 to $P_2^2 \cos 2\phi$ and Q_5 to P_2^0, then

$$P_v = P_{xx} + P_{yy} + P_{zz} + P_{(2s)^2},$$

$$Q_1 = P_{xx} - P_{yy},$$

$$Q_5 = (1/3)[P_{zz} - (1/2)(P_{xx} + P_{yy})]. \tag{15}$$

It is assumed here that $\mathcal{R}_{ss}(r) = \mathcal{R}_{pp}(r)$. (For atomic Hartree–Fock radial orbitals, the x-ray scattering by $\mathcal{R}_{ss}$ and $\mathcal{R}_{pp}$ are almost the same.[19]) Notice in Eq. (15) that there are only three multipole coefficients but four different orbital product coefficients. The spherically symmetrical component in the $2p_\alpha 2p_\alpha$ products can not be distinguished from $(2s)^2$ with the radial functions that have been used to date. This points out the ambiguity of so-called π/σ charge ratios that have been recently reported.[21] Note from Eq. (15) that for a π/σ ratio,

$$P_{zz}/(P_{xx} + P_{yy} + P_{xy}) = \left\{(1/3)[P_v - P_{(2s)^2}] + 2Q_5\right\} \Big/ \left\{(2/3)[P_v - P_{(2s)^2}] - 2Q_5 + (1/2)Q_2\right\}, \tag{16}$$

where Q_2 is the population coefficient for $P_2^2 \sin 2\phi$. The term $P_{(2s)^2}$ is not an experimental observable (or parameter) in this treatment of x-ray data so that the π/σ charge ratio rests on the choice of $P_{(2s)^2}$ as well as the diffraction data.

In the expansion [Eq. (12)], the products $\chi_{\mu a}\chi_{\nu b}$ are formally included where χ_μ is on center a and χ_ν on center b. It is convenient to represent the 2p orbitals as linear combinations of $2p\pi$, $2p\bar{\pi}$, and $2p\sigma$, where the notation of Roothaan[24] is used. There are 15 unique two-center density functions involving 2s and 2p-type orbitals, and these can be written in terms of sin $m\phi$ and cos $m\phi$ and some functions in the confocal elliptical coordinates ξ and η. For the case studied here, m is 0, 1, or 2 so that only five basis functions span the cylindrical point groups. One question, then, is "Are all 15 functions necessary, or can the charge density analyst get by with as few as five functions to effectively represent these two-center bond-type density functions?" This depends on the nature of the atomic orbitals and the internuclear distance. A measure of the linear dependency among these density functions is the projection coefficient,

$$P_{jk} = \int \rho_j \rho_k \, d\tau \Big/ \left(\int \rho_j^2 \, d\tau \int \rho_k^2 \, d\tau \right)^{1/2}, \tag{17}$$

where the integration is over all space and j and k label the two-center atomic orbital product under study. The projections are restricted to those functions that have the same trigonometric function in ϕ. In Table 1 are shown some results for projections among $2s_a 2s_b$, $2s_a 2p\sigma_b$, $2p\sigma_b 2s_b$, $2p\sigma_a 2p\sigma_b$, and $(2p\pi_a 2p\pi_b + 2p\bar{\pi}_a 2p\bar{\pi}_b)$ for standard molecular Slater-type orbitals[25] on a C–N fragment with $R = 1.37$ Å. Note that these five diatomic fragments are almost "parallel" to each other and cannot all serve as unique basis functions in a least squares analysis. The projection coefficients in Table 1 suggest that only $2p\sigma_a 2p\sigma_b$ and $(2p\pi_a 2p\pi_b + 2p\bar{\pi}_a 2p\bar{\pi}_b)$ be chosen for a representation of the cylindrically symmetrical functions. Similar conclusions can be reached from a projection analysis of the four functions that contain sinϕ and the four with cos ϕ. It is these latter functions in sin ϕ and cos ϕ that can be employed to evaluate the justification for imposing m and/or mm symmetry on the bonding density as discussed by Jones *et al.*[22]

Of further interest is the projection of the two-center atomic orbital products into the one-center multipole functions discussed above. For multipoles on the N centers of a molecule

$$\chi_{\mu_a} \chi_{\nu_b} = (1/N) \sum_{l=0}^{\infty} \sum_{q=1}^{N} C_{lq} \rho_l(\mathbf{r}_q) \tag{18}$$

as shown by Ruedenberg.[26] Note that Eq. (18) is N-fold redundant, but serves to point out that one-center multipoles are a partial representation of two-center atomic orbitals. By restricting q to centers a and b, and

TABLE 1 Projection Coefficients among Several Two-Center Density Functions for a C–N Fragment at $R = 1.37$ Å[a]

$\rho_j \backslash \rho_k$	$2s_a 2\sigma_b$	$2\sigma_a 2s_b$	$2\sigma_a 2\sigma_b$	$2\pi_a 2\pi_b + 2\bar{\pi}_a 2\bar{\pi}_b$
$2s_a 2s_b$	0.921	0.963	0.874	0.845
$2s_a 2\sigma_b$		0.869	0.953	0.706
$2\sigma_a 2s_b$			0.904	0.751
$2\sigma_a 2\sigma_b$				0.604

[a] See Eq. (17) in text.

truncating l at 3 (an octapole term) Newton[27] has done a detailed projection analysis of the expansion [Eq. (18)]. The relative root-mean-square fits vary from 5 to 60 percent for a variety of diatomic fragments. The one-center multipoles can play an important role in their partial representation of two-center charge distributions.

The discussion above is intended to emphasize that a systematic selection of density basis functions can be accomplished by projection analysis. Also note that these x-ray population coefficients are not the elements of a density matrix from a quantum chemical calculation as has been suggested.[23] To compare the x-ray results, one can project the theoretical one-electron density function by least squares into the same basis functions that were used in the x-ray analysis.

My own criterion for evaluating charge density results is to compare the electrostatic physical properties with other experiments.[28] I tried to work up the numbers reported by Matthews *et al.*[21] for a permanent dipole moment, but the information is insufficient since the local atomic axes are not reported. A coordinate system, defined with respect to the unit cell axes, should be supplied; moreover, it is convenient for the reader if all population coefficients are rotated into this defined system. Of course, if rotational invariance is violated in the representation, then there is no sense to any of the results. I was also interested in determining the local electric field gradients on the N atoms of the tetracyanoethylene oxide molecule, but the authors have prejudged the result by arbitrarily setting three of the five local quadrupole terms to zero.

From a monopole population analysis of uracil, a dipole moment was determined.[29] The result is shown in Figure 4. Note the error in the magnitude is 32 percent; the estimated standard deviation in the angle is 12°. A similar monopole analysis for cyclotriborazane leads to a dipole moment of 3.0 D, which compares favorably with a different experimental value.[30] The direction of the moment is along the threefold axis from the plane of the B atoms toward the plane of N atoms. The largest source of error in these dipole moment results is the error in the proton positions of the hydrogen atoms.

CONCLUSION

The Fourier syntheses of various residual density maps based on x-ray and neutron diffraction measurements seem to indicate that present-day diffraction data have sufficient information to pursue quantitative charge density analysis. One route is by a least squares analysis of x-ray data with generalized x-ray-scattering factors. However, published applications of the method do not lend themselves to a critical evaluation of

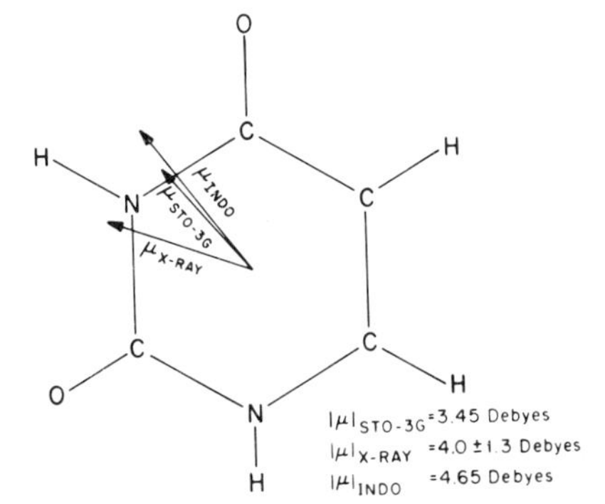

FIGURE 4 Experimental (x-ray) and theoretical dipole moments for uracil (after Stewart[29]).

charge density "results." Elementary physical principles have simply been overlooked.

This research has been supported by National Science Foundation Grant GP 22729 and the Alfred P. Sloan Foundation.

REFERENCES

1. M. Born, *Rep. Progr. Phys.*, **9**, 294 (1942–1943).
2. Bernard Borie, *Acta Cryst.*, **A26**, 533 (1970).
3. P. Coppens, *Israel J. Chem.*, **10**, 85 (1972).
4. W. H. Zachariasen, *Theory of X-Ray Diffraction in Crystals*, Wiley, New York (1945).
5. C. K. Johnson, *Acta Cryst.*, **A25**, 187 (1969).
6. R. F. Stewart and S. Hall, in *Determination of Organic Structure by Physical Method*, ed. by F. C. Nachod and J. J. Zuckerman, Academic Press, New York (1971), Vol. 3, Ch. 2.
7. A. M. O'Connell, A. I. M. Rae, and E. N. Maslen, *Acta Cryst.*, **21**, 208 (1966).
8. A. M. O'Connell, *Acta Cryst.*, **B25**, 1273 (1969).
9. R. F. Stewart, *J. Chem. Phys.*, **48**, 4882 (1968).
10. G. C. Verschoor and E. Keulen, *Acta Cryst.*, **B27**, 134 (1971).
11. P. Coppens and A. Vos, *Acta Cryst.*, **B27**, 146 (1971).
12. D. A. Matthews and G. Stucky, *J. Am. Chem. Soc.*, **93**, 5954 (1971).
13. J. C. Hanson, L. C. Sieker, and L. H. Jensen, *Acta Cryst.*, **B29**, 797 (1973).
14. G. M. Brown and H. A. Levy, *Acta Cryst.*, **B29**, 790 (1973).
15. R. F. Stewart, *Acta Cryst.*, **A24**, 497 (1968).
16. B. Dawson, *Proc. R. Soc.*, **A298**, 264 (1967).
17. K. Kurki-Suonio, *Acta Cryst.*, **A24**, 379 (1968).
18. K. Kurki-Suonio and A. Ruuskanen, *Ann. Acad. Sci. Fenn.* A VI, 358 (1971).
19. R. F. Stewart, *J. Chem. Phys.*, **51**, 4569 (1969).
20. P. Coppens, D. Pautler, and J. F. Griffin, *J. Am. Chem. Soc.*, **93**, 1051 (1971).
21. D. A. Matthews, G. D. Stucky, and P. Coppens, *J. Am. Chem. Soc.*, **94**, 8001 (1972).
22. D. S. Jones, D. Pautler, and P. Coppens, *Acta Cryst.*, **A28**, 635 (1972).

23. P. Coppens, T. V. Willoughby, and L. N. Csonka, *Acta Cryst.*, **A27**, 248 (1971).
24. C. C. J. Roothaan, *J. Chem. Phys.*, **19**, 1445 (1951).
25. W. J. Hehre, R. F. Stewart, and J. A. Pople, *J. Chem. Phys.*, **51**, 2657 (1969).
26. K. Ruedenberg, *J. Chem. Phys.*, **19**, 1433 (1951).
27. M. D. Newton, *J. Chem. Phys.*, **51**, 3917 (1969).
28. R. F. Stewart, *J. Chem. Phys.*, **57**, 1664 (1972).
29. R. F. Stewart, *J. Chem. Phys.*, **53**, 205 (1970).
30. P. W. R. Corfield and S. G. Shore, *J. Am. Chem. Soc.*, **95**, 1480 (1973).

Discussion Leader: AAFJE VOS
Reporters: HÅKON HOPE
RALPH D. NELSON JR.

Discussion

FLYGARE opened the discussion by requesting clarification concerning the status of the measurements of the anisotropy in the molecular polarizability and was informed that the measurements in question had been carried out by Bridge and had been reported in the literature. The depolarization ratio had been measured in a no-dispersion-scattering experiment with a helium–neon laser. According to BUCKINGHAM they did not look at the rotational Raman scattering, the simplifying assumption being made that the molecules could be treated as classic rotators. This remark prompted FLYGARE to suggest that one could get away from the classic quantum analog by quenching the rotation in a classic medium, i.e., a liquid, in which case there would be no rotational Raman effect, and the depolarization ratio could be calculated by hydrodynamic equations involving compressibilities. BUCKINGHAM felt that it would be difficult to correct adequately for effects connected with compressibility and interference in a liquid. He also wanted to confine his attention to isolated molecules. As a purist, he was tempted to say that there is no such thing as "*a* molecule" in a condensed phase. In a final remark FLYGARE noted that his group had measured the anisotropies of OCS and CS_2 in CCl_4, finding values close to the gas-phase results and BUCKINGHAM conceded that this was encouraging.

SCHAEFER, as a theoretician, related that he was having problems with experimentalists who wanted to know about molecular polarizabilities. He was concerned about the reliability of the data given in Hirschfelder, Curtis, and Bird (*Molecular Theory of Gases and Liquids*) and whether these data had been superseded by more recent results. BUCKINGHAM emphasized that the value of the polarizability is a function of the frequency of interest. The values in the above reference are probably for static frequencies; thus, atomic polarization is included. However, at optical frequencies the atomic polarization does not contribute, so those data have been superseded. Improved numbers have been obtained by laser techniques since 1965.

NELSON raised a question of concern to those working with condensed phases. The multipole expansion works when the distance from the point of action to the molecule is large relative to the separation of charge. When molecules approach closer than 10 Å, the multipole expansion becomes invalid and more detailed descriptions of electron distribution and bonding are needed. For polarizabilities analogous problems exist. Moments and polarizabilities derived from low-density gas measurement may not be valid in condensed states. Refraction studies indicate that the polarizability in the visible region does not change by more than about 1 percent in going from gas to condensed phase, but no extensive measurements have been made yet. The dipole moments seem to change by as much as 10 percent.

BUCKINGHAM's reply indicated that he would have liked to confine the discussion to isolated molecule properties, but the extremely interesting point of what happens to these properties when one goes to compressed gases or condensed phases did warrant discussion. He did not think there was reason for much concern over the polarizabilities. Even in water, where molecules are constantly being torn apart and recombined, the polarizability as measured by the refractivity differs by less than 1 percent from the isolated molecule value. "Polarizability really does seem to have an integrity about it. It is incorruptible."

The components of the polarizability, however, should not be expected to show the same constancy in response to external disturbances, although the information FLYGARE just provided (*vide supra*) indicates that even the components might not change too much. One should be prepared for a change by 10 percent with state.

With the dipole moments, it is necessary to consider what is actually being measured when one applies the Debye equation to the dielectric constant of a solution. One does not measure the dipole moments of single molecules, but rather the mean moment of a spherical aggregate of macroscopic dimensions embracing the single molecule carrying the dipole.

LIDE asked whether quadrupole moments derived from pressure-broadening measurements represented the same quantity as those obtained from birefringence measurements (i.e., whether they are defined in terms of the same coordinate system). BUCKINGHAM maintained that this would be the case if the lowest moment were quadrupole. LAURIE interjected that he had no faith in the reliability of pressure-broadening experiments. BUCKINGHAM agreed with LAURIE's general assessment. Pressure broadening is a dynamic property, and to determine the quadrupole moment of CO_2 by determining its effect on the line width of NH_3 involves many hazards: in particular, the assumption of a model for the intermolecular potential. Now that reliable quadrupole moments are obtainable by other methods, one can approach the intermolecular force problem more confidently and, thereby, improve the understanding of the dynamics.

WEBER was interested in the variation of the anisotropy through the visible region. BUCKINGHAM said that data were available for molecules that do not absorb in the visible range and noted the interesting observation that the anisotropy increases with frequency in all cases, although this is not required by theory. The change is typically 2–3 percent.*

WEBER further inquired about bond polarizabilities. BUCKINGHAM resisted the temptation to enter into a lengthy discussion of this important topic, but noted a few points. Bond polarizabilities give a good representation of the trace of the polarizability tensor, but do not give a very good account of the individual components. Interaction effects, as discussed earlier, can change the components without changing the trace. A bond model would be more likely to be successful for the trace than for the anisotropy.

In response to Hazony's presentation ABRAHAMS raised a question about problems that might arise from impure samples. HAZONY gave a vivid account of some of his experiences with unexpected interactions from unspecified impurities, ranging from spurious resonances to "false"-phase transitions. He felt that problems related to impurities were more widespread than was suspected by many workers and stressed the importance of extreme care in procuring pure specimens.

MOSER related an experience he had had with the use of Mössbauer spectroscopy in the determination of site geometry in SnI_2. At one time there was a hope that Mössbauer data would enable one to tell whether the Mössbauer active atoms were present in one or more site symmetries in a given structure. For Sn(II) this hope has not been realized. In SnI_2, in which tin atoms occur in two sites of unambiguously different geometry,† the Mössbauer spectrum gives no indication of the complexity of the structure. Actually, the quadrupole splitting of most Sn(II) compounds seems largely insensitive to coordination geometry and thus offers little or no help toward structure elucidation.

HAZONY was pleased to note that these results helped illustrate the most surprising aspect of the data he presented in his lecture, namely, that in the specific cases in question the quadrupole splitting is insensitive to the details of the distortion from octahedral symmetry. This makes it possible to correlate the observed QS-magnitude factors with $\langle r^{-3} \rangle$ values for fe(II) (high-spin)–Fe(III) (low-spin) compounds, whereas in crystal structure determinations of the kind Moser attempted, the method is quite insensitive.

The first question in the discussion about Stewart's presentation was asked by WILLIAMS, who was interested in optimum operating conditions for work in the charge density field. For example, would low-temperature data be useful, would there be advantages to combined x-ray and neutron data?

STEWART felt that there would be obvious advantages to low-temperature data, mentioning that the representation of the thermal motion improves with decreasing values of the thermal parameters. He also advocated the collection of diffraction data up to about 1.0–1.1 $Å^{-1}$. Positional and thermal parameters can be refined from the high-order data [$(\sin\theta)/\lambda > 0.6$ $Å^{-1}$] with the low-order data being used

*N. J. Bridge and A. D. Buckingham, *Proc. R. Soc. A*, **295**, 334 (1966).
†R. A. Howie, W. Moser, and I. C. Travena, *Acta Crystallogr., B,* **28** (10), 2965 (1972).

for valence electron distribution work. In the absence of high-order data, neutron data are very important.

TEN EYCK wanted to know the ratio of observations to variables for an extensive data set refined with "relaxed" form factors, and was informed that it would vary from a low of 5 to about 12.

FLYGARE sought details of the work on uracil. STEWART noted that the H atoms had been included in the calculations and presented the following experimental values for the numbers of valence electrons in the monopoles put on the atoms in the molecule (i.e., the L-shell populations):

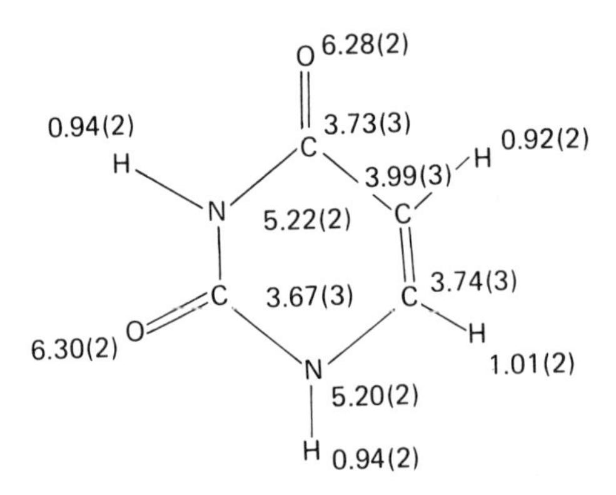

(The numbers in parentheses are estimated standard deviations in the last decimal places.) When the refinement is carried out with dipole terms included, it is rewarding to find that individual dipoles are aligned along the bond directions. For the oxygens, they point into the bond; for the nitrogens, the direction is into the ring away from the N–H bond. The carbons give virtually zero dipole moments. Also, calculations with quadrupole terms give physically realistic results. The dipole moment calculated from the x-ray results is about 3.8 D.

JOHNSON was concerned about the interpretation of the C=O results. He had in several cases found the carbonyl oxygens to show a great deal of anharmonicity in neutron diffraction. STEWART agreed that this might affect the observed local quadrupole values, but pointed out that the same analysis done on cyanuric acid had given results very similar to those obtained for uracil.

COUGHLAN thought it would be interesting to compare results from room temperature data and from low temperature (e.g., liquid N_2 temperature) data. STEWART felt that this would be a very worthwhile project. The model assumes a static charge. In principle, it should be temperature independent. It would indeed be interesting to see how insensitive the results are to changes in the Debye–Waller factors.

FLYGARE seemed quite intrigued by STEWART's results and wanted to know what would happen to the fit if the various monopoles were kept fixed at zero.

STEWART assured him that data below 0.6 $Å^{-1}$ are highly sensitive to valence electron distributions and that the idea of charge actually being transferred from one atom to another was strongly supported by diffraction data.

In connection with the discussion of the dipole moments BUCKINGHAM noted that chemists and physicists tend to use opposite sign conventions and expressed hope that a single convention would be agreed on in order to avoid confusion.

IX

Theoretical Methods

Robert Ditchfield

Studies of Molecular Properties Pertaining to Electronic Charge Distribution: A Comparison between Theory and Experiment

It is now many years since quantum mechanics was originally developed, providing a complete mathematical theory for chemical phenomena. Since that time, the actual development of techniques for extracting useful information has proceeded slowly, delayed largely by the mathematical complexity of the task of obtaining sufficiently accurate approximations to electronic wave functions. Nevertheless, some progress has been made, and in recent years there have been rapid advances in the formulation of fairly sophisticated molecular orbital methods that can be used to study electronic structure for quite large three-dimensional molecules treating all electrons equally. With the advent of such methods, the role of quantum mechanics has progressed from being purely qualitative in nature to yielding results of quantitative value.

If we neglect spin–orbit coupling and external fields, then within the Born–Oppenheimer approximation, the evaluation of the electronic energy $[E_{el}(\mathbf{R})]$ of the ground state of a molecule requires the expectation value of the Hamiltonian operator $\mathcal{H}_{el}$. Thus

$$E_{el}(\mathbf{R}) = \langle \Psi_{el}(\mathbf{r};\mathbf{R}) | \mathcal{H}_{el} | \Psi_{el}(\mathbf{r};\mathbf{R}) \rangle, \tag{1}$$

where

$$\mathcal{H}_{el} = \text{kinetic energy of electrons + potential energy of pairs of particles}$$

$$= -(1/2)\sum_j \nabla_j^2 - \sum_j \sum_N Z_N r_{jN}^{-1} + \sum_{j \neq l} \sum r_{jl}^{-1} + \sum_{M \neq N} \sum Z_M Z_N R_{MN}^{-1}. \qquad (2)$$

R and **r** denote nuclear and electronic coordinates, respectively, and Z_N is the charge on nucleus N. $\mathcal{H}_{el}$ is the usual electronic Hamiltonian for fixed positions of the nuclei, Ψ_{el} (**r**;**R**) is the many-electron ground state electronic wave function that depends parametrically on the nuclear positions, **R**, and $\sum_j$ and $\sum_N$ represent sums over electrons and nuclei, respectively. We recall from the variational principle, that improving Ψ_{el} (**r**;**R**) will result in some lowering of the energy E_{el} (**R**) such that, as Ψ_{el} (**r**;**R**) approaches the exact electronic wave function, E_{el} (**R**) approaches the experimental electronic energy.

Similarly, we can determine theoretical values for other types of first-order molecular property by evaluating expectation values of other operators, **M**, say

$$\langle M(\mathbf{R})\rangle = \langle \Psi_{el}(\mathbf{r};\mathbf{R})|\mathbf{M}|\Psi_{el}(\mathbf{r};\mathbf{R})\rangle \qquad (3)$$

However, for operators other than the Hamiltonian operator, we cannot predict how the theoretical value of a physical property will behave as Ψ_{el} (**r**;**R**) is improved.

For molecules of chemical interest it is not possible to calculate an exact many-electron wave function. As a result, we have to make certain approximations. The most commonly made approximation is the molecular orbital approximation, which is outlined in the next section. Within such a framework, it is useful to define various levels of computational method, each of which can be applied to give a unique wave function and energy for any set of nuclear positions and number of electrons. If such a model is clearly specified and if it is sufficiently simple to apply repeatedly, it can be used to generate molecular potential energy surfaces, equilibrium geometries, and other physical properties. Each such theoretical model can then be explored and the results compared in detail with experiment. If there is sufficient consistent success, some confidence can then be acquired in its predictive power. Each such level of theory therefore should be thoroughly tested and characterized before the significance of its prediction is assessed.

Using such an approach, considerable effort has been directed toward quantitatively predicting molecular structure and stabilities.[1-15] The current status of such theories for geometries of small molecules will be reviewed later. The aim of the present paper is to give a partial review of some of the commonly applied *ab initio,* single-determinant molecular orbital methods and to document their performance for calculations of physical properties. Some examples of the extent of agreement between theoretical and experimental values will be given for electric dipole moments, molecular quadrupole moments, deuterium quadrupole coupling constants, and magnetic shielding parameters.

GENERAL MOLECULAR ORBITAL THEORY

In single determinant molecular orbital methods, the many-electron wave function Ψ is approximated as a single antisymmetrized product of one-electron functions ψ_j. Thus, for a closed-shell molecule with $2n$ electrons

$$\Psi = |\psi_1\bar{\psi}_1 \ldots \psi_n\bar{\psi}_n|, \tag{4}$$

where the bar indicates that ψ_j is associated with a β-spin function. In most applications, it is usual to approximate the molecular orbitals ψ_j as linear combinations of a set of basis functions ϕ_ν

$$\psi_j = \sum_\nu c_{\nu j}\phi_\nu. \tag{5}$$

The general problem of finding the coefficients $c_{\nu j}$ in the linear combination of atomic orbitals (LCAO) by the variational method was solved by Roothaan[16] who derived the set of equations

$$\sum_\nu (F_{\mu\nu} - \epsilon_j S_{\mu\nu})c_{\nu j} = 0, \tag{6}$$

where ϵ_j are one-electron energies and $F_{\mu\nu}$ is the Fock matrix

$$F_{\mu\nu} = H_{\mu\nu} + \sum_{\lambda\sigma} P_{\lambda\sigma}\left[(\mu\nu|\lambda\sigma) - (1/2)\,(\mu\sigma|\lambda\nu)\right]. \tag{7}$$

Here $H_{\mu\nu}$ is the matrix of the one-electron Hamiltonian for motion in the field of bare nuclei, and $(\mu\nu|\lambda\sigma)$ is the two-electron integral

$$(\mu\nu|\lambda\sigma) = \int \phi_\mu^*(1)\,\phi_\nu(1)\,r_{12}^{-1}\,\phi_\lambda^*(2)\,\phi_\sigma(2)\,d\tau_1\,d\tau_2. \tag{8}$$

The density matrix $P_{\lambda\sigma}$ is defined by the sum over all occupied orbitals:

$$P_{\lambda\sigma} = 2\sum_{j}^{occ} c_{\lambda j}^{*} c_{\sigma j} \tag{9}$$

and contains all the detailed information that is implicit in the molecular orbital wave function. Since the density matrix depends on the coefficients $c\lambda_j$, Eq. (6) are not linear and have to be solved by an iterative procedure. They are frequently referred to as self-consistent field (LCAOSCF) equations. The most difficult part of LCAOSCF theory is usually the evaluation of the large number of two-electron integrals [Eq. (8)]. *Ab initio* methods evaluate all of these integrals but have to use basis functions ϕ_ν for which such integration is possible.

Once the LCAO coefficients are determined, a first-order one-electron property characterized by the operator **M** can be evaluated as indicated in Eq. (10) to Eq. (12):

$$\langle M \rangle = \langle \Psi | \mathbf{M} | \Psi \rangle \tag{10}$$

$$= 2\sum_{j}^{occ} \langle \psi_j | \mathbf{M} | \psi_j \rangle \tag{11}$$

$$= \sum_{\mu} \sum_{\nu} P_{\mu\nu} M_{\mu\nu}. \tag{12}$$

Here $P_{\mu\nu}$ is the density matrix defined above and $M_{\mu\nu}$ is given by

$$M_{\mu\nu} = \langle \phi_\mu | \mathbf{M} | \phi_\nu \rangle. \tag{13}$$

Thus, to specify the method for a given nuclear configuration, we need to prescribe the set of ϕ functions. Within the LCAO approximation, it is clear that, generally, more accurate molecular orbitals can be obtained from large basis sets, which permit increased flexibility in the representation. However, the computational effort [mainly the evaluation of $(\mu\nu|\lambda\sigma)$] increases rapidly with the number of basis functions, so applications with large sets of ϕ functions are necessarily limited to small molecules. We can arrange single determinant *ab initio* methods in an order that reflects the increasing sophistication of the basis set employed. This is shown in Table 1.

Molecular orbital theory is simplest to apply if the basis set is *minimal,* that is, consists of just enough atomic orbital type functions to describe the ground state of the corresponding atom (ls for hydrogen; 1s, 2s, $2p_x$,

TABLE 1 *ab initio* Molecular Orbital Methods

Type of Basis Set	Examples
Perfect orbitals	
Split shell + polarization	6-31G + d (6-31G*)
Split shell	4-31G, 6-31G
Minimal	STO, STO-NG

$2p_y$, and $2p_z$ for carbon, etc.). The simplest minimal set that might be used is one of Slater-type orbitals (STO),[17]

$$\phi_{1s} = (\zeta_1^3/\pi)^{1/2} \exp(-\zeta_1 r)$$

$$\phi_{2s} = (\zeta_2^5/3\pi)^{1/2}\, r \exp(-\zeta_2 r)$$

$$\phi_{2p_x} = (\zeta_2^5/\pi)^{1/2}\, x \exp(-\zeta_2 r), \tag{14}$$

with similar forms for the other 2p orbitals. The ζ parameters are scale factors that determine the size of the orbital. These may either be chosen as a standard set[18] or may be optimized by the variational method.[19] The principal difficulty in the direct application of an STO minimal basis is the evaluation of the two-electron integrals $(\mu\nu|\lambda\sigma)$. A nearly equivalent procedure is the use of a set of Slater-type functions replaced by a least squares fitted combination of N gaussians (NG). The advantage of this approach, which was first suggested by Boys[20,21] is that all two-electron integrals can be evaluated analytically. Thus we make the replacements

$$\exp(-r) \to \sum_{k=1}^{N} a_{1k} \exp(-\alpha_{1k} r^2)$$

$$r \exp(-r) \to \sum_{k=1}^{N} a_{2sk} \exp(-\alpha_{2k} r^2)$$

$$x \exp(-r) \to \sum_{k=1}^{N} a_{2pk}\, x \exp(-\alpha_{2k} r^2), \tag{15}$$

where the least squares parameters a and α are determined by techniques due to Stewart.[18,22] The resulting set of basis functions is referred to as STO-NG.[18] The value of N required to produce results close to those which would be obtained by using STO's directly depends to some extent on the property of interest. Appropriate values of N have now been determined for many types of property.[23] For molecular geometries, relative energies, electric dipole moments, and molecular quadrupole moments STO-3G is adequate and has been extensively used. For properties that derive large contributions from regions close to the nuclei, such as nuclear quadrupole-coupling constants and magnetic-shielding parameters, STO-4G or STO-5G is required.

If a larger number of ϕ functions than minimal is used, the basis set is usually described as *extended.* The first improvement is to replace each minimal basis function by a pair of ϕ_ν (inner and outer parts). Such basis sets are sometimes referred to as "double zeta" or "split shell" basis sets. At a slightly simpler level, only the valence shell functions may be split (split valence shell basis). Several basis sets of this sort, both exponential[24] and gaussian[25] have been proposed and used. We have introduced a contracted gaussian set (4-31G).[26] For example, for carbon we take

$$\phi_{1s} = \sum_{k=1}^{4} d_{1sk} \exp(-\alpha_k r^2)$$

$$\phi'_{2s} = \sum_{k=1}^{3} d'_{2sk} \exp(-\alpha'_k r^2)$$

$$\phi'_{2p_x} = \sum_{k=1}^{3} d'_{2pk}\, x \exp(-\alpha'_k r^2)$$

$$\phi''_{2s} = d''_{2s} \exp(-\alpha'' r^2)$$

$$\phi''_{2p_x} = d''_{2p}\, x \exp(-\alpha'' r^2) . \qquad (16)$$

The functions ϕ' and ϕ'' represent the inner and outer part of the valence shell in the sense that α'' is less than α'_k. For hydrogen there is no inner shell, but the valence orbital is split comparably. The d and α parameters in Eq. (16) are varied to minimize the calculated energy found by using the same basis for the atomic ground states. All valence orbitals are then rescaled in molecular calculations using a standard set of scale factors ζ' and ζ'', chosen as an average of optimized values for the ground states of some small polyatomic molecules.[26] This method has been widely applied to various properties of small organic molecules. A corresponding 6-31G set with an improved inner shell description gives very similar results.[27]

Beyond the split shell level, the next step is the addition of polarization functions with higher orbital angular quantum number (p functions on hydrogen, d functions on carbon, etc.). Hariharan and Pople[28] have recently started a systematic study of small organic molecules using a set of single gaussian polarization functions added to 6–31G. In the next section we consider applications of their 6-31G* basis, which results from the addition of a set of single d-gaussian functions to the 6-31G set for heavy atoms, with unmodified hydrogen functions. The additional six (unnormalized) functions are $(x^2, y^2, z^2, xy, yz, xz) \exp(-\alpha_d r^2)$.

The final level of basis is a very large one that approaches completion so that the molecular orbitals ψ_j become perfect. With a restricted determinantal wave function of the type given in Eq. (3), this is the Hartree–Fock limit. In practice, however, this limit can only be achieved for very small molecules, and it will be difficult to implement in an extended manner.

At any level, a calculation of first-order properties simply reduces to a determination of an expectation value using the wave function Ψ. For example, the electric dipole moment is calculated as

$$\mu_\alpha(\text{debyes}) = 2.54158\left[\sum_N Z_N (R_N)_\alpha - \sum_{\mu\nu} P_{\mu\nu} (r_\alpha)_{\mu\nu}\right], \tag{17}$$

where

$$(r_\alpha)_{\mu\nu} = \int \phi_\mu r_\alpha \phi_\nu \mathrm{d}\tau \tag{18}$$

and $\mathbf{R}_N$ is the distance vector of nucleus N from some arbitrary origin. For neutral molecules the electric dipole moment is independent of the choice of this origin. In a similar way, the molecular quadrupole moment tensor $\theta_{\alpha\beta}$ (in buckinghams)* is calculated as[29]

$$\theta_{\alpha\beta} = 1.34492(\tfrac{1}{2})\left\{\sum_N Z_N \left[3(R_N)_\alpha (R_N)_\beta - R_N^2 \delta_{\alpha\beta}\right] - \sum_{\mu\nu} P_{\mu\nu} (3r_\alpha r_\beta - r^2 \delta_{\alpha\beta})_{\mu\nu}\right\}. \tag{19}$$

It should be noted that the quadrupole moment for a neutral molecule with a permanent dipole moment is origin dependent. All values reported in this work are referred to the center of mass of the molecule. Finally, the electric field gradient at nucleus N is given (in atomic units) by

*One buckingham = 10^{-26} esu cm^2 $\approx 3.3356 \times 10^{-30}$ C m.

$$q^N_{\alpha\beta} = \left(\sum_{M \neq N} Z_M \, [3\,(R_{MN})_\alpha\,(R_{MN})_\beta - R^2_{MN}\delta_{\alpha\beta}]\, R^{-5}_{MN} \right.$$

$$\left. - \sum_{\mu\nu} P_{\mu\nu} \left\{ [3(r_N)_\alpha\,(r_N)_\beta - r_N^2\delta_{\alpha\beta}]\, r_N^{-5} \right\}^{\mu\nu} \right). \qquad (20)$$

If nucleus N has a quadrupole moment eQ_N, then there is an interaction of the nuclear quadrupole moment with the electric field gradient q^N. This is described by the nuclear quadrupole coupling constant $\chi^N = eq^N Q_N/h$, where $-e$ is the charge on the electron and h is Planck's constant. For example, for deuterium we can write

$$\chi^{\mathrm{D}}(\mathrm{Hz}) = (2.35025 \times 10^8)q^{\mathrm{D}}Q_{\mathrm{D}}, \qquad (21)$$

where the value of the deuterium quadrupole moment is taken as $Q_{\mathrm{D}} = 0.0027965$ barn.[30]

In the above equations α and β refer to directions x, y, or z. It is worth noting that the first order properties defined here are the difference between a nuclear contribution that depends solely on the molecular geometry and an electronic part that depends on Ψ.

RESULTS FOR FIRST-ORDER PROPERTIES

Electric Dipole Moments

Having set up various levels of computational method, they should then be applied extensively to molecules that are well-characterized experimentally so that the performances of the methods can be documented. In this section we present the results of some such studies for electric dipole moments. Table 2 contains results for electric dipole moments calculated at the STO-3G, 4-31G, and 6-31G* levels together with the corresponding experimental data.

It is clear from this table that there is no smooth improvement in the calculated electric dipole moments as the basis set is extended from STO-3G to 6-31G*. The minimal set of ϕ functions generally gives values that are lower than experimental. The split valence shell basis, on the other hand, somewhat overestimates electric dipole moments. In most cases, the addition of a set of single d-gaussian polarization functions to each heavy atom produces a lowering of the 4-31G values, but

TABLE 2 Electric Dipole Moments (debyes)

Molecule	STO-3G[a]	4-31G[b]	6-31G*[c]	Experiment[d]
HF	1.29	2.29	1.99	1.82
H_2O	1.69	2.52	2.19	1.85
NH_3	1.65	2.11	1.78	1.47
CH_3NH_2	1.47	1.86	1.56	1.31
CH_3OH	1.49	2.36	1.95	1.70
CH_3F	1.13	2.41	1.96	1.85
H_2CO	1.52	3.07	2.85	2.33
HCN	2.43	3.24	3.23	2.98
$CH_3CH_2CH_3$	0.03	0.07	0.08	0.084
CH_3NHCH_3	1.29	1.56	1.28	1.03
CH_3OCH_3	1.33	2.15	1.71	1.30
CH_3CHCH_2	0.25	0.32	0.29	0.37
$FCHCH_2$	0.87	2.00	1.53	1.43
CH_3CHO	1.75	3.35	3.08	2.69
FCHO	1.60	2.94	2.73	2.02
CH_3CCH	0.50	0.68	0.64	0.78
CH_3CN	3.06	4.12	4.08	3.92
FCN	1.86	1.68	2.13	2.17
CH_2[e] / CH=CH (cyclopropene)	0.55	0.54	0.57	0.45

[a] Calculated values taken from Hehre and Pople.[19]

[b] Calculated values taken from R. Ditchfield, W. J. Hehre, L. Radom, and J. A. Pople (unpublished results).

[c] Calculated values from P. C. Hariharan and J. A. Pople (unpublished data).

[d] All experimental values are taken from Nelson *et al.*[56]

[e] Calculated values taken from W. A. Lathan, L. Radom, P. C. Hariharan, W. J. Hehre, and J. A. Pople (unpublished data).

electric dipole moments at the 6-31G* level are still generally larger than experimental values.

All three levels correctly calculate the decreasing dipole moment trend in the NH_3, CH_3NH_2, $(CH_3)_2NH$ and H_2O, CH_3OH, $(CH_3)_2O$ series. Similarly, the increase in going from HCHO to CH_3CHO is handled adequately by all three methods. The dipole moments of propane, propene, and propyne are of interest since these molecules are the simplest stable polar hydrocarbons. The extended basis sets handle the dipole moments in these molecules somewhat more successfully than does STO-3G. The dipole moment in cyclopropene is calculated to have its negative end on the methylene group. From studies of the *g* values of cyclopropene and its 1,2-dideuterio derivative, Benson and Flygare[31] have found the opposite result. However, it should be noted that the experimental error involved in determining the *sign* of the electric dipole moment is quite large. The direction of the theoretical dipole appears to be mainly due to a *withdrawal* of the π electrons from the double bond into the CH_2 group.[32] It is worthy of note that this is the reverse of the usual polarity involving the interaction of saturated and unsaturated hydrocarbon fragments.

As mentioned in the previous section, the 6-31G* basis does not contain polarization functions on hydrogen. No extensive systematic study using such a basis set has been reported. However, results are available for a number of small molecules and they may be used to evaluate the effect that the addition of p functions on hydrogen has on calculated dipole moments. Neumann and Moskowitz[33,34] have obtained near Hartree–Fock wave functions for water and formaldehyde. From these studies they conclude that the inclusion of polarization functions on hydrogen results in rather small changes in the calculated dipole moment. However, it should be pointed out, that they added p functions to a split shell basis set and so their conclusions may not apply directly to the 6-31G* level of basis.

It is also worthwhile to compare the 6-31G* values with those that would be obtained if perfect molecular orbitals were used. Neumann and Moskowitz[33,34] calculated dipole moments for water and formaldehyde of 1.995 and 2.83 debyes, respectively. More recently, Dunning, Pitzer, and Aung[35] have performed a similar study for water and obtained 2.03 debyes. Near Hartree–Fock wave functions have also been reported for ammonia by Laws, Stevens, and Lipscomb.[36] Their best wave function gives a calculated dipole moment of 1.687 debyes. These limited data suggested that the 6–31G* results are probably within 0.1–0.2 debye of the Hartree–Fock limit.

It is clear from these and other results[37] that electron correlation effects can make sizable contributions to electric dipole moments. These correlation contributions appear to be of the order of 0.1–0.2 debye for first-row hydrides and somewhat larger for molecules containing more than one first-row atom.

Molecular Quadrupole Moments

The second moment of the charge distribution is a measure of the absolute size of the charge distribution in each direction. The molecular quadrupole moment, on the other hand, is a measure of the shape or deviation from spherical symmetry of the charge distribution.[29] Thus the accuracy of the expectation value of the quadrupole moment operator is a rather sensitive test of the quality of the calculated charge distribution. In Table 3 we present results for molecular quadrupole moments calculated using the minimal and the split valence shell basis sets. In general, calculated values improve as the size of the basis set is increased. Molecular quadrupole moments calculated at 4-31G are in moderately good agreement with the experimental values, the largest discrepancies appearing to be for linear molecules. The results for the three-membered

TABLE 3 Molecular Quadrupole Moments (buckinghams)[a]

Molecule	Component[b]	STO-5G[c]	4-31G[c]	Experiment
H_2O	Θ_{zz}	0.01	−0.41	−0.13[d]
	Θ_{xx}	1.33	2.60	2.63
NH_3	Θ_{11}		−2.86	−3.3 ± 0.4[e]
C_2H_6	Θ_{11}	−0.35	−0.89	−0.65[f]
$H_2C{=}CH_2$	Θ_{zz}	0.69	1.44	1.5[f]
	Θ_{xx}	0.93	1.78	
C_2H_2	Θ_{11}	4.16	6.45	8.4 ± 1.0[g]
$H_2C{=}O$	Θ_{zz}	−0.68	−0.51	−0.7 ± 0.8[h]
	Θ_{xx}	0.29	0.48	0.6 ± 1.0
HCN	Θ_{11}	0.93	1.93	3.1 ± 0.6[g]
C–C≡C	Θ_{zz}	0.65	0.71	0.6 ± 0.3[i]
	Θ_{xx}	0.74	1.96	2.9 ± 0.5
F–C≡C	Θ_{zz}		−1.12	−0.2 ± 0.2[j]
	Θ_{xx}		3.45	3.1 ± 0.2
CH_3CCH	Θ_{11}	3.50	5.59	4.82 ± 0.23[k]
FCCH	Θ_{11}		3.30	4.0 ± 0.2[l]
CH_3CN	Θ_{11}	−2.93	−2.99	−1.82 ± 1.2[m]
FCN	Θ_{11}		−5.89	−3.7 ± 1.0[n]
$H_2C{=}C{=}CH_2$	Θ_{11}	1.08	3.51	
$H_2C{=}C{=}O$	Θ_{zz}		−2.14	−0.7 ± 0.3[o]
	Θ_{xx}		3.76	3.8 ± 0.4

TABLE 3 (continued)

Molecule	Component[b]	STO-5G[c]	4-31G[c]	Experiment
O=C=O	Θ_{11}		−7.32	−4.3 ± 0.3[p]
	Θ_{yy}	2.14	2.45	
	Θ_{xx}		−4.68	−4.2 ± 0.5[q]
	Θ_{yy}		1.47	1.2 ± 0.8
	Θ_{xx}		−0.14	−0.4 ± 0.4[r]
	Θ_{yy}		−1.93	−2.0 ± 0.6

[a] 1 buckingham = 10^{-26} esu cm^2 $\approx 3.3356 \times 10^{-40}$ C m^2.

[b] The unique components of the quadrupole moment tensor are given relative to the axis system shown.

[c] Calculated values taken from R. Ditchfield, D. P. Miller, and J. A. Pople (unpublished results). These results were determined using the standard geometrical model proposed by Pople and Gordon.[39]

[d] From J. Verhoeven and A. Dymanus.[57]

[e] From S. G. Kukolich.[58]

[f] From A. D. Buckingham (private communication) quoted in Stogryn and Stogryn.[59]

[g] From Hartford *et al.*[60]

[h] From Flygare and Benson.[61]

[i] From Benson and Flygare.[62]

[j] From Rock *et al.*[63]

[k] From Shoemaker and Flygare.[64]

[l] From Shoemaker and Flygare.[65]

[m] From Pochan *et al.*[66]

[n] From Rock *et al.*[67]

[o] From Huttner *et al.*[68]

[p] From Buckingham and Disch[69] and Bridge and Buckingham.[70]

[q] From Sutter *et al.*[71]

[r] From Benson and Flygare.[31]

rings are in good agreement with the observed values. It is worth noting that the 4-31G results for the quadrupole moment of cyclopropene agree with experiment, whereas the dipole moment evaluated at 4-31G may be in conflict with what is observed.

No systematic study of molecular quadrupole moments has been made using a basis set containing polarization functions. Neumann and Moskowitz,[33,34] however, have found that inclusion of d functions on first-row atoms can produce fairly large changes in elements of the quadrupole moment tensors of water and formaldehyde. It should be noted that the addition of polarization functions leads to rather small changes in calculated second moments of the electronic charge distribution. However, these changes are magnified after applying Buckingham's[29] definition of the quadrupole moment. It is clear that more theoretical studies of molecular quadrupole moments are required to further document the performance of *ab initio* single- determinant methods for calculating this type of multipole moment.

Deuterium Quadrupole Coupling Constants

In Table 4 we present a comparison of deuterium quadrupole coupling constants calculated using the 4-31G basis set[38] with experimental values. Although in most cases calculated values are rather too large, certain experimental trends are reproduced. For example, the decreasing deuterium quadrupole coupling constant found on going from DF to CH_3D is described adequately.

These theoretical studies suggest that deuterium quadrupole coupling constants in alkanes, alkenes, and alkynes fall into three fairly distinct frequency ranges. For example, the calculated values for CH_3D, CH_3CD_3, and CD_3CHCH_2 lie between 218 and 220 kHz. In the alkenes C_2D_4, CH_3CDCH_2, and CH_3CHCD_2, the calculated results are between 229 and 231 kHz, while most alkynes have a calculated deuterium quadrupole-coupling constant that is close to 251 kHz. It should be noted, however, that the experimental data do not fall into three such clearly defined frequency regions. The values of 222 and −111 kHz reported for formaldehyde were calculated using the standard geometrical model proposed by Pople and Gordon.[39] The C–H bond length in this model is considerably shorter (1.08 Å) than the experimental value of 1.101 Å[40] The values in parentheses in Table 4 were calculated using the experimental geometry. These results illustrate the marked sensitivity of deuterium quadrupole coupling constants to the C–D bond length.[41–43] The values for CH_3D, CD_3F, and CDF_3 illustrate the effect of replacing hydrogens by a morc electronegative substituent. The resulting decrease in the deuterium quadrupole coupling constant is handled quite well at 4-31G. A striking feature of Table 4 is the formic acid result. In contrast to other molecules reported, the calculated quadrupole coupling constants for HCOOD and DCOOH are considerably *lower* than the experimental values. This point merits further study.

More sophisticated basis sets have not been used extensively to study deuterium quadrupole coupling constants. Recently, however, L. C.

TABLE 4 Deuterium Quadrupole Coupling Constants (kilohertz)

Molecule	Component[a]	4-31G[b]	Experiment
DF	$\chi_{F\text{-}D}$	412	354.24 ± 0.08[c]
HDO	χ_{xx}	353	307.95 ± 0.14[d]
	χ_{yy}	−196	−174.78 ± 0.20
	χ_{zz}	−157	−133.13 ± 0.14
NH_2D	$\chi_{N\text{-}D}$	295	290.6 ± 0.7[e]
CH_3D	$\chi_{C\text{-}D}$	220	191.48 ± 0.77[f]
CH_3CD_3	$\chi_{C\text{-}D}$	219	167 ± 12[g]
D_2CO	$\chi_{C\text{-}D}$	222(178)	171 ± 3[h]
		−111(−93)	−84 ± 3
FDCO	$\chi_{C\text{-}D}$	216	205.2 ± 4.0[i]
C_2D_4	$\chi_{C\text{-}D}$	230	–
CD_3CHCH_2	$\chi_{C\text{-}D}$	218	–
CH_3CDCH_2	$\chi_{C\text{-}D}$	229	–
CH_3CHCD_2	$\chi_{C\text{-}D}$	231	–
DCCD	$\chi_{C\text{-}D}$	251	{ 200 ± 10[j] 198 ± 7[k]
CH_3CCD	$\chi_{C\text{-}D}$	252	{ 208 ± 10[l] 199.4 ± 2.0[k]
FCCD	$\chi_{C\text{-}D}$	250	212 ± 10[l]
DCN	$\chi_{C\text{-}D}$	240	{ 199.0 ± 3.0[k] 202.18 ± 0.31[m]
CD_3CN	$\chi_{C\text{-}D}$	214	{ 171 ± 17[g] 167.5 ± 4.0[n]
CD_3F	$\chi_{C\text{-}D}$	212	–
CDF_3	$\chi_{C\text{-}D}$	197	171[i]
D_2CCO	$\chi_{C\text{-}D}$	229	240 ± 20[o]
HCOOD	$\chi_{O\text{-}D}$	340	391 ± 4[p]
DCOOH	$\chi_{C\text{-}D}$	219	249 ± 3[p]

[a] $\chi_{A\text{-}B}$ refers to the component along the bond axis A–B in the direction from A to B; χ_{xx}, etc., refers to the component in the principal field-gradient axis system.

[b] Calculated values are reported for nuclear positions specified by the standard geometrical model presented by Pople and Gordon.[39]

[c] From Muenter and Klemperer.[72]

[d] From Verhoeven *et al.*[73]

[e] From Kukolich.[74]

[f] From Wofsy *et al.*[75]

[g] From Millett and Dailey.[76]

[h] From Flygare.[77]

[i] From Kukolich.[78]

[j] From Ramsey.[79]

[k] From Caspary *et al.*[80]

[l] From Weiss and Flygare.[81]

[m] From G. Tomasevich (personal communication), quoted in Wofsy *et al.*[75]

[n] From Kukolich *et al.*[82]

[o] From Weiss and Flygare.[83]

[p] From Kukolich.[84]

Snyder (personal communication) examined some small molecules and found that the addition of polarization functions to a split shell basis set produces a considerable lowering of the calculated values.

SECOND-ORDER PROPERTIES

When a molecule is placed in an external field, its energy and wave function change in a way that reflects the perturbing influence of the external field. To determine the energy and wave function of a molecule in the total magnetic field due to a uniform external magnetic field **H** and the dipole fields arising from nuclear magnetic moments $\mu_M, \mu_N, \ldots$ situated at fixed nuclear positions $\mathbf{R}_M, \mathbf{R}_N, \ldots$ we solve the Schroedinger equation

$$\mathcal{H}'\Psi' = E'\Psi' \tag{22}$$

where

$$\mathcal{H}' = (1/2) \sum_j [-i\nabla_j + (1/c)A'(\mathbf{r}_j)]^2 - \sum_j \sum_N Z_N r_{jN}^{-1} + \sum_{j \neq l} \sum r_{jl}^{-1} + \sum_{M \neq N} Z_M Z_N R_{MN}^{-1}, \tag{23}$$

and Ψ' and E' are the wavefunction and energy, respectively, in the presence of the magnetic field. Here c is the velocity of light and the vector potential describing the total magnetic field at the position of electron j, $\mathbf{A}'(\mathbf{r}_j)$, is given by

$$\mathbf{A}'(\mathbf{r}_j) = (1/2)\mathbf{H} \times \mathbf{r}_j + \sum_M (\mu_M \times \mathbf{r}_{jM}) r_{jM}^{-3}$$

$$= \mathbf{A}(\mathbf{r}_j) + \sum_M (\mu_M \times \mathbf{r}_{jM}) r_{jM}^{-3}. \tag{24}$$

In the above equations $\mathbf{r}_j$ is the distance vector from electron j to some arbitrary origin.

Within the molecular orbital framework, the effect of the perturbation on the wave function Ψ is introduced by allowing the molecular orbitals to change. Thus we can write

$$\Psi' = |\psi_1' \overline{\psi_1'} \ldots \psi_j' \overline{\psi_j'} \ldots \psi_n' \overline{\psi_n'}|. \tag{25}$$

Within the LCAO scheme, we consider two ways of determining the perturbed molecular orbitals ψ_j'. In the first method we allow the molecular orbitals to change only via the LCAO coefficients $c_{\nu j}$. Thus we write

$$\psi_j' = \sum_\nu c_{\nu j}' \phi_\nu \tag{26}$$

and then determine $c_{\nu j}'$ using a self-consistent procedure similar to that outlined in "General Molecular Orbital Theory." However, unless complete sets of ϕ functions are used, this approach leads to calculated magnetic properties that depend on the origin with respect to which $\mathbf{r}_j$ is referred.[44] Some years ago, Stevens, Pitzer, and Lipscomb[45,46] calculated magnetic shielding constants for some diatomic molecules using this type of approach and obtained good agreement with experiment when sets of ϕ functions three to four times larger than a minimal set were employed. Since it is not possible to apply such large basis sets extensively to polyatomic molecules, we have examined[47,48] the use of minimal and split valence shell sets of ϕ functions within this approach. The results of these studies will be discussed below.

A basic defect of this type of theory is that it does not give a correct description of an isolated atom in a magnetic field. To remedy this, we follow London[49] and write each molecular orbital as a linear combination of gauge-invariant atomic orbitals* (GIAO) χ_ν,

$$\psi_j' = \sum_\nu c_{\nu j}' \chi_\nu, \tag{27}$$

where

$$\begin{aligned}\chi_\nu &= \exp\left[-(i/c)\mathbf{A}_\nu \cdot \mathbf{r}\right] \phi_\nu \\ &= f_\nu \phi_\nu .\end{aligned} \tag{28}$$

In Eq. (28), $\mathbf{A}_\nu = (1/2)\mathbf{H} \times \mathbf{R}_\nu$ is the value of the vector potential at the nuclear position $\mathbf{R}_\nu$ of atomic orbital ϕ_ν, and f_ν is an exponential gauge factor. It is clear that this approach allows *both* the LCAO coefficients and the atomic basis functions to change with the magnetic field. The important effect of using such an expansion is that calculated magnetic properties are independent of the choice of origin for any set of ϕ functions.[50]

Application of the variational principle to the energy E' leads to modi-

*This is an unfortunate phrase. The orbitals clearly do depend on the gauge via the gauge factor f_ν. Use of these functions, however, leads to calculated magnetic properties that are gauge-invariant.

fied Roothaan equations[46,48,51,52] that can be solved in a similar way to that outlined above. This leads to a perturbed wave function Ψ' and an associated density matrix $P'_{\lambda\sigma}$ that contains all the information about the diamagnetic polarization of the molecule. The electronic currents induced by the external magnetic field give rise to secondary magnetic fields at various nuclear positions in the molecule. For example, the screening field (in direction α) at nucleus N is $\sum_\beta \sigma_{N\alpha\beta} H_\beta$, where $\sigma_{N\alpha\beta}$ is a component of the magnetic shielding tensor σ_N. For either of the two approaches outlined above [Eq. (26) and (27)], these screening fields are easily calculated once $P'_{\lambda\sigma}$ is known. The methods used to determine $P'_{\lambda\sigma}$ and hence $\sigma_{N\alpha\beta}$ are presented in full elsewhere.[46,48,51,52] Thus, the total magnetic field experienced by nucleus N, which determines its NMR frequency, is given by

$$\mathbf{H}_N = \mathbf{H}(1 - \sigma_N). \tag{29}$$

In most liquid phase NMR studies only the isotropic part of the magnetic shielding tensor is obtained. In Table 5 we present calculated isotropic NMR chemical shifts, together with the corresponding experimental data for some small molecules. Calculated values obtained using minimal and split valence shell sets of *real* ϕ functions are given in the first and third columns. Values obtained with GIAO are presented in the second and fourth columns. All calculated values are for geometries in which the nuclei are fixed in the experimentally determined equilibrium configurations.[14] The theoretical results obtained with real ϕ functions [Eq. (26)] correspond to the arbitrary origin taken at the center of mass.[48]

The results in Table 5 clearly show that allowing the atomic functions to depend on the external magnetic field produces a marked improvement in the agreement between calculated and experimental values. The STO-5G results are in poor agreement with experimental in all cases. Extending the basis set to 4-31G produces a considerable improvement for ^{13}C chemical shifts but other nuclei are still poorly described. It should be noted that the agreement is still poor for the ^{13}C chemical shifts of the fluorocarbons. Although some useful qualitative information for ^{13}C shielding can be obtained from the 4-31G results,[48] the marked sensitivity of the theoretical values to the choice of origin severely limits this method. Thus at present this approach does not appear too attractive as a general theory of magnetic shielding.

The STO-5G† results are in moderately good agreement with experimental results; the improved results for protons are very noticeable.

TABLE 5 Isotropic NMR Chemical Shifts (ppm)

Molecule	Nucleus of Interest	STO-5G	STO-5G†[a]	4-31G	4-31G†[a]	Experiment[b]
C_2H_6	C	10.3	−11.8	9.0	−7.4	−8.0[c]
	H	−1.65	0.05	−2.05	−0.35	−0.75[d]
CH_3F	C	−2.5	−50.5	−41.3	−65.4	−77.5[c]
	F	65.4	49.2	99.9	74.0	76[e]
	H	0.38	−1.18	−2.62	−4.04	−4.00[d]
C_2H_4	C	−78.5	−111.6	−126.3	−130.8	−125.6[f]
	H	−14.7	−5.21	−12.6	−5.61	−5.18[d]
H_2CO	C	−116.8	−147.9	−207.0	−199.6	−197[g]
	O	−679.5	−757.6	−854.8	−858.8	−580 to −600[h]
	H	−16.6	−9.34	−14.8	−10.4	−9 to −10[i]
C_2H_2	C	−6.3	−56.9	−78.2	−75.2	−76[f]
	H	−27.8	−1.41	−20.3	−1.25	−1.35[d]
HCN	C	−55.2	−97.5	−139.2	−123.5	−120 ± 10[j]
	N	−205.4	−266.5	−306.5	−308.4	−297 ± 20[k]
	H	−28.3	−3.10	−21.5	−2.95	−2.83[d]
CH_2F_2	C	34.1	–	−33.8	−98.3	−111.3[l]
	F	95.5	–	31.3	−35.3	−57.5[e]
	H	16.8	–	11.2	−5.22	−5.22[e]
CHF_3	C	–	–	−6.4	−115.5	−118.7[l]
	F	–	–	−23.5	−108.8	−119.5[e]
	H	–	–	28.7	−6.07	−6.02[e]
CF_4	C	145.6	–	28.5	−126.2	−123.9[m]
	F	118.8	–	−54.7	−150.7	−137.6[e]

[a] † denotes that the real basis functions are modified by exponential gauge factors.

[b] ^{13}C chemical shifts are given relative to CH_4, ^{14}N shifts relative to NH_3, ^{17}O shifts relative to H_2O and ^{19}F shifts relative to HF. A negative value indicates a downfield shift. All ^{1}H shifts are given relative to CH_4.

[c] From Spiesecke and Schneider.[85]

[d] From Schneider *et al.*[86]

[e] From Frankiss.[87]

[f] From Ditchfield and Ellis.[88]

[g] From P. C. Lauterbur (personal communication), quoted in Neumann and Moskowitz.[34]

[h] Experimental range given in Emsley *et al.*[89]

[i] Experimental range given in Bovey.[90]

[j] From Appleman and Dailey.[91]

[k] From Gierke and Flygare.[92]

[l] From Ditchfield and Ellis.[93]

[m] From Motell and Maciel.[94]

This level of theory describes most experimental trends adequately but appears to underestimate many types of shielding effect. The 4-31G† results are in excellent agreement with the observed values. A number of comments can be made about individual chemical shifts. The deshielding of the methane carbon when H is replaced by CH_3 or F is well described at 4-31G†. The shielding at carbon nuclei in C–C, C=C, and C≡C is also extremely well handled by the 4-31G† level of theory, the intermediate position for the acetylenes being described very well. The calculated results for ^{19}F shielding in the fluorocarbons are also in good agreement with experiment.

The accurate calculation of ^1H NMR chemical shifts has previously proved to be extremely difficult, deviations between theory and experiment often being larger than 10 ppm. Thus the minimal and extended GIAO basis ^1H results represent a considerable improvement over previous methods. The deshielding effects of adjacent C–C, C=C, and C≡C bonds are all well described at 4-31G†. It is worth noting that the low field ^1H shift of ethane is calculated correctly only at the extended basis set level.

The methods outlined in this section calculate the full magnetic-shielding tensor σ. In Tables 6 and 7 we present 4–31G† results for magnetic-shielding anisotropies associated with first-row atoms and with protons. The calculated values are in good overall agreement with the experimental data. It is interesting to note that σ_{zz} for CH_3X compounds ($X=CH_3$, NH_2, OH, and F) remains almost constant, whereas σ_{xx} and σ_{yy} show large variations as the substituent X is changed. Experimental data for methyl carbon atoms in other types of molecule have shown similar trends.[53] The elements of the ^{19}F-shielding tensor in fluoroform are given relative to a C–F bond axis system; the *zz* element is the com-

TABLE 6 Magnetic-Shielding Anisotropies for First-Row Atoms[a] (ppm)

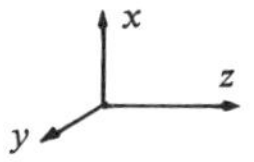	Nucleus of Interest	σ_{xx}	σ_{yy}	σ_{zz}	$\Delta\sigma(\exp)$[b,c]
C_2H_6	C	193.9	193.9	204.5	–
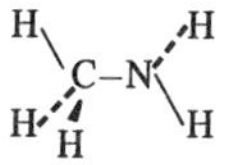	C	164.6	167.2	203.1	–
	N	285.8	232.5	256.6	–
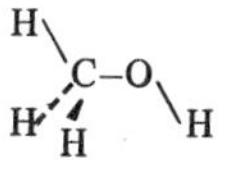	C	137.0	139.6	201.2	{ 70 ± 20
	O	400.3	318.5	303.0	63 ± 10

TABLE 6 (continued)

(axes: x, y, z)	Nucleus of Interest	σ_{xx}	σ_{yy}	σ_{zz}	$\Delta\sigma(\exp)^{b,c}$
CH_3F	C	112.4	112.4	196.5	68 ± 15
	F	509.5	509.5	438.0	−157 ± 10 −66 ± 8 −61 ± 15
$H_2C{=}CH_2$	C	−66.0	191.7	96.3	–
$H_2C{=}O$	C	−115.4	139.0	−7.9	–
	O	−644.3	447.5	−1395.2	–
C_2H_2	C	54.8	54.8	279.2	245 ± 20
HCN	C	−17.3	−17.3	278.4	282 ± 20
	N	−230.3	−230.3	338.9	577 ± 20
HF	F	376.2	376.2	482.8	102
F_2	F	−504.7	−504.7	488.4	1050 ± 50 1055
H_2O	O	355.4	308.8	320.1	–
NH_3	N	281.9	281.9	239.4	39 ± 10
$F_3C{-}H$	C	94.4	94.4	79.0	–
	F[d]	313.3	205.2	388.2	105 ± 20[e]
CF_4	F	193.5	193.5	394.1	–

[a] Calculated values were obtained using the 4-31G basis set appropriately modified by exponential gauge factors.

[b] $\Delta\sigma = \sigma_{zz} - \frac{1}{2}(\sigma_{xx} + \sigma_{yy})$.

[c] All the experimental values are taken from Appleman and Dailey.[91]

[d] These results refer to the C–F bond axis system; the z-axis points along the C–F bond and the F, C, and H atoms lie in the xz plane.

[e] This value was obtained by assuming that the ^{19}F-shielding tensor is axially symmetric about the C–F bond direction.

TABLE 7 Proton Magnetic Shielding Anisotropies (ppm)

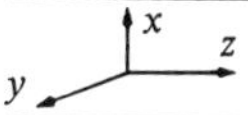 Molecule		σ_{xx}	σ_{yy}	σ_{zz}	$\Delta\sigma$(calc)[a]	$\Delta\sigma$(exp)[b]
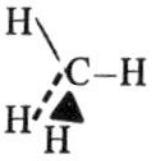 $H_3C{-}H$		29.61	29.61	38.97	9.36	
 $F_3C{-}H$		23.96	23.96	32.05	8.09	9.8 ± .5 (LC)[c]
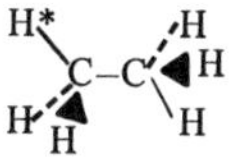 $H^*H_2C{-}CH_3$	(i)[d]	35.89	27.93	33.24	1.33	
	(ii)	30.85	27.93	38.28	8.89	
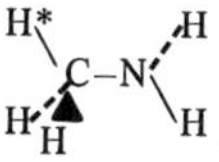 $H^*H_2C{-}NH_2$	(i)	33.40	26.87	32.75	2.61	
	(ii)	30.33	26.87	35.82	7.22	
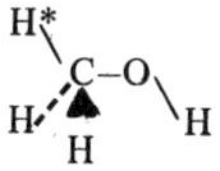 $H^*H_2C{-}OH$	(i)	31.36	25.99	31.70	3.02	
	(ii)	29.40	25.99	33.66	5.96	
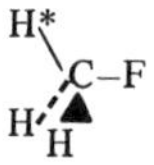 $H^*H_2C{-}F$	(i)	29.49	25.31	31.29	3.89	−0.5 ± 0.5 (LC)[c] −6.1 1.91 ± 10 (MB)[e]
	(ii)	29.20	25.31	31.57	4.31	1.8 ± 1.5(LC)[c]
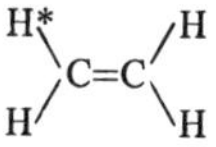 $H^*HC{=}CH_2$	(i)	23.62	27.39	30.35	4.84	
	(ii)	28.19	27.39	25.78	−2.01	
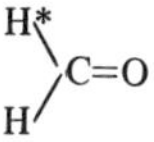 $H^*HC{=}O$	(i)	21.48	24.70	20.85	−2.24	
	(ii)	21.08	24.70	21.24	−1.65	
HC≡CH		26.76	26.76	40.91	14.15	22 ± 2 (LC)[f] 18 ± 1 (MB)[g]
HC≡N		25.08	25.08	39.19	14.07	0.0 ± 1.5 (LC)[h] −4.3 ± 1.0 (LC)[h]
HF		24.19	24.19	43.47	19.28	

[a] Calculated values were obtained using the 4-31G basis set appropriately modified by exponential gauge factors.

[b] $\Delta\sigma = \sigma_{zz} - \frac{1}{2}(\sigma_{xx} + \sigma_{yy})$.

[c] From Bernheim *et al.*[54]

[d] (i) Gives the shielding tensor elements in the molecular axis system shown; (ii) gives the shielding tensor elements in the C–H bond axis system. σ_{zz} is the component along the C–H bond, σ_{yy} is the component perpendicular to the plane of the paper.

[e] From Wofsy *et al.*[95]

[f] From Mohanty.[96]

[g] From Cederberg *et al.*[97]

[h] From Millett and Dailey.[98]

ponent along the C–F bond direction. In some studies[54] it has been assumed that the ^{19}F-shielding tensor is axially symmetric when referred to a C–F bond axis system. The calculated values presented here, however, suggest that the components of $\boldsymbol{\sigma}_F$ perpendicular to the C–F bond are far from being equal.

There are rather few experimental data for proton magnetic-shielding anisotropies. Most of the available data have been obtained from NMR in liquid crystal studies. Unfortunately, this technique appears to be beset with experimental difficulties,[55] and there are large uncertainties in proton magnetic-shielding anisotropies determined by this approach. The ^{1}H-shielding tensor has also been assumed to be axially symmetric when referred to a C–H bond axis system.[54] The results in Table 7 suggest that, in general, this is an incorrect assumption.

CONCLUSIONS

From the results presented in the previous sections we attempt to evaluate the performance of the various methods for calculations of molecular properties. Some measure of this can be obtained by considering the mean absolute deviations between theoretical and experimental values presented in Table 8.

For electric dipole moments it is clear that the agreement between theory and experiment is not improved when the basis set is extended from minimal to split valence shell. It should be noted that calculated values at the minimal level, in general, are lower than experiment, while the 4-31G results are too large. A considerable improvement is obtained when d functions are added to the first-row atoms. Limited data suggest that the 6-31G* values are close to those that would be obtained if perfect molecular orbitals were used. Sizable discrepancies, however, still remain between such Hartree–Fock values and experiment, indicating that electron correlation can make substantial contributions to electric dipole moments.

For molecular quadrupole moments the results improve steadily with improvement of the theoretical model. The calculated values at 4-31G are in moderately good agreement with experiment. Results for water and formaldehyde suggest that the addition of polarization functions could improve this agreement. Clearly, extensive studies at the 6-31G* or some comparable level are required to further document the performance of *ab initio* single determinant molecular orbital methods for calculations of this type of multipole moment.

Only split valence shell basis results have been presented for deuterium

TABLE 8 Mean Absolute Deviations[a]

Electric dipole moments (debyes)

STO-3G	4-31G	6-31G*
0.38 (19)	0.49 (19)	0.24 (19)

Molecular quadrupole moments (buckinghams)

STO-5G	4-31G
1.16 (12)	0.72 (24)

Deuterium quadrupole-coupling constants (kHz)

4-31G
31.5 (19)

Isotropic NMR chemical shifts (ppm)

	STO-5G	STO-5G†	4-31G	4-31G†
^{13}C	96.3 (8)	22.6 (6)	47.5 (9)	4.8 (9)
^{1}H	13.7 (7)	0.7 (6)	13.0 (8)	0.3 (8)

[a] Number in parentheses is the number of comparisons made.

quadrupole-coupling constants. At this level, the calculated values are generally too large by approximately 30 kHz. Preliminary studies at a level that includes polarization functions indicate that this difference between theoretical and experimental values can be reduced to about 10 kHz. More extensive theoretical work is required for this type of first-order property.

The mean absolute deviations for isotropic NMR chemical shifts show quite clearly that the atomic functions must be allowed to depend on the magnetic field if close agreement with experiment is to be obtained at the minimal or split valence shell levels. Within the GIAO framework, the results show a marked improvement as the theoretical model is improved. However, the study presented here is by no means exhaustive and further theoretical work is necessary to document more adequately the performance of these computational methods for magnetic shielding parameters.

I am grateful to Dr. P. C. Hariharan and Professor J. A. Pople for generously communicating their 6-31G* results for electric dipole moments prior to publication. I also thank Kiewit Computation Center for a generous grant of computer time.

REFERENCES

1. W. H. Fink and L. C. Allen, *J. Chem. Phys.*, **46**, 2261, 2276 (1967).
2. S. Peyerimhoff, R. J. Buenker, and L. C. Allen, *J. Chem. Phys.*, **45**, 734 (1966).
3. M. Krauss, *J. Res. Natl. Bur. Stand.*, **68A**, 635 (1964).
4. R. G. Body, D. S. McClure, and E. Clementi, *J. Chem. Phys.*, **49**, 4916 (1968).
5. R. M. Pitzer, *J. Chem. Phys.*, **46**, 4871 (1967).
6. S. Peyerimhoff, R. J. Buenker, and J. L. Whitten, *J. Chem. Phys.*, **46**, 1707 (1967).
7. R. J. Buenker and S. D. Peyerimhoff, *J. Chem. Phys.*, **48**, 354 (1968).
8. M. D. Newton, W. A. Lathan, W. J. Hehre, and J. A. Pople, *J. Chem. Phys.*, **52**, 4064 (1970).
9. W. A. Lathan, W. J. Hehre, L. A. Curtiss, and J. A. Pople, *J. Am. Chem. Soc.*, **93**, 6377 (1971).
10. W. A. Lathan, L. A. Curtiss, W. J. Hehre, J. B. Lisle, and J. A. Pople (to be published).
11. L. C. Snyder, *J. Chem. Phys.*, **46**, 3602 (1967).
12. L. C. Snyder and H. Basch, *J. Am. Chem. Soc.*, **91**, 2189 (1969).
13. R. Ditchfield, W. J. Hehre, J. A. Pople, and L. Radom, *Chem. Phys. Lett.*, **5**, 13 (1970).
14. W. J. Hehre, R. Ditchfield, L. Radom, and J. A. Pople, *J. Am. Chem. Soc.*, **92**, 4796 (1970).
15. L. Radom, W. J. Hehre, and J. A. Pople, *J. Am. Chem. Soc.*, **93**, 289 (1971).
16. C. C. J. Roothaan, *Rev. Mod. Phys.*, **23**, 69 (1951).
17. J. C. Slater, *Phys. Rev.*, **36**, 57 (1930).
18. W. J. Hehre, R. F. Stewart, and J. A. Pople, *J. Chem. Phys.*, **51**, 2657 (1969).
19. W. J. Hehre and J. A. Pople, *J. Am. Chem. Soc.*, **92**, 2191 (1970).
20. S. F. Boys, *Proc. R. Soc. (London)*, **A200**, 542 (1950).
21. J. M. Foster and S. F. Boys, *Rev. Mod. Phys.*, **32**, 303 (1960).
22. R. F. Stewart, *J. Chem. Phys.*, **50**, 2485 (1969).
23. R. Ditchfield, D. P. Miller, and J. A. Pople, *J. Chem. Phys.*, **53**, 613 (1970).
24. E. Clementi, *J. Chem. Phys.*, **39**, 1397 (1963).
25. H. Basch, M. B. Robin, and N. A. Kuebler, *J. Chem. Phys.*, **47**, 201 (1967).
26. R. Ditchfield, W. J. Hehre, and J. A. Pople, *J. Chem. Phys.*, **54**, 724 (1971).
27. W. J. Hehre, R. Ditchfield, and J. A. Pople, *J. Chem. Phys.*, **56**, 2257 (1972).
28. P. C. Hariharan and J. A. Pople, *Theor. Chim. Acta*, **28**, 213 (1973).
29. A. D. Buckingham, *Q. Rev. (London)*, **13**, 183 (1959).
30. H. Narumi and J. Watanabe, *Progr. Theor. Phys. (Kyoto)*, **35**, 1154 (1966).
31. R. C. Benson and W. H. Flygare, *J. Chem. Phys.*, **51**, 3087 (1969).
32. L. Radom, W. A. Lathan, W. J. Hehre, and J. A. Pople, *J. Am. Chem. Soc.*, **93**, 5339 (1971).
33. D. B. Neumann and J. W. Moskowitz, *J. Chem. Phys.*, **49**, 2056 (1968).
34. D. B. Neumann and J. W. Moskowitz, *J. Chem. Phys.*, **50**, 2216 (1969).
35. T. H. Dunning, R. M. Pitzer, and S. Aung, *J. Chem. Phys.*, **57**, 5044 (1972).
36. E. A. Laws, R. M. Stevens, and W. N. Lipscomb, *J. Chem. Phys.*, **56**, 2029 (1972).
37. S. Green, *J. Chem. Phys.*, **54**, 827 (1971).
38. R. Ditchfield and D. P. Miller (to be published).

39. J. A. Pople and M. S. Gordon, *J. Am. Chem. Soc.*, **89**, 4253 (1967).
40. K. Tagaki and T. Oka, *J. Phys. Soc. Japan*, **18**, 1174 (1963).
41. C. W. Kern, *J. Chem. Phys.*, **46**, 4543 (1967).
42. P. L. Olympia, I. Y. Wei, and B. M. Fung, *J. Chem. Phys.*, **51**, 1610 (1969).
43. W. C. Ermler and C. W. Kern, *J. Chem. Phys.*, **55**, 4851 (1971).
44. S. T. Epstein, *J. Chem. Phys.*, **42**, 2897 (1965).
45. R. M. Stevens, R. M. Pitzer, and W. N. Lipscomb, *J. Chem. Phys.*, **38**, 550 (1963).
46. W. N. Lipscomb, *Adv. Magn. Resonance*, **2**, 137 (1966).
47. R. Ditchfield, D. P. Miller, and J. A. Pople, *Chem. Phys. Lett.*, **6**, 573 (1970).
48. R. Ditchfield, D. P. Miller, and J. A. Pople, *J. Chem. Phys.*, **54**, 4861 (1971).
49. F. London, *J. Phys. Radium*, **8**, 397 (1937).
50. A. T. Amos and H. G. Ff. Roberts, *J. Chem. Phys.*, **50**, 2375 (1969).
51. R. Ditchfield, *J. Chem. Phys.*, **56**, 5688 (1972).
52. R. Ditchfield, *Mol. Phys.* (in press).
53. A. Pines, M. G. Gibby, and J. S. Waugh, *Chem. Phys. Lett.*, **15**, 373 (1972).
54. R. A. Bernheim, D. J. Hoy, T. R. Krugh, and B. J. Lavery, *J. Chem. Phys.*, **50**, 1350 (1969).
55. A. D. Buckingham, E. E. Burnell, and C. A. deLange, *J. Chem. Phys.*, **54**, 3242 (1971).
56. R. D. Nelson, D. R. Lide, and A. A. Maryott, *Selected Values of Electric Dipole Moments for Molecules in the Gas Phase*, National Standard Reference Series, Vol. 10, U.S. Government Printing Office, Washington, D.C. (1967).
57. J. Verhoeven and A. Dymanus, *J. Chem. Phys.*, **52**, 3222 (1970).
58. S. G. Kukolich, *Chem. Phys. Lett.*, **5**, 401 (1970); **12**, 216 (1971).
59. D. E. Stogryn and A. P. Stogryn, *Mol. Phys.*, **11**, 371 (1966).
60. S. L. Hartford, W. C. Allen, C. L. Norris, E. F. Pearson, and W. H. Flygare, *Chem. Phys. Lett.*, **18**, 573 (1973).
61. W. H. Flygare and R. C. Benson, *Mol. Phys.*, **20**, 225 (1971).
62. R. C. Benson and W. H. Flygare, *Chem. Phys. Lett.*, **4**, 141 (1969).
63. S. L. Rock, J. K. Hancock, and W. H. Flygare, *J. Chem. Phys.*, **54**, 3450 (1971).
64. R. L. Shoemaker and W. H. Flygare, *J. Am. Chem. Soc.*, **91**, 5417 (1969).
65. R. L. Shoemaker and W. H. Flygare, *Chem. Phys. Lett.*, **2**, 610 (1968).
66. J. M. Pochan, R. L. Shoemaker, R. G. Stone, and W. H. Flygare, *J. Chem. Phys.*, **52**, 2478 (1970).
67. S. L. Rock, J. C. McGurk, and W. H. Flygare, *Chem. Phys. Lett.*, **19**, 153 (1973).
68. W. Hüttner, P. D. Foster, and W. H. Flygare, *J. Chem. Phys.*, **50**, 1710 (1969).
69. A. D. Buckingham and R. L. Disch, *Proc. R. Soc.*, **A273**, 275 (1963).
70. N. J. Bridge and A. D. Buckingham, *J. Chem. Phys.*, **40**, 2733 (1964).
71. D. H. Sutter, W. Hüttner, and W. H. Flygare, *J. Chem. Phys.*, **50**, 2869 (1969).
72. J. S. Muenter and W. Klemperer, *J. Chem. Phys.*, **52**, 6033 (1970).
73. J. Verhoeven, A. Dymanus, and H. Bluyssen, *J. Chem. Phys.*, **50**, 3330 (1969).
74. S. G. Kukolich, *J. Chem. Phys.*, **49**, 5523 (1968).
75. S. C. Wofsy, J. S. Muenter, and W. Klemperer, *J. Chem. Phys.*, **53**, 4005 (1970).
76. F. S. Millett and B. P. Dailey, *J. Chem. Phys.*, **56**, 3249 (1972).
77. W. H. Flygare, *J. Chem. Phys.*, **41**, 206 (1964).
78. S. G. Kukolich, *J. Chem. Phys.*, **55**, 610 (1971).

79. N. F. Ramsey, *Am. Sci.*, **49**, 509 (1961).
80. W. J. Caspary, F. Millett, M. Reichbach, and B. P. Dailey, *J. Chem. Phys.*, **51**, 623 (1969).
81. V. W. Weiss and W. H. Flygare, *J. Chem. Phys.*, **45**, 8 (1966).
82. S. G. Kukolich, D. J. Ruben, J. H. S. Wang, and J. R. Williams, *J. Chem. Phys.*, **58**, 3155 (1973).
83. V. W. Weiss and W. H. Flygare, *J. Chem. Phys.*, **45**, 3475 (1967).
84. S. G. Kukolich, *J. Chem. Phys.*, **51**, 358 (1969).
85. H. Spiesecke and W. G. Schneider, *J. Chem. Phys.*, **35**, 722 (1961).
86. W. G. Schneider, H. J. Bernstein, and J. A. Pople, *J. Chem. Phys.*, **28**, 601 (1958).
87. S. G. Frankiss, *J. Phys. Chem.*, **67**, 752 (1963).
88. R. Ditchfield and P. D. Ellis, *Chem. Phys. Lett.*, **17**, 342 (1972).
89. J. W. Emsley, J. Feeney, and L. H. Sutcliffe, *High Resolution Nuclear Magnetic Resonance Spectroscopy*, Pergamon Press, Oxford (1966), Vol. 2, p. 1046.
90. F. A. Bovey, *Nuclear Magnetic Resonance Spectroscopy*, Academic Press, New York (1969), p. 75.
91. B. R. Appleman and B. P. Dailey, *Adv. Magn. Resonance* (in press).
92. T. D. Gierke and W. H. Flygare, *J. Am. Chem. Soc.*, **94**, 7277 (1972).
93. R. Ditchfield and P. D. Ellis, *NMR Chemical Shifts of the Fluoromethanes* (to be published).
94. E. L. Motell and G. E. Maciel, *J. Magn. Resonance*, **7**, 330 (1972).
95. S. C. Wofsy, J. S. Muenter, and W. Klemperer, *J. Chem. Phys.*, **55**, 2014 (1971).
96. S. Mohanty, *Chem. Phys. Lett.*, **18**, 581 (1973).
97. J. W. Cederberg, C. H. Anderson, and N. F. Ramsey, *Phys. Rev.*, **136A**, 960 (1964).
98. F. S. Millett and B. P. Dailey, *J. Chem. Phys.*, **54**, 5434 (1971).

Henry F. Schaefer III

Status of ab initio *Molecular Structure Predictions*

One of the most important breakthroughs of the past 5 years in theoretical chemistry has been the development of reliable quantum mechanical methods for the prediction of molecular geometries.[1] These developments were dramatized by the recent controversy concerning the structure of methylene in its triplet ground state, long thought to be linear, due to Herzberg's experimental work.[2] This conflict was resolved in favor of the theoretical predictions[3-5] by the ESR experiments of Bernheim[6] and Wasserman.[7] At present the most reliable value of the CH_2 bond angle remains a theoretical value,[8] 134°.

The thesis of the present paper is that *ab initio* calculations should be viewed as a legitimate tool for the determination of molecular structures. Theoretical geometry predictions have been made for a sufficient number of molecules, for which accurate experimental data are available, that a realistic evaluation of a variety of different *ab initio* approaches is now possible. Two factors[1] will determine the reliability of a particular *ab initio* prediction: the basis set chosen and the degree to which electron correlation is explicitly incorporated in the wave functions used.

TYPES OF BASIS SETS

Four types of basis sets (of one-electron functions) are noteworthy here. For simplicity, let us assume that Slater functions

$$A\, r^{n-1} e^{-\zeta r}$$

are used throughout, although comparable basis sets are readily constructed[9,10] in terms of gaussian functions:

1. The *minimum basis set* (MBS) includes one Slater function for each orbital in the accepted atomic electron configuration. For carbon, then, a minimum basis includes 1s, 2s, $2p_x$, $2p_y$, and $2p_z$ Slater functions.
2. The *double zeta* (DZ) basis set contains twice as many functions centered on each atom of the molecule as the minimum basis. Thus, for carbon the double zeta basis includes 1s, 1s′, 2s, 2s′, $2p_x$, $2p_x'$, $2p_y$, $2p_y'$, $2p_z$, and $2p_z'$ Slater functions.
3. The *double zeta plus polarization* (DZ + P) basis set includes one set of functions with higher l value than found in the atomic electron configuration. For carbon, this basis includes 1s, 1s′, 2s, 2s′, $2p_x$, $2p_x'$, $2p_y$, $2p_y'$, $2p_z$, $2p_z'$, $3d_{z^2}$, $3d_{x^2-y^2}$, $3d_{xy}$, $3d_{xz}$, and $3d_{yz}$ functions.
4. We refer to any basis set larger than double zeta plus polarization as an *extended basis set* (EBS).

FORM OF THE WAVE FUNCTION

Most current electronic structure calculations employ single configuration self-consistent field (SCF) wave functions. The ideal SCF wave function, obtained using a complete basis set, is the Hartree–Fock wave function for a particular system. Unfortunately, the Hartree–Fock wave function differs significantly from the exact wave function due to electron correlation, the instantaneous repulsion of pairs of electrons. Electron correlation can be taken into account using configuration interaction (CI). It is important to select configurations in a systematic manner, preferably one that depends only on the symmetry of the molecule and the size of the basis set. Two such selection procedures are the *first-order* (FO) *approach,*[11] which includes all configurations in which no more than one electron occupies an orbital beyond the valence shell; and *all single and double excitations* (S + D), in which configurations are specified with respect to the (dominant) SCF configuration. This approach includes all configurations differing by one or two orbitals from the SCF configuration.

EVALUATION OF SELF-CONSISTENT FIELD METHODS

The simplest *ab initio* procedure for which reliability has been established by an adequate number of comparisons with experiment is the floating spherical gaussian orbital (FSGO) method of Frost.[12,13] In this approach a basis set even smaller than the minimum basis is used, and the results are qualitatively reasonable, particularly for hydrocarbons.[1,12,13]

The currently most widely used method and, accordingly, that which has been most thoroughly tested is the SCF method employing the MBS. Much of the credit for this method and many of the geometry predictions to date are due to Pople and his co-workers at Carnegie-Mellon University. The results in Table 1 were obtained in this way and the accuracy is representative of that to be expected for hydrocarbons from such calculations.[14] The mean absolute deviation from experiment for bond lengths is 0.01 Å (from 23 unique bond lengths). From 17 unique bond angles, the average deviation is 0.9°. The CH bond length errors are typically less than those for CC bonds, and for carbon–carbon multiple bonds the predicted lengths are consistently about 0.03 Å too short.

For molecules other than hydrocarbons, the geometries obtained from MBS SCF calculations are usually somewhat less reliable.[15] For a set of 32 molecules, including H_2, N_2, OF_2, F_2N_2, CF_4, and C_6H_6, the average and root-mean-square (rms) errors in bond distances were 0.035 and 0.05 Å. The same errors in predicted bond angles were 1.7° and 3.4°. In general, then, we can conclude that such calculations usually provide qualitatively correct geometry predictions. The most serious known failing of the method is for the F_2O_2 molecule, where the predicted OF and OO bond lengths are 1.575 Å and 1.217 Å, compared with experimental values of[16] 1.358 and 1.392 Å.

A more reliable, but still fairly simple, level to work at is the DZ SCF level. Table 2 shows some representative results obtained in this way. There it is seen that, contrary to the MBS results, the CH bond distance errors are now greater than those for bonds between heavier atoms. For CH_3NC and CH_3CN, the CH bond distances are in error by 0.02 Å, while the CC and CN bonds have an average error of only 0.005 Å. More generally, Pople[17] finds that for a variety of simple organic molecules, SCF bond distance errors using an essentially DZ basis are only one-third to one-half as great as those resulting from the use of a minimum basis.

Since DZ SCF calculations yield more reliable geometry predictions than MBS SCF calculations, one might expect SCF calculations using even larger basis sets to yield even more accurate results. In many cases, particularly for bond distances, this is not the case. Consider, for example, the water molecule.[18] The SCF geometry predicted using an essentially DZ basis is $r(\text{OH}) = 0.951$ Å, $\theta = 112.6°$, while the DZ + P result is $r(\text{OH}) = 0.941$ Å, $\theta = 106.1°$, and experiment[19] is $r(\text{OH}) = 0.957$ Å, $\theta = 104.5°$. Here it is seen that the OH bond distance is more accurately

TABLE 1 Geometries of C_3 Hydrocarbons[a]

Molecule	Parameter	*Ab initio*[b]	Experimental
Propyne	$r(C{\equiv}C)$	1.170	1.206
	$r(C{-}C)$	1.484	1.459
	r(methyl CH)	1.088	1.105
	r(ethynyl CH)	1.064	1.056
	∠HCH	108.4	108.7
Allene	$r(C{=}C)$	1.288	1.308
	$r(C{-}H)$	1.083	1.087
	∠HCH	116.2	118.2
Cyclopropene	$r(C{=}C)$	1.277	1.300
	$r(C{-}C)$	1.493	1.515
	r(methylene CH)	1.087	1.087
	r(vinyl CH)	1.075	1.070
	∠HCH	112.5	114.7
	∠C=CH	150.3	149.9
Propene	$r(C_1{=}C_2)$	1.308	1.336
	$r(C_2{-}C_3)$	1.520	1.501
	$r(C_1{-}H_1)$	1.081	1.091
	$r(C_1{-}H_2)$	1.081	1.081
	$r(C_2{-}H_3)$	1.085	1.090
	$r(C_3{-}H_4)$	1.085	1.085
	$r(C_3{-}H_5)$	1.088	1.098
	$\angle H_1C_1C_2$	122.2	120.5
	$\angle H_2C_1C_2$	121.9	121.5
	$\angle C_1C_2C_3$	125.1	124.3
	$\angle H_3C_2C_1$	119.8	119.0
	$\angle C_2C_3H_4$	111.1	111.2
	$\angle H_4C_3H_5$	108.5	109.0
	$\angle H_5C_3H_6$	107.6	106.2
Cyclopropane	$r(C{-}C)$	1.502	1.510
	$r(C{-}H)$	1.081	1.089
	∠HCH	113.8	115.1
Propane	$r(C{-}C)$	1.541	1.526
	$r(C_1{-}H_1)$	1.086	1.091
	$r(C_1{-}H_2)$	1.086	1.091
	$r(C_2{-}H_4)$	1.089	1.096
	$\angle H_1C_1C_2$	110.7	111.2
	$\angle H_2C_1C_2$	110.7	111.2
	$\angle H_2C_1H_3$	108.2	107.7
	$\angle C_1C_2C_3$	112.4	112.4
	$\angle H_4C_2H_5$	107.2	106.1

[a] The theoretical calculations were of the self-consistent field variety, employing a minimum basis set. The distances are expressed in angstroms and the angles in degrees.

[b] From Radom and co-workers.[14]

TABLE 2 Predicted Geometries of Three Points on the Minimum Energy Path for $CH_3NC \rightarrow CH_3CN$[a]

Parameter	CH_3NC	Saddle Point	CH_3CN
θ	180° (180°)	100.8°	0° (0°)
HCX angle	110.0° (109.1)	106.2°	110.0° (109.5)
R(CH)	1.081 Å (1.101)	1.074 Å	1.082 Å (1.102)
R(CX)	1.967 Å (1.962)	1.822 Å	2.086 Å (2.081)
R(CN)	1.167 Å (1.166)	1.198 Å	1.146 Å (1.157)

[a] The theoretical geometries are from Liskow *et al.*[44] and were obtained from self-consistent field calculations using a double zeta basis set. Experimental parameters, given in parentheses, are from Costain.[45] X refers to the center of mass of the CN group.

predicted by the DZ calculation than by the one in which d functions on O and p functions on H are added.

When an EBS is used, these errors sometimes become even larger. In particular, bond distances are almost always predicted to be as much as 0.06 Å too short from SCF calculations using EBS.[1] For example, for the O_2 molecule an SCF calculation using an extended (5s, 4p, 2d, 1f) basis centered on each O atom yields a predicted bond distance r_e = 1.152 Å, more than 0.05 Å less than experiment, 1.207 Å (P. E. Cade, unpublished data). The reason for these discrepancies, of course, is that SCF wave functions, even obtained with a complete basis set, do not include the effects of electron correlation. Thus, if one has no intention of going beyond the single configuration Hartree–Fock approximation, experience suggests that the DZ SCF level may be a good stopping point for the purpose of geometry predictions.

EVALUATION OF CONFIGURATION INTERACTION METHODS

For diatomic molecules, quite a few configuration interaction calculations have been carried out using MBS.[1] In many cases, essentially full CI, which is even more extensive than the "all single and double excitations" (S + D) category described earlier, was performed. However, the results of such full CI calculations are quite comparable with the S + D level, since triple, quadruple, and higher excitations are virtually always unimportant near an equilibrium geometry. Such calculations usually result in bond lengths significantly larger than experiment. For the electronic ground states of O_2, CO, and SiO, [r_e(minimum basis, full CI) – r_e (experiment)] is 0.09, 0.11, and 0.10 Å. By using orbital exponents ζ specifically optimized for the molecule (an expensive procedure not generally followed), better results can be obtained[20]; e.g., for CO the bond distance is 0.06 Å greater than experiment. These results show, again perhaps surprisingly, that total energy and reliability of geometry predictions do not always go hand in hand. Although an MBS full CI wave function is far superior energetically to an MBS SCF wave function for the same molecule, the geometry predictions provided by the latter wave function are almost invariably the more reliable.

Again at the DZ level, the effect of CI is sometimes to make geometry predictions poorer than those at the SCF level. This is illustrated by the BH_2 and CH_2 results[21,22] seen in Table 3. In each of the four cases where the bond distance is known, the SCF result is in closer agreement with experiment than the S + D CI. However the DZ CI bond distance errors are seen to be much smaller than those expected using MBS. For bond angles, however, an opposite trend is observed, with the CI results in all cases equivalent or superior to the SCF results. A final example worth noting is the linear C_3 molecule[23] where large CI using a DZ basis yields r(CC) = 1.319 Å, as opposed to experiment,[24] 1.277 Å.

CI geometries obtained with a DZ + P basis usually come closer yet to experiment. However, even at this rather sophisticated level of theory,

TABLE 3 Predicted Geometries of the Ground and First Excited States of BH_2 and the First Three Electronic States of CH_2 [a]

Parameter		SCF	CI (all single and double excitations)	Experiment
$^2A_1$ BH_2	r(BH)	1.192	1.211	1.18[b]
	θ	128.8	129.4	131[b]
$^2B_1$ BH_2	r(BH)	1.164	1.188	1.17[b]
	θ	180	180	180[b]
$^3B_1$ CH_2	r(CH)	1.075	1.095	–
	θ	130.4	133.6	~136[c]
$^1A_1$ CH_2	r(CH)	1.103	1.133	1.11[d]
	θ	106.5	104.4	102.4[d]
$^1B_1$ CH_2	r(CH)	1.067	1.092	1.05[d]
	θ	150.5	143.8	~140[d]

[a] All calculations summarized here employed a double zeta basis set. Bond distances are given in angstroms and bond angles in degrees.
[b] From Herzberg and Johns.[46]
[c] From Wasserman *et al.*[47]
[d] From Herzberg and Johns.[48]

the results, particularly for bond distances, are sometimes less accurate than those obtained from the much simpler DZ SCF method. Table 4 illustrates these points for the HF and F_2 molecules. For both molecules the DZ SCF prediction is closer to experiment than the DZ + P FO CI result. Bond angles seem to be quite reliably predicted at the DZ + P FO CI level. Consider the ground and first excited states of the NH_2 radical.[25] For the $^2B_1$ ground state θ(DZ + P SCF) = 105.4°, θ(DZ + P FO) = 102.7°, and the experimental[26] bond angle is 103.3 ± 5°. For the $^2A_1$ state θ(DZ + P SCF) = 141.9°, θ(DZ + FO) = 144.7°, and experimentally θ = 144 ± 5°. Thus we see that the CI bond angles are in close agreement with experiment, and at this level of basis set represent an improvement over the SCF values.

CI calculations using EBS are capable of yielding quantitative molecular structure predictions. One such example is given in Table 4, where it is seen that the HF bond distance is predicted correctly to 0.003 Å using an EBS FO wave function. A second example is given by the KrF bond distance in krypton difluoride.[27] The EBS SCF value is 1.813 Å, the EBS FO value 1.907 Å, and experiment[28] 1.889 ± 0.01 Å. For bond angles, calculations at this level on the singlet states of CH_2 and H_2O suggest a reliability of 1°.[8,29] EBS S + D CI wave functions yield even more reliable results, as is seen, for example, in Table 5. For bond distances the EBS S + D CI results for six different electronic states of

TABLE 4 Predicted Bond Distances (Å) for HF and F_2 from a Range of *ab initio* Wave Functions

Basis Set	Wave Function	r_e (HF)	r_e (F_2)
Minimum	SCF	0.956[a]	1.315[a]
Double zeta	SCF	0.922[b]	1.400[c]
Double zeta	FO CI	0.951[b]	1.537[c]
Double zeta plus polarization	SCF	0.906[d]	1.344[d]
Double zeta plus polarization	FO CI	0.931[d]	1.441[d]
Extended	SCF	0.899[e]	–
Extended	FO CI	0.920[e]	–
Experiment		0.917[f]	1.417[g]

[a] From Newton *et al.*[15]
[b] From Bender *et al.*[49]
[c] From O'Neill *et al.*[50]
[d] From Bender *et al.*[51] and unpublished data.
[e] From Bondybey *et al.*[52]
[f] From Mann *et al.*[53]
[g] From Andrychuk.[54]

TABLE 5 Bond Distances r_e (Å) of the Electronic States of CH and CH^+[a]

State		*Ab initio*[b,c]	Experiment[d,e]	
CH^+	$X\ ^1\Sigma^+$	1.130	1.131	
	$a^3\Pi$	1.127	–	
	$A^1\Pi$	1.234	1.235	
			CH	CD
CH	$X^2\Pi$	1.118	1.120	1.119
	$a^4\Sigma$	1.086	–	–
	$A\ ^2\Delta$	1.102	1.102	1.103
	$B\ ^2\Sigma^-$	1.173	1.164	1.173
	$C\ ^2\Sigma^+$	1.111	1.114	1.114

[a] Extended basis sets were used in the theoretical calculations, and large configuration interaction (between 1549 and 4147 configurations) was employed.
[b] From Green *et al.*[55]
[c] From Lie *et al.*[56]
[d] From Douglas and Morton.[57]
[e] From Herzberg and Johns.[58]

CH^+ and CH differ by no more than 0.003 Å from experiment. In fact, calculations at this level are competitive with spectroscopic experiments of the highest currently attainable precision. For the excited $B\ ^2\Sigma^-$ state of CH, it is seen that the spectroscopic r_e values for CH and CD differ by 0.009 Å.

SOME RECENT PREDICTIONS

It seems clear that the ultimate goal of theoretical chemistry should not be the ceaseless generation of quantities solely to be compared with experiment. At some point we must begin to interact with experiment in a more meaningful way. Several such possibilities suggest themselves. First, theory should be used to arbitrate between conflicting interpretations of experimental data. Second, theory may be used to probe an area (which experiment suggests is interesting) where experiment is incapable of providing a complete understanding. Finally, entirely new theoretical predictions can be made. Such predictions, in turn, should suggest important new experiments. Here we describe some recent truly predictive *ab initio* molecular structure studies.

Over the past 5 years, Pimentel and co-workers[30–32] have reported the observation of the linear symmetric HCl_2, HBr_2, and HI_2 radicals and suggested that HF_2 should exist as well. These results have been challenged by Milligan and Jacox,[33] who state that the reported infrared matrix isolation spectra are due to the well-known bihalide ions HCl_2^-, HBr_2^-, and HI_2^-. Kinetic studies by Weston and co-workers[34] imply that the barrier for the exchange reaction Cl + HCl → ClH + Cl is 10 kcal/mol, a result that does not appear consistent with a bound ClHCl species. Finally, however, semiempirical BEBO calculations by Truhlar and co-workers,[35] have predicted that the neutrals HCl_2, HBr_2, and HI_2 are in fact stable. Our calculations[36] on HF_2 are summarized in Table 6. Although the four calculations yield qualitatively similar geometries, FHF is predicted to be unstable. In fact, after considering possible sources of error, we concluded that the true barrier to H atom exchange is no less than 18 kcal/mol. If this is correct, F + HF may emerge as the classic system for the study of quantum mechanical tunneling in chemical reactions. Although these results have no direct bearing on the stability of HCl_2, HBr_2, and HI_2, our intuition inclines us to believe that if HF_2 is unstable by 18 kcal/mol, then HCl_2 may also be unstable in its linear symmetric configuration.

The assignment of an emission line from the galactic sources W51 and DR21 to the HNC molecule by Snyder and Buhl[37] has aroused a great deal of interest. Metz[38] included this development in a summary of the year's "interesting, important, and exciting activities" in physics. Since HNC has heretofore been observed only in frozen inert gas matrices,[39] study of its rotational spectrum has been impossible, and little information has been available on which to make the assignment. The results of our theoretical study[40] of HNC are summarized in Table 7. First, we see in the HCN results additional evidence for the high reliability of EBS S + D geometry predictions. Thus, although the geometry of HNC is not known from experiment, we expect our predictions to be correct to ±0.003 Å. Since the line observed by Snyder and Buhl could be accounted for by a linear molecule with B_e = 45.52 GHz, the existence of HNC in interstellar space is given a fairly firm theoretical foundation. Since HNC would be the first molecule found in interstellar

TABLE 6 Predicted Geometries and Energies (Relative to F + HF) of the Linear Symmetric FHF Molecule[a]

Wave Function	r_{HF} (Å)	E(kcal/mol)
DZ SCF	1.087	53.8
DZ FO CI	1.126	21.8
DZ + P SCF	1.083	53.7
DZ + P FO CI	1.099	23.9

[a]See text for a description of the different wave functions.

TABLE 7 Geometries of HCN and HNC Predicted from EBS S+D CI Wave Functions

	HCN[a]	HNC
r_{CN} (Å)	1.153 (1.153)	1.169
r_{HC} (Å)	1.068 (1.066)	–
r_{HN} (Å)	–	0.995
Rotational constant		
B_e (GHz)	44.53 (44.51)	45.43

[a] Experimental values for HCN are given in parentheses.

space without a relatively stable counterpart in the laboratory, we are now beginning a comprehensive study of the HNC → HCN isomerization.

Some other molecular structure problems of recent interest to us have included the geometry of LiO_2 in its ground and first excited states,[41] the geometries of FeF_3 and the other first-row transition metal trihalides,[42] and the question of the existence or nonexistence of the XeF radical.[43]

Much of the research reviewed here was carried out by my colleagues Charles F. Bender, Stephen V. O'Neil, Peter K. Pearson, and Dean H. Liskow. Discussions with them have been of inestimable value in the formulation of my ideas on the subject of molecular structure.

Author is an Alfred P. Sloan Fellow. Research Supported by the National Science Foundation, Grant GP-31974.

REFERENCES

1. H. F. Schaefer, *The Electronic Structure of Atoms and Molecules. A Survey of Rigorous Quantum Mechanical Results,* Addison-Wesley, Reading, Mass. (1972).
2. G. Herzberg, *Proc. R. Soc. (London),* **A262**, 291 (1961).
3. J. M. Foster and S. F. Boys, *Rev. Mod. Phys.,* **32**, 305 (1960).
4. J. F. Harrison and L. C. Allen, *J. Am. Chem. Soc.,* **91**, 807 (1969).
5. C. F. Bender and H. F. Schaefer, *J. Am. Chem. Soc.,* **92**, 4984 (1970).
6. R. A. Bernheim, H. W. Bernard, P. S. Wang, L. S. Wood, and P. S. Skell, *J. Chem. Phys.,* **53**, 1280 (1970).
7. E. Wasserman, V. J. Kuck, R. S. Hutton, and W. A. Yager, *J. Am. Chem. Soc.,* **92**, 7491 (1970).
8. D. R. McLaughlin, C. F. Bender, and H. F. Schaefer, *Theor. Chem. Acta,* **25**, 352 (1972).
9. W. J. Hehre, R. F. Stewart, and J. A. Pople, *J. Chem. Phys.,* **51**, 2657 (1969).
10. T. H. Dunning, *J. Chem. Phys.,* **53**, 2823 (1970).
11. H. F. Schaefer, *J. Chem. Phys.,* **54**, 2207 (1971).
12. A. A. Frost, *J. Chem. Phys.,* **47**, 3707 (1967).
13. A. A. Frost and R. A. Rouse, *J. Am. Chem. Soc.,* **90**, 1965 (1968).

14. L. Radom, W. A. Lathan, W. J. Hehre, and J. A. Pople, *J. Am. Chem. Soc.*, **93**, 5339 (1971).
15. M. D. Newton, W. A. Lathan, W. J. Hehre, and J. A. Pople, *J. Chem. Phys.*, **52**, 4064 (1970).
16. R. H. Jackson, *J. Chem. Soc.*, 4585 (1962).
17. J. A. Pople, in *Computational Methods for Large Molecules and Localized States in Solids*, ed. by F. Herman, A. D. McLean, and R. K. Nesbet, Plenum Press, New York (1973), p. 11–22.
18. T. H. Dunning, R. M. Pitzer, and S. Aung, *J. Chem. Phys.*, **57**, 5044 (1972).
19. W. S. Benedict, N. Gailar, and E. K. Plylar, *J. Chem. Phys.*, **24**, 1139 (1956).
20. S. V. O'Neil and H. F. Schaefer, *J. Chem. Phys.*, **53**, 3994 (1970).
21. C. F. Bender and H. F. Schaefer, *J. Mol. Spectrosc.* **37**, 423 (1971).
22. S. V. O'Neil, H. F. Schaefer, and C. F. Bender, *J. Chem. Phys.*, **55**, 162 (1971).
23. D. H. Liskow, C. F. Bender, and H. F. Schaefer, *J. Chem. Phys.*, **56**, 5075 (1972).
24. L. Gausset, G. Herzberg, A. Lagerquist, and A. Rosen, *Astrophys. J.*, **142**, 45 (1965).
25. C. F. Bender and H. F. Schaefer, *J. Chem. Phys.*, **55**, 4798 (1971).
26. K. Dressler and D. A. Ramsay, *Philos. Trans. R. Soc. London Ser. A*, **251**, 553 (1959).
27. P. S. Bagus, B. Liu, and H. F. Schaefer, *J. Am. Chem. Soc.*, **94**, 6635 (1972).
28. C. Murchison, S. Reichman, D. Anderson, J. Overend, and F. Schreiner, *J. Am. Chem. Soc.*, **90**, 5690 (1968).
29. C. F. Bender, H. F. Schaefer, D. R. Franceschetti, and L. C. Allen, *J. Am. Chem. Soc.*, **94**, 6888 (1972).
30. P. N. Noble and G. C. Pimentel, *J. Chem. Phys.*, **49**, 3165 (1968).
31. V. Bondybey, G. C. Pimentel, and P. N. Noble, *J. Chem. Phys.*, **55**, 540 (1971).
32. P. N. Noble, *J. Chem. Phys.*, **56**, 2088 (1972).
33. D. E. Milligan and M. E. Jacox, *J. Chem. Phys.*, **53**, 2034 (1970); **55**, 2550 (1971).
34. F. S. Klein, A. Persky, and R. E. Weston, *J. Chem. Phys.*, **41**, 1799 (1966).
35. D. G. Truhlar, P. C. Olson, and C. A. Parr, *J. Chem. Phys.*, **57**, 4479 (1972).
36. S. V. O'Neil, H. F. Schaefer, and C. F. Bender, *Proc. Natl. Acad. Sci., USA* (in press).
37. L. E. Snyder and D. Buhl, *Bull. Am. Astr. Soc.*, **3**, 388 (1971).
38. W. D. Metz, *Science*, **179**, 670 (1973).
39. D. E. Milligan and M. E. Jacox, *J. Chem. Phys.*, **39**, 712 (1963); **47**, 278 (1967).
40. P. K. Pearson, G. L. Blackman, H. F. Schaefer, B. Roos, and U. Wahlgren, *Astrophys. J. Lett.*, **184**, L 19 (1973).
41. S. V. O'Neil, H. F. Schaefer, and C. F. Bender, *J. Chem. Phys.*, **59**, 3608 (1973).
42. R. W. Hand, W. J. Hunt, and H. F. Schaefer, *J. Am. Chem. Soc.*, **95**, 4517 (1973).
43. D. H. Liskow, H. F. Schaefer, P. S. Bagus, and B. Liu, *J. Am. Chem. Soc.*, **95**, 4056 (1973).
44. D. H. Liskow, C. F. Bender, and H. F. Schaefer, *J. Chem. Phys.*, **57**, 4509 (1972).
45. C. Costain, *J. Chem. Phys.*, **29**, 864 (1958).

46. G. Herzberg and J. W. C. Johns, *Proc. R. Soc. (London),* **A298**, 142 (1967).
47. E. Wasserman, V. J. Kuck, R. S. Hutton, E. D. Anderson, and W. A. Yager, *J. Chem. Phys.,* **54**, 4120 (1971).
48. G. Herzberg and J. W. C. Johns, *Proc. R. Soc. (London),* **A295**, 107 (1966).
49. C. F. Bender, P. K. Pearson, S. V. O'Neil, and H. F. Schaefer, *J. Chem. Phys.,* **56**, 4626 (1972).
50. S. V. O'Neil, H. F. Schaefer, and C. F. Bender, *J. Chem. Phys.,* **58**, 1126 (1973).
51. C. F. Bender, S. V. O'Neil, P. K. Pearson, and H. F. Schaefer, *Science,* **176**, 1412 (1972).
52. V. Bondybey, P. K. Pearson, and H. F. Schaefer, *J. Chem. Phys.,* **57**, 1123 (1972).
53. D. E. Mann, B. A. Thrush, D. R. Lide, J. J. Ball, and N. Acquista, *J. Chem. Phys.,* **34**, 420 (1961).
54. D. Andrychuk, *Can. J. Phys.,* **29**, 151 (1951).
55. S. Green, P. S. Bagus, B. Liu, A. D. McLean, and M. Yoshimine, *Phys. Rev. A,* **5**, 1614 (1972).
56. G. C. Lie, J. Hinze, and B. Liu, *J. Chem. Phys.,* **59**, 1887 (1973).
57. A. E. Douglas and J. R. Morton, *Astrophys. J.,* **131**, 1 (1960).
58. G. Herzberg and J. W. C. Johns, *Astrophys. J.,* **158**, 399 (1969).

Discussion Leader: WALTER H. STOCKMAYER

Reporters: HENRI A. LEVY
C. WELDON MATHEWS

Discussion

In reply to a query from DREIZLER, DITCHFIELD stated that Lipscomb and his group* have calculated magnetic susceptibilities for some diatomic molecules. They did not allow the atomic orbitals to depend on the magnetic field but used a large set of ϕ functions so that gauge effects were small. Their results are in good agreement with experiment. If one tries to calculate susceptibilities at the split valence shell level using the same kind of approach, the results are disappointing and there are large deviations between theory and experiment. DITCHFIELD has some preliminary results for diatomic molecules that suggest that allowing the split valence shell atomic orbitals to depend on the magnetic field leads to rather large improvements in the calculated magnetic susceptibilities.

MILLS remarked that no mention had been made of any calculations of the gradient of the dipole moment on the nuclear potential energy surface and asked whether magnetic properties, in general, are easier to calculate than dipole moments or their derivatives. DITCHFIELD responded that as far as he is aware, these kinds of methods have not been extensively used to calculate dipole moment derivatives. Thus he thinks insufficient theoretical data are available to make a good judgment of the reliability of these methods for evaluating dipole moment derivatives.

RAMSAY asked whether there is a microwave value for the dipole moment of cyclopropene. FLYGARE responded that there is a Stark effect moment whose absolute value is accurate to about 1 percent. His work did *not* determine the *sign* of the moment with that accuracy, but reptition yielded the same sign. As

* *Adv. Magn. Resonance,* **2,** 137 (1966).

DREIZLER pointed out, "some moments calculated to be zero aren't coming out zero." If measured quadrupole moments agree with the theoretical ones to within 5 percent, then certainly the dipole moment ought also to agree in sign with the theoretical value, since it has to be consistent with the observed isotope effect on the quadrupole moment.

SNYDER stated that his group had investigated the effect of polarization terms in the basis set on the calculation of quadrupole coupling constants in H_2 and CH_4. These terms decrease the calculated values by about 10 percent, thus lowering the difference from the experimental values from about 30 kHz to about 10 kHz, leaving a residual difference of the order of 5 percent.

BUCKINGHAM commented that in his recollection *all* dipole moment calculations in the Hartree–Fock limit lead to values larger than experimental results, but observed that in the table presented by Ditchfield, though most indeed were larger, several of the extended-basis moments for diatomic molecules were smaller. DITCHFIELD responded that for diatomic molecules the inclusion of correlation effects does lower the calculated dipole moment. Perhaps this lowering is due to a reduction of the somewhat exaggerated ionic components of the single-determinant molecular orbital wave function. The 6-31G* values for the simple hydrocarbons propane, propene, and propyne are in fact lower than experiment. However, for polyatomic molecules, where one has several "bond" dipoles, it is not clear what the inclusion of electronic correlation effects will do. It is also possible that the 6-31G* results may be 0.1–0.2 debye from the Hartree–Fock limit. If this amount were added to the 6-31G* values, they would be larger than experiment. Clearly, more work is required on this problem.

BUCKINGHAM stated his interest that Ditchfield obtained the correct sign of the ^{19}F chemical shift anisotropy in methyl fluoride, for on simple intuitive grounds one might have expected a sign opposite to that observed. DITCHFIELD replied that he was also somewhat surprised at the sign of the ^{19}F magnetic-shielding anisotropy in CH_3F. He recalled some work that Cornwell had published* on the ClF molecule. Cornwell found, as expected, that magnetic-field-induced mixing of occupied bonding molecular orbitals with unoccupied antibonding ones gave negative contributions to the paramagnetic component of the magnetic shielding tensor σ^p. However, the mixing of an occupied *antibonding* orbital with an unoccupied antibonding orbital leads to positive contributions to σ^p. In CH_3F a pair of the highest energy occupied orbitals are C–F antibonding. Mixing these orbitals (by the magnetic field) with unoccupied C–F antibonding orbitals leads to a positive contribution to σ^p_F. From symmetry considerations, it can be shown that this mixing can occur only when the magnetic field is perpendicular to the threefold symmetry axis. This appears to be the main reason why the fluorine nucleus is more shielded perpendicular to the C–F bond.

FLYGARE observed that Lipscomb† had mentioned some SCF calculations of magnetic susceptibilities using atomic functions without any modifications such as

* *J. Chem. Phys.*, **44**, 874 (1966).
† *Adv. Magn. Resonance*, **2**, 137 (1966).

Eq. (27) of Ditchfield's paper. The results for diatomic molecules were rather good, but they were quite poor for formaldehyde. He inquired whether this situation would be improved by using larger basis sets, or if a gauge factor would be needed to get reliable values. DITCHFIELD thought that the calculated magnetic susceptibilities would be improved by using larger basis sets. However, the size of the basis set one would have to use, even for formaldehyde, to eliminate gauge effects would be very large. Thus he felt that this type of approach did not appear too attractive as a general theory of magnetism in diamagnetic molecules. The introduction of gauge factors is simply a good way of calculating gauge-invariant magnetic properties with any type of basis set. It is apparent that magnetic shielding constants are calculated rather well by this model.

BOTHNER-BY remarked that the accomplishment in obtaining greater accuracy in the calculation of chemical shifts is really impressive and astonishing. At the same time, to make NMR a really useful tool for investigation of bond distances and angles, calculations of ^{13}C shifts to approximately 0.5 ppm or better will be needed, say, within the experimental range for a methyl group. He inquired what the prospects are for another 10- to 50-fold increase in precision. DITCHFIELD replied that there are shielding effects of the order of 10 ppm or more, for example, the "β effect," that are still not well understood. For such cases, the quantitative agreement between theory and experiment gives some hope that an analysis of the theoretical results will improve our understanding of these effects. Clearly, care will have to be taken in interpreting shifts of less than 5 ppm using the present level of theory. He would expect that agreement between theory and experiment may be further improved by augmenting the level of basis set used. However, he did not know if there would be a further tenfold increase in precision.

KUKOLICH wished to support Ditchfield's view that the fluorine-shielding tensor is not axially symmetric about the C–F bond direction. He has recently measured fluorine-shielding tensor components in CH_2F_2 and found large deviations from axial symmetry about the bond. He also found that the shielding is considerably larger in the FCF plane than in a direction perpendicular to that plane.

SUTTON asked whether it is possible, when the calculations for electric dipole and quadrupole moments have been done, to give electron charge density maps of molecules. These would be very useful to chemists, especially if the charge density due to core electrons were subtracted. DITCHFIELD stated that it is possible to do this, and several theoretical workers have reported such charge density plots. The results, in fact, are often quite informative. His group has attempted to get some idea of the charge distribution by using a Mulliken population analysis to determine the charges associated with the atoms. However, because of certain arbitrary features in the Mulliken scheme, a charge density plot is probably a more satisfactory way of looking at molecular charge distribution. A recent compilation of such electron density plots has been published by Streitwieser and Owens.*

SHOEMAKER asked SCHAEFER if calculations had been made for planar but polar (i.e., T-shaped) molecules, for example, FeF_3, which had been suggested

* *Orbital and Electron Density Diagrams,* The Macmillan Company, New York (1973).

earlier. SCHAEFER indicated that these calculations have not yet been carried out and that the use of two structural parameters rather than one would be considerably more expensive.

DITCHFIELD asked about the number and types of molecules involved in each of the categories of Schaefer's table of reliabilities of interatomic distances, especially for the higher quality calculations. SCHAEFER felt that the small number of examples using extended basis set calculations with all single and double excitations makes this portion of his table particularly speculative. Most of these studies have been of diatomic molecules along with a few triatomic molecules. The CH calculations, as a particularly good example, include results for about seven electronic states. The estimates for SCF calculations should be valid as they are based on a large number of comparisons. DITCHFIELD further inquired about the effects on the bond lengths of including correlation, since lengths tend to be too short at the Hartree–Fock limit. As an example, SCHAEFER cited the results obtained for O_2, where the bond distance is more than 0.05 Å too short at the Hartree-Fock limit. Use of additional first-order wave functions gives a bond distance that is about 0.005 Å too long. In other words, the general effect of correlation, in 90 percent of the cases, is to make the bond distance longer. In fact, the SCF and CI results obtained with extended basis sets usually are quite different.

In response to a question by PLÍVA, SCHAEFER indicated that the calculation of one geometry of HNC or HCN (with 6343 configurations included) required about 20 min on an IBM-360/195. Calculations were made for about 10 geometries each on HCN and HNC; therefore, the total problem required about 7 h of computer time. PLÍVA also inquired if second and higher derivatives of the energy with respect to internuclear distance had been obtained. SCHAEFER indicated these had been calculated for HCN and HNC since this information was needed to test the RRKM theory for the unimolecular reaction.

STOCKMAYER inquired if the optimism expressed by SCHAEFER* about the calculation of energy barriers to rotation or inversion for molecules such as NH_3 and H_2O_2 could be extended to the determination of conformational energy differences of slightly larger molecules such as CH_2FCH_2F or of polymers such as $(-OCH_2CH_2-)_n$. In the latter case, the stable conformation about the C–C bond is *gauche* in contradiction to the simplest intuitive notions. SCHAEFER felt that theoretical methods definitely could provide information about relative conformation energies for larger molecules. He felt that the prediction of rotational barriers has been perhaps the area of greatest success for these methods. Although there had been some bad results, there are a number of good cases of which ethane is a prime example. In this case any level of calculation yields a barrier correct to about 0.1–0.2 kcal/mol. He cited the large amount of work done in this area by Pople and indicated that Pople's procedures could be applied to the polymer problem with no particular difficulty. In fact, the number of parameters in the polymer problem is fairly small since the bond distances and angles normally are held constant.

HARMONY asked how well nitrogen quadrupole-coupling constants or field

* *The Electronic Structure of Atoms and Molecules: A Survey of Rigorous Quantam Mechanical Results,* Addison–Wesley, Reading, Mass. (1972).

gradients could be predicted. DITCHFIELD indicated that, at the split valence shell basis level, calculated quadrupole-coupling constants for first-row atoms show about the same percentage error as the deuterium quadrupole-coupling constants. However, there are noticeable improvements in the calculated values when polarization functions are included in the basis set. Limited data are now available for small molecules, and these indicate that calculated and experimental quadrupole-coupling constants are in reasonable agreement, provided polarization functions are used. He pointed out that one of the problems of determining nitrogen quadrupole-coupling constants is that one actually calculates electric field gradients, and in order to convert to quadrupole-coupling constants one requires the nuclear quadrupole moment. These are not known too accurately for nitrogen and oxygen.

HAZONY wished to retract his earlier suggestion (in this volume) that Mössbauer spectra showed an excellent irrelevancy to theory, since these data also indicate a covalent nature of the FeF bond. He also asked if the in-plane and out-of-plane wave functions of planar FeF_3 were treated differently. SCHAEFER indicated that they were treated the same and that, in this calculation, the d orbitals were treated at the double zeta level, which permits an independent adjustment of the inner and outer parts of the orbital. He also pointed out that the net charge on Fe was +2 rather than the expected (ionic) charge of +3 for the molecule *in the gas phase,* which may or may not be equivalent to the molecules as they are seen in a condensed phase. HAZONY was concerned about the quotation of a net charge on Fe since it necessarily includes a definition of size. He preferred instead to consider the value for r^{-3}. In results of his group, for example, significantly different values were found for bonding and nonbonding contributions for in-plane and out-of-plane orbitals. SCHAEFER indicated that field gradients had been calculated for this molecule and that other properties could be determined. He again pointed out the need for communication between experimentalists and theoreticians.

SCHOMAKER commented on the general nonacceptance of Boys's correct early conclusion on the bent nature of the triplet state of CH_2. He said that he is the more puzzled because simple qualitative theoretical considerations also indicate nonlinearity. For example, it must be supposed that if anyone had asked Pauling, even 40 years ago, the immediate answer could only have been "bent, the bond angle about tetrahedral." The argument would have been that the variation of bond energy with bond–orbital hybridization would govern, the atomic (valence state) energy being almost independent of hybridization, since each carbon valence shell orbital is singly occupied and the C-H bonds are almost purely covalent. Subsequent considerations and opinions that sp^2 bonds, rather than sp^3 bonds, are stronger and that H is a little less electronegative than C (favoring s character for the bond orbitals) would have increased the estimated angle, but surely not to the point of giving a reasonable basis for *rejecting* the bent model.

RAMSAY referred to a comment on early measurements of the gravitational constant in which the author reported that Maskelyne* (some 200 years ago) had

*Nevil Maskelyne (1732–1811), English astronomer, who conducted experiments at Mt. Schehallion, Scotland (1774–1976), on the gravitational attraction of mountains as shown by deviations of the plumb line.–Ed.

determined the constant of gravitation "not entirely inaccurately." RAMSAY felt that theoretical chemistry has now reached this happy level of acceptance. He went on to question the statement in Schaefer's paper that at present the most reliable value for the bond angle in CH_2 is from theoretical calculations, in view of the rather flat potential energy surface computed for this molecule. RAMSAY maintained that bets were, in fact, pretty well hedged since he could choose a bond angle of 150° almost as easily as one of 135°. SCHAEFER indicated that the result he had shown was of an earlier calculation, but that the most recent results did not differ much from it. In answer to a question by RAMSAY he indicated that the calculated barrier height is about 7 kcal/mol. The best bond angle estimate by calculation is 134°±2°, compared with the most recent ESR estimate of 136°± 8°. RAMSAY called attention to the more recent comments on the structure of CH_2 and CD_2 in the ground state by Herzberg and Johns,* which provides an independent estimate of the bond angle. [The authors emphasize that the 1416 Å system of CD_2 *can* be interpreted in terms of a bent lower state, but that such an interpretation does impose an additional *ad hoc* assumption of a strong predissociation of the upper state such that *only* the $K = 0$ subband is observed. This assumption had been considered, discussed, and rejected as unjustified at the time by Herzberg.† When that assumption is permitted, the vacuum ultraviolet data would suggest a bond angle of 136°, with at least ±12° uncertainty.]

FLYGARE asked about the relative energies of HNC and HCN. He was especially interested in explanations for the failures to observe HNC in the laboratory. SCHAEFER replied that at first glance the system might be compared with the conversion of CH_3NC to CH_3CN, which has been studied very thoroughly by kineticists. Preliminary predictions for the HNC–HCN system indicate an activation energy of about 30 kcal/mol with an energy difference of about 12 kcal/mol, which may be compared with an activation energy of about 38 kcal/mol and an energy difference of about 15 kcal/mol for the CH_3NC–CH_3CN system. In other words, the calculations suggest that the two should behave in a similar manner. The fact that HNC has not been observed in the laboratory is puzzling; the difficulty may be in its preparation.

BUCKINGHAM asked about the status of calculations for molecules in excited states. SCHAEFER replied that more has been done at the more accurate level of approximations and that the results are just as reliable. There are some exceptions, such as the B state of CH, which has a shallow minimum. In such a case, it obviously is difficult to establish either by experiment or by theory an accurate internuclear distance. In response to a question by BUCKINGHAM, SCHAEFER indicated that excited states of the same symmetry as the ground state are more difficult to handle and less has been done with them.

MILLS asked if calculations on open-shell molecules were distinctly less reliable than calculations on closed-shell molecules. SCHAEFER felt that there was not much difference. DITCHFIELD indicated that there is a general problem of treating

* *J. Chem. Phys.*, **54**, 2276 (1971).
† *Proc. R. Soc. (London)*, **A262**, 291 (1961).

open-shell molecules at a simple level. For example, if we wish to use a single determinant, then we must allow the alpha and beta molecular orbitals to be different. In other words, we must use a spin-unrestricted single determinant. Pople and his group have calculated geometries for a large number of open-shell molecules using this level of theory. In some cases this approximation seems to yield bond lengths that are rather too long. When a restricted open-shell method is used, these bond lengths are shortened. Consequently, DITCHFIELD would say that, in some cases, the calculated results for open-shell molecules are somewhat less reliable. Of course, a more adequate inclusion of electron correlation may change this situation.

In regard to SCHAEFER'S assertion that "the existence of HNC in interstellar space is given a fairly firm theoretical foundation," LIDE made the following qualification in a postconference comment.

"The present *ab initio* calculation of the B value of HNC shows that the claimed assignment of an interstellar line to HNC is consistent with the available theoretical evidence. In this sense it is certainly a step forward over the original assignment, which was based solely on a guessed molecular structure. However, the comparison between observed and calculated frequencies is still too crude to be regarded as strong positive evidence for the assignment. The difference between the calculated $B_e = 45.43$ GHz and the value of $B_e = 45.52$ GHz derived on the assumption that the observed interstellar line is the $J = 0 \rightarrow 1$ transition of a linear molecular implies a difference of 180 MHz (twice the difference in B_e) in observed and calculated line frequencies. However, the calculated frequency probably cannot be considered reliable to better than about ±200 MHz, and one must also allow for an uncertainty of at least ±200 MHz in the corrections for vibration/rotation interactions. The real point is that, in order to attach any meaningful confidence level to an assignment based on the coincidence between the frequency of an interstellar line and a "terrestrial value" (either from a laboratory measurement or, in principle, an *ab initio* calculation), the agreement must be within a few megahertz at least* and the terrestrial value must obviously be known to a corresponding accuracy. Thus an improvement of one to two orders of magnitude appears necessary before the calculations on HNC can provide convincing evidence for the validity of the assignment of the interstellar line."

* See, for example, D. R. Lide in *Molecules in the Galactic Environment,* ed. by M. A. Gordon and L. E. Snyder, John Wiley & Sons, New York (1973), p. 242–244.

Units and Conversion Factors

The following table gives the conversion between units used in this book for expressing structural parameters and corresponding units of the International System of Units (SI).*

Interatomic distance	1 Å = 10^{-10} m
Force constant (stretch)	1 mdyn/Å = 10^{2} N/m
Force constant (bend)	1 mdyn·Å = 10^{-18} N·m = 1 aJ
Potential constants (and other energy quantities)	1 cal/mol ≈ 6.9478 × 10^{-24} J
	1 cm^{-1} ≈ 1.9865 × 10^{-23} J
	1 eV ≈ 1.6022 × 10^{-19} J
	1 hartree ≈ 4.3598 × 10^{-18} J
Electric charge	1 esu ≈ 3.3356 × 10^{-10} C
Electric dipole moment	1 debye (=10^{-18} esu cm) ≈ 3.3356 × 10^{-30} C·m
Quadrupole moment	1 buckingham (=10^{-26} esu cm^2) ≈ 3.3356 × 10^{-40} C·m^2
Magnetic susceptibility	10^{-6} erg/Gs^2 mol ≈ 1.6606 × 10^{-29} J/T^2

*Calculated from "Recommended Consistent Values of the Fundamental Physical Constants, 1973," Report of the Task Group on Fundamental Constants, Committee on Data for Science and Technology (CODATA) of the International Council of Scientific Unions, CODATA Bulletin No. 11, Frankfurt (1973).

Participants

SIDNEY C. ABRAHAMS, Bell Laboratories, Murray Hill, New Jersey

O. WILLIAM ADAMS, Chemistry Section, National Science Foundation, Washington, D.C.

DAVID F. ANDREWS, Department of Mathematics, University of Toronto, Toronto, Ontario, Canada, and Bell Laboratories, Murray Hill, New Jersey

ALBERT E. BEATON, Office of Data Analysis Research, Educational Testing Service, Princeton, New Jersey, and Department of Statistics, Princeton University, Princeton, New Jersey

HELEN BERMAN, The Institute for Cancer Research, Philadelphia, Pennsylvania

JAMES E. BLACKWOOD, Chemical Abstracts Service, Columbus, Ohio

ROBERT BOHN, Department of Chemistry, University of Connecticut, Storrs, Connecticut

AKSEL A. BOTHNER-BY, Mellon Institute of Science, Carnegie-Mellon University, Pittsburgh, Pennsylvania

F. A. BOVEY, Bell Laboratories, Murray Hill, New Jersey

W. V. F. BROOKS, Department of Chemistry, University of New Brunswick, Fredericton, New Brunswick, Canada

A. D. BUCKINGHAM, Department of Theoretical Chemistry, Cambridge University, Cambridge, England

CRISPIN CALVO, Department of Chemistry, McMaster University, Hamilton, Ontario, Canada

CHARLES N. CAUGHLAN, Department of Chemistry, Montana State University, Bozeman, Montana
BRYCE CRAWFORD, JR., Department of Chemistry, University of Minnesota, Minneapolis, Minnesota
BENJAMIN P. DAILEY, Department of Chemistry, Columbia University, New York, New York
MICHAEL I. DAVIS, Department of Chemistry, The University of Texas at El Paso, El Paso, Texas
WILSON H. DECAMP, Department of Chemistry, University of Georgia, Athens, Georgia
ROBERT DITCHFIELD, Department of Chemistry, Dartmouth College, Hanover, New Hampshire
JERRY DONOHUE, Department of Chemistry, University of Pennsylvania, Philadelphia, Pennsylvania
HELMUT DREIZLER, Abteilung Chemische Physik im Institut für Physikalische Chemie der Universität Kiel, Kiel, Germany
DAVID J. DUCHAMP, Physical and Analytical Chemistry Research, The Upjohn Company, Kalamazoo, Michigan
JAMES W. EDMONDS, Medical Foundation of Buffalo Research Laboratories, Buffalo, New York
HOWARD T. EVANS, JR., U.S. Geological Survey, Washington, D.C.
WILLIAM H. FLETCHER, Department of Chemistry, The University of Tennessee, Knoxville, Tennessee
W. H. FLYGARE, Department of Chemistry, University of Illinois at Urbana-Champaign, Urbana, Illinois
B. M. FUNG, Department of Chemistry, The University of Oklahoma, Norman, Oklahoma
JACK D. GRAYBEAL, Department of Chemistry, Virginia Polytechnic Institute and State University, Blacksburg, Virginia
HAROLD N. HANSON, Vice President for Academic Affairs, University of Florida, Gainesville, Florida
MARLIN D. HARMONY, Department of Chemistry, University of Kansas, Lawrence, Kansas
DAVID O. HARRIS, Department of Chemistry, University of California, Santa Barbara, California
YEHONATHAN HAZONY, Computer Center, Princeton University, Princeton, New Jersey
KENNETH W. HEDBERG, Department of Chemistry, Oregon State University, Corvallis, Oregon
LISE HEDBERG, Department of Chemistry, Oregon State University, Corvallis, Oregon
F. H. HERBSTEIN, Division of Chemistry and Chemical Engineering, California Institute of Technology, Pasadena, California, on sabbatical leave from Technion-Israel Institute of Technology, Haifa, Israel
RICHARD L. HILDERBRANDT, Department of Chemistry, North Dakota State University, Fargo, North Dakota

HÅKON HOPE, Department of Chemistry, University of California, Davis, California

JOHN HOWATSON, Department of Chemistry, The University of Wyoming, Laramie, Wyoming

R. E. HUGHES, Department of Chemistry, Cornell University, Ithaca, New York

JAMES A. IBERS, Department of Chemistry, Northwestern University, Evanston, Illinois

E. JEAN JACOB, Department of Chemistry, The University of Toledo, Toledo, Ohio

L. H. JENSEN, Department of Biological Structure, University of Washington, Seattle, Washington

CARROLL K. JOHNSON, Chemistry Division, Oak Ridge National Laboratory, Oak Ridge, Tennessee

ROGER KEWLEY, Department of Chemistry, Queen's University, Kingston, Ontario, Canada

WILLIAM H. KIRCHHOFF, Measures for Air Quality Division, National Bureau of Standards, Washington, D.C.

KOZO KUCHITSU, Department of Chemistry, Faculty of Science, The University of Tokyo, Tokyo, Japan

STEPHEN G. KUKOLICH, Department of Chemistry, Massachusetts Institute of Technology, Cambridge, Massachusetts

WALTER J. LAFFERTY, Optical Physics Division, National Bureau of Standards, Washington, D.C.

RANDOLPH D. LANIER, Chemical Abstracts Service, Columbus, Ohio

ALLEN C. LARSON, Los Alamos Scientific Laboratory, Los Alamos, New Mexico

VICTOR W. LAURIE, Department of Chemistry, Princeton University, Princeton, New Jersey

IRA W. LEVIN, Physical Biological Laboratory, National Institute for Arthritis, Metabolism, and Digestive Disease, National Institutes of Health, Bethesda, Maryland

HENRI A. LEVY, Chemistry Division, Oak Ridge National Laboratory, Oak Ridge, Tennessee

DAVID R. LIDE, JR., Office of Standard Reference Data, National Bureau of Standards, Washington, D.C.

ROBERT C. LIVINGSTON, Reactor Radiation Division, National Bureau of Standards, Washington, D.C.

RICHARD C. LORD, Spectroscopy Laboratory, Massachusetts Institute of Technology, Cambridge, Massachusetts

C. WELDON MATHEWS, Department of Chemistry, The Ohio State University, Columbus, Ohio

RODERICK C. McCALLEY, Department of Chemistry, Dartmouth College, Hanover, New Hampshire

ROBIN S. McDOWELL, Los Alamos Scientific Laboratory, Los Alamos, New Mexico

BRUCE R. McGARVEY, Department of Chemistry, University of Windsor, Windsor, Ontario, Canada

SAUL MEIBOOM, Bell Laboratories, Murray Hill, New Jersey
ALAN D. MIGHELL, Inorganic Materials Division, National Bureau of Standards, Washington, D.C.
FOIL A. MILLER, Department of Chemistry, University of Pittsburgh, Pittsburgh, Pennsylvania
IAN M. MILLS, Department of Chemistry, University of Reading, Reading, England
WOLF MOSER, Department of Chemistry, University of Aberdeen, Old Aberdeen, Scotland
NORBERT MULLER, Department of Chemistry, Purdue University, Lafayette, Indiana
RALPH D. NELSON, JR., Department of Chemistry, West Virginia University, Morgantown, West Virginia
WESLEY L. NICHOLSON, Applied Mathematics Division, National Bureau of Standards, Washington, D.C., and Battelle Memorial Institute, Richland, Washington
LISE NYGAARD, Kemisk Laboratorium V, H. C. Ørsted Institutet, University of Copenhagen, Denmark
HELEN M. ONDIK, Inorganic Materials Division, National Bureau of Standards, Washington, D.C.
MARTIN A. PAUL, Division of Chemistry and Chemical Technology, National Academy of Sciences-National Research Council, Washington, D.C.
N. C. PAYNE, Department of Chemistry, University of Western Ontario, London, Ontario, Canada
JOSEF PLÍVA, Department of Physics, Pennsylvania State University, University Park, Pennsylvania
SANTIAGO R. POLO, Department of Physics, Pennsylvania State University, University Park, Pennsylvania
D. A. RAMSAY, Division of Physics, National Research Council of Canada, Ottawa, Canada
MILAN RANDIČ, Department of Chemistry, Harvard University, Cambridge, Massachusetts, on leave from The University of Zagreb, Zagreb, Yugoslavia
JEAN B. ROBERT, Laboratoire C.O.P., Université de Grenoble, Grenoble, France
ALAN G. ROBIETTE, Department of Chemistry, University of Reading, Reading, England
J. J. RUSH, Reactor Radiation Division, National Bureau of Standards, Washington, D.C.
ANTONIO SANTORO, Reactor Radiation Division, National Bureau of Standards, Washington, D.C.
HENRY F. SCHAEFER III, Department of Chemistry, University of California, Berkeley, California
VERNER SCHOMAKER, Department of Chemistry, University of Washington, Seattle, Washington
R. H. SCHWENDEMAN, Department of Chemistry, Michigan State University, East Lansing, Michigan
CHARLES C. SHERA, Chemical Abstracts Service, Columbus, Ohio

CLARA B. SHOEMAKER, Department of Chemistry, Oregon State University, Corvallis, Oregon
DAVID P. SHOEMAKER, Department of Chemistry, Oregon State University, Corvallis, Oregon
LAWRENCE C. SNYDER, Bell Laboratories, Murray Hill, New Jersey
ROBERT G. SNYDER, Western Regional Research Laboratory, U.S. Department of Agriculture, Albany, California
ROBERT F. STEWART, Department of Chemistry, Carnegie-Mellon University, Pittsburgh, Pennsylvania
WALTER H. STOCKMAYER, Department of Chemistry, Dartmouth College, Hanover, New Hampshire
LESLIE E. SUTTON, Department of Physical Chemistry, University of Oxford, Oxford, England
LYNN F. TEN EYCK, Department of Molecular Biology, University of Oregon, Eugene, Oregon
H. BRADFORD THOMPSON, Department of Chemistry, The University of Toledo, Toledo, Ohio
MARIT TRAETTEBERG, Department of Chemistry, University of Trøndheim, Trøndheim, Norway
JOHN W. TUKEY, Bell Laboratories, Murray Hill, New Jersey, and Department of Statistics, Princeton University, Princeton, New Jersey
AAFJE VOS, Laboratorium voor structuurchemie, Rijksuniversiteit Groningen, Groningen, Netherlands
ALFONS WEBER, Department of Physics, Fordham University, Bronx, New York
JACK M. WILLIAMS, Chemistry Division, Argonne National Laboratory, Argonne, Illinois
GIUSEPPE ZERBI, Istituto di Chimica delle Macromolecole del CNR, Milan, Italy

OTHER CONTRIBUTORS

BERNARD R. APPLEMAN, Department of Chemistry, Columbia University, New York, New York
ROLFE H. HERBER, School of Chemistry, Rutgers University, New Brunswick, New Jersey

Index

The listing of compounds in this index is not comprehensive. Those included have been selected because of repeated or otherwise extensive reference to them in the text.